QUANTUM GRAVITY

International School of
Cosmology and Gravitation XIV Course

80th birthday dedication to
Peter G. Bergmann

THE SCIENCE AND CULTURE SERIES — PHYSICS

Series editor: A. Zichichi, *European Physical Society, Geneva, Switzerland*

1. Perspectives for New Detectors in Future Supercolliders, 1991
2. Data Structures for Particle Physics Experiments: Evolution or Revolution?, 1991
3. Image Processing for Future High-Energy Physics Detectors, 1992
4. GaAs Detectors and Electronics for High-Energy Physics, 1992
5. Supercolliders and Superdetectors, 1993
6. Properties of SUSY Particles, 1993
7. From Superstrings to Supergravity, 1994
8. Probing the Nuclear Paradigm with Heavy Ion Reactions, 1994
9. Quantum-Like Models and Coherent Effects, 1995

THE SCIENCE AND CULTURE SERIES — PHYSICS

QUANTUM GRAVITY

International School of
Cosmology and Gravitation XIV Course

80th birthday dedication to
Peter G. Bergmann

Erice, Italy 11 – 19 May, 1995

Editors

P. G. Bergmann
New York Univ., USA

V. de Sabbata
Univ. of Bologna and Ferrara, Italy

H.-J. Treder
WIP Gravitation project, Potsdam, Germany

Series Editor
A. Zichichi

World Scientific
Singapore • New Jersey • London • Hong Kong

Published by

World Scientific Publishing Co. Pte. Ltd.
P O Box 128, Farrer Road, Singapore 912805
USA office: Suite 1B, 1060 Main Street, River Edge, NJ 07661
UK office: 57 Shelton Street, Covent Garden, London WC2H 9HE

British Library Cataloguing-in-Publication Data
A catalogue record for this book is available from the British Library.

QUANTUM GRAVITY
International School of Cosmology and Gravitation XIV Course

ISBN 981-02-2786-8

This book is printed on acid-free paper.

Printed in Singapore by Uto-Print

PREFACE

These Proceedings of the 14th Course, "Quantum Gravity", of the International School of Cosmology and Gravitation are dedicated to Peter Gabriel Bergmann's 80th birthday.

We are very glad to have in Erice Prof. Peter Gabriel Bergmann for his 80th jubilee. Most people know how deep were and are the works of Bergmann on general relativity and in developing Einstein's theory of gravitation beginning in 1936 when he went from Prague to the Princeton Institute for Advanced Study to work with Einstein. Peter came to Prague in 1933 as a refugee from Hitler's Germany where as a Jew he could not have completed his academic training. He had read some scientific publications of Einstein and he realized that he would like to do just this kind of theoretical research. So he wrote to Einstein asking to accept him as a collaborator (but he wrote at a wrong address). As after one month there was no reply, he wrote again enclosing a copy of a dissertation. This time he received an answer in which Einstein wrote that he would be glad to talk with him. Bergmann was unaware that Einstein had in the meantime written to Prof. Philipp Frank (the Director of the Institute of Theoretical Physics in Prague) asking him about Bergmann. So when Bergmann arrived in Princeton he began to work with Einstein. He in fact worked with Einstein from 1936 to 1941 at the Institute of Advanced Study in Princeton in an attempt to provide a geometrical unified field theory of gravitation and electromagnetism. Within this period they wrote two article together: A. Einstein and P. G. Bergmann in *Ann. Math.* **39**, 65 (1938); and A. Einstein, V. Bargmann and P. G. Bergmann in Th. von Karman anniversary volume 212 (1941).

In 1942, Bergmann published the book "Introduction to the Theory of Relativity" which included a foreword by Albert Einstein. This book is a reference for the subject, either as a textbook for classroom use or for individual study. Einstein said in his foreword: "Bergman's book seems to me to satisfy a definite need. ... much effort has gone into making this book logically and pedagogically satisfactory and Bergmann has spent many hours with me which were devoted to this end". He also said that if general relativity has played a rather modest role in the correlation of empirical facts, and has contributed very little to understanding of quantum phenomena, it is quite possible that some of the results of the general relativity such as, general covariance of the laws of nature and their nonlinearity may help to overcome the difficulties encountered at present in atomic and nuclear processes.

Many of the important works of Peter Bergmann are in fact concerned with the quantization of field theories which are covariant with respect to general coordinate transformations.

As to these points, Bergmann's article "General Relativity" in the volume V of the "Encyclopaedia of Physics" was pioneer work.

Of course we cannot go through all of Bergmann's works but we like to mention the great contribution that he has made to the School of Cosmology and Gravitation in Erice from its inception in 1974: being ever present and always discussing every argument in great depth, always coming to the point and clarifying every aspect of the problem under discussion. For instance in the sixth Course of the School (1979) he discussed "the fading world point" which deals with the nature of space-time and of its elements, the world points. Other important discussions were on the "unitary field theories", during the eighth Course in 1983, where he discussed various unitary theories such as Kaluza-Klein, scalar-tensor theories and projective theories. During the ninth Course (1985) devoted to "Topological properties of space-time" and during the tenth Course of School in 1987, devoted to "Gravitational Measurements" Bergmann discussed 'gravitation at spatial infinity' and the 'observables in general relativity' where he showed in a very elegant way the profound difference of the notion of observable in general relativity from the corresponding concept in special relativity (or in Newtonian physics).

So Bergmann really has been the centre, the fulcrum of all discussions in all the Courses of our International School of Cosmology and Gravitation.

We are therefore very glad that the celebration of this Peter Bergmann jubilee coincides with the Course on 'Quantum Gravity'. In fact from the early fifties onwards, one of the main goals and challenging tasks for Bergmann was the quantization of the gravitational field. The difficulties which he faced and solved in dealing with the full nonlinear properties of Einstein's theory and with nonperturbative approaches were enormous. Bergmann and his collaborators investigated the canonical (i.e. Hamiltonian) formalism of classical and quantum covariant field theories.

These covariant field theories of general type of the theory of relativity are brought into the canonical form and then quantized. These works are of the year 1948 and 1949 [*Phys. Rev.* **75**, 680 (1949) and (with Johanna H. M. Brunings) *Rev. Mod. Phys.* **21**, 480 (1949)], and constitute the basis of the modern research in this field. They are followed by other important works [*Phys. Rev.* **83**, 1018 (1951) with James L. Anderson] on the constraints in covariant field theories and the application to the cases of gauge and coordinate invariance specializing the problem by assuming a quadratic lagrangian in the differentiated quantities; moreover a beautiful work on 'Spin and Angular Momentum in G.R.' [*Phys. Rev.* **89**, 400 (1953)] in presence of a semiclassical Dirac field. The problem of origin of constraints is also considered in another important work with Irwin Goldberg [*Phys. Rev.* **98**, 531 (1954)] where there is an analysis of group-theoretical significance of the Dirac bracket in order to prepare the ground for the utilization of the Dirac bracket for the quantization of generally covariant theories.

Once again, we cannot describe all of Bergmann's work here but we would like to stress the fact that Bergmann clarified and made profound contributions in the different quantization programs based on the canonical formalism. As Bergmann himself said, "the resulting theory would give us answers to such questions as the nature of a fully quantized geometry of space-time, the role of world points in this geometry, the 'softening-up of the light cone', and the effect of this not only on the divergences associated with the gravitational field but with all other fields as well".

Peter Gabriel Bergmann is really admirable both as scientist and as a man of great culture and humanity and we are glad to convey our best wishes to him.

The reader will find in this volume an updated version of different approaches to the Quantum Gravity problem, including quantum cosmology, strings, twistors, Einstein-Cartan theory, Weyl-Cartan theory, canonical formalism, Poincaré gauge theory of gravitation, spinor-tensor manifold and other topics.

We wish to express our grateful thanks to all the authors: some have contributed review papers, and others reports on their recent research work.

Venzo de Sabbata
Hans-Jürgen Treder

Erice 12 May 1995 WELCOME

by Venzo de Sabbata

I am glad to welcome you to this fourteenth Course of the International School of Cosmology and Gravitation. Professor Zichichi, the Director of this "Ettore Majorana Centre for Scientific Culture who cannot be prsent at this moment, (maybe he will come some days) has entrusted me with the welcome address on his behalf.

First of all I will thank very much Professor Bergmann who worked hard to organize the scientific aspects of this Course. I thank Professor Treder for the same reason. Unfortunately Professor Treder cannot be present, because right now he is not in good health. I thank all lecturers, who have accepted our invitations to come to Erice to give lectures (whitout any renumeration, as is usual in this School), from their different points of view, on Quantum Gravity. We all are attempting to reconcile with each other two important conceptual structures of physics, quantum theory and general relativity, which at first sight appear to be far from each other. I thank all lecturers also, because they accepted as far as they were able, to be present the whole time and to participate in all discussions, which certainly will arise among all participants. In fact this is one of the important tasks of these Courses. As perhaps many participants already know, because they were present in previous Courses, the atmosphere of the School is very friendly, and exchanges of ideas can occur quite freely among all participants including lecturers.

As you know from the poster, lectures will cover a large range of different approaches and strategies employed in recent years toward that argument. So it will include principal mathematical, physical and epistemological aspects characteristic of relativistic quantum physics in Minkowski space-time and in curved space-time, ranging through various canonical and covariant approaches to quantum gravity, including quantization of Einstein's general relativity, quantum cosmology, strings, Einstein--Shrödinger theory, Einstein-Cartan theory, Weyl-Cartan geometry, twistors and other topics.

Moreover among the participants there are some outstanding scientists as Professor Bidyut Kumar Datta, former Director of the Mathematical Institute of Calcutta University and Professor Xin Yu (alias Alfred Yu), Director of Department of Applied Mathematics of Hong Kong Polytechnic. They will present some particularby interesting arguments.

So we'll have a very broad spectrum of subjects, of which at first sight each has sufficient material to constitute a Course by itself. We have considered it more important to exhibit the various theories that attempt to achieve the goal of quantization, and to exchange information on the latest developments discussing future prospects.

I really hope that this Course with all these different topics serves to make some further progress on this fascinating subject.

Before asking Professor Bergmann to address the meeting, some few points regarding the organization of the lectures:

as you can see from the tentative time-table, we will have six lectures each day, three in the morning and three in the afternoon (with a break, both in the morning and in the afternoon, for coffee). Every lecture will last no more than fifty minutes, and after every lecture there will be a few minutes of discussion. For lunch and dinner you can choose any of the restaurants approwed for the School (you will see the list near the entrance), signing a list marked "Quantum Gravity" after every meal and either "lunch" or "dinner". You have to pay only for beverages. Now Professor Bergmann will address the meeting.

By Peter Bergmann

Thank you, Venzo.

Let me remind you that Venzo de Sabbata is wearing three hats. He represents the Director of the Centre, Prof. Zichichi; he is the permanent director of the International School of Cosmology and Gravitation, of which this is the fourteenth Course; and he is the co-director of this Course, Quantum Gravity, together with Prof. Treder of Potsdam, and myself.

Schedule: There will be six working days, Friday, 12 June, Saturday, Sunday, and Tuesday through Thursday, 18 June. Monday, 15 June, will be devoted to a full-day excursion to Greek and Roman architectural monuments in Western Sicily.

As the schedule is crowded, I urge all speakers to stay within their allotted time limits. Remember: If you run over your time, you are depriving a fellow ~~speaker~~ lecturer of hers or his. I am equally urging the chairs to enforce time limits.

To make these limits tolerable, I suggest that you concentrate on the essential ideas, leaving all details to manuscripts. If you request fifty copies of your materials, they will suffice to supply all participants.

« ETTORE MAJORANA »
INTERNATIONAL CENTRE FOR SCIENTIFIC CULTURE

A reasonable run above that number will cover your personal needs.

And now, let us begin!

By Peter Bergmann

Thank you, Venzo.

Let me remind you that Venzo de Sabbata is wearing three hats. He represents the Director of the Centre, Prof.Zichichi; he is the permanent director of the International School of Cosmology and Gravitation, of which this is the fourteenth Course; and he is the co-director of this Course, Quantum Gravity, together with Prof. Treder of Potsdam, and myself.

Schedule: There will be six working days, Friday, 12 june, Saturday, Sunday and Tuesday through Thursday, 18 june. Monday, 15 june, will be devoted to a full-day excursion to Greek and Roman architectural monuments in Western Sicily.

As the schedule is crowded, I urge all speakers to stay within their allowed time limits. Remember: If you run over your time, you are depriving a fellow lecturer of hers or his. I am equally urging the chairs to enforce time limits.

To make these limits tolerable, I suggest that you concentrate on the essential ideas, leaving all details to manuscripts. If you request fifty copies of your materials, they will suffice to supply all participants. A reasonable run above that number will cover your personal needs.

And now, let us begin!

Bergmann and Einstein working in Einstein's study at Princeton in early [illegible], when they were actively engaged in attempts at constructing unitary field theories. (Photo by Lotte Jacobi)

CONTENTS

Gravity Without the Metric and Torsion

Pratul Bandyopadhyay
Physics and Applied Mathematics Unit
Indian Statistical Institute
Calcutta-700 035, INDIA

Abstract

The role of θ-term in gravity without the metric formulation of Capovilla, Jacobson and Dell has been investigated when the gauge group is taken to be SL(2,C) and it is shown that this corresponds to the introduction of torsion and as such represents the Einstein-Cartan action. Moreover, as this term is related to chiral anomaly, this helps us to formulate the Einstein- Hilbert action as a symmetry breaking efect in quantum field theory. The relevance of θ-term in the Lagrangian gets reflected in the loop space formalism when we utilise the equivalence of the loop space with the space of gauge potentials modulo gauge transformations satisfying certain properties and Mandelstam constraints. The topological feature associated with the gauge orbit space of a non-Abelian gauge theory when the topological θ-term is introduced in the Lagrangian corresponds to a vortex line and the gauge orbit space appears to be multiply connected in nature. This has inplication in the loop space formalism in the sense that the latter involves nonlocality and there is no way we could arrive at a corresponding continuum limit. This suggests that in the construction of solution of functionals annihilated by the Hamiltonian constraint any regularization procedure which destroys the gauge invariance of the loop space variables destroys the topology of the gauge orbit space and the continuum limit can only be achieved by removing the vortex line. Thus the constraint equations of canonical quantization of gravity can be achieved in the limit of torsion tending to zero. This provides the link between covariant and canonical quantization of gravity and demonstrates explicitly the role of the arrow of time in nonperturbative quantum geometry also when we take the gauge invariant holonomy variables as the fundamental entity.

1 Introduction

Quantum Gravity is generally analysed in covariant and canonical quantization schemes. The covariant approach deals with a Lorentzian manifold $(M, g^B_{\mu\nu})$ where $g^B_{\mu\nu}$ is a Lorentzian metric generally called "background metric" over a differentiable manifold M. The components of the quantum metric field $g^Q_{\mu\nu}$ are then defined as

$$g^Q_{\mu\nu} = g^B_{\mu\nu}(x) + h_{\mu\nu}(x)$$

$$x \in M$$

$$\mu, \nu = 0, 1, 2, 3$$

where $h_{\mu\nu}(x)$ is viewed as quantum fluctuation. Though it allows us to use the techniques of formal perturbation theory, the main difficulty arises due to the nonrenormalizibility of the theory. In the canonical approach the classical space-time is described by a globally hyperbolic Lorentzian manifold (M, g) which is foliated into a family of three dimensional hypersurfaces Σ_t which depends on a timelike parameter t. The line element of the Lorentzian metric g is then decomposed into $3+1$ form

$$ds^2 = (N^2 - N_a N_b \gamma^{ab})dt^2 + 2N_a dt dx^a - \gamma_{ab} dx^a dx^b$$

$$(a, b = 1, 2, 3)$$

where γ_{ab} represent the components of a Riemannian 3-metric along each Σ_t where N and N_a are the ADM lapse and shift functions. In the Hamiltonian formulation, this gives rise to certain constraint equations where the role of the configuration variable is played by γ^{ab}. The canonical approach has attained significant importance recently due to the introduction of the Ashtekar variables. Ashtekar(1) has rewritten Einstein's equations using SO(3) vector potentials and its canonically conjugate momentum as fundamental variables. The phase space variable appears as a standard triad though there is caveat that these triads are complex-valued defined on a real Lorentzian manifold. The constraint equations now become much simplified. Recently Capovilla, Jacobson and Dell(2) have formulated a theory of gravitation without metric which constitutes a Lagrangian formulation of Ashtekar's theory in which the metric of the triad has been completely eliminated in favour of the connection. It is equivalent to Ashtekar's formalism though this equivalence breaks down unless a certain matrix is invertible. The connection carries SO(3) index. This suggests that the metric is not a fundamental entity. This metric-free action for General Relativity implies a link between covariant quantization of gravity with canonical quantization formulated by Ashtekar. Besides, as emphasized by Capovilla, Jacobson and Dell, the fact that the Einstein equations can be rewitten in terms of the spin connection may imply a twistor theoretic construction of the theory. It may be recalled that Newman and Penrose (3) formulated the gravitation theory involving spinorial variable. Carmeli and Malin(4) have formulated the SL(2,C) gauge theory of gravitation which is closely related to the Newman-Penrose formalism. It has been shown in a recent paper (5) that SL(2,C) gauge theory of gravitation may be taken to lead to Einstein-Cartan theory incorporating torsion which appears as a quantum effect. In fact, the torsion term may be taken to be originated from the geometry of microlocal space-time which is associated with the quantization of a fermion and as such may be taken to be an effect of quantum gravity. The torsion may be introduced in the formulation of Capovilla, Jacobson and Dell by introducing a θ-term in the action(6). Indeed, if we take into account the SL(2,C) gauge theory, this θ-term gives rise to the Pontryagin index which contributes to the action as the component of the torsion. This topological term can be taken to be the fundamental entity and the standard CJD action which corresponds to the Einstein-Hilbert action for pure gravity can be introduced through the incorporation of chiral anomaly in the matter field Lagrangian, where the coupling of gravity with the matter field is neglected. In this sense, this is analogous to the contention of Adler(7) that the Einstein-Hilbert gravitational action is obtained as a symmetry breaking effect in quantum field theory. The relevance of θ-term in the CJD action gets reflected in the loop space formalism when we utilise the equivalence of the loop space with the space of gauge potentials modulo gauge transformations satisfying certain properties and Mandelstam constraints Giles(8) has studied the reconstruction of gauge potentials from loop space holonomy variables and it is observed that this can be achieved when these satisfy certain algebraic constraints. Loll(9) has utilised this equivalence relation between loop space and gauge orbit space in the study of nonabelian gauge theories from the point of view that the gauge invariant traced holonomies are the fundamental variables. The topological feature associated with the gauge orbit space of a nonabelian gauge theory when the topological θ-term is introduced in the Lagrangian corresponds to a vortex line and the gauge orbit space appears to be multiply-connected in nature. This has implication in the loop space formalism in the sense that the latter involves nonlocality and there is no way we could arrive at a corresponding continuum result. In the gravity without the metric formalism of CJD, the θ-term in the Lagrangian which corresponds to torsion is effectively associated with the vortex line. This indicates that in the construction of solution of functionals annihilated by the Hamiltonian constraint we have to take care of this point in the regularization procedure. Indeed, any regularization procedure which destroys the gauge invariance of the holonomy variables destroys the topology of the gauge orbit space and hence the effect of the vortex line. When we arrive at the solution of the Hamiltonian constraint in loop space formalism through such a regularization, we are in a sense destroying the effect of torsion. This can be interpreted by arguing that canonical formalism of quantum gravity can only be derived by the limiting effect of torsion tending to zero. This may be taken to provide a link between the covariant and canonical quantization of gravity.

2 Gravity Without the Metric,Torsion and Pontryagin Index

As it is well known, in its $3+1$ form, the action in terms of Ashtekar's variable, can be written as

$$
\begin{aligned}
S &= \int \dot{A}_{ai} E^a_i - NH - N^a H_a - \Lambda_i G_i \\
H &= \frac{1}{2} i f_{ijk} E^a_i E^b_j F_{abk} \quad = \frac{1}{2} i \epsilon_{abc} f_{ijk} E^a_i E^b_j B^c_k \\
H_a &= E^b_i F_{abi} \quad = \epsilon_{abc} E^b_i B^c_i \\
G_i &= D_a E^a_i \quad = \partial_a E^a_i + i f_{ijk} A_{aj} E^a_k
\end{aligned} \tag{1}
$$

Here $a, b, c,$ are spatial indices, $i, j, k,$ are SO(3) indices, F_{abi} is an SO(3) curvature,$B^a_i = \frac{1}{2}\epsilon^{abc} F_{bci}$ is the corresponding "magnetic" field. The action yields three kinds of constraint Viz., the Hamiltonian constraint H, the vector constraint H_a and SO(3) vector's worth G_i of "internal" constraints which here takes the form of Gauss's law for the "electric" field E_i^a. The tensor $q^{ab} = gg^{ab}$, where g^{ab} is the metric tensor on the foliating hypersurface and g is its determinant, is given by

$$q^{ab} = E_i^a E_i^b \tag{2}$$

Capovilla, Jacobson and Dell(2) found a very elegant Lagrangian formulation of the above. The CJD action is

$$S^{CJD} = \frac{1}{8}\int \eta(\Omega_{ij}\Omega_{ij} + a\Omega_{ii}\Omega_{jj}) \tag{3}$$

$$\Omega_{ij} = \epsilon^{\alpha\beta\gamma\delta} F_{\alpha\beta i} F_{\gamma\delta j}$$

Here α, β are space-time indices.

The Lagrangian multiplier η is a scalar density of weight one, and $F_{\alpha\beta i}$ is an SO(3) field strength. They showed that a $3+1$ decomposition of this action yields Ashtekar's action directly provided that the parameter $a = -\frac{1}{2}$ and provided that the determinant of the "magnetic" field B_i^a is non-zero(10). In this way the equivalence to Einstein's theory is established. Indeed, with $a = -\frac{1}{2}$, the equivalence to Einstein's theory may also be shown directly in a manifestly covariant way. This demonstration hinges on the tetrad formalism and the space-time metric may be given directly in terms of the curvature when we can write(2)

$$(-g)^{\frac{1}{2}} g^{\alpha\beta} = -(\frac{2i}{3\eta})k^{\alpha\beta} = -(\frac{2i}{3\eta}) f_{ijk} \epsilon^{\alpha\gamma\delta\rho} \epsilon^{\beta\mu\nu\sigma} F_{\gamma\delta i} F_{\rho\sigma j} F_{\mu\nu k} \tag{4}$$

The constraint that is obtained when the CJD action is varied with respect to the Lagrange multiplier η is actually the Hamiltonian constraint in disguise.

$$\psi = \Omega_{ij}\Omega_{ij} - \frac{1}{2}\Omega_{ii}\Omega_{jj} = i(2\eta^2 detB)^{-1} H \tag{5}$$

We can now introduce the matrix ψ_{ij} defined by

$$E_i^a = \psi_{ij} B_j^a \tag{6}$$

where we note that such a matrix always exists provided that the magnetic field is non-degenerate. If we insert this expression in the vector constraint $H_a \approx 0$, we find that the vasishing of the vector constraint is equivalent to the statement that the matrix is symmetric. We also note that as long as $detB \not\models 0$, $\psi \approx 0$ and $H \approx 0$ are equivalent statements. Gauss' law $G_i \approx 0$ follows when the action is varied with respect to A_{0i}. Bengtsson and Peldan (11) have shown that if we perform the canonical transformation

$$A_{ai} \rightarrow A_{ai}, \; E_i^a \rightarrow E_i^a - \theta B_i^a \tag{7}$$

the expression for the Hamiltonian constraint changes though the remaining constraints are unaffected. This corresponds precisely to the addition of a "CP violating" θ-term to the CJD Lagrangian when the new action is given by

$$S = \frac{1}{8}[\int \theta\Omega_{ii} + \eta(\Omega_{ij}\Omega_{ij} - \frac{1}{2}\Omega_{ii}\Omega_{jj})] \tag{8}$$

It has been shown that when the gauge group is taken to be SL(2,C) when i, j corresponds to the SL(2,C) indices, this θ-term effectively corresponds to torsion(6). In fact, the θ-term now corresponds to the Lagrangian

$$L = -\frac{1}{4}\theta Tr\epsilon^{\mu\nu\alpha\beta}F_{\mu\nu}F_{\alpha\beta} \tag{9}$$

with

$$F_{\mu\nu} = \partial_\mu A_\nu - \partial_\nu A_\mu + [A_\mu, A_\nu] \tag{10}$$

where $A_\mu = \vec{A}_\mu \vec{g}$, g^1, g^2, g^3 being the SL(2,C) generators. This gives rise to the current

$$\vec{J}^\mu = \epsilon^{\mu\nu\alpha\beta}\vec{A}_\nu X \vec{F}_{\alpha\beta} = \epsilon^{\mu\nu\alpha\beta}\partial_\nu \vec{F}_{\alpha\beta} \tag{11}$$

which in turn is responsible for torsion.Indeed normalising $\vec{J}^\mu = \frac{1}{\chi}\epsilon^{\mu\nu\alpha\beta}\vec{A}_\nu X \vec{F}_{\alpha\beta}$ with $\chi = \frac{-8\pi G}{c^4}$ and taking

$$\vec{A}_\nu X \vec{F}_{\alpha\beta} = k^2 S_{\nu\alpha\beta}\vec{n} \tag{12}$$

k being the planck length, $\vec{n}$ being unit vector,we note that the action

$$S = A\int \vec{J}_\mu \vec{J}_\mu d^4x \tag{13}$$

with

$$A = \frac{1}{k^2} = \frac{1}{16\pi G}$$

can be written as

$$S = \frac{-4}{k^2} \int S_{\nu\alpha\beta} S^{\nu\alpha\beta} d^4x \tag{14}$$

leading to torsion. The θ-term in the action(8) effectively corresponds to the Pontryagin index when the group structure is taken to be SL(2,C) and this is the case also for the torsion term given by the coupling. To study this aspect, we consider that the θ-term is essentially associated with the chiral anomaly which arises when chiral currents interact with a gauge field. Indeed, to describe a matter field in the geometry associated with the Lagrangian(9),we note that in the background of the SL(2,C) gauge fields, the Lagrangian for a Dirac spinor field may be written as (neglecting the mass term),

$$L = -\bar{\psi}\gamma^\mu D_\mu \psi - \frac{1}{4} Tr \epsilon^{\alpha\beta\gamma\delta} F_{\alpha\beta} F_{\gamma\delta} \tag{15}$$

where D_μ is the SL(2,C) gauge covariant derivative defined by $D_\mu = \partial_\mu - igA_\mu$ where g is some coupling strength and $A_\mu \epsilon$ SL(2,C) . However, in the Lagrangian(15), if we split the Dirac massless spinor in chiral forms and identify the internal helicity $+\frac{1}{2}(\frac{-1}{2})$ with left (right) chirality we can write,

$$\begin{aligned} \bar{\psi}\gamma^\mu D_\mu \psi &= \bar{\psi}\gamma^\mu \partial_\mu \psi - ig\bar{\psi}\gamma_\mu A^a_\mu g_a \psi \\ &= \bar{\psi}\gamma^\mu \partial_\mu \psi - \tfrac{ig}{2}\{\bar{\psi}_R \gamma^\mu A^1_\mu \psi_R - \bar{\psi}_R \gamma^\mu A^2_\mu \psi_R \\ &\quad + \bar{\psi}_L \gamma^\mu A^2_\mu \psi_L + \bar{\psi}_L \gamma^\mu A^3_\mu \psi_L\} \end{aligned} \tag{16}$$

Then the three SL(2,C) gauge field equations give rise to the following three conservation laws(12)

$$\begin{aligned} \partial_\mu[\tfrac{1}{2}(-ig\bar{\psi}_R\gamma_\mu\psi_R) + j^1_\mu] &= 0 \\ \partial_\mu[\tfrac{1}{2}(-ig\bar{\psi}_L\gamma_\mu\psi_L + ig\bar{\psi}_R\gamma_\mu\psi_R) + j^2_\mu] &= 0 \\ \partial_\mu[\tfrac{1}{2}(-ig\bar{\psi}_L\gamma_\mu\psi_L) + j^3_\mu] &= 0 \end{aligned} \tag{17}$$

These three equations represent a consistent set of equations, if we choose

$$j^1_\mu = -\tfrac{1}{2} j^2_\mu \ , \ J^3_\mu = \tfrac{1}{2} j^2_\mu \tag{18}$$

which evidently guarantees the vector current conservation. Then we can write

$$\begin{aligned} \partial_\mu(\bar{\psi}_R \gamma_m u \psi_R + j^2_\mu) &= 0 \\ \partial_\mu(\bar{\psi}_L \gamma_\mu \psi_L - j^2_\mu) &= 0 \end{aligned} \tag{19}$$

From these we find

$$\partial_\mu(\bar{\psi}\gamma_\mu\gamma_5\psi) = \partial_\mu J^5_\mu = -2\partial_\mu j^2_\mu \tag{20}$$

Thus the anomaly is expressed here in terms of the second SL(2,C) component of the gauge field current J^2_μ

The term $\epsilon^{\alpha\beta\gamma\delta} Tr F_{\alpha\beta} F_{\gamma\delta}$ in the Lagrangian can actually be expressed as a four divergence of the form $\partial_\mu \Omega^\mu$. In fact Ω^μ is the Chern-Simons secondary characteristic class given by

$$\Omega^\mu = -\frac{1}{16\pi^2}\epsilon^{\mu\alpha\beta\gamma} Tr[A_\alpha F_\beta \gamma - \frac{2}{3}(A_\alpha A_\beta A_\gamma)] \tag{21}$$

and we recognize that the gauge field Lagrangian is related to the Pontryagin density

$$P = -\frac{1}{16\pi^2} Tr^* F_{\mu\nu} F_{\mu\nu} = \partial_\mu \Omega^\mu \tag{22}$$

The Pontryagin index

$$q = \int P d^4 x \tag{23}$$

is then a topologocal invariant.

The axial vector current is now modified as

$$\tilde{J}^5_\mu = J^5_\mu + 2h\Omega_\mu \tag{24}$$

and though $\partial_\mu J^5_\mu \neq 0$, we have $\partial_\mu \tilde{J}^5_\mu = 0$. From the relation(24), it is noted that Ω_μ is associated with j^2_μ and the Pontryagin index can be expressed as(12)

$$q = \int j^2_0 d^3 x = \int \partial_\mu j^2_\mu d^4 x \tag{25}$$

Since $j^2_\mu j^2_\mu$ term gives rise to the torsion,we note that the Pontryagin index is associated with the torsion term in the Einstein-Cartan action as observed by Drechsler(13) and is related to gravitational anomaly when we deal only with the Einstein part of the gravitational action. Indeed from the relation $Tr\ {}^* F_{\mu\nu} F_{\mu\nu} \sim \partial_\mu \Omega^\mu$ we note that the θ-term in the action (26) corresponds to the Pontryagin index. In fact, from the relation (24), we can write

$$j_\mu^2 = -\frac{1}{2}(J_\mu^5 + J_\mu^v) \qquad (26)$$

where J_μ is any arbitrary vector current which is conserved. Taking the particular solution,we can write

$$j_\mu^2 = -\frac{1}{2}J_\mu^5 = -\frac{1}{2}\bar{\psi}\gamma_\mu\gamma_5\psi \qquad (27)$$

so that

$$\begin{aligned} \partial_\mu j_\mu^2 &= -\tfrac{1}{2}\partial_\mu(\bar{\psi}\gamma_\mu\gamma_5\psi) \\ &= -\tfrac{1}{2}[(\partial_\mu\bar{\psi})\gamma_\mu\gamma_5\psi + \bar{\psi}\gamma_\mu\gamma_5(\partial_\mu\psi)] \\ &= -im\bar{\psi}\gamma_5\psi \end{aligned} \qquad (28)$$

This suggests that we can write

$$\partial_\mu j_\mu^2 I = im\gamma_\mu\bar{\psi}\gamma_\mu\gamma_5\psi = -2im\gamma_\mu j_\mu^2 \qquad (29)$$

where I is the identity matrix.
Thus we find

$$-4m^2\int j_\mu^2 j_\mu^2 d^4x = \int(\partial_\mu j_\mu^2)(\partial_\mu j_\mu^2)d^4x = qc \qquad (30)$$

wher q is the Pontryagin index given by $q = \int \partial_\mu j_\mu^2 d^4x$ as discussed earllier and c is any arbitrary constant. Thus we note that the torsion term given by $\int j_\mu^2 j_\mu^2 d^4x$ actually corresponds to the pontryagin index when various constants are normalized to be 1. In view of this, we find that the net efect of the θ-term in the action is just to introduce a torsion term and thus corresponds to the Einstein-Cartan action.

3 Loop Approach, Gauge Orbit Space and Vortex Line

By a loop γ we mean a continuous map from the unit interval into a d-dimensional manifold $\sum$, $\gamma : [0,1] \to \sum$ such that $\gamma(0) = \gamma(1)$. Given a loop γ and a connection (gauge potential) A, we can form the holonomy of γ

$$U(\gamma) = Pexp\int A_\mu(\gamma(t))\dot{\gamma}_\mu dt \qquad (31)$$

where P denotes the path ordering along γ. The traced holonomy defined by

$$T(\gamma) = TrU(\gamma) \tag{32}$$

is an object independent of the base point on γ and is invariant under gauge transformations of the connection A.

It has been pointed out by Loll(9) that there is an equivalence between the set of gauge potentials modulo gauge transformations and the space of complex- valued functions on loop space which turn out to be traced holonomies satisfying certain properties and Mandelstam constraints

$$\frac{U}{G} \simeq \{T(\gamma), \gamma \in loop\ \ space\}_{Mandelstam\ \ constraints} \tag{33}$$

where U is the space of all connections A and G is the space of all local gauge transformations. The constraints imply that loop variables $\{T(\gamma)\}$ where γ runs through all the loops in the manifold $\sum$ must satisfy very special properties in order that local gauge potentials (modulo gauge) may be reconstructed from them. Giles(8) has studied the problem of reconstruction of gauge fields from loop variables which involves certain algebraic constraints. If $P_{x,y}$ is a space-time curve oriented from y to x, parallel transport along $P_{x,y}$ is characterized by the path ordered phase

$$U(P_{x,y}) = Pexp \int_y^x A_\mu(x)dx^\mu \tag{34}$$

If $P_{x,y}$ and $P^1_{y,z}$ are two paths, their composite $P_{x,y}0P^1_{y,z}$ is a path from z to x whose associated path ordered phase is related to those of $P_{x,y}$ and $P^1_{y,z}$ by the product rule

$$U(P_{x,y}P_{y,z}) = U(P_{x,y})UU(P^1_{y,z}) \tag{35}$$

This is a principal algebraic rule which must be satisfied. As an example, we may cite that for a pair of loops γ_1 and γ_2 intersecting at a point x and gauge group $G = SU(2)$, we have

$$T(\gamma_1)T(\gamma_2) = T(\gamma_1 0_x \gamma_2) + T(\gamma_1 0_x \gamma_2^{-1}) \tag{36}$$

in the two dimensional representation. This becomes an identity when we define $T(\gamma)$ by (32) but if we start abstractly from the set of all loop functions, these relations have to be imposed as constraints. Since the gauge potential and also the associated field strength in nonabelian theories have a nontrivial behaviour under gauge transformation, they are not physical observables of the theory. In view of this attempts have been made to reformulate the theory in terms of gauge invariant variables. A natural set of gauge ivariant variables are the Wilson loops associated with parallel transport around closed space-time curves. These variables are nonlocal and overcomplete relative to the potentials and hence are constrained.

When these constraints are satisfied we can have an equivalence relation between loop space and gauge orbit space when a loop can be visualized as an orbit in the latter space. The paths characterizing different Wilson loops are distinguished from each other not by their local coordinates but by their homotopy classes. Now we may point out that the inclusion of the topological Lagrangian ($\theta - term$) in non-Abelian gauge theories introduces certain topologically nontrivial Abelian background gauge field in the configuration space associated with the vortex line and the field theory may be regarded as a particle whose position is indicated by A (non-Abelian gauge potential) moving in the space U of non-Abelian gauge potentials under the influence of an Abelian gauge potential A. In the language of differential forms, we can write

$$A = g^{-1}\partial g + g^{-1} a g \tag{37}$$

where $A = A_i dx^i$, $a = a_i dx^i$ The space of gauge orbits U/G where G denotes the space of local gauge transformations $g(x)$ consists of the points $a(x)$. Following Wu and Zee(14) we can consider a homotopic analysis of this gauge orbit space. Recalling that $\pi_3(g) = Z$ for all simple non-Abelian groups G and $\pi_2(G) = 0$, $\pi_n(U) = 0$ for all n, we can write

$$\pi_n(U/G) = \pi_{n-1}(G) \tag{38}$$

for $n > 1$

In $3+1$ dimension

$$\pi_1(U/G) = \pi_0(G) = \pi_3(G) = Z \tag{39}$$

The equality of $\pi_0(G) = \pi_3(G)$ follows from the condition that the gauge transformation $g(x)$ approaches a constant independent of the direction of x

as $x \to \alpha$. Thus U/G is multiply connected and has the topology of a ring and the corresponding field strength corresponds to a vortex line. In the previous section, we have observed that in $3+1$ dimension, the Pontryagin term ($\theta - term$) arises out of SL(2,C) invariance in spinor affine space where the gauge field belong to SL(2,C). These SL(2,C) gauge fields may be traced back to be generated from the spinorial variable $\theta(\bar{\theta})$ in the metric corresponding to the direction vector (vortex line) attached to the space-time point. This gives rise to the Pontryagin index associated with the charge corresponding to the gauge field part(13)

$$q = \int j_0^2 d^3x = \int_{surface} \in^{ijk} d\sigma^i F_{jk}^{(2)} \tag{40}$$

which suggests $F_{jk}^{(2)}$ to be the magnetic field like components for the background vector potential $B^{(2)}$ and so q is actually associated with the magnetic pole strength. In this way, the introduction of a vortex line attached to a space-time point effectively associates a background magnetic monopole(15). The fact that in $3+1$ dimension, the gauge orbit space U/G has the topology of a ring indicates that there is a hole in it. So the magnetic flux through the hole in the gauge orbit space is non-zero.In view of this the θ-vacuum may be taken to arise from the Bohm-Ahranov type of effect in ordinary space. Indeed, the situation is more general in this formalism. The Pontryagin term is associated with the Berry phase through the relationship with the chiral anomaly and the Berry phase may be viewed as a generalized version of Bohm- Ahranov phase. In fact, we have the relation for the Berry phase(16)

$$e^{i\phi} = e^{i2\pi\mu} \tag{41}$$

where

$$2\mu = q = \int j_0^2 d^3x = \int \partial_\mu j_\mu^2 d^4x = -\frac{1}{2} \int \partial_\mu J_\mu^5 d^4x \tag{42}$$

Thus the background Abelian gauge field when the θ-term is introduced in a non-Abelian gauge theory is effectively responsible for the Berry phase. Simon(17) has pointed out that the Berry phase is essentially given by the holonomy when a particle moves in a closed parameter space. Indeed, the loop space variable(holonomy) is associated with this topological phase factor. The equivalence between the holonomy variables and the gauge potentials modulo gauge transformations satisfying certain constraints now reveals that the topology of the gauge orbit space also manifests itself in the loop space formalism and the nonlocality of this loop space variables is associated with a vortex line. Thus there is no way we could arrive at a corresponding continuum limit

4 Torsion and the constraint Equation

Bengtsson and Peldan(10) have pointed out that the Hamiltonian constraint gets modified by the θ-term representing torsion though the other constraonts remain unaffected. It bears the seed of the fact that the Hamiltonian constraint may be taken to arise when the torsion term tends to be vanishing. This is so when we consider the loop approach also. Jacobson and Smolin(18) considered an infinite dimensional family of quantum states of the gravitational field using self-dual representation based on Ashtekar's variables and constructed from Wilson loops. Any holomorphic gauge invariant functional of A_a^i depends on the connection only via the trace of its holonomy map $H : \alpha_\Sigma \to C$ where α_Σ is the loop space of Σ. On a loop $\gamma : S^1 \to \Sigma$, H is defined by

$$H(\gamma) = Tr Pexp \oint ds A_a^i(\gamma(s))\dot{\gamma}^a(s)\tau^i \tag{43}$$

Where τ^i are the Pauli matrices. For each γ, $H(\gamma)$ is a gauge invariant holomorphic functional of A_a^i. A space of states can be constructed out of functions of the $H(\gamma)'s$. The state

$$|0>= 1 \tag{44}$$

can be called the ground state. Given N closed curves γ_α, $\alpha = 1........N$ in Σ we can define an N loop state

$$|\{\gamma_\alpha\} >= \pi_{\alpha=1}^{N} \; H(\gamma_\alpha) \tag{45}$$

The totality of all states formed by linear combinations of states of the form(44) and (45) can be called the "Fock space" associated to the loop space of Σ. Any gauge invariant functional $\psi[A_a^i]$ of the complexified SU(2) connection A_a^i on Σ can be regarded equivalently as a functional of the trace of the holonomy map of the connection. Now to study the action of the Hamiltonian constraint on the loop variables, we take the simplest operator representation for the "electric field"

$$E_i^a = \frac{\delta}{\delta A_a^i(x)} \tag{46}$$

which is determined by the commutation relations

$$[A_a^i(x) \ , \ E_j^b(y)] = -\sqrt{2}h\delta_a^b\delta_j^i\delta^3(x,y)$$

$$[E_i^a(x) \ , E_j^b(y)] = 0 \tag{47}$$

This helps us to write the Hamiltonian constraint as

$$C(x) = \epsilon^{ijk} F_{ab}^k(x) \frac{\delta^2}{\delta A_a^i(x) \delta A_b^j(x)} \tag{48}$$

with a proper choice of ordering and the constraint equation becomes

$$C(x)\psi[A_a^i] = \epsilon^{ijk} F_{ab}^k \frac{\delta^2}{\delta A_a^i(x)\delta A_b^j(x)} \psi[A] = 0 \tag{49}$$

The other constraint equations take the form

$$C_a(x)\psi[A_a^i] = F_{ab}^i(x) \frac{\delta}{\delta A_b^i(x)} \psi[A] = 0 \tag{50}$$

$$G^i(x)\psi[A_a^i] = D_a \frac{\delta}{\delta A_a^i(x)} \psi[A] = 0 \tag{51}$$

with a particular choice of operator ordering. The choice of the ordering can be done from the following considerations. The ordering of the constraint G^i is determined by the requirement that it generates infinitesimal gauge transformations. Similarly the ordering of the diffeomorphism constraint is determined by the requirement that it generates on gauge invariant functionals infinitesimal deffeomorphisms. The ordering of the Hamiltonian constraint may be considered from the requirement that the algebra of constraints be consistent so that it has no anomalous c-number term. Jacobson and Smolin have shown that there exists a regularization procedure such that

$$C(N)|\{\gamma_\alpha\} >= 0 \tag{52}$$

when the curves γ_α are differentiable and nonintersecting and C(N) is the Hamiltonian constraint smeared with any smooth density

$$C(N) = \int_\Sigma N(x)C(x) \tag{53}$$

As mentioned above, the gauge constraint (51) means that $\psi[A_a^i]$ is a gauge invariant functional of the complexified SU(2) connection A_a^i on Σ, Now the holonomy map

$$h_A : \alpha_\Sigma \to SL(2, C)$$

where α_Σ is the loop space of Σ and

$$h_A(\gamma) = Pexp \oint ds A_a^i(\gamma(s))\dot{\gamma}^a(s)\tau^i \tag{54}$$

takes into account the gauge invariant information in A_a^i . Therefore any gauge invariant functional $\psi[A_a^i]$ can be regarded as a functional of h_A. However h_A is not actually gauge invariant since it transforms at the base point of the loop. Thus functionals of h_A will not quite be annihilated by the gauge constraint G_i. This is remedied by taking the functionals of the trace of the holonomy map so that we take

$$H(\gamma) = Trh_A(\gamma)$$

Now to evalute

$$C(N)|\{\gamma_\alpha\} >$$

where

$$|\{\gamma_\alpha\} >= \pi_{\alpha=1}^{N} H(\gamma_\alpha)$$

we note that γ is a differentiable loop parametrized by the unit interval [0,1]. The first functional derivative involved in the unregulated calculation of $C(N)h_A(\gamma)$ yields (18)

$$\frac{\delta}{\delta A_b^j(x)} h_A(\gamma) = \int_0^1 ds \delta^3(x, \gamma(s))\dot{\gamma}^b(s)U(1, s)\tau^j U(s, 0) \tag{55}$$

where

$$U(t,s) = Pexp \int_s^t du A_a^i(\gamma(u))\dot{\gamma}^{\sigma}(u)\tau^i \tag{56}$$

After a second functional differentiation we obtain

$$\begin{aligned} C(N)h_A(\gamma) &= \textstyle\int d^3x N(x) F_{ab}^{ij}(x) \int ds \int dt \delta^3(x;\gamma(t))\delta^3(x,\gamma(s))\dot{\gamma}^a(s)\dot{\gamma}^b(t) \\ &\quad [\theta(s-t)U(1,s)\tau^i U((s,t)\tau^j U(t,0) + \theta(t-s)U(1,t) \\ &\quad \tau^j U(t,s)\tau^i U(s,0)] \\ &= \{-\tfrac{1}{4}\textstyle\int ds \int dt \delta^3(\gamma(s),\gamma(t)) \\ &\quad N(\gamma(s))F_{ab}^{ij}(\gamma(s))\dot{\gamma}^a(s)\dot{\gamma}^b(s)\delta^{ij}\}h_A(\gamma) \\ &= \{-\tfrac{1}{4}\textstyle\int ds \frac{\delta^2(0)}{J(s)} N(\gamma(s))F_{ab}^{ij}(\gamma(s))\dot{\gamma}^a(s)\dot{\gamma}^b(s)\delta^{ij}\}h_A(\gamma) \end{aligned} \tag{57}$$

where we have used the relation $\tau^{(i}\tau^{j)} = -\frac{1}{4}\delta^{ij}$ ($\tau^i = \frac{1}{2i}$*times the Pauli matrics*) and $J(s)$ is the Jacobian factor. Here we note that the contraction of F_{ab}^{ij} with $\dot{\gamma}^a(s)\dot{\gamma}^b(s)$ leads to zero but the difficulty lies with the factor $\delta^2(0)$ which reflects the δ - function present in the first functional derivative (55). This needs certain regularization procedure. Jacobson and Smolin considered the "smearing" the states into expressions involving finite "flux tubes" rather than holonomy elements. The regulated functions are taken as follows. A two dimensional smearing density function $f(\sigma)$ is chosen which is taken to be Gaussian. They considered the expression

$$B(s) = \int d^2\sigma f(\sigma) A_a^j(\gamma(s);\sigma)\dot{\gamma}^a(s;\sigma)\tau^j \tag{58}$$

where

$$\dot{\gamma}^a(s;\sigma) = [\frac{\partial\gamma(s;\sigma)}{\partial s}]^a \tag{59}$$

The regulated holonomy functional is then defined by

$$h_A^f(\gamma) = Pexp \int ds B(s) \tag{60}$$

It is noted that h_A^f is not gauge covariant. The gauge covariance is restored only in the limiting case

$$\lim_{f\to\delta^2} h_A^f(\gamma) = h_A(\gamma) \tag{61}$$

To evalute the Hamiltonian constraint on $h^f_A(\gamma)$, we note that the first functional derivative yields

$$\begin{aligned}\frac{\delta}{\delta A^j_b(x)}h^f_A(\gamma) &= \int dsU^f(1,s)[\frac{\delta}{\delta A^j_b(x)}B(s)]U^f(s,0) \\ &= \int ds\int d^2\sigma f(\sigma)\delta^3(x,\gamma(s;\sigma)) \\ &\quad \dot{\gamma}^b(s;\sigma)U^f(1,s)\tau^j U^f(s,0)\end{aligned} \tag{62}$$

where

$$U^f(t,s) = Pexp\int_s^t duB(u)$$

This, in the coordinate system (s,σ) gives

$$\frac{\delta}{\delta A^j_b(x)}h^f_A(\gamma) = f(x)\dot{\gamma}^b(x)U^f(1,x)\tau^j U^f(x,0)$$

where

$$f(x) = f(x^2,x^3); \gamma(x) = \gamma(x^1,x^2,x^3); U^f(1,x) = U^f(1,x^1)$$

Taking another derivative

$$C(x)h^f_A(x) = -\frac{1}{4}f^2(x)F^{ij}_{ab}\dot{\gamma}^a(x)\dot{\gamma}^b(x)\delta ij h^f_A = 0 \tag{63}$$

Thus $h^f_A(\gamma)$ provides an exact solution to the Hamiltonian constraint C for any smooth $f(\sigma)$ and for any differentiable nonintersecting $\gamma(s,\sigma)$. However this has been achieved at the expense of gauge invariance, $B(s)$ does not transform as a gauge covariant quantity. Jacobson and Smolin have pointed out that in the limit$f(\sigma)$ approaches a δ -function, gauge invariance is restored and we can write

$$C(N)h_A(\gamma) = \lim_{f\to\delta^2} C(N)h^f_A(\gamma) \tag{64}$$

With this definition, we have

$$C(N)h_A(\gamma) = 0 \tag{65}$$

It is argued that since $C(N)h_A(\gamma)$ is not well defined as it involves the product of divergent and a vanishing quantity we can safely take that $h_A(\gamma)$ is on the boundary of domain of functionals of the form $h^f_A(\gamma)$ on which the action of the operator $C(N)$ is welldefined. We can then extend the domain to the boundary by continuity. In this way we can achieve the Hamiltonian constraint equation in loop space.

However as we have pointed out in the previous section that the loop space may be made equivalent to the gauge orbit space provided certain algebraic constraints are satisfied, and the θ-term in the action suggests that the topology of the gauge orbit space incorporates a vortex line, we cannot arrive at a corresponding continuum llimit. In essence, the loop space is nonlocal in nature. This forbids us to take the limit $f \rightarrow \delta^2$ when the θ-term is present in the Lagrangian.

We have discussed above that the θ-term is associated with torsion in the gravitational action and so the above regularization procedure is valid when there is no torsion. That means, the Hamiltonian constraint can only be achieved when we take the limit torsion tending to zero. This is in conformity with the results obtained by other considerations also(6). However, the other constraints remain unaffected by the torsion term. Thus we see that the canonical quantization along with the constraint equations can be taken to arise when the torsion term in the covariant quantization scheme is taken to be vanishing.

5 Discussion

The Lagrangian formulation of Capovilla, Jacobson and Dell (CJD) avoids the metric and its equivalence with Ashtekar's formulation in $3+1$ decomposition under a certain condition suggests that the metric is not a fundamental entity. The equivalence of the loop space variables (holonomy) with the set of gauge potentials modulo gauge transformations when certain algebraic constraints are satisfied suggests gravitation can be well described in terms of gauge invariant objects. However this gauge invariance is lost when we evalute the Hamiltonian constraint equation using certain regularization procedure. This regularization procedure is valid when the topology of the gauge orbit space does not involve any vortex line and we can arrive at a continuum limit. As the vortex line is associated with the torsion term, the canonical quantization can be consistently achieved when there is no torsion. The vanishing of the Hamiltonian when the torsion term tends to zero effectively leads to the Hamiltonian constraint equation which corresponds to the

Wheeler-Dewitt equation in the metric formulation of the canonical quantization scheme of gravity. Indeed, in the metric formulation of covariant quantization of gravity, the torsion term is found to be associated with the metric fluctuation when we introduce an anisotrophy in the fluctuating component(19). This anisotropic feature of the internal space is associated with the quantization of a fermion when a massive elementary fermion appears as a stochastically extented particle(20). Thus the fact that canonical quantization along with the constraint equations arise when the torsion term in the covariant quantization scheme is taken to be vanishing is analogous to the result that nonrelativistic quantum mechanics arises when we tke the sharp point limit of a stochastically extended relativistic particle as pointed out by Prugovecki(21). This provides the link between covariant and canonical quantization of gravity and the constraints associated with the Hamiltonian formulation of classical relativity are anologous to those occurring in the theory of a relativistic particle. It is to be noted that the origin of torsion in the CJD action is the θ- term and this term is a four-divergence indicating that it bears no effect classically but contribute at the quantum mechanical level. In view of this torsion appears as a quantum mechanical effect. As the θ-term is T-violating it suggests an arrow of time. This θ-term induces a vortex line in the gauge orbit space and effectively corresponds to the essential nonlocality of loop space. This nonlocality is thus associated with torsion. We can arrive at the continuum limit only when we take the vortex line vanishing suggesting that we "forget" about time. That is there is no time, no before, no after. In view of this, we observe that the problem of time in canonical quantization of gravity has its bearing in nonperturbative quantum geometry also when we take the gauge invariant holonomy variables as the fundamental entity.

Acknowledgement : The author is grateful to **Mrs.BINA ROY** for LATEX production of the manuscript.

References

[1] A.Ashtekar : Phys.Rev.D36,1587 (1987) New Perspectives in Canonical Gravity (Bibliopolis, Naples 1988)

[2] R. Capovilla, T. Jacobson and J. Dell : Phys.Rev.Lett.63, 2325, (1989) Class Quant. Grav. 8, 59 (1991)

[3] E.T.Newman and R.Penrose :J.Math.Phys.3, 566 (1962)

[4] M.Carmeli and S.Malin : Ann. Phys.103, 208 (1977) M.Carmeli : Group Theory and General Relativity (McGraw Hill , New York (1984) M.Carmeli, E.Leibowitz and N. Nissani : Gravitation, SL(2,C) Gauge Theory and Conservation Laws (World Scientific 1990)

[5] A. Bandyopadhyay, P.Chatterjee and P.Bandyopadhyay Gen. Rel. and Grav. 18, 1193 (1986) P.Bandyopadhyay : Festschrift volume devoted to the 85th Jubilee of D.D.Ivanenko (World Scientific 1991) p.239

[6] L.Mullick and P.Bandyopadhyay : J. Math.Phys 36, 370 (1995)

[7] S.L.Adler : Rev.Mod.Phys 54, 729 (1982)

[8] R.Giles : Phys.Rev. D24, 2160 (1981)

[9] R.Loll : Proceedings of the XIX international Colloquium on Group Theoretical Methods in Physics, Salamanca, Spain (1992)

[10] I.Bengtsson and P.Peldan : Int. J.Mod.Phys A7, 1287 (1992)

[11] I.Bengtsson and P.Peldan : Phys.Lett. B244, 261 (1990)

[12] A.Roy and P.Bandyopadhyay : J.Math.Phys. 30, 2366 (1989)

[13] W.Drechsler : Gen.Rel.Grav.15, 703 (1983)

[14] Y.Wu and A.Zee : Nucl.Phys. B 258, 157 (1985)

[15] K.Sen and P.Bandyopadhyay : J.Math.Phys. 35, 2270 (1994)

[16] D.Banerjee and P.Bandyopadhyay : J.Math.Phys. 33, 990 (1992)

[17] B.Simon : Phys. Rev. Lett. 51, 2167 (1983)

[18] T.Jacobson and L.Smolin : Nucl.Phys.B 299 295 (1988)

[19] P.Mahato and P.Bandyopadhyay : Nuovo Cimento B 98, 53 (1987)

[20] P.Bandyopadhyay and K.Hajra : J.Math.Phys. 28, 711 (1989) K.Hajra and P.Bandyopadhyay : Phys. Lett . A 155, 7 (1991)

[21] E.Prugovecki : Stochastic Quantum Mechanics and Quantum Space-time, (Reidel, Dordrecht, The Netherlands 1985).

RELATIVISTIC THEORIES OF HEAT

Peter G. Bergmann

Attempts to modify the classical theories of heat so as to render them compatible with relativity are frustrated by the fundamental incompatibility of thermodynamics and statistical mechanics, exemplified by the absence or presence, respectively, of an arrow of time. One possible approach is a model theory that interpolates between the two "pure" theories. The model that is presented here is tried out on the notions of performance of work, the transfer of heat, and entropy.

In the presence of a gravitational potential the concept of temperature as a criterium of thermal equilibrium is split into a global temperature, which must have the same value throughout a spatial domain at equilibrium even if the gravitational potential varies with location, and a local temperature, whose variability is required by the principle of equivalence. Examples presented are a relativistic perfect gas and a radiation field. A thermometer measures the local temperature.

Finally a Lorentz-invariant formulation of thermodynamics suggests that the reciprocal of the temperature in classical thermodynamics be replaced by a four-vector, which is not necessarily proportional to the velocity vector for such systems as a radiation pencil.

RELATIVISTIC THEORIES OF HEAT

Peter G. Bergmann
Syracuse University and New York University

Relativistic thermodynamics as a program goes back to the early years of relativity, with A. Einstein and M. Planck as the initiators. Thermodynamics and statistical mechanics had formed part of the foundations of (classical) physics. It was only natural that attempts were made to adapt these theories to the novel concepts of space and time that define special and general relativity. The work of the pioneers was partly continued, partly attacked by a number of workers, among whom Ott and Møller are probably best known.[1] More recently A. Komar and I have devoted ourselves once more to these ancient topics. We believe that we have shed some new light on these subjects; but we have encountered serious difficulities, not all of which we have overcome.

In order to develop a more general theory of heat, one might start either with classical thermodynamics or with classical statistical mechanics. One would hope that either way one would obtain similar results. But one realizes very quickly that fundamentally these two theories are not compatible. Thermodynamics deals with the dynamics of a single physical system, whereas statistical mechanics is concerned with ensembles of such objects. Statistical mechanics is basically time-symmetric, whereas thermodynamics emphasizes, in the Second Law, the distinction between past and future. And whereas in thermodynamics the state of an object is described fully by a finite set of data, the Gibbsian ensemble requires for its complete specification an infinite data set, a function of the phase space coordinates.

Before one can hope to construct relativistic analogs to the classical theories of heat, one needs to decide how to bridge their contradictions. This is not a task with a single clear-cut solution. Komar and I have tried diverse approaches, and our attempts will be presented in different and independent talks.

CONTINUOUS THERMALIZATION.

In the classical theory of heat one outstanding variable is the entropy. Though the thermodynamic and the statistical entropy are defined differently, and obey different laws (one increases monotonically with time, the other is a constant of the motion), they have enough in common that their similarities are

---

[1]For references see, for instance, V. H. Hamity, Phys. Rev. 187, 1745-1752 (1969).

conspicous. One can construct a model of a theory that interpolates, as it were, the two entropies.

This model works in the phase space of statistical mechanics, but drastically reduces the degree of arbitrariness of the probability density: Give a Hamiltonian function of the canonical coordinates, H(x), where x stands for the canonical coordinates , the distribution $\sigma(x)$ depends on H and on a single parameter β (representing the "reciprocal temperature") as follows:

$$\sigma(x) = \frac{1}{Z}e^{-\beta H}, \qquad (1.1)$$

$$Z = \int e^{\beta H}\, dx.$$

The integral that fixes Z is to extend over the whole of phase space. Obviously, the distribution is non-negative, and the integral $\int \sigma\, dx$ equals unity.

With the help of the expression (1.1) one can form the "mean" of any function F(x) over phase space,

$$\langle F\rangle \equiv \int F\sigma\, dx \qquad (1.2)$$

In particular the mean of the Hamiltonian, U, is the expression

$$U \equiv \langle H\rangle \equiv \int H\sigma\, dx\,. \qquad (1.3)$$

If the Hamiltonian and the parameter β are changed infinitesimally, the resulting change in U is

$$\delta U = \int H\, \delta\sigma\, dx + \int \sigma\, \delta H\, dx. \qquad (1.4)$$

The two integrals on the right may be identified with the transfer of heat and with the performance of work on the object, respectively, on account of the relationship

$$\delta U = \delta H \qquad (1.5)$$

of classical mechanics, which suggests the identification of the second integral. If this identification is accepted, that of the first integral follows as a matter of course.

Entropy.

The first integral on the right of Eq. (1.4), $\int H\, \delta\sigma\, dx$, is not an exact differential. It becomes exact if multiplied by β:

$$\beta \int H\delta\sigma\, dx = -\delta\langle \ln \sigma\rangle = \delta(\ln Z + \beta U)\,. \qquad (1.6)$$

This result suggests the further identification of the entropy,

$$S = -\langle \ln \sigma \rangle \tag{1.7}$$

and of β as the reciprocal of the Kelvin temperature. Reversible adiabatic processes are those in which the value of S remains constant.

In this model on an "interpolating theory" one can visualize the restriction to distributions of the form (1.1) as a device that restores the canonicity of the distribution after every sampling, or, in other words, a device that "thermalizes" the ensemble sufficiently often, so that its deviation from canonicity (which is required by Liouville's equation) is not observed. Even so the Second Law of thermodynamics cannot be deduced from the foundations of that formalism, but it can be added without leading to inconsistencies.

TEMPERATURE AND GRAVITATIONAL POTENTIAL

In what follows it is argued that thermal equilibrium of spatially extended systems is characterized by the presence of a "global" temperature, but that there also exists a "local" temperature, which is sensitive to the value of the gravitational potential. It is further asserted that thermometers read out the local temperature.

With any reasonable definition of "thermodynamic equilibrium" it would appear that both matter and radiation fields in an extended spatial domain can be in equilibrium only if the temperature (and other relevant thermodynamic potentials) is well defined and has the same numerical value throughout that domain. But if the gravitational potential has different values at different locations, then general relativity predicts a "gravitational red shift" even if there is thermodynamic equilibrium.

Suppose, for instance, that chunks of Fe_{57} interact with each other by emitting and absorbing Mössbauer x-rays. Then theory predicts, and experiment confirms, that resonance can be established only by offsetting the gravitational red shift by a compensating Doppler shift. If the sharp Mössbauer line is replaced by black-body thermal radiation, we must extrapolate the behavior of the single sharp line to each small frequency segment, resulting in a shift of the broadband spectrum as a whole, which will be perceived as a change in temperature associated with the change in gravitational potential.

Seemingly, we are confronted by a paradox. On the one hand, the very notion of temperature as a thermodynamic potential appears to call for a single temperture in an extended spatial domain as a condition for thermodynamic equilibrium, regardless of a spatially variable gravitational potential. On the other, the phenomenon of gravitational red shift inexorably involves a dependence of the temperature on the local value of the gravitational potential if equilibrium is to be achieved.

This problem is resolved by the introduction of two different quantities, both exhibiting characteristics ordinarily associated with the notion of temperature. One of these quantities will need to be constant throughout the spatial domain in question as a precondition for thermal equilibrium; it will be called the "global temperature". The other will exhibit the dependence on the gravitational potential that accounts for the gravitational red shift; it will be referred to as the "local" temperature. These two temperature concepts will be connected with each other in examples whose properties are well known from statistical mechanics.

Consider first a three-dimensional extended region, with time forming the fourth dimension, which is endowed with properties that enable thermodynamic equilibrium to be formed, and which otherwise is kept as simple as possible. This four-dimensional manifold will serve as a stage for the construction of microdynamic models of macroscopic systems.

Thermodynamic equilibrium requires that the background space-time be stationary, which is another way of saying that our manifold must possess a time-like Killing field. If the Killing field is surface-forming, then the metric tensor can be given the form

$$g_{00} = 1+2V(\vec{x}), g_{ok} = 0, k = 1,2,3$$
$$g_{ik} = -h_{ik}(\vec{x}), i,k = 1,2,3. \qquad (2.1)$$

All components are functions of the three spatial cooordinates only; they are independent of x^0. For the present investigation the spatial structure is of no interest. Accordingly we shall specialize the metric further by requiring

$$h_{ik} = \delta_{ik}. \qquad (2.2)$$

As a final simplification of our model of space-time we shall disregard any effect of the dynamic system back into the metric. Thus the gravitational potential $V(\vec{x})$ will not be regarded as a dynamical variable but is to be considered given ab initio. This, then, is to be our model for a background space-time, on which equilibrium physical systems are to be examined.

Perfect Relativistic Non-Quantum Gas. Let N mass points of equal masses m travel on time-like geodesics in space-time defined in the preceding section. Along each curve the line element will be given by the expression

$$d\tau^2 = (1+2V)\ dt^2 - d\vec{x}^2, \tag{2.3}$$

hence the action for one "molecule" will be:

$$S = \int L dt\,, \quad L = -m(1+2V(\vec{x}) - \dot{\vec{x}}^2)^{1/2} \tag{2.4}$$

The Hamiltonian of a molecule is obtained by the usual Legendre transform. It turns out to have the form:

$$H = [(1+2V(\vec{x})\ (m^2+\vec{P}^2)]^{1/2}, \quad \vec{P} = (1+2V-\dot{\vec{x}}^2)^{-1/2} m\dot{\vec{x}}. \tag{2.5}$$

Incidentially, the canonical momentum is independent of the choice of the variable of integration in the action integral. The choice of $t=x^0$ does not destroy the essentially relativistic character of the result, which establishes H as the constant of the motion, the one that generates Killing displacements along the time axis. The Hamiltonian of the whole gas consists of the N-fold repetition of Eq. (2.5).

Consider now the perfect gas in a Gibbs canonical distribution. In the non-relativistic gas the Hamiltonian is the sum of two terms, one of which involves only the configuration coordinates $\vec{x}$, the other only the momentum coordinates $\vec{p}$. As a result, the momentum distribution at a fixed location does not depend on the gravitational potential; only the density varies. For the relativistic gas the Hamiltonian does not decompose that simply. At a fixed location the momentum spectrum is

$$\sigma = \frac{1}{Z'}\exp(-\sqrt{m^2+\vec{p}^2})/kT')\,, \quad T' = \frac{T}{\sqrt{1+2V(\vec{x})}}\,. \tag{2.6}$$

Z' is a normalizing coefficient, T the global temperature, which by assumption is independent of location, and T' the local temperature, which is sensitive to the local value of the gravitational potential $V(\vec{x})$.

At a location with large negative potential T' will be greater than T. That is to say, if an observer stationed at a location where V=0 monitors black-body radiation that has its origin near a large mass, he will find that to maintain radiative equilibrium he must permit the incoming radiation to interact with a body at the temperature T<T'.

Radiation Fields. Consider a radiation field of arbitrary integral spin that propagates at the speed of light, in an arbitrary but finite region, as in the preceding sections, but with some definite boundary conditions. Such a field can be decomposed into normal modes, each with a sinusoidal dependence on the time coordinate t. For the n-th mode the characteristic frequency will be designated as ϖ_n.

If the radiation is bosonic, the probability of there being l_n quanta in the n-th normal mode will be

$$p_n = (1-e^{-\hbar\omega_n/kT})e^{-l_n\frac{\hbar\varpi_n}{kT}} . \qquad (2.7)$$

This number cannot depend on the location of the observer, as the outcomes of any measurements of the numbers l_n are discrete. But an observer who uses his own clock to determine the characteristic frequency of the n-th normal mode will obtain the result

$$\varpi_n' = (1+2V)^{-1/2}\varpi_n \qquad (2.8)$$

The probabilities (2.7), associated with the locally determined frequencies (2.8), are consistent with a temperature T',

$$T' = (1+2V)^{-1/2}T \qquad (2.9)$$

the same outcome as in the preceding section.

Thermometers. A thermometer, having a small heat capacity, upon made to interact with a system or subsystem that is in thermodynamic equilibrium with its surroundings, will itself tend to attain equilibrium conditions without unduly changing the other system's conditions, and will provide a reading of the temperature. But which temperature, global or local?

Once equilibrium has been achieved, the thermometer must be at the same temperature as its surroundings, which is T. But there remains a distinction between the thermometer being at the temperature T and its exhibiting (on its read-out) that, or any other temperature. The location of the thermometer remains characterized by the value of g_{00}=1+2V. According to the principle of equivalence one must introduce a locally Minkowski coordinate system if ordinary laws of physics are to apply.

The preceding two examples demonstrate that for the laws of statistical mechanics to take their "ordinary" form, that is to say the form prevailing in the absence of gravitation, one needs to introduce the temperature T' of Eqs. (2.6) and (2.9), and a Minkowski coordinate system. The conclusion is that the thermometer will exhibit local temperature, T'.

Lorentz-Invariant Formulation. In thermodynamics we are usually concerned with objects that are at thermal equilibrium. In the absence of time-varying external influences an object at equilibrium does not experience changes in the course of time. Obviously, by that definition thermal equilibrium is not a Lorentz-invariant property. That definition can, however, be generalized easily. Thermal equilibrium shall be defined as any state that can be reduced to equilibrium in the older sense by a suitable Lorentz transformation. By this definition we have also introduced the notion of rest frame.

As long as a rest frame exists, the introduction of a manifestly Lorentz-covariant notation does not involve new physics. But what about a situation in which several objects exist side by side, and each has its own distinct frame? That is, of course, possible obly if these objects do not interact with each other, - otherwise each object experiences time-varying external influences. We can conform to all requirements by specifying that in the infinite past and in the infinite future there are no interactions, but at finite times there are.

To simplify matters, consider two objects, identified by labels A and B. In the infinite past they have, respectively, the four-dimensional linear momenta ${}^{A}p_{\rho}$ and ${}^{B}p_{\rho}$. In the future their momenta will be labeled by ${}^{C}p_{\rho}$ and ${}^{D}p_{\rho}$, respectively. By the First Law we must have

$$A_{p_\rho} + B_{p_\rho} = C_{p_\rho} + D_{p_\rho} \tag{3.1}$$

or

$$C_{p_\rho} - A_{p_\rho} = -(D_{p_\rho} - B_{p_\rho}) = \epsilon_\rho . \tag{3.2}$$

ϵ_ρ is the net linear four-momentum transferred in the course of interaction from the second to the first body.

Both objects have well-defined rest masses, before and after the period of interaction: rest masses obey no conservation law. They also have definite entropies (which are additive). The sum of the entropies of the two bodies cannot decrease as the result of their interactions, according to the Second Law, and that means that in general their interaction is not time-symmetric.

Originally I had believed that for a large class of physical objects rest mass and entropy increase and decrease together, and that therefore for any given set of parameters in Eq. (3.2) the interaction can take place only in one direction. But Art Komar has constructed an example that appears to be capable of moving in either direction. One way the two rest masses increase, the other way they decrease. Apparently my original analysis has been too simplistic.

We should welcome comments and suggestions.

A different approach to relativistic heat theory starts from statistical mechanics. The Boltzmann factor for a perfect gas at rest is $e^{-\beta E}$. If that gas is looked at from a different Lorentz frame, the same Boltzmann factor takes on the form $e^{-\beta^\rho p^\rho}$. Thus the four-vector β^ρ appears naturally[2], as a generalization of the single parameter β.

Originally this four-vector is the product of the scalar β, the reciprocal temperature in the rest frame, and of the unit four-velocity of the rest frame relative to the chosen coordinate system. Generalizations suggest themselves readily, for instance in the case of a collimated pencil of radiation that travels at the speed of light. Such a pencil has no rest frame. Even in non-relativistic so-called grand canonical ensembles there occur various poetentials, each of which makes a contribution to the Boltzmann factor. Thus, the introduction of additional potentials, $\beta^1, \ldots \beta^3$, represents but a small step in the generalization of the theory of heat. What is new is that their appearance is related to the existence of an invariance group, such as the Lorentz or the Poincaré group.

---

[2]D. Van Dantzig, Physics 6, 673 (1939); Proc. Koninkl. Ned. Akad. Wetenschap 42, 601 (1939); 42, 601 (1939); 42, 608 (1939).

CLASSICAL GRAVITY AND QUANTUM MATTER FIELDS IN UNIFIED FIELD THEORY

HORST-HEINO v. BORZESZKOWSKI
Technische Universität Berlin, Institut für Theoretische Physik,
„Gravitationsprojekt“, An der Sternwarte 16, D-14482 Potsdam
and
HANS-JÜRGEN TREDER
Rosa-Luxemburg-Str. 17a, D-14482 Potsdam

ABSTRACT

The Einstein-Schrödinger purely affine field theory of the non-symmetric field provides canonical field equations without constraints. These equations imply the Heisenberg-Pauli commutation rules of quantum field theory. In the Schrödinger gauging of the Einstein field coordinates $U^i_{kl} = \Gamma^i_{kl} - \delta^i_l \Gamma^m_{km}$, this unified geometric field theory becomes a model of the coupling between a quantized Maxwellian field in a medium and classical gravity. Therefore, independently of the question as to the physical truth of this model, its analysis performed in the present paper demonstrates that, in the framework of a quantized unified field theory, gravity can appear as a genuinely classical field.

1. Introduction: The Compatibility of Quantum Field Theory with Classical Gravitational Fields

The question of the compatibility of classical, i. e., non-quantized, Einsteinian gravity and the quantum structure of matter fields has been rediscussed in some recent papers. The answer given by Drechsler [1] says that quantized material source terms do not necessarily force also the Einsteinian (metric) part of geometry but only additional, non-metric, geometrical fields, coming into play at small distances, to develop quantum aspects. The physical and geometrical arguments supporting this statement are there formulated in the framework of a gauge theory originally proposed by Drechsler and quantized by using a geometro-stochastic quantization method developed by Prugovecki [2, 3, 4]. - In the present, paper we shall demonstrate that already the purely affine field theory of Einstein [5] and Schrödinger [6] provides a unified field theory in which the quantum matter fields are coupled to a classical Einsteinian gravitational field.

To show this we explore the fact that the Einstein-Schrödinger theory can be written in a canonical form with Hamiltonian equations without constraints. (That there are Hamiltonian features of the purely affine theory was already shown by Einstein [7, 8]). In its canonical formulation, the field variables, i. e. the configuration variables, are Einstein's affine tensors $U^i_{kl} = \Gamma^i_{kl} - \delta^i_l \Gamma^m_{km}$, and the canonically conjugate momenta are given by contractions of the curvature tensor Then the resulting Poisson brackets lead to the Heisenberg-Pauli commutation rules with corresponding uncertainty relations of the field variables and momenta satisfied in usual relativistic quantum field theory [9].

However, by using the Schrödinger gauging of the affine connection $\Gamma^{*i}_{kl} = \Gamma^{i}_{kl} - \delta^{\,i}_{\,k}\Gamma_l$, one can split the geometrical quantities in a Maxwell-like field described by the field variables $\Gamma_l = \frac{1}{2}(\Gamma^{m}_{em} - \Gamma^{m}_{ml})$ and the homothetic curvature taking the part of canonically conjugate momenta. Then, for the unified field, the Heisenberg-Pauli commutation rules and the corresponding canonical field equations remain valid. For the field quantities Γ_l this implies the validity of quantum electrodynamics. But the relations satisfied by the asterisk quantities resulting from the laws for the unified field say that the there are no measurable quantum effects related to the asterisk field Γ^{*i}_{kl}. Indeed, they turn out to be of a form that implies Rosenfeld's inequality relations, $\Delta g_{ik}\,\Delta x^i\,\Delta x^k \geq \kappa\hbar c$, stating a relation between the measurement fluctuations of the metric and of the coordinates.

The Einstein-Schrödinger theory thus is a unified field theory that incorporates Einsteinian gravity. This theory can, in principle, be quantized by the rules of canonical quantization. However, this quantization is physically meaningless for the gravitational constituents of the unified field, while the quantization remains meaningful for all the other (non-metric) field parts. (In this sense, gravity described by this theory is called "classical".)

In this paper, we do not want to discuss the "physical truth" of Einstein's "relativistic theory of the non-symmetric field". The Einstein-Schrödinger theory here serves us only as self-consistent model of a quantizable unified field theory describing quantum matter fields and classical gravity. It is used to show that the classical nature of gravity which is a consequence of its universality expressed by Einstein's strong principle of equivalence does not necessarily contradict the quantum features of matter.

To prepare our arguments, in Sec. II we shall discuss some aspects of purely metric general relativity theory by comparing it to the purely affine theory, and in Sec. III we describe the interpretative frame to be used in order to introduce Einsteinian gravity into a unified geometrical theory. Finally, we shall consider the Einstein-Schrödinger theory (Sec. IV).

2. Purely Metric and Purely Affine Theories

In a purely metric theory of general relativity, a Riemannian structure of space-time is assumed so that the metrical tensor $g_{ik}(x^l) = g_{ki}(x^l)$ is the (only) fundamental geometrical quantity. If one is interested in second-order differential equations, then, up to the cosmological term, one is uniquely led to the Einstein-Hilbert Langrange density $\mathbf{L}_{EH}$ which, in the holonomic representation, reads as

$$\mathbf{L}_{EH} = \mathbf{R} = g^{1/2} R = g^{1/2} g^{kl} R_{kl}, \tag{1}$$

where the Ricci tensor R_{kl},

$$R_{kl} = \delta^{\,m}_{\,n} R^{n}_{klm} = \delta^{\,m}_{\,n}\left(-\left\{ {n \atop kl} \right\}_{,m} + \left\{ {n \atop km} \right\}_{,l} - \left\{ {p \atop kl} \right\}\left\{ {n \atop pm} \right\} + \left\{ {p \atop km} \right\}\left\{ {n \atop pl} \right\} \right), \tag{2}$$

is formed via contraction of the Riemann-Christoffel tensor from the Christoffel affine connection

$$\left\{ {i \atop kl} \right\} = \left\{ {i \atop lk} \right\} = \frac{1}{2} g^{ip}\left(-g_{kl,p} + g_{lp,k} + g_{pk,l} \right). \tag{3}$$

$\mathbf{L}_{EH}$ is the only coordinate-covariant and Lorentz-invariant expression that one can construct in this geometry. From (1) one can deduce the first-order Lagrangian $\mathbf{L}_E$ introduced by Einstein [10]. $\mathbf{L}_E$ is Lorentz-invariant but not coordinate-covariant.

A similar situation arises when one considers the anholonomic representation of the Einstein-Hilbert Lagrangian using, instead of the g_{ik}, tetrads h_i^A (and the metrical spinors $\sigma_i^{\mu\nu'}$, respectively). In this case Moller's Lagrangian $\mathbf{L}_M$ [11] turns out to be the only first-order Lagrangian that one can construct. $\mathbf{L}_M$ is a scalar density under coordinate transformations, but it is not Lorentz-invariant.

Thus, with, $\mathbf{L}_{EH}$, $\mathbf{L}_E$, and $\mathbf{L}_M$ there exist Lagrangians leading via the variation by g_{ik}, h_i^A, and $\sigma_i^{\mu\nu'}$, respectively, to Einstein's vacuum equations. They do not however, provide the usual canonical field formalism. Indeed, $\mathbf{L}_{EH}$ is not of the necessary first-order form, and the first-order Lagrangians $\mathbf{L}_E$ and $\mathbf{L}_M$ provide, due to the missing invariance properties, physically uninterpretable results (Borzeszkowski and Treder [12]).[a] The physical reason for this is that the realization of general relativity within a metric theory implies the validity of the strong principle of equivalence, and, in virtue of the latter, the notion of energy localization is not meaningful.

On the quantum level, this circumstance is reflected by Rosenfeld's inequalities [13] what can be seen as follows. In the theory here under consideration, a matter tensor $T_{ik}(x^l) = T_{ki}(x^l)$ can be defined by Einstein's equations

$$R_{ik} = - \kappa (T_{ik} - \frac{1}{2} g_{ik} T) = - \kappa \tilde{T}_{ik} \qquad (4)$$

By taking the average of these equations over the volume $L_0 \cong (\Delta x)^3$ of the measurement region (given by the extent of the measurement body) one gets for the uncertainties (or inaccuracies) ΔR and $\Delta \Gamma$ of the quantities R_{ik} and T_{ik} the relation[b]

$$\Delta R \approx \kappa \Delta T \approx \kappa \Delta E (\Delta x)^{-3} \qquad (5)$$

and thus for the uncertainties Δg of g_{ik},

$$\Delta g \approx \Delta R (\Delta x)^2 \approx \kappa \Delta E (\Delta x)^{-1}, \qquad (6)$$

where ΔE is the uncertainty of E. Hence, Heisenberg's fourth uncertainty relation

$$\Delta E \Delta t \geq \hbar \qquad (7)$$

and Bohr's condition of a classical measurement body[c]

$$\Delta x \geq c \Delta t \qquad (8)$$

yield

$$\Delta g (\Delta x)^2 \approx \kappa \Delta E \Delta x \geq \kappa c \Delta E \Delta t \geq \kappa \hbar c \qquad (9)$$

and thus the Rosenfeld inequalitiy

$$\Delta g_{ik} \Delta x^i \Delta x^k \geq \kappa \hbar c, \qquad (10)$$

where $(\kappa \hbar c)^{1/2}$ is Planck's fundamental length.[d]

---

[a] This coincides with the fact discussed by Mielke [13] according to which attempts to construct a gravitational Hamiltonian are accompanied by a transition from GRT to its "teleparallelism equivalent" which differs from GRT.

[b] Here and in the following the $\approx$ signs relate the usual (stochastic) measurement errors, while $\geq$ denotes inequalities arising from fundamental relations of general relativity, quantum field, and measurement theories.

[c] In the Bohr-Rosenfeld paper [15] and the Rosenfeld paper [14] this condition is shown to be imposed in order to guarantee optimal measurements.

Eq. (10) states a relation between the uncertainties of the basic quantity g_{ik} of Riemannian geometry (and, consequently, of the Christoffel connection and the Riemann-Christoffel curvature tensor) and the position in space-time. This relation holds true for each method of quantization if gravity is determined by Einstein's equations. For, it is a simple consequence of these equations and of quantum mechanics applied to the matter source of Einsteinian gravity. Therefore, (10) must be compatible with each quantization procedure applied to such fields.

If one assumes Einstein's condition that the Lorentz signature of space-time has to be maintained, i. e., that $\Delta g \leq 1$, then Rosenfeld's relation (10) says that $(\Delta x) \geq \hbar\kappa c$, i. e., that Planck's length is the smallest measurable distance. According to Rosenfeld, relation (10) (possibly [e]) shows that, at distances of the order of magnitude of Planck's length, the quantum fluctuations induced by the measuring procedure are so violent that the quantum gravity region is not accessible to measurement. This interpretation is especially plausible when one follows Rosenfeld's own "measurement-theoretical" derivation of (10) (cf. also, Borzeszkowski and Treder [16]).

The Einstein-Schrödinger theory assumes space-times which are characterized by a non-symmetric affine connection Γ^i_{kl}. The simplest coordinate-covariant Lagrangian one can form of Γ^i_{kl} is given as

$$\mathfrak{H} = \frac{2}{\lambda}(-det\, P_{kl})^{1/2}, \tag{11}$$

where

$$P_{kl} = \delta_n{}^m P^n_{klm}. \tag{12}$$

The constant λ and the negative sign in (11) are conventions which do not influence the result, and P^n_{klm} is the curvature tensor constructed from Γ^i_{kl}:

$$P^n_{klm} = -\Gamma^n_{kl,m} + \Gamma^n_{km,l} - \Gamma^p_{kl}\Gamma^n_{pm} + \Gamma^p_{km}\Gamma^n_{pl} \tag{13}$$

As was shown by Einstein [5], in this theory the quantities

$$U^i_{kl} = \Gamma^i_{kl} - \delta^i_l \Gamma^p_{kp} \tag{14}$$

behave like canonical "field coordinates", and the components of the non-symmetric tensor P_{kl} take the part of the momenta. Therefore, here one finds some fundamentals of the canonical formalism (for details cf., Treder [17]). If it should be possible to complete the canonical scheme, on the quantum level, one has to use the commutation rules and uncertainty relations of the canonical formalism of quantization founded by Heisenberg and Pauli [9]. According to this procedure, generally one finds the uncertainty relations

$$\Delta\varphi\Delta(\partial(\sqrt{-g}L)/\partial\dot{\varphi})(\Delta x)^3 \geq \hbar \tag{15}$$

or

$$\Delta\varphi\Delta\pi(\Delta x)^3 \geq \hbar, \tag{15a}$$

---

[d] As the derivation shows, (10) is a consequence of the (strong) equivalence of passive and active gravitational masses to the inertial mass (and thus to energy) of bodies. This equivalence lies on the basis of Einstein's purely metric theory of general relativity. In virtue of this, the energy-momentum tensor of matter is the source of gravitational fields, and only therefore the above line of arguments can be given.

[e] Rosenfeld used this cautious formulation because he considered it necessary to examine his conclusions by a quantized theory of Einsteinian gravity or by a more general theory embracing gravity. In particular, he required to derive relation (10) from such a theory. In the following we shall see that one can deduce (10) from the Einstein-Schrödinger theory.

where L is the Lagrangian of the system considered and φ and π denote symbolically the field variables and momenta, respectively (the dot denotes the time derivative). Correspondingly, we find here the dispersion relation

$$\Delta U \Delta P \geq (\Delta x)^{-3} . \tag{15b}$$

This means, considering a (purely) affine theory of general relativity one is led to a situation which, to some extent, is similar to usual field physics. First, one finds canonically conjugate field quantities, and, second, the uncertainty relations that one can formally write down do not impose fundamental limitations on the measurability of fields (cf. for this, the discussion in [15]). Moreover, the fact that for $(\Delta x) \geq \hbar\kappa c$ the inequality (10) recovers the relation satisfied in the (purely) metric general relativity theory, namely

$$\Delta R \Delta \Gamma \geq \kappa\hbar c(\Delta x)^{-5} , \tag{16}$$

even signals that the affine and the metric theories are possibly compatible at distances greater than Planck's length. (Relation (16) is equivalent to (10) since Γ and R are of the order of magnitude g/L and g/L^2, respectively, where L is the characteristic length over which g changes significantly.)

Hence, the Einstein-Schrödinger affine field theory satisfies a requirement imposed by quantum mechanics on a general-relativistic theory of gravitation, namely the validity of Rosenfeld's uncertainty relations (10) and (16), respectively. These relations hold true in consequence of the usual canonical uncertainty relations valid for the unified field.

To consider our question in more physical terms in the following we shall discuss it in the framework of Einstein-Cartan theories providing the interpretative scheme for the unified theory.

3. On Einstein-Cartan Theories

According to Eisenhart [18], in non-Riemannian geometry the affine connection can always be written as

$$\Gamma^i_{kl} = \left\{ {i \atop kl} \right\} + \Omega^i_{kl} , \tag{17}$$

where $\left\{ {i \atop kl} \right\}$ is again the Christoffel affinity defined by the requirement to be a symmetric connection which determines a second rank tensor g_{ik} via Ricci's lemma,

$$g_{ik;l} = g_{ik,l} - \left\{ {p \atop il} \right\} g_{pk} - \left\{ {p \atop kl} \right\} g_{ip} = 0 , \tag{18}$$

and Ω^i_{kl} is a tensor without any index symmetry.

The curvature tensor (13) which can also be written as

$$P^i_{klm} = R^i_{klm} - \Omega^i_{kl;m} + \Omega^i_{km;l} + \Omega^r_{kl} \Omega^i_{rm} - \Omega^r_{km} \Omega^i_{rl} \tag{19}$$

(R^i_{klm} denotes the Riemann-Christoffel curvature tensor formed of {}) possesses the symmetry $P^i_{klm} = -P^i_{kml}$ so that one can construct from Eq. (19) by index contraction the Ricci tensor (12)

$$P_{kl} := R_{kl} + \Omega_{kl} = R_{kl} - \Omega^r_{kl;r} + \Omega^r_{kr;l} - \Omega^r_{kl} \Omega^n_{rn} + \Omega^r_{kn} \Omega^n_{rl} \tag{20}$$

which can be splitted into its symmetric and antisymmetric parts, $P_{(kl)}$ and $P_{[kl]}$,

$$P_{(kl)} := \frac{1}{2}(P_{kl} + P_{lk}) = R_{kl} + \Omega_{(kl)}$$
$$P_{[kl]} := \frac{1}{2}(P_{kl} - P_{lk}) = \Omega_{[kl]} \qquad (21)$$

and by, another contraction, the tensor of homothetic curvature,

$$S_{lm} = P^r_{rlm} = -\Omega^r_{rl;m} + \Omega^r_{rm;l} \qquad (22)$$

Following Einstein and Cartan, one can use this framework to generalize metric general relativity theory by interpreting the purely Riemannian parts $\left\{ {i \atop kl} \right\}$, R_{lm} and R^i_{klm} as quantities describing the gravitational field, while the other parts stemming from Ω^i_{kl} introduce new degrees of freedom which describe matter. Assuming now the strong principle of equivalence for gravity we can again define the matter tensor T_{kl} by relation (4). We shall call such a non-Riemannian framework requiring additionally the definition of matter by condition (4) an Einstein-Cartanian one.

Generally, the Einstein-Cartan theories, using Einstein's and Weyl's expression, are given as "mixed" and respective "neutral" theories. In these theories the metrical tensor g_{ik} and the Levi-Civita connection Γ^i_{kl} (and respectively the tensor Ω^i_{kl}) are the independent field variables. Accordingly, the field equations read

$$\frac{\delta \mathbf{L}}{\delta g^{ik}} = 0 \text{ and } \frac{\delta \mathbf{L}}{\delta \Gamma^i_{kl}} = 0 \text{ (and, respectively, } \frac{\delta \mathbf{L}}{\delta \Omega^i_{kl}} = 0\text{)}. \qquad (23)$$

In the purely affine theory proposed by Einstein (1923) and Schrödinger (1944, 1950) the Γ^i_{kl} (and respectively, their combinations U^i_{kl} defined in (14) are the only field variables so that the equations $\delta\mathbf{L}/\delta U^i_{kl} = 0$ result. We shall work with Schrödinger's Lagrangian since one then arrives at an immediately manifest model which does not require to conceive new (classical or quantum) methods.

4. Einstein-Schrödinger Theory

A classical example of an Einstein-Cartan theory is the Einstein-Schrödinger theory in the gauging of Schrödinger working with the so-called asterisk connection Γ^{*i}_{kl} .In this theory one can isolate gravitational and electromagnetic parts which respectively satisfy Einstein's and Maxwell's equations. Therefore, the Schrödinger gauging takes the purely affine theory discussed above into an Einstein-Cartanian form. To see this, let us summarize some features of this theory.

The field equations resulting from

$$\frac{\delta \mathbf{H}}{\delta P_{mn}} \frac{\delta P_{mn}}{\delta \Gamma^i_{kl}} = 0 \qquad (24)$$

read in Schrödinger's version,

$$P_{ik,l} - P_{rk}\Gamma^r_{il} - P_{ir}\Gamma^r_{lk} - \frac{2}{3}(P_{ik}\Gamma_l + P_{il}\Gamma_k) = 0\,, \qquad (25)$$

where Γ_l denotes the torsion vector,

$$\Gamma_l = \delta^m_n \Gamma^n_{[lm]} = \frac{1}{2}\delta^m_n(\Gamma^n_{lm} - \Gamma^n_{ml}) \tag{26}$$

and the variation of $\mathbf{H}$ by Γ_l yields

$$\frac{\delta \mathbf{H}}{\delta \Gamma_l} = 2(\sqrt{-g}\, P^{[lk]})_{,k} = 0\,. \tag{27}$$

Regarding that the reduced minor

$$\frac{2}{\lambda \mathbf{H}}\frac{\partial \mathbf{H}}{\partial P_{ik}} = g^{ik} = \frac{1}{\lambda} g^{im} g^{kn} P_{nm} = \frac{1}{\lambda} P^{ik} \tag{28}$$

defines the fundamental tensor g_{ik} satisfying the relation $g_{ir}g^{kr} = g_{ri}g^{rk} = \delta^k_i$ so that the equation $P_{ik} = \lambda g_{ik}$ holds, these equations can be rewritten in the form given by Einstein

$$\frac{\partial \mathbf{H}}{\partial U^l_{ik}} = (\sqrt{-g}\, g^{ik})_{,l} + (\sqrt{-g}\, g^{mk})(U^i_{ml} - \frac{1}{2}\delta^i_l U^r_{mr})$$

$$+ \sqrt{-g}\, g^{im}(U^k_{lm} - \frac{1}{2}\delta^k_l U^r_{rm}) := N^{ik}_l = 0 \tag{25a}$$

and

$$(\sqrt{-g}\, g^{[ik]})_{,k} = 0, \tag{27a}$$

where

$$U^l_{ik} = \Gamma^l_{ik} - \delta^l_k \Gamma^r_{ir} \tag{29}$$

is the Einstein affine tensor and

$$\Gamma^l_{ik} = U^l_{ik} - \frac{1}{3}\delta^l_k U^r_{ir} \tag{29a}$$

the Levi-Civita connection.

The equations (25) represent 60 conditions on the 64 components of Γ^l_{ik} so that the 4 components of the torsion vector remain undetermined. The relations (27) are conditions for the existence of solutions Γ^l_{ik} of (25). They are an expression of the fact that this purely affine theory is invariant under Einstein's super-gauge transformations according to which with each solution Γ^l_{ik} of (25) also

$$\Gamma'^l_{ik} = \Gamma^l_{ik} + \delta^l_i \Phi_k \tag{30}$$

is a solution of Eq. (25).

In order to discuss the physical meaning of the purely affine field theory Schrödinger had chosen the special gauging $\Phi_k = \frac{2}{3}\Gamma_k$ so that the connection is given by the following asterisk affinity

$$\Gamma^{*l}_{ik} = \Gamma^l_{ik} + (2/3)\delta^l_k \Gamma^r_{ir}\,. \tag{31}$$

Then the relation $\Gamma^*_l = \Gamma^{*n}_{ln} = 0$ is satisfied and equation (25) can be written as the "plus-minus lemma":

$$P_{ik,l} - P_{rk}\Gamma^{*r}_{il} - P_{ir}\Gamma^{*r}_{lk} = \lambda(g_{ik,l} - g_{rk}\Gamma^{*r}_{il} - g_{ir}\Gamma^{*r}_{lk}) = 0, \tag{32}$$

where

$$P_{ik} = P^*_{ik} - \frac{2}{3}(\Gamma_{i,k} - \Gamma_{k,i}). \tag{33}$$

The tensor $F_{ik} (= -F_{ki})$,

$$F_{ik} = P^*_{ik} - P_{ik} = P^*_{[ik]} - P_{[ik]} = \frac{2}{3}(\Gamma_{i,k} - \Gamma_{k,i}), \tag{34}$$

is connected with the Hermitian part S^H_{ik} of the tensor of homothetic curvature via the relation,

$$S^H_{ik} := -\frac{1}{2}(S_{ik} + \bar{S}_{ik}) = \frac{3}{2} F_{ik} \tag{35a}$$

$$S_{ik} = \delta^m_r P^r_{mik} = -\Gamma^r_{(ir),k} + \Gamma^r_{(kr),i} + \Gamma_{i,k} - \Gamma_{k,i}. \tag{35b}$$

As a consequence, the tensor F_{ik} satisfies the first group of the Maxwell equations identically,

$$F_{ik,l} + F_{kl,i} + F_{li,k} = 0. \tag{36}$$

Furthermore, according to (27) the following relations are valid:

$$\lambda(\sqrt{-g}\, g^{[ik]})_{,k} = (\sqrt{-g}\{P^{*[ik]} + F^{ik}\})_{,k} = 0. \tag{37}$$

These relations define a current density $\sqrt{-g}\, J^i$ with $(\sqrt{-g}\, J^i)_{,i} = 0$ by the second group of Maxwell-like equations

$$(\sqrt{-g}\, F^{ik})_{,k} = \sqrt{-g}\, J^i \quad \text{with} \quad \sqrt{-g}\, J^i := (\sqrt{-g}\, P^{*[ki]})_{,k}. \tag{38}$$

The current Γ_l is independent of the functions Γ_l (i. e., of the vector potential). Since however the g_{ik} are given by the P_{ik} one finds vacuum polarization.

Due to the covariance of $\mathbf{H}$, for each variation δx^l of the coordinates x^l the relation

$$\delta\mathbf{H} = N^{ik}_l U^l_{ik} - (\sqrt{-g}\, g^{ik}\, \delta U^r_{ik})_{,r} = 0 \tag{39}$$

is satisfied. In virtue of the field equations (25a), this identity provides Einstein's conservation law,

$$(\sqrt{-g}\, g^{ik}\, \delta U^r_{ik})_{,r} = 0 \tag{40}$$

or, via equation (28),

$$\frac{1}{\lambda}(\sqrt{-g}\, P^{ik}\, \delta U^r_{ik})_{,r} = 0. \tag{40a}$$

In the case of translations, where $(\delta x^i)_{,l} = 0$, the conservation law (40a) can be written as

$$-\lambda \frac{\delta\mathbf{H}}{\delta x^k} = \{\sqrt{-g}\, P^{mn}(\Gamma^r_{mn} - \delta^r_n \Gamma^s_{ms})_{,k}\}_{,r} = 0. \tag{41}$$

The affine tensor density

$$\vartheta^i_k = \sqrt{-g}\, P^{mn}(\Gamma^i_{mn} - \delta^i_n \Gamma^s_{ms})_{,k} = \lambda\sqrt{-g}\, g^{mn} U^i_{mn,k} \tag{42}$$

satisfying the conservation law (41) is a canonical energy-momentum complex ($\sqrt{-g}\, g^{mn} U^i_{mn,k}$ is the Einstein affine complex).

As was shown by P. G. Bergmann [19], the principles of general relativity (i. e., of coordinate covariance) and of equivalence (i. e., of minimal coupling) always imply the existence of a covariant vector density

$$\vartheta^i = \mathbf{L}\xi^i - \frac{\delta \mathbf{L}}{\delta \Gamma^l_{mn,i}} \bar{\delta} \Gamma^l_{mn} - \frac{\delta \mathbf{L}}{\delta \psi_{A,i}} \bar{\delta} \psi_A \tag{43}$$

satisfying the equation

$$\vartheta^i_{,i} = 0 \tag{44}$$

where the divergence is a scalar density ($\bar{\delta}K$ denotes the change of a quantity K with fixed values of the coordinates). In the case of Einstein's 1915 theory starting from the covariant Lagrangian $\mathbf{L} = \mathbf{L}_{EH} + \mathbf{L}_{mat}[g,\psi]$ one is led to Moller's expression reading for pure gravity [20]:

$$\vartheta^i_k = \{\sqrt{-g}(g_{kn,m} - g_{km,n})g^{im}g^{ln}\}_{,l} \qquad (\vartheta^k = \vartheta^k_i \alpha^i, \ \ \xi^i = \alpha^i = const.) \tag{45}$$

And in the present case, where one starts from the Lagrangian (11), one obtains (40).

Returning now to the discussion of the affine tensor density (41), with the asterisk affinities the expression $\sqrt{-g}\,\vartheta^i_k$ splits up into four terms,

$$\frac{1}{\sqrt{-g}}\vartheta^i_k = P^{*mn}U^{*i}_{mn,k} - \frac{1}{3}S^{mn}U^{*i}_{mn,k} + \frac{4}{3}P^{*[in]}\Gamma_{n,k} - \frac{2}{3}S^{in}\Gamma_{n,k}\,. \tag{46}$$

Regarding that, according to (36) and (38), Γ^{*i}_{kl} can be considered as a background field and $\Phi_l = (2/3)\Gamma_l$ as gauge field (with the field strength $(1/2)S_{ik}$) the relation (46) can be interpreted as follows. The second and the third terms of the right-hand side describe the interaction between gauge field and background, while the fourth term demonstrates that the homothetic curvature S^{in} and the torsion vector (of the unified field Γ^i_{kl}) behave like canonically conjugate quantities.

Thus, in the gauging of Schrödinger one arrives at Maxwell-like equations (36) and (38) in a polarized and charged medium. Therefore, the torsion vector $\Gamma_l = \Gamma^n_{[ln]}$ can be considered as vector potential $A_i = (2/3)\Gamma_i$ and the curvature tensor $F_{ik} = P^*_{ik} - P_{ik}$ plays the role of the field-strength, and the torsion vector and the homothetic curvature are canonically conjugate variables in the same sense as the vector potential and the field strength are such variables in the electrodynamics of charged and polarized media.

If one now regards that

$$P^i_{klm} = P^{*i}_{klm} - \frac{2}{3}\delta^i_k(\Gamma_{l,m} - \Gamma_{m,l}) \tag{47}$$

such that

$$P_{(kl)} = P^*_{(kl)} = R_{kl} + \Omega_{(kl)} \tag{48}$$

then one can rewrite relation (24) or, equivalently,

$$P_{kl} = \lambda g_{kl} \tag{49}$$

in the equation

$$R_{kl} = -\Omega_{(kl)} + \lambda g_{(kl)}\,, \tag{50}$$

where $\Omega_{(kl)}$ can be interpreted as the energy-momentum tensor of the electromagnetic field F_{kl} moving in a medium.

Therefore, in the gauging of Schrödinger the unified field Γ^i_{kl} is splitted into a field Γ_l satisfying Maxwell's equations and a Riemannian part which, via (50), defines the matter tensor :

$$\tilde{T}_{kl} = -\frac{1}{\kappa}\Omega_{(kl)}. \tag{51}$$

This version of a unified theory is thus an Einstein-Cartan theory.

Turning now to commutator and uncertainty relations one may state the following. In accordance with Eq. (46) there are three types of relations, namely relations between asterisk/asterisk, asterisk/asterisk-free and asterisk-free/asterisk-free quantities. The commutators of asterisk and asterisk-free quantities of course vanish because they are, due to the splitting procedure, logically independent, while the asterisk-free quantities describing the electromagnetic field satisfy commutator and uncertainty relations known of quantum electrodynamics. And as to the asterisk quantities describing gravity, one can derive uncertainty relations by utilizing the Einstein-Cartan structure of this theory. Indeed, from (48) - (50) it follows that

$$\Delta P^*_{(kl)} \approx \Delta P_{(kl)} \approx \Delta R_{kl} \approx \kappa \cdot \Delta T_{kl}, \tag{52}$$

and hence with (7)

$$\Delta P_{00} \approx \Delta R_{00} \approx \kappa \cdot \Delta E(\Delta x)^{-3}. \tag{53}$$

Further, the symmetric part of the affine connection satisfies the relation

$$\Delta \Gamma \approx \Delta\{\} \approx (\Delta x)^{-1}, \tag{54}$$

where the latter part of (54) results from that an uncertainty of the coordinates provides an uncertainty of the affinities of the order of magnitude of $(\Delta x)^{-1}$ (it stems from the inhomogeneous part of the transformation formula for affinities). Thus, for the gravitational field (now described by asterisk quantities), from (7), (8), (52), and (54) one finally finds again relation (16), namely

$$\Delta P^* \Delta \Gamma^* \geq \kappa\hbar c(\Delta x)^{-5} \tag{55}$$

which is equivalent to Rosenfeld's relation (10).

Thus, the theory considered here unifies Maxwellian fields in a medium with Einstein's gravity, where both satisfy the uncertainty relations respectively known of quantum electrodynamics and Einstein's metric general relativity. This means, gravity itself whose quantization, as it was argued in Sec. 1, does not lead to measurable local quantum effects and usual quantum matter fields moving in a background determined by gravity are consistently unified.

5. References

1. W. Drechsler, *Found. Phys.* **23** (1993) 261.
2. W. Drechsler, *Forschr. Phys.* **23** (1976) 607; W. Drechsler, W. *Found. Phys.* **7** (1977) 629.
3. W. Drechsler and E. Prugovecki, *Found. Phys.* **21** (1991) 23.
4. E. Prugovecki, *Found. Phys. Lett.* **2** (1989) 81, 163, 403; E. Prugovecki,. *Found. Phys.* **21** (1990) 93.
5. A. Einstein, *The Meaning of Relativity*, 5th ed., App. II (Princeton U. P., Princeton,1950).
6. E. Schrödinger, *Space-Time Structure* (Cambridge U. P., Princeton, 1950).
7. A. Einstein,, *Ber. Preuss. Akad. Wiss.*, (1923) pp. 32, 76.

8. A. Einstein, „Eddingtons Theorie und das Hamiltonsche Prinzip“, Appendix to: A. S. Eddington, *Relativitätstheorie in mathematischer Behandlung* (Springer, Berlin, 1925).
9. W. Heisenberg and W. Pauli, *Zeitschr. Phys.* **56** (1929) 1.
10. A. Einstein, *Ber. Preuss. Akad. Wiss.*, (1916) p. 1111.
11. C. Moller,. *Mat. Fys. Skr. Dan. Vid. Selskab.* **1** (1961) no. 10.
12. H.-H. v. Borzeszkowski and H.-J. Treder, H.-J., *Gen. Rel. Grav.* **25** (1993) 291.
13. E. W. Mielke, E.W., *Ann. Phys.* N. Y. **219** (1992) 78.
14. L. Rosenfeld, „Quantentheorie und Gravitation“ in *Entstehung, Entwicklung und Perspektiven der Gravitationstheorie*, ed. H.-J. Treder, (Akademie-Verlag, Ber lin, 1966).); English translation in *Selected Papers of Léon Rosenfeld*, eds. R. S. Cohen
and J. Stachel, (Reidel, Dordrecht, 1971).
15. N. Bohr and L. Rosenfeld, *Det. Kgl. Dan. Vid. Selskab., Mat.-fys. Medd.* **XII** (1933) no. 8.
16. H.-H. v. Borzeszkowski, and H.-J. Treder, *The Meaning of Quantum Gravity* (Reidel, Dordrecht, 1988).
17. H.-J. Treder, *Astron. Nachr.* **315** (1994) 1.
18. L. P. Eisenhart, *Non-Riemannian Geometry* (Amer. Math. Soc. Colloquium Publications, New York, 1994).
19. P. G. Bergmann,. in *Encyclopedia of Physics*, vol. IV, ed. S. Flügge (Springer, Berlin etc.. 1962).
20. C. Moller,. *Ann. Phys.* **4** (1958) 347.

MIXED FIELD THEORIES AND WEYL-CARTAN GEOMETRY

HORST-HEINO v. BORZESZKOWSKI
Technische Universität Berlin, Institut für Theoretische Physik,
„Gravitationsprojekt", An der Sternwarte 16, D-14482 Potsdam
and
HANS-JÜRGEN TREDER
Rosa-Luxemburg-Str. 17a, D-14482 Potsdam

ABSTRACT

Einstein's purely metric theory of general relativity (GRT) is based on Riemannian geometry, where both the Ricci and Weyl lemmata are satisfied. Generalizations of GRT generally lead to a violation of these lemmata. In particular, we argue that, performing an independent variation of the metric (or the tetrads) and the connection, one is automatically led to Weyl-Cartan-like generalizations of GRT for which the Weyl lemma is violated. In this context, we discuss the model first considered by Weyl 1950 and show that it can be interpreted as a Weyl-Cartan theory implying an additional Fermi-like interaction with coupling constant $(8\pi G / c^3)\hbar$.

1. Introduction

One can put forward sound arguments for imposing certain *a priori* conditions on the relation between the metrical basic quantities (the metric g_{ik} and the orthonormal tetrads h_i^A, respectively) and the connection of space-time. Such conditions are the Ricci lemma stating that the non-metricity of space-time,

$$\phi_{ikl} = g_{ik;l} \tag{1}$$

vanishes, $\phi_{ikl} = 0$, and second the Weyl lemma according to which the expression

$$D_l h_i^A = h_{i,l}^A + \Gamma_{Bl}^A h_i^B + \Gamma_{il}^k h_k^A \tag{2}$$

is equal to zero, $D_l h_i^A = 0$.[a] They represent relations between the metric, the tetrads, the tensor connection, and the spin connection. However, these *a priori* assumptions are problematic if one starts with an action principle and considers, besides the metrical basic quantities g_{ik} (or h_i^A), the coefficients of the connection of space-time as independent variables.[b]

In the course of generalizing Einstein's general relativity involving only Riemannian space-time it is generally accepted that, notwithstanding all physical problems arising then,

---

[a] The anholonomic indices *A, B,...* and the holonomic ones *i, k, ...* run both from *0* to *3*.
[b] A theory arising by such a procedure was called by Einstein [1] a „mixed theory" and by Weyl [2] a „neutral theory".

the metricity condition $\phi_{ikl} = 0$ can possibly, at least partly, be given up. The meaning of a non-vanishing covariant derivative of the metric was visualized by Hecht and Hehl [3] whereby arriving at a Weyl-Cartan space-time and, more generally, at a metric-affine space-time with nonvanishing torsion and nonmetricity.[c]

As far as the relation $D_l h_i^A = 0$ is concerned, the situation, however, seems to be more complicated. On the one hand, it is difficult to regard it as a separate condition for it only transforms the coefficients of the connection from a holonomic to an anholonomic frame or vice versa (cf., Treder and Borzeszkowski [5]; the same was stressed in Hehl et al. [4]). Therefore, one cannot drop it as long as one wants to work with *one* connection. On the other hand, one has to be cautious when one assumes, besides g_{ik} (or h_i^A), the existence of independent connection components. Indeed, requiring $D_l h_i^A = 0$, with respect to its coupling to gravity, one assumes a definite relation between tensorial and spinorial matter. Even if one considers this relation as physically meaningful in discussing generalizations of Einstein's general relativity it is more stringent, as in the case of the condition $g_{ik;l} = 0$, first to perform the independent variation of g_{ik} (or h_i^A) and Γ^i_{kl} (or Γ^A_{Bl}), and then, dependently on physical requirements, to formulate relations between g (or h) and Γ.

Generally speaking, when one works with a Lagrangian constructed within a metric-affine framework carrying metric (or tetrads and spinvectors, respectively) and connection as completely independent structures one has to vary by g (or h)and Γ without postulating any relations between g (or h)) and Γ *a priori.*[d] Such conditions can and should be imposed after the variation, namely in the course of the physical interpretation of the resulting equations. For, only then one knows those parts of the geometry on which, for reasons of measurement or other physical grounds, one can or has to impose certain additional conditions. [To avoid misunderstandings, in parentheses, it should be mentioned the following: To transform the connection calculated in this manner from holonomic to anholonomic coordinates and vice versa, of course, the equation $D_l h_i^A = 0$ must be used.]

To demonstrate this point and to discuss generalizations of Einstein's purely metric theory of general relativity (GRT) in this context, in the present paper, we shall perform the mixed variation of the Einstein-Hilbert Lagrangian coupled to tensorial and spinorial matter. This approach first considered by Weyl [2] leads to a Weyl-Cartan generalization of GRT satisfying the Ricci lemma but not the Weyl lemma.

To prepare this discussion, in Sec. 2, in the case of Weyl-Cartan space-times that result by mixed variation of the Einstein-Hilbert Lagrangian, we consider requirements to be imposed on geometry for measurement and other physical reasons. (Here we call, in con-

---

[c] This framework was used by Hehl et al. [4] to formulate a gauge theory of gravity.

[d] Here we are talking about metric-affine theories, where Einstein's purely metric general relativity appears as a degenerate case with merely g_{ik} as basic variables. It should, however, be mentioned that there is still another degenerate case, where the components of the connection are the only basic variables. These so-called purely affine theories were considered by Eddington [6], Einstein [7] ,and Schrödinger [8] in order to arrive at a unified field theory of gravity and matter. Then the field equations read $(\delta\mathbf{H}/\delta R_{ik})(\delta R_{ik}/\delta\Gamma) = 0$, where $\mathbf{H}$ now is a Lagrange density formed of Γ^i_{kl} and R^i_{ikl} is the non-symmetric Ricci tensor.

trast to the convention used by Hecht and Hehl [3], a space-time Weyl-Cartanian when either the Ricci lemma or/ and the Weyl lemma is/are violated. This is justified because both possibilities were considered by Weyl [2, 9, 10], and since, second, both lemmata show some similarity.)[e] The Weyl approach of 1950 reconsidered in Sec. 3 will then serve as an example of a Weyl-Cartan theory which, interpreted as Einstein-Cartan theory, implies an additional Fermi-like interaction with the coupling constant, $\kappa\hbar c$, where $\kappa = 8\pi G/c^4$ is Einstein's gravitational constant, $\hbar$ Planck's quantum, and c the speed of light [11].

2. The "Einstein-Hilbert Case" of a Weyl-Cartan Geometry

Let us consider a generalization of GRT that is free of additional physical hypotheses and ask for the measurement implications it brings about. This generalization results from the mixed variation of the Einstein-Hilbert Lagrangian and yields second-order differential equations, where the coupling between gravity and matter is determined by the requirements of general covariance and gauge invariance.

To start with, some remarks on GRT from the point of view of measurement. In GRT, three requirements are satisfied:

(i) *The Ricci Lemma*

$$g_{ik;l} = 0. \tag{3}$$

It specifies the relation between metric g and connection Γ so that Γ is given as (cf., Schouten [12]),

$$\Gamma^i_{kl} = \left\{ {i \atop kl} \right\} + g^{ir} g_{km} \Gamma^m_{[rl]} + g^{ir} g_{lm} \Gamma^m_{[rk]} + \Gamma^i_{[kl]}, \tag{4}$$

where

$$\left\{ {i \atop kl} \right\} = \frac{1}{2} g^{im} (g_{mk,l} - g_{kl,m} + g_{lm,k}). \tag{5}$$

Here the skew-symmetric part $\Gamma^i_{[kl]}$ can be chosen arbitrarily, and the symmetric part consisting of the first three terms on the right is then uniquely determined by g and Γ [12]. In GRT, where $\Gamma^i_{[kl]} = 0$, Eq. (4) leads to $\Gamma^i_{kl} = \left\{ {i \atop kl} \right\}$. As was argued by Schrödinger [8], Eq. (3) is a sufficient condition for $ds^2 = g_{ik} dx^i dx^k$ to be in accordance with the affine measure of distances along every geodesic. It guarantees that metric and distance measurement, on the one hand, and connection and direction comparison, on the other, are compatible.

From the physical point of view, the major meaning of Eq. (3) consists in that it ensures the telecomparison of length. Indeed, from (3) it follow

$$(A^i A^k g_{ik})_{;l} \frac{dx^l}{d\tau} = 0 , \tag{6}$$

---

[e] In reply to a critical comment of Friedrich Hehl we would like to add the following: In our terminology we follow the geometers who denote investigations of the compatibility of metric and connection in this manner. This notation goes back to Weyl's papers from 1918 until 1923 concerning the so-called *Mathematische Analyse des Raumproblems* and to Cartan's 1923 paper *Sur un theoreme fondamental de M. H. Weyl* where, with respect to Weyl, Cartan extends these considerations to spaces with torsion.

where A^i denotes an arbitrary vector parallely transported along the curve $x^i = x^i(\tau)$ The physical meaning of Eq. (6) and thus of Eq. (3) was clarified by Einstein's criticism [13] of Weyl's 1918 paper [9], wherein Weyl had the idea to give up the relation (3).

(ii) *The Weyl Lemma* [14]

$$h^A_{i,l} + \Gamma^A_{Bl} h^B_i - \Gamma^k_{il} h^A_k = 0 . \tag{7}$$

It specifies the relation between the tetrads h^A_i, replacing g_{ik} via $g_{ik} = \eta_{AB} h^A_i h^B_k$, and the connection. Otherwise, if one does not look at the tetrads as the "root" of the metric g but as anholonomic coordinates, replacing the coordinates x^i, then (7) has to be interpreted as rule for the transformation of the connection from holonomic coordinates x^i to anholonomic coordinates h^A_i, and vice versa.

From the first point of view, Eq. (7) resembles (3), and it can be derived from the Einstein-Hilbert Lagrangian $\mathbf{L}_{EH}$ by an independent variation of h and Γ, similarly as (3) can be obtained from $\mathbf{L}_{EH}$ by a variation of g and Γ. Viewed from the second standpoint, (7) means that genuine tensorial matter and tensorial matter occurring as the result of the fusion of spinorial matter are parallely transported in the same manner.

(iii) *The Levi-Civita condition* [15]

$$[A_i A^k_{;l} - A^k A_{i;l}] \frac{dx^l}{d\tau} = 0 \tag{8}$$

Unlike the conditions (3) and (7), it is a requirement on Γ merely. It defines a parallel transport along the curve $x^i = x^i(\tau)$. Because in GRT the metric is symmetric, the general solution of (8) reads (cf. Eisenhart[16])

$$\Gamma^i_{kl} = \left\{ {i \atop kl} \right\} + \delta^i_k \Phi_l \tag{9}$$

where Φ_l is an arbitrary vector. This means, the definition of the Levi-Civita transport allows a non-symmetric connection Γ which can be considered as resulting from the Christoffel connection $\left\{ {i \atop kl} \right\}$ by regauging. To make the condition of telecomparison of lengths given by (3) and of parallel transport of vectors defined by Eq. (9) compatible one has, however, to assume the gauge $\Phi_l = 0$.

Physically speaking, Eq. (8) says that there are gyroscopes which, when transported along a world line $x^i = x^i(\tau)$, only measure the rotation caused by $u^i = dx^i/d\tau$ (this means, there are no "cosmic" Coriolis forces).

In Einstein's GRT with minimally coupled matter based on Riemann space-times (with vanishing torsion and nonmetricity) all these three requirements are satisfied. If one, however, turns to Weyl-Cartan space-times (with non-vanishing torsion and nonmetricity), then one arrives at theories; where not all of these conditions hold true.

To get an idea about the implications such generalizations of GRT have let us start again with the Einstein-Hilbert Lagrangian plus matter term and vary the metric g (or the tetrad h) and Γ independently.[f] Physically, this means to assume that gyroscopes are independent measurement devices whose data are not reducible to distance measurements.

[f] Following Levi-Civita, this was proposed by Einstein [18] in 1925.

Accordingly, we assume

$$L = \sqrt{-g}(R[g,\Gamma] + 2\kappa\, L_{mat}[g,\Gamma,\Phi]) \tag{10a}$$

or

$$L = |h|(R[h,\Gamma] + 2\kappa\, L_{mat}[h,\Gamma,\Psi]), \tag{10b}$$

where $R = g^{ik} R_{ik}$, $\kappa = 8\pi G/c^4$, $g = det(g_{ik}$, $|h| = det(h_i^A)$, and Φ and Ψ denote tensorial and spinorial matter fields, respectively. Here always

$$g_{ik} = g_{ki} = h^A{}_i h^B{}_k \eta_{AB.} \tag{11}$$

is satisfied, and $R[g, \Gamma]$ and $R[h, \Gamma]$ are the holonomic and anholonomic representations of the Ricci scalar, respectively.

As a consequence of Eq.(11), one has

$$g_{ik\|l} = h_i^A h_{Ak\|l} + h^A_{i\|l} h_{Ak} \tag{11a}$$

so that Ricci's lemma requires

$$h_i^A h_{Ak\|l} = -h^A_{i\|l} h_{Ak}\,. \tag{11b}$$

(Here $\|$ denotes the covariant derivative with respect to holonomic and anholonomic indices.)

Using Palatini's rule [17]

$$\delta R^i_{klm} = -\ (\delta\Gamma^i_{kl})_{;m} +\ (\delta\Gamma^i_{km})_{;l} + 2\Gamma^n_{[ml]}\delta\Gamma^i_{kn} \tag{12}$$

the variation of L with respect to Γ provides field for the quantities equations.

$$g_{ikl} = g_{\overset{+-}{ik;l}} - \frac{2}{3}(g_{ik}\Gamma_l + g_{il}\Gamma_k) \tag{13}$$

with

$$\Gamma_l = \frac{1}{2}(\Gamma^r_{kr} - \Gamma^r_{rk}) = \Gamma^r_{[kr]} \tag{14}$$

and g_{ikl} as terms resulting from $\delta L_{mat}/\delta\Gamma$. (Here we use Einstein's plus-minus symbol defined as $g_{\overset{+-}{ik;l}} = g_{ik,l} - \Gamma^r_{il} g_{rk} - \Gamma^r_{lk} g_{ir}$.)

In the case, where

$$\frac{\delta L_{mat}}{\delta\Gamma} = 0 \tag{15}$$

holds, one finds from (13) the equations

$$g_{\overset{+-}{ik;l}} = \frac{2}{3}(g_{ik}\Gamma_l + g_{il}\Gamma_k)\,. \tag{16}$$

These are 60 equations for the 64 components of Γ so that the vector Γ_l remains unspecified. [In parentheses, it should be mentioned that, for non-symmetric g_{ik}, it was shown by Schrödinger [8] that one arrives at (16), too.]

For symmetric g and symmetric Γ, Eq.(16) goes over into the Ricci lemma (3) leading to $\Gamma^i_{kl} = \left\{ {i \atop kl} \right\}$. But for symmetric g and non-symmetric Γ the general solution of (16) reads (cf., Eisenhart [16])[g]

[g] It coincides with Eq. (9) so that it defines a Levi-Civita parallel transport.

$$\Gamma^i_{kl} = \left\{ {i \atop kl} \right\} + \delta^i_l \Phi_l \tag{17}$$

so that one has

$$\Gamma_l = \frac{1}{2}(\Gamma^r_{kr} - \Gamma^r_{rk}) = -\frac{3}{2}\Phi_l \tag{18}$$

and thus

$$g_{\underset{+}{i}\underset{-}{k};l} = g_{ik}\Phi_l + g_{il}\Phi_k \text{ and } g_{ik;l} = 2g_{ik}\Phi_l . \tag{19}$$

Considering now Bose fields coupled to gravity via (10a) then one is led to the same relation between g and Γ as in the vacuum case, namely to (17). This is due to the fact that g_{ikl} in (13) vanishes because of

$$\frac{\delta(\sqrt{-g}\, L_{mat}[g,\Gamma,\Phi])}{\delta\Gamma} = 0 . \tag{20}$$

Indeed, for scalar and for Maxwell-Proca matter, L_{mat} does not depend on Γ : In the scalar case one has $\Phi_{i;k} = \Phi_{i,k}$, and in the other case only the rotation of the four-potential $\Phi_{i,k} - \Phi_{i,k}$ occurs.

For spinorial fields, however, one finds

$$\frac{\delta(|h|\, L_{mat}[h,\Gamma,\Psi])}{\delta\Gamma} \neq 0 . \tag{21}$$

As it will be shown in the next section, for Dirac matter this particularly leads to an equation stating a relation between h and Γ. This relation means a violation of the Weyl lemma. It provides a non-symmetric connection of the form (4), i.e., a connection satisfying the Ricci lemma.

But before doing this, we shall discuss some aspects of this procedure under the assumption that the Ricci lemma is satisfied from the very beginning.

For this reason, let us return to the relations (3) and (4), respectively. With the rotation tensor (often also denoted as contorsion tensor)

$$\omega_{ikl} = \Gamma_{k[il]} + \Gamma_{l[ik]} + \Gamma_{i[kl]} = -\omega_{kil} \tag{22}$$

Eq. (4) can be written as

$$\Gamma^i_{\ kl} = \left\{ {i \atop kl} \right\} + \omega^i_{\ kl} \quad , \tag{23}$$

and the covariant derivative of a vector A^i defined by (23) is given by

$$A^i_{;l} = A^i_{,l} + A^r\Gamma^i_{rk} = A^i_{,l} + A^r \left\{ {i \atop rk} \right\} + A^r\omega^i_{rk} = A^i_{\perp k} + A^r\omega^i_{rk} , \tag{24}$$

where $A^i_{\perp k}$ means the Christoffel derivative of A^i. From the skew symmetry of ω_{ikl} in the first two indices it follows that the Ricci lemma

$$g_{ik;l} = g_{ik\perp l} - \omega_{ikl} - \omega_{kil} = 0 \tag{25}$$

implies also $g_{ik\perp l} = o$.

This means: The plus-minus relations (25) (64 equations!) are much stronger than the Ricci lemma (40 equations!). But Eqs. (25) and the Weyl lemma are equivalent. Indeed, with (11) and (22), Eqs. (25) can be written as

$$\begin{aligned} g^{+-}_{ik;l} &= (h^A_{i,l} - h^A_r \Gamma^r_{il}) h_{Ak} + h^A_i (h_{Ak,l} - h^A_i \Gamma^r_{lk}) \\ &= \omega_{kil} + \omega_{ikl} - 2\Gamma_{i[kl]} \\ &= -2\omega_{ikl} = 0, \end{aligned} \tag{26}$$

and the only solution of Eq. (26) is Weyl's lemma $\omega^i_{\ kl} = 0$.

In this notation, the Riemann-Cartan curvature tensor can be written as (cf. Eisenhart [16] and Cartan [19])

$$R^i_{klm} = -\Gamma^i_{\ kl,m} + \Gamma^i_{\ km,l} + \Gamma^i_{\ nl}\Gamma^n_{\ km} - \Gamma^i_{\ nm}\Gamma^n_{\ kl} = R^{\{\}i}_{klm} - \omega^i_{\ kl\perp m} + \omega^i_{\ km\perp l} + \omega^i_{\ nl}\omega^n_{\ km} - \omega^i_{\ nm}\omega^n_{\ kl}\ , \tag{27}$$

where $R^{\{\}i}_{klm}$ denotes the Riemann-Christoffel tensor. According to the Ricci lemmata (25) and (26), the homothetic curvature $R^n_{nlm} = -\omega^n_{\ nl;m} + \omega^n_{\ nm;l}$ vanishes: because of

$$\omega^n_{\ nl} = 0, \tag{28}$$

while the nonvanishing Ricci tensor and Ricci scalar read,

$$R^i_{klm} = R_{kl} = R^{\{\}}_{kl} - \omega^m_{\ kl\perp m} + \omega_{k\perp l} + \omega^m_{\ nl}\omega^n_{\ km} - \omega_n\omega^n_{\ kl} \tag{29}$$

and

$$R = R^{\{\}} - g^{kl}\omega^m_{\ kl\perp m} + \omega^k_{\ \perp k} + \omega_{mnl}\omega^{lnm} - \omega_n\omega^n_{\ kl}g^{kl} \tag{30}$$

with $\omega_l = \omega^n_{\ ln}$.

The Euler derivatives of the Einstein-Hilbert Lagrangian,

$$\mathbf{L}_{EH} = \sqrt{-g}R = \sqrt{-g}\,g^{ik}R_{ik}\,, \tag{31}$$

provides

$$\begin{aligned} \frac{1}{\sqrt{-g}}\frac{\delta\mathbf{L}_{EH}}{\delta(\sqrt{-g}\,g^{kl})} &= R_{(kl)} = \frac{1}{2}(R_{kl} + R_{lk}) = R^{\{\}}_{kl} - \omega^n_{\ kl\perp n} + \frac{1}{2}(\omega_{k\perp l} + \omega_{l\perp k}) \\ &\quad + \frac{1}{2}(\omega^m_{\ nk}\omega^n_{\ kl} + \omega^m_{\ nl}\omega^n_{\ km}) - \omega_n\omega^n_{\ (kl)} \end{aligned} \tag{32}$$

or

$$\frac{1}{\sqrt{-g}}\frac{\delta\mathbf{L}_{EH}}{\delta g^{kl}} = R_{(kl)} - \frac{1}{2}g_{kl}R \tag{32a}$$

and

$$\frac{1}{\sqrt{-g}}\frac{\delta\mathbf{L}_{EH}}{\delta\omega^{ikl}} = = -2\omega_{ikl} + g_{il}\omega_k - g_{kl}\omega_i\,. \tag{33}$$

We shall now consider the general-relativistic Lagrange density

$$L = \sqrt{-g}(R + L_{mat}) \tag{34}$$

where L_{mat} is a functional of the matter field ϕ, the metric g, and the rotation tensor $\omega^i_{\ kl}$ (respectively, the torsion tensor $\Gamma^i_{\ [kl]}$). Then the field equations obtained by the independent variation with respect to these variables are

$$\frac{\delta\mathbf{L}}{\delta\Phi} = 2\kappa\,\frac{\delta(\sqrt{-g}\,L_{mat})}{\delta\Phi} = 0 \quad \rightarrow \quad \text{matter field equations} \tag{35a}$$

$$\frac{\delta \mathbf{L}}{\delta g^{ik}}+\frac{2\kappa}{\sqrt{-g}}\frac{\delta(\sqrt{-g}\,L_{mat})}{\delta g^{ik}}=0 \quad\rightarrow\quad R_{(kl)}-\frac{1}{2}g_{kl}R=-\kappa\,T_{ik} \tag{35b}$$

and

$$\frac{1}{\sqrt{-g}}\Big(\frac{\delta \mathbf{L}_{EH}}{\delta\,\omega^{ikl}}+2\kappa\,\frac{\delta(\sqrt{-g}\,L_{mat})}{\delta\,\omega^{ikl}}\Big)=0$$

$$\rightarrow\quad -2\,\omega_{ikl}+g_{il}\,\omega_k-g_{kl}\,\omega_i.=-2\kappa\,\frac{\delta(\sqrt{-g}\,L_{mat})}{\delta\,\omega^{ikl}}=S_{ikl} \tag{35c}$$

For $S_{ikl}=0$, the solution of Eq. (35c) is $\omega_{ikl}=0$. Hence, in the case of Einstein-Bose matter, where L_{mat} does not depend on Γ^{i}_{kl}, the Einstein equations of GRT with $\omega_{ikl}=0$, i.e., with vanishing torsion, $\Gamma^{i}_{[kl]}=0$, are recovered. However, in the case of Fermi-Dirac matter S_{ikl} is proportional to the Dirac current, $S_{ikl}\sim\eta_{iklm}J^m$, and a nonvanishing torsion, $\Gamma^{i}_{[kl]}\neq 0$, results. Then one is led to a Weyl-Cartan theory considered in Sec. 3 in more detail.

3. The "Einstein-Hilbert Case" of a Weyl-Cartan Geometry

Now let us consider (10a) and (10b), where (11) is assumed, as starting point of a mixed field theory and vary L with respect to the metric (or tetrads) and the components of the connection without assuming any a priori conditions. This leads to the equations

$$\frac{\delta \mathbf{L}}{\delta g^{ik}}=\frac{\delta(\sqrt{-g}R)}{\delta g^{ik}}+2\kappa\,\frac{\delta(|h|\,L_{mat})}{\delta g^{ik}}=0 \qquad \text{(grav. field equations)} \tag{36}$$

$$\frac{\delta \mathbf{L}}{\delta\Phi}=2\kappa\,\frac{\delta(|h|\,L_{mat})}{\delta\Phi}=0 \qquad \text{(matter field equations)} \tag{37}$$

and

$$\frac{\delta \mathbf{L}}{\delta\Gamma}=\frac{\delta(\sqrt{-g}R)}{\delta\Gamma}+2\kappa\,\frac{\delta(|h\,L_{mat})}{\delta\Gamma}=0 \qquad \text{(field equations for the connection field).} \tag{38}$$

Here the field equations (38) supersede the Ricci lemma and provide in general algebraic equations for the connection Γ^{i}_{kl} or Γ^{A}_{Bl} which are different from Ricci's and Weyl's lemmata.

However, in the case of a tensorial matter field Φ (38) becomes identical with Ricci's lemma. Indeed, then one obtains $\delta(|h|\,L_{mat}/\delta\Gamma)=0$ so that (38) reduces to the relation

$$\frac{\delta(\sqrt{-g}R)}{\delta\Gamma}=0, \tag{39}$$

which becomes identical with $g_{ik;l}=0$. Hence, for Bose-Einstein matter the mixed theory provides the same field equations as the purely metric general relativity.

As opposed to this, in the case of a spinorial matter field, e.g., for a Dirac field ψ , one is led to other Euler equations. With the Lagrangian $L_{mat} = L_{mat}[h, \Gamma, \Psi]$ for Dirac matter one gets [9, 11]:

$$\frac{\delta(\sqrt{-g}R)}{\delta\Gamma} + 2\kappa \frac{\delta(|h| L_{mat})}{\delta\Gamma} = 0 \text{ with } \frac{\delta(|h| L_{mat})}{\delta\Gamma} \neq 0 . \tag{40}$$

This relation defines the conformal connection

$$\Gamma^i_{kl} = \left\{ {i \atop kl} \right\} + \sqrt{2}\kappa\hbar c(-g_{kl} S^i + \delta^i_l S_k) = \left\{ {i \atop kl} \right\} + S^i_{kl}, \tag{41}$$

where

$$S_{ikl} + S_{kil} = 0 \tag{41a}$$

($\hbar$ is the Planck quantum, c the vacuum velocity of light, and $\psi^* \gamma_i \psi$ the vector of Dirac's matter current).

With the conformal connection (41) one can rewrite the Lagrange density (10b) in an Einstein-Cartan theory [11],

$$\begin{aligned} \mathbf{L} &= |h|(R[h] + 2\kappa \{ L_{mat/0}[h,\psi] + \kappa \hbar^2 c^2 S_i S^i \}) \\ &= |h|(R[h] + 2\kappa L_{mat/0}[h,\psi] + \kappa^2 \hbar^2 c^2 S_i S^i) \end{aligned} \tag{42}$$

depending only on h^A_i and Ψ . Here $L_{mat/0}$ is the usual Dirac Lagrangian and $|h|(R + L_{mat/0})$ thus is a purely metric expression. This Lagrangian leads to Einstein's equations of GRT with an additional self-interaction of the matter field having the form of a Fermi coupling of fourth order. This coupling is a pendant of the universal Fermi coupling of weak interaction; its coupling constant is given as,

$$g_W = \kappa (\hbar c)^2 = mc^2 l^3 = \hbar c l^2 . \tag{43}$$

Here $mc^2 = (\hbar c / \kappa)^{1/2}$ is the rest energy of a planckion and $l = (\kappa \hbar c)^{1/2} = \kappa mc^2$ is Planck's elementary length. Instead of the Dirac equation, the variation by Ψ,

$$\frac{1}{\kappa}\frac{\delta \mathbf{L}}{\delta\psi} = 2\frac{\delta(|h| \ L_{mat/0})}{\delta\psi} + (\hbar c)^2 \kappa \ \frac{\delta(S_i S^i)}{\delta\psi} = 0, \tag{44}$$

provides

$$i\hbar c\sqrt{2}\, \sigma^{\,l}_{\mu\nu'}(\psi^{\mu}_{,l} + \Lambda^{\{\}\mu}_{\alpha l} \psi^{\alpha}) + im\gamma_{\beta'\nu'}\psi^{\beta'} + 2\kappa \hbar^2 c^2 \sigma^{\,l}_{\alpha\nu'}\psi^{\alpha} \sigma_{l\mu\lambda'} \psi^{\mu} \psi^{\lambda'} = 0, \tag{45}$$

where $\Lambda^{\{\}\mu}_{\alpha l}$ is the spinor connection of GRT satisfying the relation

$$\Lambda^{\{\}\mu}_{\alpha l} = -\frac{1}{2}\sigma^{k\mu\nu'} \sigma_{k\alpha\nu' \perp l} + \frac{1}{2}\delta^{\mu}_{\alpha} \Lambda^{\{\}\nu}_{\nu l} \tag{46}$$

(the index $\perp$ denotes again the covariant derivative performed with respect to $\{\}$ and $\sigma^{\,i}_{\mu\nu'} = h^i_A \sigma^{\,A}_{\mu\nu'}$ is forme of the tetrads h^i_A and Pauli's spin matrices $\sigma^{\,A}_{\mu\nu'}$).

Turning to the question of the validity of the Ricci and of the Weyl lemma, we state that the Ricci lemma is satisfied, while the Weyl lemma is violated in such a manner that there arise no problems of mathematical consistency.

As to the lemma of Ricci, it holds because of the relations (41) and (41a) They say that Γ is of the form (23) with (22) and thus it satisfies (3) and (4), respectively. However, due to $\delta(|h| L_{mat})/\delta\Gamma \neq 0$, the Weyl lemma is violated. But, as was argued in Treder and Borzeszkowski [5], the so-called contracted Weyl lemma,

$$h^{\ l}_{A,l} + \Gamma^{B}_{Al} h^{l}_{B} - \Gamma^{l}_{kl} h^{k}_{A} = 0 \tag{47}$$

and, respectively,

$$\sigma^{\ l}_{\mu\nu',l} - \Lambda^{\rho'}_{\mu l} \sigma^{\ l}_{\rho\nu'} - \Lambda^{\rho}_{\nu' l} \sigma^{\ l}_{\mu\rho'} + \Gamma^{l}_{kl} \sigma^{\ k}_{\mu\nu'} = 0 \tag{47a}$$

is satisfied. It guarantees the self-consistency of the theory. Indeed, starting with the Lagrangians

$$L_{mat} = \psi^{\nu'} \sigma^{\ l}_{\mu\nu'} \psi^{\ \mu}_{;l} + \frac{mc}{\hbar} \sigma^{\ l}_{\mu\nu'} \psi^{\mu} \psi^{\nu'}, \tag{48}$$

$$\overline{L}_{mat} = (\psi^{\nu} \sigma^{\ l}_{\mu'\nu} \psi^{\ \mu'}_{;l} + \frac{mc}{\hbar} \sigma^{\ l}_{\mu'\nu} \psi^{\mu'} \psi^{\nu}) \tag{48a}$$

and

$$L'_{mat} = \frac{1}{2}(L_{mat} - \overline{L}_{mat}) \tag{48b}$$

the variation with respect to ψ^{μ} and $\psi^{\mu'}$ leads to equations which are only compatible if Eq. (47a) is satisfied. The physical meaning of the contracted lemma is that it "saves" the duality between the Lorentz covariant and the coordinate covariant representations of the equation of continuity following by fusion from the Dirac spinor field equations (cf. for this, Treder and Borzeszkowski [5]).

4. Conclusion

The method of mixed variation of the Einstein-Hilbert Lagrangian coupled to Fermi-Dirac matter leads to an "Einstein-Cartan-Weyl" theory with matter not minimally coupled to gravity. This coupling implies a violation of Weyl's lemma, namely $h^{A}_{i\|l} h_{Ak} = \omega_{ikl} \neq 0$, and introduces, according to (41), a tensor $\Gamma^{i}_{kl} - \left\{ {i \atop kl} \right\}$ given by the Dirac current.

In the case, where the spinorial field is a Weyl spinor χ, a purely metric approach varying L only with respect to h^{A}_{i} provides only Einstein's equations plus the covariantly written Weyl equations, $\sigma^{\ l}_{\mu\nu'} \chi^{\ \mu}_{\perp l} = 0$. The mixed variation, however, gives a non-linear Heisenberg equation [20] of the type,

$$i \sigma^{l} \chi_{\perp l} + l^{2} \sigma^{l} \chi [\overline{\chi} \sigma_{l} \chi] = 0 \tag{49}$$

where $l = \kappa\hbar c$. One knows from Heisenberg's theory that, from the measurement-theoretical point of view, the length l has to be interpreted as "smallest distance". Hence the mixed variation of the Einstein-Hilbert Lagrangian coupled to spinorial matter shows Planck's length to be the smallest distance accessible to measurement (this statement is independent of Rosenfeld's uncertainty relations $\Delta g(\Delta x)^2 \geq \kappa\hbar c$).

The mixed variation of the Einstein-Hilbert Lagrangian coupled to spinorial matter leads to a generalization of GRT. Because the magnitude of additional effects arising then is automatically fixed by the gravitational constant G to be extremely small, this theory cannot be considered as a unification of GRT gravity and weak (or electroweak) interaction. It rather is to interpret as a possible generalization of Einsteinian GRT gravity to Weyl-Cartan gravity that cannot be excluded from consideration by a priori assumptions concerning the Weyl lemma.

The framework considered above can of course be extended by adding other terms to the Lagrangian, in particular, by adding suitable Riemann-Cartan terms, e.g., Weitzenboeck-like terms,

$$A\omega_i\omega^i + B\omega_{ikl}\omega^{ikl} + C\omega_{ikl}\omega^{lki} \qquad (50)$$

(where A, B; and C are constants which should be chosen such that the theory is not underdetermined) or, more generally, the class of terms assumed by Hehl et al. [4]

5. References

1. A. Einstein, *The Meaning of Relativity*, 5th.ed., App. II (Princeton, 1955).
2. H. Weyl, *Phys. Rev.* **77** (1950) 699; *Naturwiss.* **38** (1950) 73.
3. R. D. Hecht and F. W. Hehl, "A Metric-Affine Framework for a Gauge Theory of Gravitaty", in *Proceedings of the 9th Conference on General Relativity and Gravitational Physics, Capri (Napoli), 25-28 Sept. 1990*, R. Cianci, R. de Ritis, M. Francaviglia, G. Marmo, C. Rubano, P. Scudellaro, eds. (World Scientific, Singapore 1991) pp. 246 - 291.
4. F. W. Hehl, *Found. Phys.* **15** (1985) 451; F. W. Hehl, J. D. McCrea, E. W. Mielke, Y. Ne'eman, "Metric-Affine Gauge Theory of Gravity: Field Equations, Noether Identities, World Spinors, and Breaking of Dilation Invariance", *Physics Reports* **258** (1995) pp. 1 - 171.; cf. also the literature cited therein.
5. H.-J. Treder and H.-H. v. Borzeszkowski, *Int. Jour. Theor. Phys.* **8** (1973) 319.
6. A. S. Eddington, *The Mathematical Theory of Relativity*, (Cambridge, 1922).
7. A. Einstein, "Eddingtons Theorie und das Hamiltonsche Prinzip", in the German translation of Ref. 6.: *Relativitätstheorie in mathematischer Behandlung* (Berlin, 1925), p. 367; "Zur allgemeinen Relativitätstheorie", *Ber. Preuss. Akad. Wiss.* 1923, p. 32.
8. E. Schrödinger, *Space-Time Structure* (Cambridge, 1950).
9. H. Weyl, "Gravitation und Elektrizität", *Ber. Preuss. Akad. Wiss.* 1918, p. 465.
10. H. Weyl, *Raum, Zeit, Materie*, 1th ed. Berlin 1918.
11. H.-J. Treder, *Found. Phys.* **6** (1976) 527.
12. J. A. Schouten, *Ricci-Calculus* (Berlin etc., 1954).
13. A. Einstein, "Über eine naheliegende Ergänzung des Fundaments der allgemeinen Relativitätstheorie", *Ber. Preuss. Akad. Wiss.* 1921, p. 261.
14. H. Weyl, *Mathematische Analyse des Raumproblems* (Berlin, 1922).
15. T. Levi-Civita, *The Absolute Differential Calculus* (London, Glasgow, 1926).
16. L. P. Eisenhart, *Non-Riemannian Geometry* (New York, 1926).
17. A. Palatini, *Rend. Circ. Mat. Palermo* **43** (1919) 203.
18. A. Einstein, „Einheitliche Feldtheorie von Gravitation und Elektrizität", *Ber. Preuss. Akad. Wiss.* 1925, p. 414.
19. E. Cartan, *Leçons sur la géométrie des éspaces de Riemann* (Paris, 1928).
20. W. Heisenberg, *Einführung in die einheitliche Feldtheorie der Elementarteilchen* (Stuttgart,1967).

PHYSICAL THEORIES IN SPACE-TIME ALGEBRA

BIDYUT KUMAR DATTA
Istituto di Fisica dell' Universita,
Via Irnerio 46, 40126 Bologna, ITALIA

and

Department of Mathematics, Tripura University,
Agartala 799004, INDIA

ABSTRACT

A comprehensive review of space-time algebra as required for the development of physical theories in space-time is given. Expressing the Dirac equation without complex numbers, the geometrical significance of the wave function Ψ is pointed out. For this purpose certain bilinear functions of Ψ are expressed as tensors which are interpreted as observables. Finally, it is demonstrated that the bivectors $\gamma_\alpha \wedge \gamma_\beta$ play the role of generators of rotations in space-time. The $\sqrt{-1}$ of the Dirac theory can be interpreted geometrically as the generator of rotations in the $e_2 \wedge e_1$-plane with $\frac{\hbar}{2} \gamma_2 \wedge \gamma_1$ representing the intrinsic spin in real Dirac algebra in conventionally used labelling. Furthermore, it is pointed out that supersymmetry may be introduced in Einstein-Cartan theory by using space-time algebra.

1. Review of Space-Time Algebra

We give a brief resumé of the space-time algebra as developed by Hestenes (1966, 1971, 1982). The space-time algebra is the Clifford algebra of real 4-dimensional space-time with a thoroughgoing geometric interpretation. It is built out of objects with direct geometric interpretations; the properties of these objects are specified by introducing algebraic operations which directly determine their interrelations. Furthermore, space-time algebra derives its potency from the fact that both the elements and the operations of the algebra are endowed with direct geometric interpretation.

The geometric product of a generic proper vector a with itself is a scalar quantity describing the metric of space-time :

$$a^2 > 0 \text{ iff } a \text{ is a \underline{timelike} vector,} \tag{1.1a}$$

$$a^2 = 0 \text{ iff } a \text{ is a \underline{lightlike} vector,} \tag{1.1b}$$

$$a^2 < 0 \text{ iff } a \text{ is a \underline{spacelike} vector.} \tag{1.1c}$$

The geometric product ab of two proper vectors a and b can be understood geometrically by separating it into symmetric and antisymmetric parts :

$$ab = a.b + a\wedge b , \tag{1.2a}$$

where

$$a.b \equiv \frac{1}{2}(ab + ba) = b.a , \tag{1.2b}$$

$$a \wedge b \equiv \frac{1}{2}(ab - ba) = -b \wedge a. \tag{1.2c}$$

(1.1) tells that a.b is a scalar quantity, the usual <u>inner product</u> of space-time vectors. Here "scalar" means "real number". The quantity $a \wedge b$, called the <u>outer product</u> of a,b, is a (proper) bivector or 2-vector. The geometric product obeys <u>associative rule.</u>

The inner and outer product of a vector a and a bivector B are defined respectively by

$$a.B \equiv \frac{1}{2}(aB - Ba) = -B.a, \tag{1.3a}$$

$$a \wedge B \equiv \frac{1}{2}(aB + Ba) = B \wedge a. \tag{1.3b}$$

Then we have

$$aB = a.B + a \wedge B. \tag{1.3c}$$

Eqs. (1.3a) and (1.2b,c) give

$$a.(b \wedge c) = a.bc - a.cb = -(b \wedge c).a. \tag{1.4}$$

<u>Note</u>

(i) a.B is a vector.

(ii) $a \wedge B$ is a trivector or 3-vector.

(iii) Every trivector in space-time algebra can be factored into an outer product of three vectors.

The inner and outer product of a vector a and a trivector T are defined respectively by

$$a\cdot T \equiv \frac{1}{2}(aT + Ta) = T\cdot a \,, \tag{1.5a}$$

$$a\wedge T \equiv \frac{1}{2}(aT - Ta) = -T\wedge a \,. \tag{1.5b}$$

Thus

$$aT = a\cdot T + a\wedge T \,. \tag{1.5c}$$

Eqs. (1.5a), (1.3a) and (1.2) give

$$a\cdot(b\wedge B) = a\cdot bB - b\wedge(a\cdot B) \,, \tag{1.6}$$

where a,b are vectors and B is a bivector. $a\wedge T$ is a 4-vector or <u>pseudoscalar</u>. Unit pseudoscalar, denoted by i, assigns an orientation to space-time. Every pseudoscalar is a scalar multiple of i. It can be shown that

$$i^2 = 1 \,, \tag{1.7a}$$

$$ai = -ia \,. \tag{1.7b}$$

From the above it follows that

$$a\wedge i \equiv \frac{1}{2}(ai + ia) = 0 \,, \tag{1.8a}$$

$$a\cdot i \equiv \frac{1}{2}(ai - ia) = ai \,. \tag{1.8b}$$

ai is a trivector, called the dual of a. Every trivector T is dual of some vector t :

$$T = ti \,. \tag{1.9a}$$

Multiplying on the right by i and using (1.7a), we get

$$Ti = -t \,. \tag{1.9b}$$

So, the dual of a trivector T is a unique vector. This establishes an isomorphism of the linear space of all trivectors to the space of all vectors. So, trivectors are often called pseudovectors.

Multivectors

A generic element of the space-time algebra is called a (proper) multivector. Every proper multivector M is a sum of Clifford objects of arbitrary grade (grade 0 = scalar, grade 1 = vector, grade 2 = bivector or 2-vector, grade 3 = trivector or 3-vector, grade 4 = 4-vector or pseudoscalar). Thus M can be expressed as

$$M = \langle M\rangle_0 + \langle M\rangle_1 + \langle M\rangle_2 + \langle M\rangle_3 + \langle M\rangle_4 , \qquad (1.10)$$

where $\langle M\rangle_k$ is a Clifford object of grade k or denotes the k-vector part of M.

A multivector M in space-time algebra is said to be even if

$$\langle M\rangle_1 = \langle M\rangle_3 = 0 .$$

The even multivectors constitutes a subalgebra of the full space-time algebra.

Reversion

The reverse of M, denoted by $\tilde{M}$, is defined by

$$\widetilde{M} = \langle M\rangle_0 + \langle M\rangle_1 - \langle M\rangle_2 - \langle M\rangle_3 + \langle M\rangle_4 \qquad (1.11)$$

The reverse of a product equals the product of the reverse :

$$\widetilde{(AB)} = \widetilde{B}\,\widetilde{A} \quad . \qquad (1.12)$$

Lorentz Rotation $\mathcal{R}$

Any Lorentz rotation $\mathcal{R}$ which maps a generic proper vector a into the vector a' can written in the canonical form :

$$a \rightarrow a' = \mathcal{R}(a) = R\,a\,\widetilde{R} \quad . \qquad (1.13a)$$

Here R is an even multivector, unique except for sign, with the property

$$R\,\widetilde{R} = 1 \,. \qquad (1.13b)$$

R is called a spinor.

Two Special Classes of Lotentz Rotations : Boosts and Spatial Rotations.

A Lorentz rotation $\mathcal{L}(a) = L\,a\,\widetilde{L}$ which takes a unit timelike vector u into the vector v is said to be a boost of u into v if it leaves the vectors orthogonal to the $v \wedge u$ - plane invariant.

Any vector a can be expressed as the sum

$$a = a_{\parallel} + a_{\perp} \qquad (1.14a)$$

with the component $a_{\|}$ in the $v \wedge u$ - plane given by

$$a_{\|} = a\cdot(v \wedge u)\,(v \wedge u)^{-1}\,, \tag{1.14b}$$

and the component $a_{\perp}$ orthogonal to it

$$a_{\perp} = a \wedge (v \wedge u)\,(v \wedge u)^{-1}\,. \tag{1.14c}$$

A Lorentz rotation $\mathcal{U}(a) = U a \widetilde{U}$ is said to be spatial rotation if it leaves a timelike vector u invariant :

$$U u \widetilde{U} = u \tag{1.15a}$$

or equivalently,

$$U U^{\dagger} = 1, \qquad U^{\dagger} \equiv u \widetilde{U} u\,. \tag{1.15b}$$

The set of all Lorentz rotations satisfying (1.5) is the group of spatial rotations in the spacelike hypersurface with normal u, called the little group of u.

Any Lorentz rotation can be uniquely expressed as a spatial rotation followed by a boost of a given timelike vector u :

$$R = L\,U\,. \tag{1.16}$$

Magnitude

The magnitude (or modulus) $|M|$ of a multivector M is determined by

$$|M| = \left[\langle M \widetilde{M}\rangle_0\right]^{1/2}\,. \tag{1.17}$$

The inverse M^{-1} of a multivector M is defined by

$$M\, M^{-1} = 1\,. \tag{1.18}$$

If $M\widetilde{M} = |M|^2$, the inverse of M exists and can be expressed as

$$M^{-1} = \widetilde{M} / |M|^2\,. \tag{1.19}$$

It follows that every vector has an inverse, so it is possible to divide by vectors. The vector division is made possible by (1.2). So, the space-time algebra is an associative division algebra.

2. The Algebra of a Euclidean Plane

The Clifford algebra for 2-dimensional Euclidean space is generated by two orthonormal vectors $\{\sigma_k\}$, $(k = 1,2)$, and is spanned by

$$1,\ \{\sigma_k\}\,,\ i'\,, \tag{2.1}$$

where $i' = \sigma_1\sigma_2$ is the pseudoscalar (highest grade multivector) for the space.

The basic properties of i' are

$$i'^2 = -1 \implies i' = \sqrt{-1}\,, \tag{2.2a}$$

$$\widetilde{i'} = -\,i'\,, \tag{2.2b}$$

$$i'\, M = M\, i'\,, \tag{2.2c}$$

for every multivector M in the space and

$$a \wedge b = \lambda\, i' \tag{2.2d}$$

for any vectors in the Euclidean plane. The scalar λ is positive if and only if the vectors make up a right handed set in the order given.

Besides the property (2.2a) ascribed to the traditional unit imaginary, our i' is a bivector. So, it has geometric and algebraic properties :

(i) it is the unit of directed area and

(ii) also the generator of rotations in the plane.

In view of (2.1) any multivector M can be expressed as

$$M = \alpha_0 + \alpha_1\sigma_1 + \alpha_2\sigma_2 + \alpha_3 i' \tag{2.3}$$

with scalar coefficients $\alpha_0, \alpha_1, \alpha_2, \alpha_3$. The algebra (2.1) is the geometric algebra of the i'-plane and is denoted by $\zeta_2(i')$. Obviously, the ζ_2 algebra is a 4-dimensional linear space. We can write (2.3) as

$$M = \langle M\rangle_+ + \langle M\rangle_- \quad , \tag{2.4}$$

where the even multivector part $\langle M\rangle_+$ represents the plane spinor

$$\langle M\rangle_+ = \alpha_0 + i'\alpha_3 \tag{2.5}$$

and the odd multivector part $\langle M\rangle_-$ represents the vector in the i'-plane

$$\langle M\rangle_- = \alpha_1\sigma_1 + \alpha_2\sigma_2 \quad . \tag{2.6}$$

So, ζ_2 can be expressed as the sum of two linear spaces

$$\zeta_2 = \zeta_2^+ + \zeta_2^- , \tag{2.7}$$

where ζ_2^+ is the 2-dimensional linear space of spinors and ζ_2^- is the 2-dimensional vector space. The algebra (2.5) is an even subalgebra ζ_2^+ of ζ_2 or spinor algebra. Likewise, the algebra (2.6) is an odd subalgebra ζ_2^- of ζ_2 or vector algebra. Every spinor in ζ_2^+ represents a rotation-dilation in 2-dimensional plane.

3. The Algebra of Euclidean 3-Space

The Clifford algebra for 3-dimensional Euclidean space is generated by three orthonormal vectors $\{\sigma_k\}$ and is spanned by

$$1, \{\sigma_k\} , \{\hat{i}\,\sigma_k\} , \hat{i} , \tag{3.1}$$

where $\hat{i} = \sigma_1\sigma_2\sigma_3$ is the pseudoscalar (highest grade multivector) for the space. The basic properties of $\hat{i}$ are

$$\hat{i}^2 = -1 , \tag{3.2a}$$

$$\hat{i}^\sim = -\hat{i} , \tag{3.2b}$$

$$\hat{i}\,M = M\,\hat{i} \tag{3.2c}$$

for every multivector M in the space, and

$$a \wedge b \wedge c = \lambda\,\hat{i} \tag{3.2d}$$

for any vectors in the Euclidean 3-space. The scalar λ is positive if and only if the vectors make up a righthanded set in the order given. $\hat{i}$ is the dextral unit pseudoscalar (trivector) giving the unit oriented cube.

In view of (3.1) any multivector M of the space can be expressed as

$$M = \langle M\rangle_0 + \langle M\rangle_1 + \langle M\rangle_2 + \langle M\rangle_3 \tag{3.3}$$

with

$\langle M\rangle_0$ = α, the scalar part of M,

$\langle M\rangle_1$ = a, the vector part of M,

$\langle M\rangle_2$ = $\hat{i}$ b, the bivector part of M expressed as a dual of a vector, and

$\langle M\rangle_3$ = $\hat{i}\,\beta$, the pseudoscalar part of M.

So, (3.3) can be put in the form

$$M = \alpha + a + \hat{i}\,b + \hat{i}\,\beta \tag{3.4}$$

with scalar coefficients α, β. The algebra (3.1) is the geometric algebra of the $\hat{i}$-space and is denoted by ζ_3. It is also called the Pauli algebra, but in geometric algebra the three Pauli σ_k are no longer viewed as three matrix-valued components* of a single isospace vector, but as three

---

* $\hat{\sigma}_1 = \begin{bmatrix} 0 & 1 \\ 1 & 0 \end{bmatrix}$, $\hat{\sigma}_2 = \begin{bmatrix} 0 & -i' \\ i' & 0 \end{bmatrix}$, $\hat{\sigma}_3 = \begin{bmatrix} 1 & 0 \\ 0 & -1 \end{bmatrix}$

independent basis vectors for real space. Obviously, the ζ_3 algebra or the Pauli algebra is a 8-dimensional linear space. We can write (3.4) as

$$M = \langle M\rangle_+ + \langle M\rangle_- \tag{3.5}$$

where the even multivector part $\langle M\rangle_+$ represents the spinor space

$$\langle M\rangle_+ = \alpha + \hat{i}\, b \tag{3.6}$$

and the odd multivector part $\langle M\rangle_-$ is given by

$$\langle M\rangle_- = a + \hat{i}\, \beta \,. \tag{3.7}$$

So, ζ_3 can be expressed as the sum of an even part ζ_3^+ and an odd part ζ_3^- :

$$\zeta_3 = \zeta_3^+ + \zeta_3^- \,. \tag{3.8}$$

ζ_3^+ is closed under multiplication, so it is a sub-algebra of ζ_3, but ζ_3^- is not. ζ_3^+ may be referred to as spinor algebra or even subalgebra of ζ_3 (Pauli) algebra. Every spinor in ζ_3^+ represents a rotation-dilation in 3-dimensional space.

We have three linearly independent bivectors given by

$$i_1' = \sigma_2\sigma_3 = \hat{i}\,\sigma_1\,,\quad i_2' = \sigma_3\sigma_1 = \hat{i}\,\sigma_2,\quad i_3' = \sigma_1\sigma_2 = \hat{i}\,\sigma_3 . \tag{3.9}$$

So, any bivector B in ζ_3 can be expressed as

$$B = \mathcal{B}_1\, i_1' + \mathcal{B}_2\, i_2' + \mathcal{B}_3\, i_3' \tag{3.10}$$

with scalar coefficients $\mathcal{B}_k$. Thus the set of all bivectors in ζ_3 is a 3-dimensional linear space with basis $\{i_1', i_2', i_3'\}$. By using (3.9), eq. (3.10) can be expressed as

$$B = \hat{i}\,(\mathcal{B}_1 \sigma_1 + \mathcal{B}_2 \sigma_2 + \mathcal{B}_3 \sigma_3) = \hat{i}\, b\,, \tag{3.11}$$

where b is a vector defined by

$$b = \mathcal{B}_1\sigma_1 + \mathcal{B}_2\sigma_2 + \mathcal{B}_3\sigma_3 \quad . \tag{3.12}$$

The bivector B is called the dual of the vector b. By using (3.10) and (3.11) one can express (3.6) as

$$\langle M \rangle_+ = \alpha + \mathcal{B}_1\, i_1' + \mathcal{B}_2\, i_2' + \mathcal{B}_3\, i_3' \quad . \tag{3.13}$$

This shows that $\{1, i_1', i_2', i_3'\}$ make up a basis for ζ_3^+. Thus ζ_3^+ is a linear space of 4-dimensions. As the elements of ζ_3^+ were called quaternions by Hamilton, ζ_3^+ may also be referred to as quaternion algebra.

4. The Algebra of Space-Time (STA)

The Clifford algebra of real 4-dimensional space-time is generated by four orthonormal vectors $\{\gamma_\mu\}$ and is spanned by

$$\underset{\text{(scalar)}}{1},\ \underset{\text{(vectors)}}{\{\gamma_\mu\}}\ ,\ \underset{\text{(bivectors)}}{\{\sigma_k, i\,\sigma_k\}}\ ,\ \underset{\text{(pseudovectors)}}{\{i\,\gamma_\mu\}}\ ,\ \underset{\text{(pseudoscalar)}}{i} \tag{4.1}$$

$$(\mu = 0,1,2,3\ ;\ \ k = 1,2,3)$$

where i is the pseudoscalar (4-vector) for the space-time

$$i \equiv \gamma_0 \gamma_1 \gamma_2 \gamma_3 = \underset{\sim}{\sigma}_1 \underset{\sim}{\sigma}_2 \underset{\sim}{\sigma}_3 = \gamma_5 , \quad (4.2a)$$

$$\underset{\sim}{\sigma}_k \equiv \gamma_k \gamma_0 . \quad (4.2b)$$

The even elements of the basis (4.1) for the space-time

$$1 , \{ \underset{\sim}{\sigma}_k , i \underset{\sim}{\sigma}_k \} , i \quad (4.3)$$

coincide with Pauli algebra because of the relation (4.2b). Thus, vectors in Pauli algebra become bivectors as viewed from (real) Dirac algebra. In STA the four Dirac γ_μ are no longer viewed as four matrix-valued components of a single isospace vector, but as four independent basis vectors for real space-time. The pseudoscalar i anti-commutes with the space-time vectors γ_μ.

In STA the reversion changes the sign of bivectors leaving scalars and vectors unchanged. So, the reversion of an even multivector can be obtained by changing the sign of bivectors.

The algebra (4.1) is the space-time algebra or real Dirac algebra having sixteen components and is a 16-dimensional linear space. The algebra (4.3) is an even subalgebra of STA or real Dirac algebra with respect to the selection of the timelike vector γ_0, and is a 8-dimensional linear space of spinors.

5. The Dirac Equation Without Complex Numbers

A Dirac spinor Ψ is represented by

$$\Psi = \begin{bmatrix} \psi_1 \\ \psi_2 \\ \psi_3 \\ \psi_4 \end{bmatrix} = \begin{bmatrix} \alpha_1 + i'\beta_1 \\ \alpha_2 + i'\beta_2 \\ \alpha_3 + i'\beta_3 \\ \alpha_4 + i'\beta_4 \end{bmatrix} , \tag{5.1}$$

where α's and β's are real numbers, and $i' = \sqrt{-1}$ is the unit imaginary of the matrix algebra. The representation (5.1) in terms of the components $\psi_1, \psi_2, \psi_3, \psi_4$ presumes a specific representation of the Dirac matrices :

$$\hat{\gamma}_o = \begin{bmatrix} I & 0 \\ 0 & I \end{bmatrix} , \quad \hat{\gamma}_k = \begin{bmatrix} 0 & -\hat{\sigma}_k \\ \hat{\sigma}_k & 0 \end{bmatrix} , \tag{5.2a}$$

where I is a 2×2 unit matrix and $\hat{\sigma}_k$ are the Pauli matrices, which are traceless Hermitian matrices satisfying

$$\hat{\sigma}_1 \hat{\sigma}_2 \hat{\sigma}_3 = i' I . \tag{5.2b}$$

We introduce in the spinor space a basis (Hestenes 1975):

$$u_1 = \begin{bmatrix} 1 \\ 0 \\ 0 \\ 0 \end{bmatrix} , \quad u_2 = \begin{bmatrix} 0 \\ 1 \\ 0 \\ 0 \end{bmatrix} , \quad u_3 = \begin{bmatrix} 0 \\ 0 \\ 1 \\ 0 \end{bmatrix} , \quad u_4 = \begin{bmatrix} 0 \\ 0 \\ 0 \\ 1 \end{bmatrix} , \tag{5.3}$$

such that

$$\gamma_0 u_1 = u_1 , \tag{5.4a}$$

$$i \underset{\sim}{\sigma}_3 u_1 = \gamma_2 \gamma_1 u_1 = i' u_1 , \tag{5.4b}$$

$$u_2 = -i \underset{\sim}{\sigma}_2 u_1 , \quad u_3 = \underset{\sim}{\sigma}_3 u_1 , \quad u_4 = \underset{\sim}{\sigma}_1 u_1 \tag{5.4c}$$

Supposing (5.1) refers to this representation, one can write any Dirac spinor Ψ as

$$\Psi = \psi u_1 \ , \tag{5.5}$$

where

$$\psi = \underset{\text{(scalar)}}{\alpha_1} + \underset{\text{(bivectors)}}{(\alpha_4\gamma_1\gamma_0 + \beta_4\gamma_2\gamma_0 + \alpha_3\gamma_3\gamma_0 + \beta_2\gamma_3\gamma_2 + \alpha_2\gamma_3\gamma_1 + \beta_1\gamma_2\gamma_1)} + \underset{\text{(pseudoscalar)}}{\gamma_5\beta_3} \ . \tag{5.6}$$

Here the unit imaginary i' has been eleminated. The ψ given by (5.6) is an even multivector and can be expressed as an element of STA by interpreting the γ's as vectors.

Dirac equation for an electron with charge e and mass m in an external electromagnetic field is expressed as

$$(i'\hbar\Box - \frac{e}{c}A)\Psi = mc\Psi \ , \tag{5.7}$$

where

$$\Box \equiv \gamma^{\mu}\delta_{\mu} \ , \quad \delta_{\mu} \equiv \frac{\partial}{\partial x^{\mu}} \ , \tag{5.8a}$$

$$A = A_{\mu}\gamma^{\mu} = A^{\mu}\gamma_{\mu} \ . \tag{5.8b}$$

Replacing $i' = \sqrt{-1}$ by the bivector $\gamma_2\gamma_1$ and using (5.5), we put (5.7) in the form

$$(\hbar\Box\,\psi\gamma_2\gamma_1 - \frac{e}{c}A\psi)\gamma_0 u_1 = mc\,\psi u_1 \ , \tag{5.9}$$

with the help of (5.4a).

The coefficient of u_1 in (5.5) is an even multivector. (5.4a) implies that even multivector operating on u_1 generates a complete basis for Dirac spinors. So, the coefficients of u_1 in (5.9) can be equated, although u_1 does not have an inverse. Thus, eq.(5.9) yields

$$(\hbar \,\Box\, \psi \gamma_2\gamma_1 - \frac{e}{c} A\psi)\, \gamma_0 = mc\psi \tag{5.10a}$$

or equivalently

$$\hbar \,\Box\, \psi \gamma_2\gamma_1 - \frac{e}{c} A\psi = mc\psi\gamma_0 \,. \tag{5.10b}$$

This is the Dirac equation in STA without unit imaginary $i' = \sqrt{-1}$. By using eq. (5.5), a solution of the one equation can be expressed as a solution of the other.

6. Observables and the Wave Function

In order to determine the geometrical significance of the wave function Ψ in Dirac's theory, it is required that certain bilinear functions of Ψ be tensors. Interpretation of these tensors as observables determines the physical significance of Ψ.

As $\tilde{i} = i$, $\tilde{\psi}$ can be obtained from ψ by changing the sign of its bivector part. As ψ is an even multivector, so is $\psi\tilde{\psi}$. But $\psi\tilde{\psi}$ is invariant under reversion. So, its bivector part must vanish. Furthermore, since $i^2 = -1$, $\psi\tilde{\psi}$ can be put in the "polar form"

$$\psi\tilde{\psi} = \rho \exp(i\beta) = \rho\cos\beta + i\rho\sin\beta \,, \tag{6.1}$$

where ρ and β are scalars.

Also we know that

$$R\tilde{R} = 1 . \tag{6.2a}$$

Thus, one can define R by the equation

$$R = \left[\rho \exp(i\beta)\right]^{-1/2} \psi \tag{6.2b}$$

or, may write

$$\psi = \left[\rho \exp(i\beta)\right]^{1/2} R . \tag{6.2c}$$

The quantity R in (6.2) determines a proper Lorentz transformation of a frame $\{\gamma_\mu\}$ into a frame $\{e_\mu\}$ according to the equation

$$e_\mu = R\gamma_\mu\tilde{R} . \tag{6.3}$$

$R = R(x)$ is a differentiable function of the space-time point x. Thus, eq. (6.3) specifies a differentiable set of four vector fields with values $e_\mu = e_\mu(x)$ at each point x determined by a proper Lorentz transformation of a fixed frame $\{\gamma_\mu\}$. This completely describes the geometrical significance of R. By virtue of eq. (6.3) the spinor R may be regarded as a representation of a Lorentz transformation.

Eqs. (6.2c) and (6.3) give

$$\psi\gamma_\mu\tilde{\psi} = \rho e_\mu . \tag{6.4}$$

Treating (6.4) as a generalization of the transformation (6.3), one may interpret the multiplication of e_μ by ρ as a dilation.

The spinor ψ represents a Lorentz transformation because, by eq. (6.4) it determines a "rotation-dilation" of the frame $\{\gamma_\mu\}$ into the frame $\{\rho e_\mu\}$.

It is to be noted that ψ does not operate in some "abstract spin space" detached from space-time, it transforms space-time vectors into space-time vectors.

The physical interpretations of ρ and R are fixed by specifying interpretations for the e_μ.

The quantity (Hestenes 1975)

$$\rho v \equiv \rho e_0 = \psi \gamma_0 \tilde{\psi} \tag{6.5a}$$

may be identified with the probability current of the Dirac theory. So, the timelike vector

$$v \equiv e_0 = R \gamma_0 \tilde{R} \tag{6.5b}$$

can be interpreted as the (local) "world velocity" of a Dirac particle, while ρ is the probability density in the local rest frame determined by v.

The tensor components of the probability current with respect to the frame $\{\gamma_\mu\}$ are

$$\rho v^\mu = \rho v \cdot \gamma^\mu = (\psi \gamma_0 \tilde{\psi} \gamma^\mu)_0 . \tag{6.5c}$$

The conservation of probability is given by

$$\Box \cdot (\rho v) = \partial_\mu (\rho v^\mu) = 0 . \tag{6.6}$$

The spacelike vector

$$s \equiv \frac{\hbar}{2}\, e_3 = \frac{\hbar}{2}\, R\,\gamma_3\,\tilde{R} \tag{6.7a}$$

can be identified as the (local) spin vector of the Dirac theory. The corresponding "current"

$$\rho s \equiv \rho\,\frac{\hbar}{2}\, e_3 = \frac{\hbar}{2}\,\psi\,\gamma_3\,\tilde{\psi} \tag{6.7b}$$

has components

$$\rho s_{\mu} = \rho s\cdot\gamma_{\mu} = \frac{\hbar}{2}\,(\psi\,\gamma_3\,\tilde{\psi}\,\gamma_{\mu})_o \quad . \tag{6.7c}$$

The "proper spin density" of the electron is ρS, where S is the (local) spoin bivector :

$$S \equiv \frac{\hbar}{2}\, e_2 e_1 = \frac{\hbar}{2}\, R\,\gamma_2\,\gamma_1\,\tilde{R}$$

$$= \frac{\hbar}{2}\, R\, i\,\sigma_3\,\tilde{R} = i s v \quad . \tag{6.8a}$$

The tensor components of S are given by

$$S^{\alpha\beta} = (S\gamma^{\beta}\gamma^{\alpha})_o = i s\wedge v\wedge\gamma^{\beta}\wedge\gamma^{\alpha}$$

$$= s_{\mu}\, v_{\nu}\,\epsilon^{\mu\nu\alpha\beta} \quad , \tag{6.8b}$$

where the "alternating tensor" $\epsilon^{\mu\nu\alpha\beta}$ is defined by

$$\epsilon^{\mu\nu\alpha\beta} = -i\,\gamma^{\mu}\wedge\gamma^{\nu}\wedge\gamma^{\alpha}\wedge\gamma^{\beta} = -(\gamma_5\gamma^{\mu}\gamma^{\nu}\gamma^{\alpha}\gamma^{\beta})_o \; . \tag{6.9}$$

So, we may write

$$S = \frac{1}{2}\, S^{\alpha\beta}\,\gamma_{\alpha}\,\gamma_{\beta} = \frac{1}{2}\, S^{\alpha\beta}\,\sigma_{\alpha\beta} \quad , \tag{6.8c}$$

where $\sigma_{\alpha\beta}$ are the generators of rotations :

$$\sigma_{\alpha\beta} = \frac{1}{2}(\gamma_\alpha\gamma_\beta - \gamma_\beta\gamma_\alpha) \quad . \tag{6.8d}$$

7. Generators of Rotations in Space-time : Intrinsic Spin

Equations (6.1), (6.2a) and (6.8a) yield

$$S\psi = \frac{\hbar}{2}\,\psi\,\gamma_2\gamma_1 \quad . \tag{7.1}$$

Interpreting this as a matrix equation, we can write by using (5.4a,b) (Hestenes 1975)

$$S\psi u_1 = i'\frac{\hbar}{2}\,\psi\,u_1 \,, \tag{7.2}$$

or in terms of Dirac spinor Ψ (by relation (5.5))

$$S\,\Psi = i'\frac{\hbar}{2}\,\Psi \quad . \tag{7.3}$$

This implies that $i'\frac{\hbar}{2}$ is an eigenvalue of the bivector S describing the spφin. So, $i'\frac{\hbar}{2}$ is a representation of the spφin bivector S by one of its eigenvalues.

In two-dimensional i'-plane of vectors, the unit imaginary $i' = \sqrt{-1}$ is a unit bivector. It is a generator of rotations in the plane as well as the "intrinsic" rotation represented by $\sigma_1 \wedge \sigma_2$ in the spinor i'-plane. The spinor plane is not detached from the real vector plane as depicted figures (i) and (ii). There is a one-to-one correspondence between them. This feature gives geometric algebra a great advantage over matrix algebra in which abstract spin space

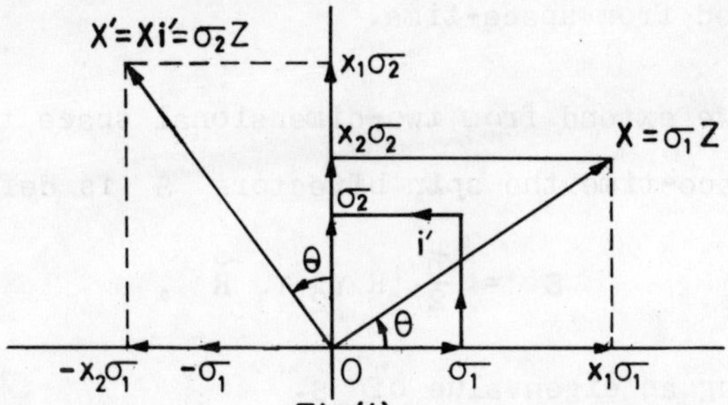

Fig.(i)

Diagram of the i′plane of vectors ("real plane")

unit imaginary $i' = \sqrt{-1}$

Also i' is a unit bivector: $i' = \sigma_1\sigma_2 = \sigma_1 \wedge \sigma_2$.

i' as an operator on vectors:

$\sigma_1 i' = \sigma_2 , \sigma_2 i' = -\sigma_1$.

i' is the generator of rotations in the i′-plane.

$\sigma_1 i'^2 = -\sigma_1$ provides a geometric interpretation for $i'^2 = -1$,when i and −1 are both regarded as operators on vectors.

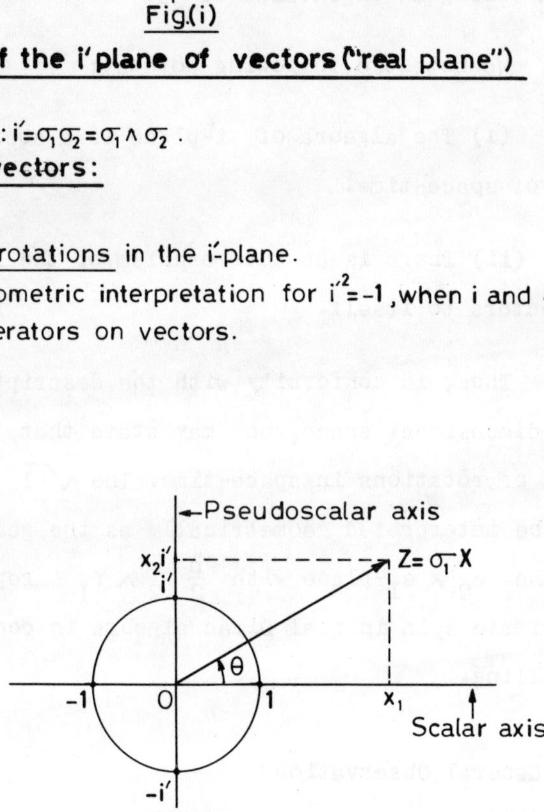

Fig.(ii)

Diagram of the Spinor i′-plane ("Argand diagram")

X, σ_1 -vectors, Z-Spinor; $Z = \sigma_1 X = x_1 + i'x_2$, if $X = x_1\sigma_1 + x_2\sigma_2$.

Each point in the Spinor plane represents a rotation-dilation.

is detached from space-time.

If we extend from two-dimensional space to four-dimensional space-time the spin bivectors S is defined by

$$S = \frac{\hbar}{2} R \gamma_2 \gamma_1 \tilde{R} ,$$

$i' \frac{\hbar}{2}$ being an eigenvalue of S.

We note the following points :

(i) The algebra of i'-plane is a subalgebra of the algebra of space-time.

(ii) There is an automorphism of the linear space of all bivectors to itself.

Thus, in conformity with the description of bivectors in two-dimensional space, one may state that $\gamma_\alpha \wedge \gamma_\beta$ are generators of rotations in space-time. The $\sqrt{-1}$ of the Dirac theory can be interpreted geometrically as the generator of rotations in the $e_2 \wedge e_1$-plane with $\frac{\hbar}{2} \gamma_2 \wedge \gamma_1$ representing the intrinsic spin in real Dirac algebra in conventionally used labelling.

8. General Observation

Interpreting the γ_μ in the Riemann-Cartan space-time as vectors instead of matrices, the geometric product $\gamma_\mu \gamma_\nu$ can be understood geometrically by separating it into symmetric and antisymmetric parts (Hestens 1967)

$$\gamma_\mu \gamma_\nu = \gamma_\mu \cdot \gamma_\nu + \gamma_\mu \wedge \gamma_\nu , \tag{8.1a}$$

where

$$\gamma_\mu \cdot \gamma_\nu = \frac{1}{2} (\gamma_\mu \gamma_\nu + \gamma_\nu \gamma_\mu) \equiv g_{\mu\nu} , \tag{8.1b}$$

$$\gamma_\mu \wedge \gamma_\nu = \frac{1}{2} (\gamma_\mu \gamma_\nu - \gamma_\nu \gamma_\mu) \equiv \sigma_{\mu\nu} . \tag{8.1c}$$

De Sabbata et al (1985) have pointed out that in gauge theory of gravity as well as in supergravity if super symmetric action contains the graviton $(e_\mu{}^k)$ and the gravitino $(\psi_\mu{}^a)$, the affine connection is nonsymmetric because of the presence of a fermion field which is a source of torsion according to the Einstein-Cartan theory.

The equation

$$\gamma_\mu \gamma_\nu = g_{\mu\nu} + \sigma_{\mu\nu} \tag{8.2}$$

suggests, as conjectured by de Sabbata (1995), that supersymmetry is inherited from it, because of the simultaneous existence of the commutator (Fermionic field) and anticommutator (Bosonic field) : $g_{\mu\nu}$ is connected with Bosons and $\sigma_{\mu\nu}$ with Fermions.

Next, the covariant differential of a spinor field as derived by De Sabbata et al (1985)

$$D\psi = dx_k \partial_k \psi + \frac{1}{2} w_{ik} \gamma^{[i} \gamma^{k]} \psi \tag{8.3}$$

yields dilation and rotation simultaneously, which, in turn indicates the presence of both curvature and torsion.

Furthermore, De Sabbata (1993, 1994) has shown that curvature R and torsion Q may play the role of conjugate variables of the geometry giving the commutation relation

$$\left[Q, R\right] = (\hbar G/c^3)^{-3/2} \quad . \tag{8.4}$$

All the above facts strongly indicate that supersymmetry may be introduced in Einstein-Cartan theory by using space-time algebra.

Acknowledgements

The author places on record his grateful indebtedness to Professor A. Zichichi, President, ICSC WORLD LABORATORY, LAUSANNE, for financial support under E-8 GRAVI PROJECT. The author also expresses his gratitude to Professor Venzo de Sabbata for his encouragement and fruitful suggestions and thankfully acknowledges the contributions of Professor D. Hestenes and his school in preparing this lecture-note.

References

De Sabbata, V. and Gasperini, M. - "Introduction to Gravitation" (World Scientific, Singapore, 1985).

De Sabbata, V. - Comm. Theor. Phys. **2** (1993) 217.

De Sabbata, V. - Nuovo Cimento **107A** (1994) 363.

De Sabbata, V. - in "Proc. of Int. School on Quantum Gravity", eds. P.G. Bergmann and V. de Sabbata (World Scientific, Singapore, 1995).

Hestenes, D. - "Space-Time Algebra" (Gordon and Breach, New York, 1966).

Hestenes, D. - J. Math. Phys. 8 (1967) 798-808.

Hestenes, D. - Amer. J. Phys. 39 (1971) 1013-1027.

Hestenes, D. - J. Math. Phys. 16 (1975) 556-572.

Hestenes, D. and Sobczyk, G. - "Clifford Algebra to Geometric Calculus" (D. Reidel, Dordrecht, 1984).

TWISTORS, TORSION, AND TENSOR-SPINORS SPACE-TIME

Venzo de Sabbata
World Laboratory, Lausanne, Switzerland
Dept.of Physics, University of Bologna and Ferrara, Italy
Istituto di Fisica Nucleare, sezione di Ferrara, Italy

ABSTRACT

The analogies between twistors and torsion are developed. The SL(2,C) group of transformations is used to suggest a unified link between the two approaches. Both twistors and torsion transform invariantly under SL(2,C) group structure. The role of space-time defects in the quantization of space-time is clarified and commutation relations between conjugate twistor variables and between curvature and torsion as conjugate variables are suggested. At last we make some considerations on tensor-spinors spacetime algebra.

1. Introduction

I will speak of torsion, spinors, twistors and some words on tensor-spinor space-time. First of all torsion. Why speak of torsion in a course of quantum gravity? because we think that torsion constitutes a way to go towards quantization of gravity.

In a previous paper [1], [2], we have shown that curvature R and torsion Q may play the role of conjugate variables of the geometry, enabling us to write commutation relations like

$$[Q, R] = (\hbar G/c^3)^{-3/2} \tag{1}$$

or, more explicitly,

$$[Q^{\alpha}{}_{\mu\nu}\, \Delta x^{\mu} \wedge \Delta x^{\nu}, R_{\beta\mu\nu\rho}\, \Delta x^{\mu} \wedge \Delta x^{\nu} \wedge k^{\rho}] = i(\hbar\, G/c^3)\delta^{\alpha}_{\beta} \tag{2}$$

where Δx^{μ}...... refer to changes in coordinates and k^{ρ} is a vector joining neighboring particles. The relation (1) may also be written as

$$\Delta Q\, \Delta R \geq L_{Pl}^{-3} \tag{3}$$

where L_{Pl} is the Planck length.

We can arrive in a very easy way at the relation (1); consider in fact that torsion has an intrinsic geometrical meaning: if we take a small closed circuit and write $l^{\alpha} = \oint Q^{\alpha}{}_{\mu\nu}\, dA^{\mu\nu}$ where $dA^{\mu\nu} = dx^{\mu} \wedge dx^{\nu}$ is the area element enclosed by the loop, then l^{α} represents the so-called "closure failure", so torsion represents the failure of the loop to close, l^{α} having the dimension of length ($Q^{\alpha}{}_{\mu\nu}$ has the dimension of inverse length).

In the above relation, we know that torsion can be related to the instrinsic spin $\hbar$, and, as the spin is quantized, we can say that the defect in space-time topology should occur in multiple of the Planck length, $(\hbar G/c^3)^{1/2}$ i.e.

$$\oint Q^{\alpha}_{.\beta\gamma} dx^{\beta} \wedge dx^{\gamma} = n\,(\hbar\, G/c^3)^{1/2}\, k^{\alpha} \tag{4}$$

(n is an integer and k^{α}=unit point vector).

Now (4) is analogous to the well known $\oint pdq = n\hbar$, i.e. to the Bohr-Sommerfeld relation, and would define a minimal fundamental length, i.e. the Planck length entering through the minimal unit of spin or action $\hbar$. So $\hbar$ has to deal with the intrinsic defects built into the torsion structure of space-time through $l^{\alpha} = \oint Q^{\alpha}_{.\beta\gamma} dx^{\beta} \wedge dx^{\gamma}$. Then, through eq.(4), we are led to commutation relations such as:

$$[Q^{\alpha}_{.\beta\gamma}, dx^{\beta} \wedge dx^{\gamma}] = (\hbar\, G/c^3)^{1/2}\, n^{\alpha} \tag{5}$$

and to corresponding uncertainty relations:

$$\Delta Q^{\alpha}_{.\beta\gamma}\,(\Delta x^{\beta} \wedge \Delta x^{\gamma}) \geq (\hbar\, G/c^3)^{1/2}\, n^{\alpha} \tag{6}$$

Therefore, the Einstein-Cartan theory of gravitation should provide, in contrast to Einstein's General Relativity Theory, genuine quantum-gravity effects.

We can also observe that from relation $l^{\alpha} = \oint Q^{\alpha}_{\mu\nu} dA^{\mu\nu}$, considering the fourth component, we can define time in the quantum geometric level through torsion as

$$t = (1/c) \oint Q\, dA = n(\hbar\, G/c^5)^{1/2}. \tag{7}$$

In that way we have quantized the torsion field but not the background. We believe that there is the possibility to include also the metrical background (the curvature).

In order to show this possibility, we can go a little further, trying to see if there are conjugate variables for which we can write down commutation relations. We will see that R and Q can be such conjugate variables. Consider in fact the equation of geodesic deviation of a free particle with non-zero spin (in which coupling of spin to the space-time curvature occurs):

$$\frac{d^2x^{\mu}}{ds^2} + \left\{ {\mu \atop \alpha\beta} \right\} \frac{dx^{\alpha}}{ds}\frac{dx^{\beta}}{ds} = K_{\alpha\beta.}^{\ \ \ \mu} \frac{dx^{\alpha}}{ds}\frac{dx^{\beta}}{ds} \tag{8}$$

or:

$$\frac{d^2x^{\mu}}{ds^2} + \left\{ {\mu \atop \alpha\beta} \right\} \frac{dx^{\alpha}}{ds}\frac{dx^{\beta}}{ds} = -2\, Q^{\mu}_{.\alpha\beta} \frac{dx^{\alpha}}{ds}\frac{dx^{\beta}}{ds} \tag{9}$$

(being $K_{\alpha\beta.}^{\ \ \ \mu} = -Q_{\alpha\beta.}^{\ \ \ \mu} - Q^{\mu}_{.\alpha\beta} + Q_{\beta.\alpha}^{\ \mu}$ the contortion tensor). Following the above discussion, in the quantum picture the right hand side is proportional to multiples of the Planck length. In the limit of vanishing torsion, we have the usual geodesic equation. So torsion in eq.(9) can be identified with the quantum correction (in multiples of $\hbar$) to the classical equation of motion.

Unlike electromagnetism, where a single particle suffices to measure the components of the field tensor, in the case of gravitation we require the relative acceleration between two particles to specify the components of the curvature tensor, i.e. $R^{\alpha}_{.\beta\gamma\delta}$ is defined through the acceleration:

$$d^2x^\alpha/ds^2 = R^{\alpha}_{.\beta\gamma\delta}\,(dx^\beta/ds)\,(dx^\gamma/ds)\,k^\delta \tag{10}$$

(k^δ= unit length vector in direction of the line joining the two particles).

As many approaches lead to the notion of "maximal acceleration", for instance in the case of gravitation, this is of $\sim m_{Pl}\,c^3/\hbar = c^{7/2}/(\hbar\,G)^{1/2}$ (m_{Pl} = Planck mass), then the maximal value for d^2x^α/ds^2 would also imply a maximal curvature which turns out to be $\approx c^3/\hbar\,G \approx 10^{66}\,\mathrm{cm}^{-2}$. In the classical limit, i.e. $\hbar \Rightarrow 0$, we have the usual curvature singularity, i.e. the maximal value is infinite, the relation between acceleration and curvature being of the form: $a = c^2\sqrt{R}$, R= curvature scalar. This would imply a relation of the type

$$\Delta R_{\alpha\beta\gamma\delta}(\Delta x^\alpha \wedge \Delta x^\beta)\,(\Delta x^\gamma \wedge \Delta x^\delta) \geq \hbar\,G/c^3 \tag{11}$$

Now we can try to go a little further: if we wish to connect the initial and final position of one and the same particle, we cannot avoid uncertainty associated with torsion, i.e. for a sufficiently small area element dS, uncertainty in distance between initial and final position would be $\Delta l^\mu = Q^\mu\,dS$ and this will induce fluctuations in distance in the metric through $\Delta l = \sqrt{g_{\mu\nu}dx^\mu dx^\nu}$. So what is important is not the point themselves but the "fluctuations" in their position. Now the fluctuations in position strike the idea of twistors where the points (i.e. intersections points between two twistors) become fuzzy or fluctuate quantum-mechanically. Points are secondary in both approaches.

Since curvature causes relative acceleration between neighbouring test particles, we have momentum uncertainty related to curvature as:

$$ma^\mu ds = \Delta p^\mu = m\,R^\mu{}_{\alpha\beta\gamma}\frac{dx^\alpha}{ds}\,dx^\beta k^\gamma = m\,R\,\frac{dS}{ds}\,k^\mu \tag{12}$$

So as position fluctuations are given by torsion, momentum fluctuations are due to curvature, and we can interpret quantum effects (and then uncertainty principle) as consequences of space-time deformation i.e.

$$\Delta p^\mu \cdot \Delta x_\mu \geq i\,\hbar \tag{13}$$

where

$$\Delta x^\mu = Q\,c\,ds\,k^\mu \qquad \text{and} \qquad \Delta p^\mu = m\,R\,\frac{dS}{ds}\,k^\mu \tag{14}$$

(i.e. uncertainty in initial and final position and relative acceleration between them due to space-time deformations), Q (torsion) and R (curvature) thus both playing simultaneous roles as conjugate variables of the geometry (gravitational field), thus enabling us to write commutation relations between curvature and torsion (analogous to $[x,p] = i\hbar$) as:

$$[Q, R] = (\hbar G/c^3)^{-3/2} \tag{15}$$

i.e. the relation (1).

2. Twistors

We can notice that in the case of twistors (see [3],[4]) the space-time point can be written as 2 × 2 complex matrix having SL(2,C) group structure. The result is that twistors transform invariantly when a local Lorentz transformation is performed. In fact it is well known that there is a homomorphism between the SL(2,C) group of complex unimodular (2 × 2) matrix and the Lorentz O(1,3) group.

In the torsion picture we know very well that the spinor gauge fields described by tetrads transform invariantly under SL(2,C). So we have the same group of transformation SL(2,C) for torsion and for twistors.

One fundamental feature of twistor theory is that we operate in complex space; if we take account of helicity, we have six dimensions, and this means 3-D complex space and with a complex space we are close to the description of quantum theory; so with twistor theory we represent space-time structure and, at the same time, we are close to the description of quantum theory. Moreover a point in space-time is represented by a sphere in twistor complex space, and this non-local feature in twistors is also present in quantum theory.

We think that this fact can be fruitful because when we go to quantization, in general we use a complex Hilbert space and also because it happens that spinor structure can be, in both cases of torsion and twistors, the basis of a space-time structure.

At this point we like to notice that with the introduction of Hestenes geometric algebra [5] we can operate in real space-time. We will discuss this point in a moment.

In the twistors view, the space-time 'direction of motion' of twistors (a twistor representing a massless particle in motion) remains null, the points of space-time intersection between twistors become subject to quantization, i.e. the points become 'fuzzy' or fluctuate quantum-mechanically.

In the torsion view, we have seen that fluctuations in position due to torsion and fluctuation in momentum due to curvature enabled us to write quantum uncertainty relations for the conjugate variables of curvature and torsion.

The points are derived or secondary concepts in both cases of torsion and twistors.

In the twistors case, the physical world is pictured as a sort of network of line segments, each segment is thought of as an 'object' having a well defined angular momentum of $n\hbar/2$. Intersections of segments give rise to space-time points; these points are then derived objects: the twistors themselves are more basic.

So while in the twistor case complex spin vectors are regarded as primary, in the torsion case the spinor gauge fields are basic entities: the metric is a derived concept, and is given by the expectation value of a tetrad product, i.e.

$$\langle h_a^b\, h_c^d\, \eta_{bd}\rangle = g_{ac} \tag{17}$$

Curvature tensors and other geometric quantities are built from these tetrads. Analogously in the twistor case at every point the metric is defined as

$$g_{ab} = \langle \varepsilon_{AB}\ \varepsilon_{A'B'}\rangle \tag{18}$$

in terms of complex spin vectors.

There are many ways to formally introduce the twistor concepts. Remembering that a twistor represents a massless particle in motion, first of all we can observe that the most important physical attributes in a system in special relativity are its four-momentum p_a and its six-angular momentum M^{ab} $(= -M^{ba})$, defined relative to some origin 0. A twistor Z^α is an object which can be described, in a suitable coordinate frame, by four complex numbers

$$(Z^0\,, Z^1\,, Z^2\,, Z^3) \tag{19}$$

where the complex conjugate twistor is described by

$$(\bar{Z}_0\,, \bar{Z}_1\,, \bar{Z}_2\,, \bar{Z}_3) = (\overline{Z^2}\,, \overline{Z^3}\,, \overline{Z^0}\,, \overline{Z^1}) \tag{20}$$

(the symbol $\overline{Z^2}$ denoting the complex conjugate of the complex number Z^2, etc.).

For the massless particle the 4-momentum and angular momentum components, in a standard Minkowski frame, are to be related to those of Z^α by

$$\begin{aligned} p^0 &= (1/\sqrt{2})(Z^3\,\bar{Z}_1 + Z^2\,\bar{Z}_0)\\ p^1 &= (1/\sqrt{2})(Z^3\,\bar{Z}_1 - Z^2\,\bar{Z}_0)\\ p^2 &= (1/\sqrt{2})(-Z^2\,\bar{Z}_1 - Z^3\,\bar{Z}_0)\\ p^3 &= (1/\sqrt{2})(Z^2\,\bar{Z}_1 - Z^3\,\bar{Z}_0) \end{aligned} \tag{21}$$

and

$$\begin{aligned} M^{01} &= -M^{10} = (1/2)(Z^0\,\bar{Z}_0 - Z^1\,\bar{Z}_1 - Z^2\,\bar{Z}_2 + Z^3\,\bar{Z}_3)\\ M^{02} &= -M^{20} = (1/2)(Z^0\,\bar{Z}_1 + Z^1\,\bar{Z}_0 - Z^2\,\bar{Z}_3 - Z^3\,\bar{Z}_2)\\ M^{03} &= -M^{30} = (1/2)(Z^0\,\bar{Z}_1 - Z^1\,\bar{Z}_0 - Z^2\,\bar{Z}_3 + Z^3\,\bar{Z}_2)\\ M^{12} &= -M^{21} = (1/2)(Z^0\,\bar{Z}_1 - Z^1\,\bar{Z}_0 + Z^2\,\bar{Z}_3 - Z^3\,\bar{Z}_2)\\ M^{32} &= -M^{23} = (1/2)(Z^0\,\bar{Z}_0 - Z^1\,\bar{Z}_1 + Z^2\,\bar{Z}_2 - Z^3\,\bar{Z}_3)\\ M^{13} &= -M^{31} = (1/2)(Z^0\,\bar{Z}_1 + Z^1\,\bar{Z}_0 + Z^2\,\bar{Z}_3 + Z^3\,\bar{Z}_2) \end{aligned} \tag{22}$$

$$M^{00} = M^{11} = M^{22} = M^{33} = 0$$

The most directly physical one, relevant for our purpose, is to use angular momentum of zero rest mass particle. The total momentum 4-vector $\bar{p}_a$ and angular momentum bivector $M^{ab} = -M^{ba}$, with respect to some origin 0, correspond to those of a zero rest mass particle of helicity h. We have

$$\bar{S}_a = h\,\bar{p}_a \tag{23}$$

$\bar{S}_a$ being the Pauli-Lubanski vector, being given by

$$\bar{S}_a = (1/2)\varepsilon_{abcd}\,\bar{p}^b\, M^{cd} \tag{24}$$

$\bar{p}^a$ is null and future pointing i.e. $p_a p^a = 0$ and $p^0 > 0$. Equation (23) is automatically satisfied by (21), (22) with

$$h = (1/4)(Z^\alpha\,\bar{Z}_\alpha + \bar{Z}_\alpha\,Z^\alpha) \tag{25}$$

In obtaining (25), it has not been assumed that Z^α and $\bar{Z}_\alpha$ commute but that either all components commute (classical twistor variables) or that the twistor commutation rules

$$[Z^\alpha\,, Z^\beta] = 0, \qquad [\bar{Z}_\alpha\,, \bar{Z}_\beta] = 0 \qquad [Z^\alpha\,, \bar{Z}_\beta] = \hbar\delta^\alpha_\beta \tag{26}$$

hold. We see the formal analogy with the commutation relation (1).

3. Twistor and commutation relations

To see better this analogy, we consider the two-spinor expression $p_{AA'}$ for p^a in the form

$$p_{AA'} = \bar{\pi}_A\,\pi_{A'} \tag{27}$$

(for same dual conjugate spin vector π_A).

The angular momentum in this notation is defined as [3]

$$M^{AA'BB'} = i\omega^{(A}\bar{\pi}^{B)}\varepsilon^{A'B'} - i\varepsilon^{AB}\bar{\omega}^{(A'}\pi^{B')} \tag{28}$$

The conjugate pair $(\omega^A\,, \pi_A)$ uniquely defines momentum p_a and angular momentum M^{ab}.

The gauge freedom in ω^A , π_A is overall phase change, i.e. invariance under

$$\omega^{A'} \Longrightarrow e^{i\vartheta}\,\omega^A, \qquad \pi_{A'} \Longrightarrow e^{i\vartheta}\,\pi_A \tag{29}$$

ω^A , π_A are 2 spinor parts of twistors Z^a, written as:

$$Z^\alpha \Longleftrightarrow (\omega^A\,, \pi_A) \tag{30}$$

The complex conjugate of the twistor Z^a is the dual $\bar{Z}_\alpha$, i.e.

$$\bar{Z}^\alpha \iff (\bar{\pi}_{A'}, \bar{\omega}^{A'}) \tag{31}$$

The scalar product $Z^a\bar{Z}_\alpha$ can be now formed as:

$$Z^\alpha \bar{Z}_\alpha = \omega^A \bar{\pi}_A + \pi_{A'} \bar{\omega}^{A'} \tag{32}$$

This so called twistor norm, i.e. the invariant Hermitian form $Z^\alpha\bar{Z}_\alpha$, has an interesting physical interpretation: it mesures the helicity of zero mass particle, i.e. (from (23), (24), (27), (28)):

$$Z^\alpha \bar{Z}_\alpha = 2h \tag{33}$$

The helicity (or spin) is thus a conformal invariant. When $h = 0$ we have:

$$Z^\alpha \bar{Z}_\alpha = 0 \tag{34}$$

and Z^α (or $\bar{Z}_\alpha$) is referred to as a null twistor. We can think of a complexified space-time coordinate as:

$$Z_\mu = x_\mu + ih_\mu \tag{35}$$

where ih_μ crresponds to the 'internal helicity' project vector attached to the space-time point x_μ.

In fact the anisotropic feature of the h-space, leads us to consider it as an attached 'projection vector' to the space-time point x_μ, and with twistor geometry this 'projection vector' is automatically interpreted as an internal helicity.

Space-time point x_μ can be written as a 2×2 complex matrix having SL(2,C) group structure. Physically the two opposite orientations of the projection vector give rise to two opposite internal helicities which correspond to particle and antiparticle in isospin space and which can easily be formulated in terms of the extended metric $g_{\mu\nu}(x, \vartheta, \bar{\vartheta})$ (see below) where ϑ and $\bar{\vartheta}$ are two component spinorial variables (see eq.(36) below).

Considering a complex manifold, the coordinates in this extended space, where we identify the coordinate in the complex manifold as $Z^\mu = x^\mu + ih^\mu$, where $h^\mu = (1/2)\eta^\mu_a \vartheta^a$, can be written as

$$Z^\mu = x^\mu + (i/2)\eta^\mu_a \vartheta^a \tag{36}$$

Here ϑ^μ (=1,2) is a two component spinor or, in matrix representation:

$$Z^{BB'} = x^{BB'} + (i/2)\eta^{BB'}_a \vartheta^a \tag{37}$$

where

$$X^{BB'} = (1/\sqrt{2}) \begin{pmatrix} x^0 - x^1 & x^2 + ix^3 \\ x^2 - ix^3 & x^0 + x^1 \end{pmatrix} \tag{38}$$

$$\eta_a^{BB'} \in SL(2, C) \tag{39}$$

The twistor equation in terms of the complex conjugate spinors $\bar{\pi}_B(\pi_{B'})$ is:

$$\bar{Z}_\alpha Z^\alpha + \eta_a^{BB'} \vartheta^a \bar{\pi}_B \pi_{B'} = 0 \tag{40}$$

The four momentum corresponds to:

$$\bar{p}^{BB'} = \bar{\pi}^B \pi^{B'} \tag{41}$$

$$Z^\alpha = (\omega^B \, , \pi_{B'}); \qquad \bar{Z}^\alpha = (\bar{\pi}_B \, , \bar{\omega}^{B'})$$

with

$$\omega^B = i \left(x^{BB'} + (i/2)\eta_a^{BB'} \vartheta^a \right) \pi_{B'} \tag{42}$$

The helicity operator is then:

$$h = Z^\alpha \bar{Z}_\alpha = -\eta_a^{BB'} \vartheta^a \bar{\pi}_B \pi_{B'} \tag{43}$$

$\bar{p}^\mu$ is the canonical conjugate of x^μ in $Z^\mu = x^\mu + (i/2)\eta_a^\mu \vartheta^a$

$$\bar{p}^{BB'} = \bar{\pi}^B \pi^{B'}, \quad \bar{p}_\mu^2 = 0 \tag{44}$$

Defining

$$\beta = i\vartheta^{A'} \pi_A, \qquad \bar{\beta} = -i\bar{\vartheta}^A \bar{\pi}_{A'} \, , \tag{45}$$

$$h = -\vartheta^A \bar{\vartheta}^{A'} \bar{\pi}_A \pi_{A'} = \beta\bar{\beta} \tag{46}$$

The metric is then:

$$g_{\mu\nu}(x, \vartheta, \bar{\vartheta}) = g_{\mu\nu}^{BB'}(x) \bar{\vartheta}_B \vartheta_{B'} \tag{47}$$

One can now construct connections (both symmetric and antisymmetric) from this metric and its derivatives. Torsion and curvature corresponding to this complexified space can in turn be constructed from these connections.

In fact in terms of tetrads ($h_\mu^a \, h_\nu^b \, \eta_{ab} = g_{\mu\nu}$) we can define the fundamental 1-forms ϑ^i:

$$\vartheta^i = h_\mu^i dx^\mu \tag{48}$$

and the connection 1-form $\omega^i{}_j$:

$$\omega^i{}_j = \gamma^i{}_{jk} \vartheta_k \tag{49}$$

where $\gamma^i_{.jk}$ are Ricci rotation coefficients ($\gamma^i_{.jk} = -h^i_{\mu;\nu} h_j^\mu h_k^\nu$) and, as is well known, we have the relations:

$$h_\alpha^a x^\beta - x^\beta h_\alpha^a = \delta_\alpha^\beta \, (hG/c^3)^{1/2} n^a \tag{50}$$

or

$$\Delta h_\alpha^a \Delta x^\beta \geq \delta_\alpha^\beta (\hbar G/c^3)^{1/2} n^a \tag{51}$$

(where n^a is a direction vector). We can take covariant exterior derivatives of eqs.(48) and (49) obtaining curvature and torsion 2-forms:

$$\Theta^i = h^i_\alpha dx^\alpha = D\,\vartheta^i = d\vartheta^i + \omega^i{}_k \wedge \vartheta^k = (1/2)Q^i{}_{jk}\vartheta^j \wedge \vartheta^k \tag{52}$$

$$\Omega^i{}_j = D\,\omega^i{}_j = d\omega^i{}_j + \omega^i{}_k \wedge \omega^k{}_j = (1/2)R^i{}_{jkl}\vartheta^l \wedge \vartheta^k \tag{53}$$

so that we have the conjugate relation between torsion and curvature forms:

$$[D\,\vartheta^i\,, D\,\omega_m{}^j] = [\Theta^i\,, \Omega_m{}^j] = \delta^i_m(\hbar G/c^3)^{1/2} n^j \tag{54}$$

and

$$(1/2)[Q_{kl}.{}^i\,\vartheta^k \wedge \vartheta^l\,, R_{mjkl}\,\vartheta^j \wedge \vartheta^k \wedge k^l] = (\hbar G/c^3)\delta^i_m \tag{55}$$

One can define the radius vector field which can be integrated along a curve and when this is done for a loop (of radius r), one finds that with torsion, it does not return to initial value, i.e. it gives rise to a local translation in the internal space of the radius vector field. So the change in the intersection loop of two twistors, when taken along such a loop, is given by:

$$\delta Z^\alpha = \frac{1}{r^2}\int (d\omega^B_{ji} + \omega^B_{ki} \wedge \omega^{Bk}_j) Z^\alpha dx^i dx^j \geq \hbar^{1/2} n^\alpha \tag{56}$$

The fluctuation in the norm $\langle \delta Z^\alpha\,\delta \bar{Z}_\alpha \rangle \geq \hbar$ then gives a twistor geometric definition of Planck's constant. Thus this automatically leads to the quantization of the background geometry. This may provide a basis for the twistor commutation relation:

$$[Z^\alpha\,, Z^\beta] = 0, \qquad [\bar{Z}_\alpha\,, \bar{Z}_\beta] = 0, \qquad [Z^\alpha\,, \bar{Z}_\beta] \geq \hbar\delta^\alpha_\beta \tag{57}$$

(see eqs.(26) and (63)).

Eqs.(52, 53) and (56) can constitute a first step towards quantization of space-time through twistors in terms of spinors θ and complex conjugate variable $\bar{\pi}$ or $\bar{\omega}$ which are connected with torsion and curvature; in this context twistor theory seems to constitute a mathematical-geometrical basis for the commutation relation (1).

The twistor two-form $d\bar{Z}_\alpha \wedge dZ^\alpha$ which defines Penrose's symplectic structure, agrees, apart from the factor i, with the natural two form $\omega_j{}^i$.

Remember now that a null twistor may be represented by a pair of two-spinors. This pair of spinors has a simple geometrical interpretation: Let $(\omega^A\,, \pi_{A'})$ represent the null twistor Z^α. Then, if p_a is the null contangent vector corresponding to $\bar{\pi}_A \pi_{A'}$, $g^{ab}p_b$ is the tangent vector to the null geodesic γ defined by Z^α, and points x on γ are determined by the equation $\omega^A - ix^{AA'}\pi_{A'} = 0$.

Moreover, following Crampin and Pirani, we have [6]

$$\begin{aligned} \bar{Z}_\alpha dZ^\alpha &= i\bar{\pi}_A(\pi_{A'}dx^{AA'} + x^{AA'}d\pi_{A'}) - ix^{AA'}\bar{\pi}_A d\pi_{A'} \\ &= i\bar{\pi}_A\pi_{A'}dx^{AA'} = ip_a dx^a \end{aligned} \tag{58}$$

and then we have

$$d\bar{Z}_\alpha \wedge dZ^\alpha = idp_a \wedge dx^a \tag{59}$$

where $x^{AA'}$ is the spinor equivalent of x^a. The null geodesic is well defined if and only if $\omega^A \pi_A$ is real, which is the condition that the corresponding twistor be null. The twistor $\bar{Z}_\alpha$ (the complex conjugate of Z^α) is represented by the spinor pair $(\bar{\pi}_A, \bar{\omega}^{A'})$; if moreover $(\lambda^A, \mu_{A'})$ represents a twistor W, the scalar product $Z_\alpha W^\alpha$ is $\bar{\pi}_A \lambda^A + \bar{\omega}^{A'} \mu_{A'}$. In particular, $\bar{Z}_\alpha dZ^\alpha = \bar{\pi}_\alpha d\omega^\alpha + \bar{\omega}^{A'} d\pi_{A'}$, so for a null twistor we have eq.(58). Then we have

$$d\bar{Z}_\alpha \wedge dZ^\alpha = idp_a \wedge dx^a \tag{60}$$

and, as shown before, dp_a and dx^a can be associated, respectively, with fluctuations ΔR and ΔQ in the curvature and torsion of the background geometry. Thus

$$[Q, R] = L_{Pl}^{-3} \tag{61}$$

Therefore analogous to the well-known quantum commutation relations

$$[x^\alpha, x^\beta] = 0, \qquad [p_\alpha, p_\beta] = 0, \qquad [x^\alpha, p_\beta] = \hbar\delta^\alpha_\beta \tag{62}$$

we have the twistor commutation relations for the geometry as:

$$[Z^\alpha, Z^\beta] = 0, \qquad [\bar{Z}_\alpha, \bar{Z}_\beta] = 0, \qquad [Z^\alpha, \bar{Z}_\beta] = \hbar\delta^\alpha_\beta \tag{63}$$

(see also (26) and (57)).

Thus the uncertainty relations for $Z^\alpha, \bar{Z}_\beta$ take the form:

$$[Z^\alpha, \bar{Z}_\beta] = (1/2)\hbar\delta^\alpha_\beta \tag{64}$$

So far the discussion above has been done for massless particles, the twistors norm $Z^\alpha \bar{Z}_\beta$ being physically associated with the helicity of a zero mass particle.

We can try to extend these cosiderations to particles with mass and charge.

4. Supersymmetry in the spacetime geometric algebra?

We think, as we have said, that a deeper common basis for twistor and torsion would be a spinorial space-time and this concept of a spinor structure of a space-time has to be developed.

Yu Xin [7] shows that the concept of 'internal' and 'external' spaces are just two different representations of a single primitive structure: the spinorial space-time, or in other words the spinor structure, represents the 'internal' abstract space and 'external' space time, depending on its representation.

In fact we have seen that in order to take into account both mass and spin it seems at first sight that we have to do with two different spaces: a real space-time where we we describe the curvature, due to the mass, with tensors and complex space-time where we describe torsion, due to the spin, with spinors. But this is not completely satisfactory: one would like to describe these two fundamental physical properties, mass and spin, in a unique manifold, the real space-time and this can be done through Hestenes algebra [5].

In other words we can describe at the same time bosons and fermions: curvature and torsion must be given in the same real space-time. We know that when we introduce Dirac equation in Riemann-Cartan space-time (and this is necessary if we like to describe bosons and fermions), we find that contorsion tensor is completely antisymmetric. In fact if we consider the term in the Lagrangian for the Dirac equation in U_4 that contains interaction between spinor and torsion, we find $K_{abc}\bar{\psi}\gamma^{[a}\gamma^{b}\gamma^{c]}\psi$ (see [8]) (where K_{abc} is the contorsion tensor defined through torsion tensor as $= -Q_{abc} - Q_{cab} + Q_{bca}$) and spin density tensor is given by

$$S^{abc} = (1/\sqrt{-g})\,[\partial\sqrt{-g}L_m/\partial K_{bca}] = -(i/4)\bar{\psi}\gamma^{[a}\gamma^{b}\gamma^{c]}\psi = S^{[abc]}$$

that is the spin density tensor is totally antisymmetric.

Moreover we know that the transformation law for spinor field is written as

$$\psi'(x') = U(\Lambda)\psi(x) \tag{65}$$

where $U(\Lambda)$ is the usual 4×4 constant matrix representing the Lorentz transformation (the spinor indices are not written explicity). Λ is the Lorentz matrix involved in the vector Lorentz transformation in the flat tangent space $x'^i = \Lambda^i_k x^k$.

Dirac equation $(i\gamma^k\partial_k\psi - m\psi = 0)$ is transformed as

$$i\gamma^i(\partial\psi'(x')/\partial x'^i) - m\psi'(x') = i\gamma^i\Lambda_i{}^k\partial_k U\psi - mU\psi = 0 \tag{66}$$

It is well known that multiplying the left by U^{-1} and imposing on the Dirac eqation to be invariant in form under a Lorentz transformation, we obtain the condition for the matrix U

$$U^{-1}\gamma^i U = \gamma^k\Lambda^i{}_k \tag{67}$$

Considering an infinitesimal transformation

$$\Lambda_{ik} \simeq \eta_{ik} + \omega_{ik} \tag{68}$$

(where $\omega_{ik} = \omega_{[ik]}$), we have

$$U = 1 + (1/2)\omega_{ik}S^{ik} \tag{69}$$

The ω_{ik} are six constant infinitesimal parameters and S^{ik} are the generators of the infinitesimal Lorentz transformation which in order that (67) be fulfilled must satisfy

$$S^{ik} = \gamma^{[i}\gamma^{k]} = (1/2)(\gamma^i\gamma^k - \gamma^k\gamma^i) \tag{70}$$

Here we can observe that considering the Dirac equation in Riemann-Cartan space-time we have the relation

$$g_{\mu\nu} = (1/2)(\gamma_\mu\gamma_\nu + \gamma_\nu\gamma_\mu) \tag{71}$$

where γ are the Dirac matrices. But we have also

$$\sigma_{\mu\nu} = (1/2)(\gamma_\mu\gamma_\nu - \gamma_\nu\gamma_\mu) \tag{72}$$

We see that eqs.(70) and (72) are formally identical, and it seems that we can pass from one to another with the help of the Hestenes algebra [5].

We have also

$$\gamma^\mu\gamma^\nu = g^{\mu\nu} + \sigma^{\mu\nu} \tag{73}$$

Eq.(73) seems to include automatically supersymmetry because we have commutator (fermonic field) and anticommutator (bosonic field): $g_{\mu\nu}$ is connected with bosons and $\sigma_{\mu\nu}$ is connected with fermions, and they are given simultaneously. Notice that eq.(73) follows directly from Hestenes' geometric product [5]

$$\gamma_\mu\gamma_\nu = \gamma_\mu \cdot \gamma_\nu + \gamma_\mu \wedge \gamma_\nu \tag{74}$$

where

$$\gamma_\mu \cdot \gamma_\nu = g_{\mu\nu} \tag{75}$$

and

$$\gamma_\mu \wedge \gamma_\nu = \sigma_{\mu\nu} \tag{76}$$

In other words the Hestenes geometric algebra [5] with the concept of multivectors (see B.K.Datta, these Proceedings [9]) and with the precise geometrical interpretation of imaginary numbers, seems to be very important: the imaginary unit i appearing in the Dirac, Pauli and Schrödinger equations has a geometrical interpretation in terms of rotation in real space-time [10]: thus one has to do with real space-time; spinors and tensors are treated in a unified way (and this allows us to write eq.(73)), and the antisymmetric terms in the stress-energy tensor are clearly related to spin [11].

These concepts (related to eqs.(70),(72),(73),(74)) will be explored in future works.

In conclusion I like to give the following Table which shows the main analogies between Einstein-Cartan and twistors space-time:

Table 1.

Einstein-Cartan space-time	Twistor space-time
$D\vartheta^i = (1/2)Q^i_{jk}\vartheta^j \wedge \vartheta^k$	
	$\delta Z^\alpha = \frac{1}{r^2}\int(d\omega^B_{ji} + \omega^B_{ki} \wedge \omega^{Bk}_j)Z^\alpha dx^i dx^j$
$D\omega^i_j = (1/2)R^i{}_{jkl}\vartheta^j \wedge \vartheta^k$	
or	
$\Delta x^\mu = Q\, c\, ds\, k^\mu$	$\bar{Z}_\alpha dZ^\alpha = i\bar{\pi}_A \pi_{A'} dX^{AA'} = ip_a dx^a$
$\Delta p^\mu = m\, R\, \frac{dS}{ds}\, k^\mu$	$\bar{Z}_\alpha dZ^\alpha = \bar{\pi}_A d\omega^A + \bar{\omega}^{A'} d\pi_{A'}$
	then
$\Delta p^\mu . \Delta x_\mu \geq i\hbar$	$d\bar{Z}_\alpha \wedge dZ^\alpha = idp_a \wedge dx^a$
then	
$[Q^\alpha{}_{\mu\nu}\Delta x^\mu \wedge \Delta x^\nu\, , R_{\beta\mu\nu\rho}\Delta x^\mu \wedge \Delta x^\nu \wedge k^\rho]$ $= i(\hbar G/c^3)\delta^\alpha_\beta$	
or	then
$[Q\, , R] = (\hbar G/c^3)^{-3/2} = L^{-3}_{Pl}$	$[Z^\alpha, \bar{Z}_\beta] = (1/2)\hbar\delta^\alpha_\beta$
and	or
$\Delta Q\, \Delta R \geq L^{-3}_{Pl}$	$\Delta Z^\alpha\, \Delta\bar{Z}_\beta \geq (1/2)\hbar\delta^\alpha_\beta$

5. References

1. V. de Sabbata, *Il Nuovo Cimento* **107A**, 363 (1994).
2. V. de Sabbata, *Communications in Theoret. Phys.* **2**, 217 (1993).
3. Roger Penrose, *"Twistor theory, its aims and achievements"* in 'Quantum Gravity' ed. by C.J. Isham, R. Penrose and D.W. Sciama, Clarendon Press, Oxford 1975.
4. Roger Penrose, *"Twistor Theory"* in *'Cosmology and Gravitation- Spin, Torsion, Rotation and Supergravity'* ed. by Peter G. Bergmann and Venzo de Sabbata, Plenum Press, N.Y. 1980.
5. D. Hestenes, *"Clifford Algebra to Geometric calculus"* D. Reidel Publ. Co. Dordrecht, Holland 1982.
6. M. Crampin and F.A.E. Pirani, in *"Relativity and Gravitation"* ed. by Charles G. Kuper and Asher Peres, Gordon and Breach, N.Y.1971, 105.
7. X. Yu, *Astrophys.Space Sci.* **154**, 321 (1989), see also *"General Relativity on Spinorial Space-time"* preprint; see also these Proceedings.
8. V. de Sabbata and M. Gasperini, *"Introduction to Gravitation"* World Scientific Publishing, Singapore 1985.
9. B.K. Datta, *"Physical theories in spacetime algebra"* these Proceedings.
10. D. Hestenes, *J. Math. Phys.* **16**(3), 556 (1975).
11. see for instance A.N. Lasenberg, C.J.L. Doran and S.F. Gull, *Found. Phys.* **23**(10), 1295 (1993).

QUANTIZED MATTER IN A DE SITTER GAUGE THEORY WITH CLASSICAL METRIC AND AXIAL TORSION

W. DRECHSLER

Max-Planck-Institut für Physik

– Werner-Heisenberg-Institut –

P.O.Box 40 12 12, 80712 Munich (Fed. Rep. of Germany)

ABSTRACT

The dualism between classical matter and geometry is extended to include elementary hadronic matter described in quantized form in terms of a gauge theory based on the group $SO(4,1)$ containing a built-in fundamental length parameter R of geometric origin of the order of a Fermi. Quantized matter at subnuclear distances is described in terms of generalized quantum mechanical wave functions (first quantized geometro-stochastically quantized fields) which are identified with sections on a soldered Hilbert bundle $\mathcal{H}$ with Cartan connection constructed over a curved Riemann-Cartan space-time base. The bundle $\mathcal{H}$ carries a unitary irreducible spin zero phase space representation of the de Sitter group, i.e. a system of covariance of $SO(4,1)$ describing elementary objects extended at a scale of length given by R. The Lorentz subgroup of $SO(4,1)$ is related to a vierbein formulation of gravitation. Two sets of source equations for the underlying Riemann-Cartan geometry are considered: (i) Einstein-type equations and (ii) current-curvature equations of Yang-Mills type relating the energy-momentum and $SO(4,1)$-gauge current distributions of classical *as well as* quantized matter to the underlying geometry. It is shown that Einstein's metric theory of gravitation is contained in this scheme as an unquantized classical theory embedded in a geometric framework involving –besides classical macrophysical sources – quantized microphysical material sources determining quantum effects for the additional nonmetric fields of the geometry (torsion) which are induced at small distances by that part of matter described in a quantum mechanical manner. Axial vector torsion is considered explicitly and a set of nonlinear field equations for the axial vector torsion field is derived from the current-curvature equations.

1. Introduction

General relativity is a classical theory of the gravitational interaction intended to describe the physics of gravity at a macroscopic[1)] or even cosmological scale of length. General relativity is not a theory of matter: it treats a classical macroscopic distribution of matter in a rather rudimentary way through its energy-momentum content and relates a source distribution of energy and momentum to the metric geometry of the underlying space (or rather space-time). Elementary matter has much more properties than given by the deposition of energy and momentum at a certain place in space-time. It has internal rotational properties (spin); it has isospin, charge and hypercharge properties – to name just a few – which all could influence the geometry of the underlying space. In fact, elementary particle physics has gone a long way to show, starting from atomic physics and the explanation of optical, infrared, X- and γ-ray spectra of atoms, molecules and nuclei that the atomic world – the microphysical world at atomic and subatomic or even nuclear and subnuclear distances – is governed by a dynamics described by quantum theory. Quantum mechanics, indeed, *is* a theory of matter; more exactly, a theory of matter at an atomic scale of length. Also nuclear physics and particle physics seems to be described by quantum theory. Although the final theory of hadrons and leptons is not yet known, it is clear that quantum mechanical aspects play a fundamental rôle in the description and understanding of elementary matter at subatomic and subnuclear distances.

On the other hand, we have learned from general relativity that it is a very attractive idea to consider a feed-back mechanism between matter, on the one side, and the underlying geometry, on the other side. The question in this context is whether the dualism between matter and the ambient geometry can be extended to the subatomic or even subnuclear level. Formulated differently one may ask: *Does quantized matter again leave an imprint in the underlying geometry by inducing certain quantum effects in the geometry at microphysical distances?* Let me point out

---

1) When the words "macroscopic" and (see below) "microscopic" or "microphysical" are used, we mean a description characterized by a certain length scale. The division line may roughly be taken to be the radius of an atom, i.e. 10^{-8} cm. In our later discussion a fundamental length scale R of the order of $R \approx 10^{-13}$ cm characterizing hadron physics will be introduced.

already here that I don't want to discuss the question of quantizing a particular geometry *as such*, or pose this as a mathematical problem or exercise. I would rather like to answer the physically motivated question: What are the effects on the underlying geometry – the geometric framework or "stratum" assumed to exist in order to formulate a particular theory – if the material sources for this geometry are treated quantum mechanically? Is there again an extended dualism at work corresponding to Einstein's vision, however now not necessarily a dualism relating the energy-momentum of matter to the metric of the underlying space via the contracted metric curvature tensor, but a similar dualism (at least in spirit) relating other, in fact, *quantum mechanical* properties of matter to certain characteristic *nonmetric* features of the geometry? So our basic thesis will be that the geometry acquires quantum properties only through the quantum nature of matter showing up at small distances. This is a quite different starting point than the one taken by postulating that Einstein's theory is to be quantized as such by quantizing the metric or the connection of a pseudo-Riemannian space-time (even in the absence of material sources) and trying to give a meaning to such a procedure. It may be wise to be much more cautious at this point and, in fact, much more modest and ask in this context the following question: *Is it, indeed, necessary to quantize Einstein's metric theory of gravitation if elementary matter which, necessarily, is to be treated in a quantum mechanical manner, is introduced into the theory?* I shall try to present arguments (probably as the devil's advocate at this school) that it is very well possible, and certainly physically reasonable, to leave the metric of Einstein's theory classical, i.e. unquantized, and ask what *additional nonmetric geometric fields* may acquire a quantum structure at small subnuclear distances if the quantum nature of the sources is taken into consideration. We shall thus consider classical macroscopic sources *as well as* quantized microscopic sources of the underlying geometry side by side – without ideolgocal preference of one over the other – and determine what the possible modifications of Einstein's theory at small distances are.

The philosophical idea behind this proposal is that of a *multilayered structure of the reality of the outside world.* I don't want to enter at this point into a general philosophical discussion about the reality of the world as perceived by us. As physicists we can be neither naive realists nor pure epistemological idealists. Our theories are rational images of certain domains of reality with limiting properties being satisfied by the theories on the border of two such domains. We don't have one unique overall description of nature – although this may be the ultimate goal. We have various partial descriptions valid in limited ares. Why do I say this? I don't want to base my following arguments on the monistic thesis that the world around us is from top to bottom a quantum world being quantum mechanically structured and organized throughout. I think we don't know enough yet to make such a statement. In the present situation it is, in my view, a better strategy to start from an interplay of several antagonistic principles or entities involved in a basic *dualism* between *geometry* and *matter* : here the *geometry* (the arena for all the phenomena), there *matter* (a space filling agent affecting the underlying geometry and being affected by it). In this dualism both classical *and* quantum aspects are assumed to be present at the same time.[2)] So I don't want to take an ideological position concerning the ultimate description of matter (classical versus quantum). All that is assumed is that there is *one* geometry of the outside world and matter leaves an imprint on it. This imprint may have different facetts (classical, quantum mechanical, and further yet unknown ones). But I don't want to be forced right at the start of formulating a problem to take a complicated limit (provided it exists) which turns, for example, 10^α protons, 10^β electrons, 10^γ-photons etc. etc. – all described quantum mechanically - into a gas ball of the mass and the properties of the sun. So I admit – even in a theory which claims to treat quantum effects – that there are huge accumulations of matter in nature, which are best described collectively in a classical manner when certain long range features of the underlying geometry are studied. I shall thus not hesitate to introduce as part of the geometric structure a purely classical reference geometry (the base space of the bundle used below).

There is one further point worth to be remembered. There is no field without a testparticle measuring the field. We should demand in this context following Ehlers, Pirani and Schild[1] that the underlying geometry – or, to be more precise, its kinematical part – should be measurable by freely falling testparticles. The theoretical

---

[2)]Such a dualistic point of view may be appropriate even if – ultimately – the world would, indeed, be found to be constructed in a monistic manner with *one* fundamental principle acting behind the scene determining both geometry and matter.

formalism must be such that ultimately quantum testparticles may be used and *can* be used in order to determine operationally the quantum kinematics of elementary material objects within the geometry chosen as a framework and determine parts of this geometry by measurement. However, the full geometry (including the fibers, see below) do contain short range quantum information which lead to interactions among several quantum testparticles when they scatter from each other. This would also have an image in the geometry determining the dynamics between individual quantum testparticles which could even lead to the production of further new such "testparticles". Since this is usually not considered when one speaks about testparticles and "testparticle approximation" (i.e. considering a rigid particle which does not contribute to the fields being measured), we made a distinction above between the quantum kinematics of elementary quantum testparticles and their interaction if they collide or rather, get into their mutual range of interaction, for example, the strong interaction when they are hadrons.

In the following sections we shall develop a description for quantized matter which will allow us to formulate the quantum kinematics of elementary matter in the presence of gravitation and establish covariant source equations for the underlying geometry. Due to the gauge formulation used we shall, however, at the same time give a geometric formulation of another interaction besides gravity which has, so far, not been related to geometry, namely the strong interaction among hadrons in particle physics. So in this context the rôle played by the metric of Eintein's theory – its classical or quantum nature in a theory with quantized elementary sources – appears to be connected with the understanding of the geometric and quantum aspects of another interaction characterizing matter when regarded at small distances: the strong interaction.

2. Description of Quantized Matter

The next question concerns the particular choice of the underlying geometric framework for the description of quantized hadronic matter as well as classical macroscopic matter. We shall use as underlying gemoetric arena a fiber bundle geometry raised over a 4-dimensional classical curved space-time manifold $B = U_4$ (Riemann-Cartan base with metric and metric compatible torsion). The standard fiber of the bundle on which we describe quantized matter is a one-particle Hilbert space, $\mathcal{H}_{\tilde{\eta}}$, yielding a so-called first quantized Hilbert bundle. Fock space bundles with many-particle Hilbert spaces as fibers have been constructed and investigated but will, for simplicity, not be used in the following discussion. The bundle

$$\mathcal{H} = \mathcal{H}(B, \mathcal{F} = \mathcal{H}_{\tilde{\eta}}, U(g)) \tag{2.1}$$

with fiber $\mathcal{H}_{\tilde{\eta}}$, being associated to a G-frame bundle $P(B,G)$, carries a system of covariance (a generalized system of imprimitivity) of the group G, given by a unitary irreducible representation $U(g)$ of G, with $g \in G$, and a *positive operator valued* measure, $E(\Delta_j)$, on a family $\{\Delta_j\}$ of Borel sets in the phase space variables over which the standard fiber $\mathcal{H}_{\tilde{\eta}}$ will be constructed a a resolution kernel Hilbert space with resolution generator $\tilde{\eta}$. (Compare the work of Prugovečki[2,3], and Ali and Prugovečki[4].)

As structural group G of the bundle $\mathcal{H}$ and P we shall use here the de Sitter group $G = SO(4,1)$ which is related to the Poincaré group (the group of motion in Minkowski space) in a analogous way as the Lorentz group is related to the Galilei group, namely by group contraction. We shall construct a one-particle phase space representation, $U(g) = U^{(\rho)}(g)$, of $SO(4,1)$ for spin zero particles[3)] related to the principal series of unitary irreducible representations (UIR) of the $(4,1)$-de Sitter group labelled by the parameter ρ; $0 \leq \rho < \infty$. This yields a system of covariance of the de Sitter group defined on a Hilbert space $\mathcal{H}_{\tilde{\eta}} = \mathcal{H}_{\tilde{\eta}}^{(\rho)}$, as indicated above, which determines the stochastic localization properties of particles together with

---

[3)] There is a formulation for arbitrary spin in terms of scalar functions on the homogeneous spaces of the group $SL(2, C)$ [compare Bacry and Kihlberg[5] and Drechsler[6] and a forthcoming paper by Drechsler and Tuckey[7]]. To simplify the following discussion we, however, treat here only the spinless case.

their conjugate momenta in terms of generalized wave functions, $\Psi_x^{(\rho)}$; $x \in B$ which are given as sections on the bundle $\mathcal{H}$ with fiber $\mathcal{H}_{\tilde{\eta}}^{(\rho)}$. This stochastic phase space description in terms of scalar de Sitter gauge fields will briefly be summarized below (for further details compare Drechsler and Prugovečki[8]).

An essential point in this gauge theory based on the group $G = SO(4,1)$ is that the Lorentz subgroup $H = SO(3,1) \equiv O(3,1)^{++}$ will be considered as the gauge group of gravity in a vierbein formulation à la Cartan. The Lorentz frame bundle $P_L(B, H = SO(3,1))$ appearing in general relativity is thus regarded as a subbundle of the de Sitter frame bundle $P(B, G = SO(4,1))$ with the de Sitter group playing the rôle of the gauge group for extended strongly interacting particles[4)]

Hadrons, which are the building blocks of matter as we know it, are definitely not pointlike particles. Already at the lowest level of compositness observed in nature they appear as *extended structures* characterized by a "radius" of $< r > \approx 0.80$ Fermi. This was established in the Hofstadter experiments carried out in the fifties by scattering electrons off hadrons $(p, n(d), H^3, He^3)$. To take this experimental fact into account we introduce a fundamental length parameter R of the order of a Fermi into our geometric description. However, the idea is not to introduce this length in the space-time base of the bundles considered (which would yield a lattice space) but to associate this length with the fibers.

Let us thus consider a generalized tangent bundle with "curled up" fiber F given by the homogeneous space $G/H = SO(4,1)/SO(3,1)$ of the de Sitter group isomorphic to a (4,1)-de Sitter space V_4' (being a space of constant curvature with curvature radius R) and structural group G,

$$E = E(U_4, F = G/H \equiv V_4', G), \tag{2.2}$$

which is associated to the principal de Sitter frame bundle

$$P = P(U_4, G = SO(4,1)) \, . \tag{2.3}$$

We call E the de Sitter bundle over space-time and identify the homogeneous fiber

---

4) We shall not discuss electromagnetic or weak interaction effects. To include them, one would have to introduce a further $U(1) \otimes SU(2)$ fiber in the principal bundles regarded in the text.

with a de Sitter space V_4' of *fixed* radius $R \approx 10^{-13}$ cm typical for hadron physics.[5)] E is a soldered bundle[9,10] with the property that at each $x \in U_4$ the fiber over x, F_x, and the space-time base have a first order contact and the group G acts transitively on the fiber. The stability group of the point of contact is the Lorentz subgroup H which is to be linked, as mentioned, with gravitation. In the contraction limit $R \to \infty$ the bundle (2.3) goes over into the affine frame bundle $P_A(U_4, ISO(3,1))$ having the Poincaré group as structural group, and the bundle E in (2.2) goes over into the affine tangent bundle $T_A(U_4)$.

The $(4,1)$-de Sitter space V_4' may be viewed as a single-shell hyperboloid of radius R embedded in a 5-dimensional Lorentzian space $R_{4,1}$ with metric $\eta_{ab} = diag(1,-1,-1,-1,-1)$. The coordinates of a point $\xi \in V_4' \subset R_{4,1}$ are denoted by $\xi^a; a = 0,1,2,3,5$. Then the one-particle de Sitter phase space is given by

$$\mathcal{N}^{\pm} = V_4' \times \mathcal{C}^{\pm}, \tag{2.4}$$

where

$$\begin{aligned} V_4' &: [\xi,\xi] = \xi^a \xi^b \eta_{ab} = -R^2, \\ \mathcal{C}^{\pm} &: [\zeta,\zeta] = \zeta^a \zeta^b \eta_{ab} = 0; \quad \zeta^5 = \frac{1}{R}. \end{aligned} \tag{2.5}$$

Here the summations over twice appearing indices run over $a,b = 0,1,2,3,5$. $\mathcal{C}^{\pm}$ is the intersection of the light cone $C^{\pm}$ in $R_{4,1}$ (representing infinity in de Sitter space) with the surface $\zeta^5 = \frac{1}{R}$. The superscript $\pm$ of $\mathcal{C}^{\pm}$ stands for sign $\zeta^0 = \pm$ with the vector $\zeta^a = (\zeta^i, \zeta^5 = \frac{1}{R})$; $i = 0,1,2,3$ characterizing a so-called horosphere or horocycle[11,12] through the origin $\overset{\circ}{\xi} = (0,0,0,0,-R)$ of V_4'. The vector $\zeta \in \mathcal{C}^{\pm}$ plays the rôle of the wave vector or momentum vector for a wave phenomenon in de Sitter space (i.e. a space of constant curvature with curvature radius R and group of motion $SO(4,1)$). Moreover, we introduce a six-dimensional horospherical submanifold of $\mathcal{N}^{\pm}$, denoted by

$$\tilde{\Sigma}^{\pm} = h \times \mathcal{C}^{\pm}, \tag{2.6}$$

[5)] Hence R is a fundamental length parameter of geometric origin characterizing the fiber of E in a similar manner as the velocity of light characterizes the space-time base. The group contraction described by the limit $c \to \infty$ links a Lorentz invariant to a Galilei invariant description. Correspondingly, in the limit $R \to \infty$ the de Sitter group $SO(4,1)$ contracts to the Poincaré group $ISO(3,1)$ being the kinematic group for relativistic point particles i.e. leptons.

which is composed of a horosphere h, being a space-like hypersurface in V_4', and the cone $\mathcal{C}^{\pm}$.

Let us now introduce the de Sitter phase space bundle

$$\tilde{E} = \tilde{E}(U_4, \tilde{F} = V_4' \times \mathcal{C}^{\pm}, G = SO(4,1)) \tag{2.7}$$

which is a soldered bundle associated to P given in (2.3). The soldering is performed here through the local subspace[6)] $V_4'(x)$ of $\tilde{F}_x = \mathcal{N}_x^{\pm}$ being tangent to the space-time base at every $x \in U_4$ as in the bundle E defined in (2.2), which, in a natural way, is a subbundle of $\tilde{E}$. We denote the hyperbolic rotation in $R_{4,1}$ leaving the hyperboloid and the cone in $R_{4,1}$ invariant by $A_g \in SO(4,1)$, which is the defining 5×5 representation of the de Sitter group.

Let us now construct a phase space representation of $SO(4,1)$ related to the principal series of UIR's of the de Sitter group for spinless particles characterized by the parameter ρ and denote it by $U^{(\rho)}(A_g)$ or simply by $U(g)$. The value of ρ determines the mass of the particle in question. The Laplace-Beltrami operator, $\Box_\xi$, on V_4' has eigenvalues[14] $\kappa(\kappa+3)/R^2$ with $\kappa = -\frac{3}{2} + i\rho$; $0 \le \rho < \infty$, leading to the following relation between ρ, the radius R of de Sitter space, and the mass m of the particle: $\left(\frac{mc}{\hbar}\right)^2 R^2 = \rho^2 + \frac{1}{4}$.

$\mathcal{H}_{\tilde{\eta}}^{(\rho)}$ is the Hilbert space $L^2(\tilde{\Sigma}^{\pm})$ of square-integrable functions in the variables $(\xi, \zeta) \in \tilde{\Sigma}^{\pm} \subset \mathcal{N}^{\pm}$ with respect to the G-invariant measure[8,7)] obeying $d(\tilde{\Sigma}(A_g\xi, A_g\zeta) = d\tilde{\Sigma}(\xi,\zeta)$, given by

$$d\tilde{\Sigma}(\xi,\zeta) = \frac{1}{R^2}\frac{1}{[\xi,\zeta]^2}\delta(\|[\xi,\zeta]\| - c)d\mu(\xi)\delta([\zeta,\zeta])d^4\zeta, \tag{2.8}$$

where $d\mu(\xi) = \frac{R}{|\xi^5|}d\xi^0 d\xi^1 d\xi^2 d\xi^3$ is the invariant measure on V_4'.

One can now construct a coherent state basis for the Hilbert space $\mathcal{H}_{\tilde{\eta}}^{(\rho)} = L^2(\tilde{\Sigma}^{\pm})$ with measure (2.8) in two steps. First one considers a "momentum space"

---

6) Compare in this context Ch. Ehresmann[13].

7) In (2.8) c is a positive constant determining a particular horosphere h_ζ^c in V_4' characterized by ζ which is parallel to the horosphere h_ζ^1, for $c = 1$, through the origin $\overset{\circ}{\xi}$ characterized by the same vector ζ (compare Ref. 14).

representation $\hat{U}(g)$ of $SO(4,1)$, realized on a space $L^2(\mathcal{C}^\pm)$ with respect to the measure $d\Omega(\zeta) = \delta([\zeta,\zeta])d^4\zeta$, and constructs a coherent state basis in $L^2(\mathcal{C}^\pm)$ in terms of horospherical waves (which are analogous to plane waves in flat space) from an $SO(3)$ invariant resolution generator $\tilde{\eta}(\zeta')$ by using de Sitter boost and Lorentz boost transformations parametrizing thereby the basis states by the coset $SO(4,1)/SO(3)$ [8]. Then, in a second step, an integral transform is considered providing a map $\mathcal{W}_{\tilde{\eta}} : L^2(\mathcal{C}^\pm) \to L^2(\tilde{\Sigma}^\pm)$ to go over to the space $\mathcal{H}_{\tilde{\eta}}^{(\rho)}$ with its corresponding coherent state basis (for details see Ref. 8). $\mathcal{H}_{\tilde{\eta}}^{(\rho)}$ is called a resolution kernel Hilbert space with resolution generator $\tilde{\eta}$ possessing a decomposition into $\mathcal{H}^+ \oplus \mathcal{H}^-$, where $\mathcal{H}^+$ (with sign $\zeta^0 > 0$) denotes the one-particle Hilbert space, and $\mathcal{H}^-$ (with sign $\zeta^0 < 0$) denotes the one-antiparticle Hilbert space. $\mathcal{H}_{\tilde{\eta}}^{(\rho)}$ carries a unitary irreducible spin zero phase space representation of the de Sitter group associated to the principal series of UIR's characterized by ρ, and the map $\mathcal{W}_{\tilde{\eta}}$ is an isometry between Hilbert spaces obeying the intertwining relation

$$U(g)\mathcal{W}_{\tilde{\eta}} = \mathcal{W}_{\tilde{\eta}}\hat{U}(g) \; . \tag{2.9}$$

Having introduced the Hilbert space $\mathcal{H}_{\tilde{\eta}}^{(\rho)}$ we now consider as the geometric arena for the description of spinless quantized matter the first quantized Hilbert bundle

$$\mathcal{H} = \mathcal{H}(U_4, \mathcal{F} = \mathcal{H}_{\tilde{\eta}}^{(\rho)}, U(A_g)) \tag{2.10}$$

associated to (2.3) which carries a system of covariance of the $(4,1)$-de Sitter group. The variables (ξ,ζ) in each local fiber $\mathcal{N}_x^\pm$ of $\tilde{E}$ play here the rôle of local geometro-stochastic variables on $\mathcal{H}$ determining the quantum kinematical localization properties of spinless quantum particles (with internal de Sitter gauge degrees of freedom)

---

[8] The resolution generator $\tilde{\eta}(\zeta')$ contains, morover, a length parameter l which determines the stochastic smearing of the phase space points (compare in this context the discussion presented in Ref. 2). Having thus introduced the resolution generator together with a definite choice for the regularizing length parameter l (which may, for example, be taken to be the Planck length) we have made the transition from classical de Sitter phase space with sharp points (ξ,ζ) to *stochastic de Sitter phase space* with fuzzy, i.e. stochastically smeared points. It was shown by Prugovečki[2,3] for G being the Poincaré group that in the sharp point limit, $l \to 0$, the stochastic phase space formalism develops singularities reminiscent of the singularities appearing in the conventional field theory defined on configuration space only. We shall thus take l to be nonzero and fixed obeying $l << R$.

on curved space-time U_4. The main idea here is to represent the quantum mechanical properties and the ensuing stochastic smearing (i.e. fuzziness) of the states in the phase space variables (ξ,ζ), *in the local fibers* of $\mathcal{H}$, and study the quantum propagation of this local quantum phenomenon on the curved space-time base[8].

Denoting now the generalized coherent state basis in the local fibers $\mathcal{H}_{\tilde{\eta}}^{(\rho)}(x)$ of $\mathcal{H}$, which is adapted to a particular choice of gauge[9)] $\sigma = s(x)$ on P, by $\mathbf{\Phi}_{\xi,\zeta}^{s(x)}$, where $x \in U_4$ and $(\xi,\zeta) \in \tilde{\Sigma}_x^{\pm}$, one has the following resolution of the identity in $\mathcal{H}$ at the point $x \in U_4$:

$$\int_{\tilde{\Sigma}_x^{\pm}} \mid \mathbf{\Phi}_{\xi,\zeta}^{s(x)}\rangle d\tilde{\Sigma}(\xi,\zeta)\langle \mathbf{\Phi}_{\xi,\zeta}^{s(x)} \mid = \mathbf{1}_x^{\pm}. \tag{2.11}$$

Here $d\tilde{\Sigma}(\xi,\zeta)$ is the measure (2.8) on the local (horospherical) hypersurface $\tilde{\Sigma}_x^{\pm}$ in $\mathcal{N}_x^{\pm}$. Any state vector $\mathbf{\Psi}_x^{(\rho)\pm}$ belonging to the principal series of UIR's of $SO(4,1)$ with zero spin may now be expanded with respect to the *local coherent quantum frame basis*, $\mathbf{\Phi}_{\xi,\zeta}^{s(x)}$, of $\mathcal{H}_{\tilde{\eta}}^{(\rho)}(x)$ at $x \in U_4$ according to

$$\mathbf{\Psi}_x^{(\rho)\pm} = \int_{\tilde{\Sigma}_x^{\pm}} \Psi_x^{(\rho)}(\xi,\zeta)\mathbf{\Phi}_{\xi,\zeta}^{s(x)} d\tilde{\Sigma}(\xi,\zeta). \tag{2.12}$$

The coefficient function $\Psi_x^{(\rho)}(\xi,\zeta)$ in (2.12) is the *scalar de Sitter coordinate wave function*, called for short the *generalized wave function*, which may be regarded as the (phase space) component of a section on the first quantized bundle $\mathcal{H}$ with respect to a base provided by $\mathbf{\Phi}_{\xi,\zeta}^{s(x)}$. *The function $\Psi_x^{(\rho)}(\xi,\zeta)$ will represent first quantized matter in the theory.* We refer to this gauge dependent function $\Psi_x^{(\rho)}(\xi,\zeta)$ as to a section on $\mathcal{H}$ in the above sence.

One can adopt a convenient bracket notation for the $SO(4,1)$-invariant integration with measure (2.8) over the space-like hypersurface $\tilde{\Sigma}_x^{\pm}$ at $x \in U_4$ and solve Eq. (2.12) for $\Psi_x^{(\rho)}(\xi,\zeta)$ yielding

$$\Psi_x^{(\rho)}(\xi,\zeta) = \langle \mathbf{\Phi}_{\xi,\zeta}^{s(x)} \mid \mathbf{\Psi}_x^{(\rho)\pm}\rangle_{\tilde{\Sigma}_x^{\pm}} . \tag{2.13}$$

[9)] We assume that $P(U_4, SO(4,1))$ admits a global section, i.e. that P is a trivial bundle. This corresponds to the usual assumption made in general relativity that a spin structure exists on space-time necessary to represent fermionic matter on space-time.

$\Psi_x^{(\rho)}(\xi,\zeta)$ has the following transformation property under de Sitter gauge transformations (i.e. changes of section on $\mathcal{H}$ corresponding to a change of gauge on P) (compare Ref. 8)

$$\left(U(A_{g(x)})\Psi_x^{(\rho)}\right)(\xi,\zeta) = \Psi_x^{(\rho)}(A_{g(x)}^{-1}\xi, A_{g(x)}^{-1}\zeta). \tag{2.14}$$

A G-invariant scalar product of two sections $\Psi_{1,x}^{(\rho)}(\xi,\zeta)$ and $\Psi_{2,x}^{(\rho)}(\xi,\zeta)$ is defined by

$$\langle\Psi_{1,x}^{(\rho)} \mid \Psi_{2,x}^{(\rho)}\rangle_{\tilde{\Sigma}_x^{\pm}} = \int_{\tilde{\Sigma}_x^{\pm}} \Psi_{1,x}^{(\rho)*}(\xi,\zeta)\Psi_{2,x}^{(\rho)}(\xi,\zeta)d\tilde{\Sigma}(\xi,\zeta). \tag{2.15}$$

The covariant derivative of a section $\Psi_x^{(\rho)}(\xi,\zeta)$ on $\mathcal{H}$ is given by

$$\begin{aligned} D^R\Psi_x^{(\rho)}(\xi,\zeta) &= [d + i\Gamma^R(x)]\Psi_x^{(\rho)}(\xi,\zeta) \\ &= [d + \frac{i}{2}[\omega^R(x)]_{ab}\tilde{M}^{ab}]\Psi_x^{(\rho)}(\xi,\zeta) \end{aligned} \tag{2.16}$$

where $[\omega^R(x)]_{ab} = -[\omega^R(x)]_{ba}$ is the pull back of a connection on P by the map $\sigma = s(x)$, and $\tilde{M}_{ab}$, with $\tilde{M}_{ab} = \eta_{ac}\eta_{bd}\tilde{M}^{cd}$, denote the generators of the spin zero phase space representation $U(A_g)$ of $SO(4,1)$ given by

$$\tilde{M}_{ab} = -\tilde{M}_{ba} = L_{ab}(\xi) + L_{ab}(\zeta), \tag{2.17}$$

with

$$L_{ab}(\xi) = i\left(\xi_a\frac{\partial}{\partial\xi^b} - \xi_b\frac{\partial}{\partial\xi^a}\right); \quad L_{ab}(\zeta) = i\left(\zeta_a\frac{\partial}{\partial\zeta^b} - \zeta_b\frac{\partial}{\partial\zeta^a}\right). \tag{2.18}$$

We, finally, introduce the kernel for the propagation from (ξ,ζ) to (ξ',ζ') in the local fiber over $x \in U_4$ in $\mathcal{H}$ which is determined by the following overlap of the coherent state basis $\mathbf{\Phi}_{\xi,\zeta}^{s(x)}$ at x:

$$\tilde{K}_{\tilde{\eta},x}^{(\rho)}(\xi',\zeta';\xi,\zeta) = \langle\mathbf{\Phi}_{\xi',\zeta'}^{s(x)} \mid \mathbf{\Phi}_{\xi,\zeta}^{s(x)}\rangle_{\tilde{\Sigma}_x^{\pm}}\,. \tag{2.19}$$

Eq. (2.19) defines a reproducing kernel in $\mathcal{H}_{\tilde{\eta}}^{(\rho)}(x)$ with the reproducing property following from (2.11)

$$\tilde{K}_{\tilde{\eta},x}^{(\rho)}(\xi',\zeta';\xi,\zeta) = \int_{\tilde{\Sigma}_x^{\pm}} \tilde{K}_{\tilde{\eta},x}^{(\rho)}(\xi',\zeta';\xi'',\zeta'')\tilde{K}_{\tilde{\eta},x}^{(\rho)}(\xi'',\zeta'';\xi,\zeta)d\tilde{\Sigma}(\xi'',\zeta''), \tag{2.20}$$

obeying, moreover,

$$\left[\tilde{K}^{(\rho)}_{\tilde{\eta},x}(\xi',\zeta';\xi,\zeta)\right]^* = \tilde{K}^{(\rho)}_{\tilde{\eta},x}(\xi,\zeta;\xi',\zeta'). \tag{2.21}$$

The kernel $\tilde{K}^{(\rho)}_{\tilde{\eta},x}(\xi',\zeta';\xi,\zeta)$ determines the propagation of the generalized wave function $\Psi^{(\rho)}_x(\xi,\zeta)$ in the local fiber variables. For the problems related to a physically reasonable definition of the geometro-stochastic quantum propagation on $\mathcal{H}$, i.e. of propagating excitations in the local fiber at x to point of the base different from x and its relation to parallel propagation on $\mathcal{H}$ (with respect to a connection) as well as the Feynman path integral formulation (in the flat space limit and in the absence of a stochastic smearing, i.e. in the sharp point limit) we have to refer to the literature[3,15,16,17;10)]

In the present context we shall require that the propagation of the generalized wave function $\Psi^{(\rho)}_x(\xi,\zeta)$ on $\mathcal{H}$ is governed by a de Sitter gauge covariant and U_4-covariant second order wave equation with real eigenvalue α[18]:

$$(\Box_{\mathcal{H}} + \alpha)\Psi^{(\rho)}_x(\xi,\zeta) = 0. \tag{2.22}$$

Specializing to axial vector torsion in the Riemann-Cartan base (as will be done below) the operator $\Box_{\mathcal{H}}$ for scalar functions defined on $\mathcal{H}$ may be written as

$$\Box_{\mathcal{H}} = \frac{1}{\sqrt{-g}} D^R_\rho \sqrt{-g} g^{\rho\sigma} D^R_\sigma \tag{2.23}$$

with $dx^\rho D^R_\rho = D^R$ as given by (2.16) and $g = det\ g_{\mu\nu}(x)$ with $g_{\mu\nu}(x)$ denoting the metric in the space-time base U_4. The constant α in (2.22) is of dimension L^{-2} (L=length) characterising the wave motion on $\mathcal{H}$.

We are now in a position to define a set of de Sitter gauge covariant source currents constructed by G-invariant integration over the local fiber at x of a hermitean bilinear expression in $\Psi^{(\rho)}_x(\xi,\zeta)$ and its adjoint for a fixed value of ρ:

$$J^{(\rho)}_{kab}(x) = \frac{i}{2} \int_{\tilde{\Sigma}^\pm_x} \Psi^{(\rho)*}_x(\xi,\zeta) \left[\vec{\tilde{M}}_{ab} \vec{D}^R_k - \overleftarrow{D}^R_k \overleftarrow{\tilde{M}}_{ab} \right] \Psi^{(\rho)}_x(\xi,\zeta) d\tilde{\Sigma}(\xi,\zeta) \tag{2.24}$$

---

10) The literature quoted does not treat the de Sitter case considered here but discusses quantum propagation on the Hilbert bundle in the Poincaré case arising in the limit $R \to \infty$. However, the problems involved in determining the quantum propagation are anaglogous in both cases.

with $\overset{\rightarrow R}{D_k} = \lambda^\mu_k(x) \overset{\rightarrow R}{D_\mu}$, where $\lambda^\mu_k(x)$ are the vierbein fields and $\lambda^k_\mu(x)$ their inverse, obeying $g_{\mu\nu}(x) = \lambda^i_\mu(x)\lambda^k_\nu(x)\eta_{ik}$, with $\eta_{ik} = diag(1,-1,-1,-1)$ being the Minkowski metric. Moreover, one has in (2.24)

$$\overset{\rightarrow R}{D_k} = \overset{\rightarrow}{\partial}_k + i \overset{\rightarrow R}{\Gamma_k}(x); \quad \overset{\leftarrow R}{D_k} = \overset{\leftarrow}{\partial}_k - i \overset{\leftarrow R}{\Gamma_k}(x), \tag{2.25}$$

and analogously for $\overset{\rightarrow}{\tilde{M}}_{ab}$ and $\overset{\leftarrow}{\tilde{M}}_{ab} = \overset{\rightarrow}{\tilde{M}}_{ab}{}^\dagger$. As a result of (2.22) and the reality of α the currents $J^{(\rho)}_{kab}(x)$ are covariantly conserved.[18]. The equations (2.24) and (2.16)-(2.18) show that the currents $J^{(\rho)}_{kab}(x)$ result from an averaged internal motion taking place in the local fibers on $\mathcal{H}$. For $a, b = i, j = 0,1,2,3$ it is an internal rotational motion (Lorentz rotation); for $a, b = i, 5$; $i = 0,1,2,3$ it is a generalized translation (de Sitter boost) in the fiber. It is thus apparent that our formalism describes quantized material objects possessing internal gauge degrees of freedom and extension. We shall use the currents (2.24) as source currents for the bundle geometry tying thereby the quantized motion in the fiber to the geometry of the entire space.

3. The Nonlinear Gauge

In order to recover gravitation in a G-invariant manner as a gauge theory of the Lorentz subgroup $H = SO(3,1)$ of $G = SO(4,1)$ we introduce a new Higgs-type field in the formalism given as a section, $\xi(x)$, of the soldered bundle E defined in (2.2) obeying $\xi^a(x)\xi^b(x)\eta_{ab} = -R^2$ [compare (2.5)]. Global sections on E always exist. The "zero section", $\xi(x) = \overset{\circ}{\xi}$, may be identified with the space-time base of E. The field $\xi(x)$ acts, on the one hand, *as a symmetry reducing field* and, on the other hand, as an *extension field* in the bundle framework: if $\xi(x)$ is *parallel* with respect to $\omega^R(x)$, i.e. satisfies

$$D^R\xi^a(x) = d\xi^a(x) + [\omega^R(x)]_b{}^a\xi^b(x) = 0, \tag{3.1}$$

the $SO(4,1)$ gauge symmetry reduces to the $SO(3,1)$ gauge symmetry describing pure (metric) gravitation. We assume that the full de Sitter gauge symmetry does not reduce everywhere to the Lorentz subsymmetry, but that this reduction of symmetry indeed occurs far outside the quantized material sources present in the geometry. There are thus regions in space-time, denoted by $D_{(i)}; i = 1, \ldots N$, where the G-symmetry does *not* reduce, i.e. where $D^R\xi^a(x) \neq 0$.[11] In these regions the connection ω^R (or W^R, see (3.2) below) takes values in the Lie algebra $\mathbf{g}$ of $SO(4,1)$ while for regions where (3.1) is true ω^R (or W^R) reduces to a set of one-forms with values in the Lie algebra $\mathbf{g}'$ of the subgroup $SO(3,1)$. In general $\mathbf{g}$ may be decomposed as $\mathbf{g}=\mathbf{g}'\oplus\mathbf{t}$, where $\mathbf{t}$ is a vector space generating the homogeneous space G/H isomorphic to V_4'.

One may now consider de Sitter boost transformations, $A(\xi(x)) \in SO(4,1)$, transforming the origin $\overset{\circ}{\xi}$ in $V_4'(x)$ into $\xi(x)$ at $x \in U_4$ and use these transformations to go over to a nonlinear realization of the de Sitter transformations in terms of Lorentz transformations in $\overset{\circ}{\xi}$ yielding for ω^R the following form [$a = (i,5)$ row index, $b = (j,5)$ column index]:

$$[\omega^R(x)]^a{}_b \overset{A^{-1}(\xi(x))}{\longrightarrow} [W^R(x,\xi(x))]^a{}_b = \begin{pmatrix} [W^R(x,\xi(x))]^i{}_j & [\theta^R(x,\xi(x))]^i \\ [\theta^R(x,\xi(x))]_j & 0 \end{pmatrix}, \tag{3.2}$$

with

$$[W^R(x,\xi(x))]^i{}_j = [\bar{\omega}(x)]^i{}_j + [\tau^R(x,\xi(x))]^i{}_j, \tag{3.3}$$

and

$$[\theta^R(x,\xi(x))]^i = [W^R(x,\xi(x))]^i{}_5 = \frac{1}{R}[A^{-1}(\xi(x))]^i{}_a D^R\xi^a(x). \tag{3.4}$$

We call the form $[W^R(x,\xi(x))]^a{}_b$ in (3.2) the nonlinearly transforming form of the connection on P (i.e. the nonlinear gauge denoted by N.L.). It transforms under

---

[11] The section $\xi^a(x); a = 0,1,2,3,5$, discriminates between the "inside" ($D^R\xi^a(x) \neq 0$) and the "outside" ($D^R\xi^a(x) = 0$) of the domains $D_{(i)}$. In this respect $\xi^a(x)$ could be called an *extension field* representing the extension of elementary objects described by the field $\Psi_x^{(\rho)}(\xi,\zeta)$ as a function of $x \in U_4$. There are thus two types of extensions present in the theory: *firstly*, the stochastic extension l (see footnote 8 above) determining the basic fuzzines in the stochastic phase space fiber variables (ξ,ζ) in the proper description of the *quantum kinematics* of particles reflecting the limitation in the measurement of conjugate variables ξ and ζ corresponding to position and momentum, and, *secondly*, the extension in the space-time base introduced through the symmetry reducing section $\xi^a(x)$ depending on R. This latter aspect is essential in recovering gravitation in a G-invariant manner as part of the extended gauge symmetry based on the bigger group $G = SO(4,1)$ containing the gauge group $H = SO(3,1)$ of gravitation as a subsymmetry. Moreover, a wave equation for $\xi^a(x)$ has to be specified guaranteeing that the dynamical behaviour of the $\xi(x)$-field is coupled to the equation of motion for the actual matter field $\Psi_x^{(\rho)}(\xi,\zeta)$ on $\mathcal{H}$ (compare Ref. 18).

gauge transformations, $\xi'(x) = A_{g(x)}\xi(x)$, with a matrix $A(\Lambda(\xi'(x), \xi(x))) \in H$ leaving the form of the r.-h. side of (3.2) unchanged (for details see Ref. 18). The first term on the r.-h. side of (3.3) is the metric part defining a connection on P_L. However, the Lorentz part (3.3) of (3.2) has a torsion addition denoted by $\tau^R(x, \xi(x))$. Eq. (3.4) defining the soldering forms $[\theta^R(x, \xi(x))]^i$ of the de Sitter connection shows explicitly that (3.2) is Lorentz-valued (i.e. g'-valued) for $D^R\xi^a(x) = 0$, i.e. outside the domains $D_{(i)}$ where the G-gauge symmetry reduces to the H-gauge symmetry. A form analogous to (3.2) is obtained for the curvature two-forms $[\Omega^R(x, \xi(x))]^a{}_b$ in the N.L. gauge which is given by

$$[\omega^R(x)]^a{}_b \overset{A^{-1}(\xi(x))}{\longrightarrow} [\Omega^R(x, \xi(x))]^a{}_b = \begin{pmatrix} [\Omega^R(x, \xi(x))]^i{}_j & [\Theta^R(x, \xi(x))]^i \\ [\Theta^R(x, \xi(x))]_j & 0 \end{pmatrix}, \tag{3.5}$$

where

$$[\Theta^R(x, \xi(x))]^i = [\Omega^R(x, \xi(x))]^i{}_5 = \overset{\mathrm{W}}{D}\,[\theta^R]^i = d[\theta^R]^i + [W]_k{}^i \wedge [\theta^R]^k \tag{3.6}$$

is called the torsion of the Cartan connection in $P(U_4, SO(4,1))$ (given in the N.L. gauge) being, actually, the $(i, 5)$ component of curvature.

The Lorentz part in (3.5) is given, with the index i lowered, by

$$\Omega^R(x, \xi(x))_{ij} = d[W^R]_{ij} + [W^R]_{ik} \wedge [W^R]_j{}^k - [\theta^R]_i \wedge [\theta^R]_j\,, \tag{3.7}$$

where, for simplicity, we have suppressed the arguments $(x, \xi(x))$ on the r.-h. sides of (3.6) and (3.7). By $\overset{\mathrm{W}}{D}$ in (3.6) we denote the covariant derivative with respect to the N.L. form $[W^R(x, \xi(x))]_{ij}$ of the connection involving the Lorentz part only.

In the transformation of (3.2) and (3.5) under changes of the section, $\xi(x) \to \xi'(x) = A_{g(x)}\xi(x)$ (de Sitter gauge transformations), $A(\Lambda(\xi'(x), \xi(x)))$ plays the rôle of the Wigner rotation of the de Sitter group defined by[6]

$$A_{g(x)}A(\xi(x)) = A(\xi'(x))\; A(\Lambda(\xi'(x), \xi(x))) \tag{3.8}$$

where $\Lambda(\xi'(x), \xi(x)) = \Lambda(\xi(x), g(x))$ is a transformation of the stability subgroup $H = SO(3,1)$ of the point $\overset{\circ}{\xi}$ in V_4'. The transformations $A(\Lambda(\xi'(x), \xi(x)))$ provide a nonlinear realization of $SO(4,1)$ on the stability subgroup H of $\overset{\circ}{\xi}$ leaving, as mentioned, the form of the r.-h. sides of the equations (3.2) and (3.5) invariant, respecting thus the obtained separation into de Sitter boots $[(i,5)$ and $(5,j)]$ and Lorentz $[(i,j)]$ parts.

The G-invariant form (3.2) of the connection allows us to split the Lorentz part $[W^R(x,\xi(x))]_{ij}$ according to (3.3) into a metric part pertaining to a V_4, which is related to gravitation as in general relativity, and a rest which is considered to be a torsion contribution originating from the embedding of the Lorentz gauge symmetry into a bigger (here de Sitter) gauge symmetry. The torsion field will be coupled to the quantum aspect of matter as described in Sect. 2, and a set of covariant field equations involving the torsion field will be introduced in Sect. 4 below.

Let us, for later use, expand the r.-h. side of (3.3) in a basis of one-forms θ^k (local Lorentz coframe):

$$[\bar{\omega}(x)]_{ij} + [\tau^R(x,\xi(x))]_{ij} = \theta^k\left(\bar{\Gamma}_{kij}(x) + K^R_{kij}(x,\xi(x))\right), \tag{3.9}$$

where $\bar{\Gamma}_{kij}(x)$ are the Ricci rotation coefficients and $K^R_{kij}(x,\xi(x))$ are the components of the torsion tensor with respect to a moving Cartan frame (local Lorentz frame). As mentioned above, we shall consider the case of a *completely antisymmetric* torsion tensor (axial vector torsion) given by four independent components.[12)]

Correspondingly, we write for the de Sitter boost part (3.4) of (3.2)

$$[\theta^R(x,\xi(x))]_i = W^R(x,\xi(x))_{i5} = \theta^k\Gamma^R_{ki5}(x,\xi(x)). \tag{3.10}$$

For the Lorentz indexed curvature tensor one finds the following decomposition from (3.7) and (3.9):

$$R^R_{ijkl}(x,\xi(x)) = \bar{R}_{ijkl}(x) + P^R_{ijkl}(x,\xi(x)) + Q^R_{ijkl}(x,\xi(x)), \tag{3.11}$$

where $\bar{R}_{ijkl}(x)$ is the Riemannian curvature tensor of a V_4, P^R_{ijkl} is the torsion contribution to curvature in a U_4, and Q^R_{ijkl} is the 5,5 de Sitter boost contribution corresponding to the last term on the r.-h. side of (3.7). Finally, from (3.6) and (3.10) we obtain a tensor expressible in terms of the Γ^R_{ki5}:

$$R^R_{ijk5}(x,\xi(x)) = T^R_{ijk}(x,\xi(x)). \tag{3.12}$$

---

12) The torsion tensor of a U_4 has in general 24 independent components belonging to three irreducible classes: axial vector (4), vector (4), and tensor (16).

For the explicit forms of the expressions appearing on the r.-h. sides of (3.11) and (3.12) in terms of the connection coefficients introduced in (3.9) and (3.10), as well as for a discussion of the cyclic identities and the Bianchi identities for these curvature quantities, we refer to the appendices or Ref. 19.

We conclude this section by pointing out that the condition $D^R\xi^a(x) = 0$ introduced in (3.1) above – allowing, with the help of a section on E, to reduce the connection and curvature forms from a g-valued to a g′-valued (Lorentz-valued) form – is a typical *symmetry braking condition.* If this condition is satisfied the de Sitter symmetry reduces to the linearly realized Lorentz symmetry which, as we know, is related to gravitation. For $x \in D_{(i)}$ (i.e. $D^R\xi^a(x) \neq 0$) the de Sitter symmetry does *not* reduce and is realized nonlinearly. This will be assumed to occur inside extended hadrons where the strong interaction force dominates the scene. Far outside the range of the strong forces one has $D^R\xi^a(x) = 0$, and all that is left in the geometry – as we shall see – is the long range gravitational interaction described in terms of the geometry of a V_4 coupled to a distant classical matter distribution. There is, however, one remark necessary here: in some intermediate range of distances outside the domains $D_{(i)}$ there may also be torsion effects present in the geometry. It is still unclear how fast torsion decays outside the domaines characterized by a strong nonlinear de Sitter gauge interaction of matter described in a quantum mechanical manner.

4. Field Equations

As field equations for the bundle geometry we now introduce, besides Eq. (2.22) for the generalized quantum mechanical wave function $\Psi_x^{(\rho)}(\xi,\zeta)$, the following two sets of de Sitter gauge covariant and U_4-covariant nonlinear source equations[13]. The first is of Einstein type, the second is of current-curvature Yang-Mills type:

$$R^R_{ik}(x,\xi(x)) - \frac{1}{2}\eta_{ik}R^R(x,\xi(x)) = \kappa \overset{N.L.}{T_{ik}}(x,\xi(x)) \,, \tag{4.1}$$

$$\overset{N.L.}{D^i}\, R^R_{ijab}(x,\xi(x)) = \bar{\kappa}\, \overset{N.L.}{J_{jab}}(x,\xi(x)) \,. \tag{4.2}$$

Here κ and $\bar{\kappa}$ are two independent coupling constants; κ is Einstein's gravitational

---

[13] A further equation for the reducing section $\xi(x)$ is introduced and discussed in Ref. 18.

constant, and $\bar{\kappa}$ is a new coupling constant characterizing the quantum fiber dynamics (QFD), i.e. the dynamical relation between quantized matter described on $\mathcal{H}$ and the full *uncontracted* bundle curvature tensor. The constant $\bar{\kappa}$ is supposed to be related to strong interaction effects. The operator $\overset{N.L.}{D^i}$ in (4.2) denotes the full covariant derivative of the Lorentz *and* de Sitter indices [the latter taken with respect to the N.L. form, $W^R(x,\xi(x))$, of the connection]. The current on the r.-h. side of (4.2) is the total current tranformed to N.L. form with the help of the boost $A^{-1}(\xi(x))$ as described in Sect. 3. This source current is composed of a classical and a quantum part (see below). $R^R_{ijab}(x,\xi(x))$ is the full curvature tensor with the Lorentz part [for $a,b=k,l$] containing metric, torsion, and quadratic de Sitter boost contributions (3.11)], and with the de Sitter boost part (3.12) [for $a,b=i,5$]. Eqs. (4.1) are of Einstein type involving the contracted Lorentz curvature tensor, $R^R_{ik}(x,\xi(x))=\eta^{jl}R^R_{ijkl}(x,\xi(x))$, and the corresponding curvature scalar $R^R(x,\xi(x))$ again composed of three parts (metric, torsion and boost). On the r.-h. side of (4.1) appears the total energy-momentum tensor decomposing into the classical symmetric part, $\bar{T}_{ik}(x)$, of general relativity representing classical matter, and a quantum part induced by $\Psi^{(\rho)}_x(\xi,\zeta)$ possessing no symmetries in the indices i,k :

$$\overset{N.L.}{T_{ik}}(x,\xi(x))=\bar{T}_{ik}(x)+\overset{N.L.}{T_{ik}}(\Psi). \tag{4.3}$$

Here $\overset{N.L.}{T_{ik}}(\Psi)$ is a short-hand notation for $T_{ik}(\overset{N.L.}{\Psi_x}{}^{(\rho)})$ with the *N.L.* form of the generalized wave function $\Psi^{(\rho)}_x(\xi,\zeta)$ defined by

$$\overset{N.L.}{\Psi_x}{}^{(\rho)}(\xi,\zeta)=\Psi^{(\rho)}_x(\xi,\zeta;\xi(x))=\left(U(A^{-1}(\xi(x)))\Psi^{(\rho)}_x\right)(\xi,\zeta). \tag{4.4}$$

$\bar{T}_{ik}(x)$ in (4.3) is the source in Einstein's equations

$$\bar{R}_{ik}(x)-\frac{1}{2}\eta_{ik}\bar{R}(x)=\kappa\bar{T}_{ik}(x), \tag{4.5}$$

where we continue to denote purely metric quantities pertaining to a V_4 by a bar.

The original de Sitter gauge covariant field equations (4.1) of Einstein type relate the energy-momentum tensor (4.3) of classical *and* quantum matter to the contracted full Lorentz indexed curvature tensor $R^R_{ik}(x,\xi(x))$ derived from (3.11) by contraction. Using the classical equations (4.5) in (4.1) yields the following relation between the energy-momentum tensor $\overset{N.L.}{T_{ik}}(\Psi)$ of quantized matter and the nonmetric quantities $\tilde{R}^R_{ik}(x,\xi(x)) = P^R_{ik}(x,\xi(x)) + Q^R_{ik}(x,\xi(x))$, and $\tilde{R}^R(x,\xi(x)) = \eta^{ik}\tilde{R}^R_{ik}(x,\xi(x)) = P^R(x,\xi(x)) + Q^R(x,\xi(x))$:

$$\tilde{R}^R_{ik}(x,\xi(x)) - \frac{1}{2}\eta_{ik}\tilde{R}^R(x,\xi(x)) = \kappa \overset{N.L.}{T_{ik}}(\Psi). \tag{4.6}$$

However, we do *not* define a tensor $\overset{N.L.}{T_{ik}}(\Psi)$ for quantized matter constructed in terms of the fields (4.4) to consider it as the driving source term in the field equations for the nonmetric quantities since this dynamical feedback is already established by the new field equations (4.2), the current-curvature equations of Yang-Mills type, involving the *uncontracted* full curvature tensor of the de Sitter theory. We shall thus not specify the symmetric and antisymmetric parts of $\overset{N.L.}{T_{ik}}(\Psi)$ explicitly but, instead, treat Eqs.(4.6) *as identities* determining the r.-h. side in terms of the fields on the l.-h. side which are themselves determined by the current-curvature equations (4.2). (Compare the discussion given in Ref. 18 and Ref. 20.)

Before we turn to the form these equations ultimately assume in using the symmetries of the quantities involved and by taking the Bianchi identities into account, let us note that it would be the tensor $\overset{N.L.}{T_{ik}}(\Psi)$ which would show quantum fluctuations and that this energy-momentum tensor is related to a combination of *nonmetric* geometric fields constructed with torsion and de Sitter boost contributions. The metric determined by the classical equations (4.5), indeed, remains a classical quantity despite the presence of quantized material sources in the theory. At short distances, however, a further interaction comes into play which is characterized by a new coupling constant $\bar{\kappa}$ measuring the coupling strength of quantized matter to torsion and de Sitter boost fields appearing here on the scene together with gravitation. Since the coupling constants κ and $\bar{\kappa}$ are unrelated in their strengths reflecting the fact that two different sets of dynamical equations, (4.1) and (4.2), are involved, the additional nonmetric fields may very well mediate a strong interaction inside the domaines $D_{(i)}$, i.e. inside extended hadrons, coupling torsion and de Sitter boost degrees of freedom in a nonlinear manner. This may be concluded despite the extreme weakness of the classical gravitational interaction which is regarded in this context as a residual interaction reaching far outside the domains $D_{(i)}$, and being described there geometrically in terms of the classical metric with Lorentzian signature.

Let us now turn to the current-curvature equations (4.2) which represent 40 nonlinear partial differential equations constrained by the 40 Bianchi identities [compare Ref. 19, App.1]:

$$\overset{N.L.}{D}_{\{i}\, R^{R}_{jk\}ab}(x,\xi(x)) = S^{R\;\;l}_{\{ij}(x,\xi(x)) R^{R}_{k\}lab}(x,\xi(x)). \tag{4.7}$$

Here $\{ijk\}$ denotes the cyclic sum of the indices ijk, and S^{Rl}_{ij} is Cartan's U_4-torsion tensor related to K^{Rl}_{ij} appearing in (3.9) by [we suppress the arguments $(x, \xi(x))$]

$$K^{Rl}_{ij} = \frac{1}{2}\left(S^{R}_{ij}{}^{l} + S^{Rl}{}_{ij} - S^{R}{}_{j}{}^{l}{}_{i}\right). \tag{4.8}$$

For a completely antisymmetric torsion tensor we have $K^{R}_{ijk} = \frac{1}{2} S^{R}_{ijk}$. We assume, furthermore, that the boost fields, Γ^{R}_{ij5}, defined in (3.10) are antisymmetric in i and j. We thus have a completely antisymmetric torsion quantity, with $a = (k, 5)$, transforming under $SO(4,1)$, defined by[19]

$$S^{R}_{ija}(x,\xi(x)) = \left(2K^{R}_{ijk}(x,\xi(x)), 2\Gamma^{R}_{ij5}(x,\xi(x))\right). \tag{4.9}$$

We now split the source current appearing in (4.2) into a purely classical and a quantum part according to

$$\bar{\kappa}\; \overset{N.L.}{J^{(\rho)}_{jab}}(x,\xi(x)) = \begin{cases} \kappa \bar{J}_{jkl}(x) + \bar{\kappa}\; \overset{N.L.}{J^{(\rho)}_{jkl}}(\Psi) & \text{for } a,b = k,l, \\ \bar{\kappa}\; \overset{N.L.}{J^{(\rho)}_{jk5}}(\Psi) & \text{for } a,b = k,5. \end{cases} \tag{4.10}$$

The quantized N.L. currents depending on Ψ are obtained from (2.24) with help of the de Sitter boost $A(\xi(x))$:

$$\overset{N.L.}{J^{(\rho)}_{jab}}(\Psi) = \overset{N.L.}{J^{(\rho)}_{jab}}(x,\xi(x)) = J^{(\rho)}_{jcd}(x)\, [A(\xi(x))]^{c}{}_{a}\, [A(\xi(x))]^{d}{}_{b}. \tag{4.11}$$

The current $\bar{J}_{jkl}(x)$ in (4.10) is the classical current contribution which would appear in a purely metric theory. In Einstein's theory, however, the $\bar{J}_{jkl}(x)$ are *derived quantities* expressible in terms of $\bar{T}_{ik}(x)$ and $\bar{T}(x) = \eta^{ik}\bar{T}_{ik}(x)$ and their metric covariant derivatives[18] with $\bar{J}_{jkl}(x)$ obeying the cyclic identities

$$\bar{J}_{\{jkl\}}(x) = 0. \tag{4.12}$$

In a purely metric theory no further source equations of the type (4.2) for $(a,b) = (k,l)$ would thus have to be considered. Due to the Ricci identities such equations would contain no new information which is not already contained in (4.5). In a theory with torsion, or when the Lorentz quantities are embedded in a theory with a bigger gauge group, the situation is different and the field equations (4.2) do contain additional information.

We shall not repeat here the details of the derivation which allows one to deduce from (4.2) two sets of field equations, i.e. for $(a,b)=(k,l)$ and for $(a,b)=(k,5)$, and show how they turn – together with the corresponding Bianchi identities (4.7) and considering the symmetries of the quantities involved – into two sets of nonlinear field equations for the $(4+6)$ field quantities K^{R}_{ijk} and Γ^{R}_{ij5} coupled to the quantized sources (4.11) (compare Ref. 18 and the appendices of Ref. 19). After several steps involving as the first one the taking of the cyclic sum of the indices jkl in (4.2), for $(a,b)=k,l$, in order to eliminate the metric curvature contained in R^{R}_{ijkl} by using the cyclic identities of the Riemann tensor together with (4.12), one obtains

$$\overset{\mathrm{W}}{D^i}\,(P^{R*}_{ik}+3Q^{R*}_{ik})-4({}^*K^{Ri}Q^{R}_{ik}-{}^*K^{R}_{k}Q^{R})=-3\bar{\kappa}\,{}^*\overset{N.L.}{J^{(\rho)}_{k}}(\Psi)\,, \tag{4.13}$$

and

$$\overset{\mathrm{W}}{D^i}\,T^{R}_{ikj}-\Gamma^{Ril5}R^{R}_{ikjl}=\bar{\kappa}\,\overset{N.L.}{J_{kj5}}(\Psi)\,, \tag{4.14}$$

where we have introduced the dual quantities ($\varepsilon_{0123}=+1$)

$${}^*K^{R}_{s}=-\frac{1}{6}\varepsilon_s{}^{ijk}K^{R}_{ijk}, \tag{4.15}$$

$${}^*\overset{N.L.}{J^{(\rho)}_{s}}(\Psi)=-\frac{1}{6}\varepsilon_s{}^{ijk}\,\overset{N.L.}{J^{(\rho)}_{ijk}}(\Psi), \tag{4.16}$$

and

$$P^{R*}_{ijkl}=\frac{1}{2}P^{R}_{ijpq}\varepsilon^{pq}{}_{kl}\quad;\quad P^{R*}_{ik}=\eta^{jl}P^{R*}_{ijkl}\,, \tag{4.17}$$

$$Q^{R*}_{ijkl}=\frac{1}{2}Q^{R}_{ijpq}\varepsilon^{pq}{}_{kl}\quad;\quad Q^{R*}_{ik}=\eta^{jl}Q^{R*}_{ijkl}\,. \tag{4.18}$$

For a discussion of the covariant current conservation implied by[14] (4.2) and (2.22) for real α, as well as for an investigation of the contracted Bianchi identities

---

[14] The axial current (4.16), being the source current in (4.13), is not conserved (see below) despite the fact that the r.-h. side of (4.2) is covariantly conserved. This difference is due to the cyclic sum in the indices jkl being taken in deriving (4.13).

yielding an expression for the nonmetric quantities $\overset{\mathrm{w}}{D^i}\left[\tilde{R}^R_{ki} - \frac{1}{2}\eta_{ki}\tilde{R}^R\right]$, with the quantities $\tilde{R}^R_{ki}$ and $\tilde{R}$ as defined in (4.6), we refer to Ref. 18.

Let us make some concluding remarks concerning the current-curvature equations (4.2):

For $K^R_{ijk} = 0$ and $\Gamma^R_{ij5} = 0$ all source currents $\overset{N.L.}{J^{(\rho)}_{kab}}(\Psi)$ vanish, as do the $\overset{N.L.}{T_{ik}}(\Psi)$, and the $SO(4,1)$ symmetry reduces everywhere to a $SO(3,1)$ symmetry. The field equations reduce to Einstein's equations (4.5) determining a classical metric from a classical distribution of energy and momentum. Furthermore, the Eq. (2.22) for $\Psi^{(\rho)}_x(\xi,\zeta)$ reduces to a Klein-Gordon-type equation with only Lorentz degrees of freedom appearing in $\Box_{\mathcal{H}}$ through D^R_ρ. Finally, the bundle E may, for the zero section $\xi(x) \equiv \overset{\circ}{\xi}$ be identified with the base $B = V_4$.

If the de Sitter symmetry does not reduce everywhere, i.e. if $D^R\xi(x) \neq 0$ in some domain $D_{(i)}$ of space-time, we have within this domain a complicated nonlinear interrelation between axial vector torsion and de Sitter boost fields, ${}^*K^R_s$ and Γ^R_{ij5}, governed by Eqs.(4.13) and (4.14) together with various constraints[18], and, of course, Einstein's equations (4.5) for the classical background metric $g_{\mu\nu}(x)$. An interesting question now is, how fast axial torsion decays outside the domains $D_{(i)}$, where $\Gamma^R_{ij5} = 0$ [remember that the regions $D_{(i)}$ are defined in terms of de Sitter boost quantities only], or whether there may be propagating torsion possible in the geometry disconnected from the material sources. For $\Gamma^R_{ij5} = 0$ all the $(i,5)$-components of the currents (4.11) vanish, as does T^R_{ijk}, and the residual field equations for torsion together with the Bianchi identities take the form[15)]

$$\overset{\mathrm{w}}{D^i}\, P^{R*}_{ik} \equiv \bar{\nabla}^i P^{R*}_{ik} + K^{Rij}_k P^{R*}_{[ij]} = -3\bar{\kappa}\; {}^*\overset{N.L.}{J^{(\rho)}_k}(\Psi)\,, \tag{4.19}$$

$$\overset{\mathrm{w}}{D^i}\, P^R_{ik} \equiv \bar{\nabla}^i P^R_{(ik)} + K^{Rij}_k P^R_{[ij]} = \frac{1}{2}\partial_k P^k - K^{Rijl} P^R_{kijl}\,, \tag{4.20}$$

with $\bar{\nabla}^i$ denoting the metric covariant derivative and with $P^R = -6^* K^{R*}_s K^{Rs}$. Eqs.(4.20) are identically satisfied as the result of the contracted Bianchi identities.

Let us focus the attention on Eq.(4.19) in the metrically flat case with $\bar{\nabla}_i$ re-

15) As usual (ij) and $[ij]$ denote the symmetric and antisymmetric index combination, respectively.

placed by ∂_i, i.e. neglecting classical gravitation; then one has

$$\Box^* K_s^R - \frac{1}{12}\partial_s{}^* P^R - \varepsilon_s{}^{ijl}\, {}^*K_i^R \partial_j\, {}^*K_l^R \overset{N.L.}{=} -3\bar{\kappa}\, {}^*J_k^{(\rho)}\,(\Psi) \tag{4.21}$$

with ${}^*P^R = P^{R*} = -6\partial^j {}^*K_j^R$. For the divergence of the axial vector source current one derives (correcting a sign error in Ref. 18)

$$\Box^* P^R + 4\varepsilon^{kijl}(\partial_k {}^*K_i^R)(\partial_j {}^*K_l^R) \overset{N.L.}{=} -6\bar{\kappa}\partial^k {}^*J_k^{(\rho)}\,(\Psi)\,. \tag{4.22}$$

For finding and classifying solutions of Eq.(4.21) it may be helpful to relate the ε-selfcoupling term in (4.22) to the Pontrajagin invariant K (compare Ref. 21). Solutions to Eq.(4.21) or nontrivial solutions to the source-free equation are not yet known. [For the trivial solution ${}^*K_s^R = k_s \exp(\pm ik \cdot x)$, with $k \cdot k = 0$, of the source-free equation, each term on the l.-h. side of (4.21) is separately zero.]

In the gauge theory presented here, which is based on the group $SO(4,1)$ and the introduction of a fundamental length parameter R of geometric origin typical for hadron physics, with the description involving, furthermore, a back-reaction of classical as well as quantized matter on the underlying bundle geometry [called quantum fiber dynamics (QFD)], one observes a dichotomy of the following kind. Classical matter determines the classical, i.e. metric, long range gravitational field reaching far outside the classical material source distributions. Matter treated in a quantum mechanical manner does not influence the metric at all; it relates to other, non-metric, aspects of the underlying geometry (torsion, de Sitter boost contributions) which come into play at short distances and describe an interaction which, so far, has been interpreted in a totally different manner as a force of nongeometric origin between elementary hadrons extended at the scale of R, i.e. the strong interaction. The idea thus is to regard the strong short range interaction between hadrons as an interaction obtained in an endeavour to embed the gauge group of the classical theory of gravitation into a bigger group (here the ten parameter de Sitter group) and keep, as an essential element, Einstein's dualism between classical *and* quantum matter, on the one side, and the underlying enlarged geometry, on the other side. Whether this proposal provides, indeed, a reasonable framework for the joint understanding of all the aspects of the strong and gravitational interactions has still to be seen.

REFERENCES

1. J. Ehlers, F.A.E. Pirani and A. Schild, in *General Relativity*, L.O'Raifeartaigh, ed., Clarendon Press, Oxford, 1972.
2. E. Prugovečki, *Stochastic Quantum Mechanics and Quantum Spacetime*, Reidel, Dordrecht 1984; corr. printing 1986.
3. E. Prugovečki, *Quantum Geometry*, Kluwer, Dordrecht, 1992.
4. S.T. Ali and E. Prugovečki, Acta Applicande Math. **6**, 1 (1986).
5. H. Bacry and A. Kihlberg, J. Math. Phys. **10**, 2132 (1969).
6. W. Drechsler, J. Math. Phys. **26**, 41 (1985).
7. W. Drechsler and P.A. Tuckey, *Geometro-Stochastically Quantized Fields with Internal Spin Variables*, preprint MPI-PhT/95-38, 1996.
8. W. Drechsler and E. Prugovečki, Found. Phys. **21**, 513 (1991).
9. S. Kobayashi, Can. J. Math. **8**, 145 (1956).
10. W. Drechsler, Fortschr. Phys. **38**, 63 (1990).
11. I.M. Gel'fand, M.I. Graev and N.Ya. Vilenkin, *Generalized Functions*, Vol.5, Academic Press, London 1966. Chapt.V.
12. S. Helgason, *Topics in Harmonic Analysis on Homogeneous Spaces*, Birkhäuser, Basel 1981.
13. Ch. Ehresmann, Colloque de Toplogie (espaces fibrés), Bruxelles 1950, p.29.
14. W. Drechsler and R. Sasaki, Nuovo Cimento **46**, 527 (1978).
15. E. Prugovečki, *Principles of Quantum General Relativity*, World Scientific, Singapore, 1994.
16. E. Prugovečki, Class. Quantum Grav. **11**, 1981 (1994).
17. W. Drechsler and P.A. Tuckey, *On Quantum and Parallel Transport in a Hilbert Bundle over Spacetime*, preprint MPI-PhT/95-37, 1995.
18. W. Drechsler, Found. Phys. **22**, 1041 (1992).
19. W. Drechsler, Class. Quantum Grav. **6**, 623 (1989).
20. W. Drechsler, Found. Phys. **23**, 261 (1993).
21. W. Drechsler, Gen. Relativ. Grav. **15**, 703 (1983).

QUANTUM GRAVITY AND BLACK HOLE ENTROPY

VALERI FROLOV

CIAR Cosmology Program; Theoretical Physics Institute,
*University of Alberta, Edmonton, Canada T6G 2J1**

1 Black-Hole Entropy

According to the thermodynamical analogy in black hole physics, the entropy of a black hole in the Einstein theory of gravity is

$$S^{BH} = A_H/(4l_P^2), \tag{1.1}$$

where A_H is the area of a black hole surface and $l_P = (\hbar G/c^3)^{1/2}$ is the Planck length [1, 2]. In black hole physics the Bekenstein-Hawking entropy S^{BH} plays essentially the same role as in the usual thermodynamics. In particular it allows to estimate what part of the internal energy of a black hole can be transformed into work. Four laws of black hole physics that form the basis in the thermodynamical analogy were formulated in [3]. According to this analogy the entropy S is defined by the response of the free energy F of the system containing a black hole to the change of its temperature:

$$dF = -SdT. \tag{1.2}$$

The generalized second law [1, 2, 4] (see also [5, 6, 7, 8] and references therein) implies that when a black hole is a part of the thermodynamical system the total entropy (i.e. the sum of the entropy of a black hole and the entropy of the surrounding matter) does not decrease.

The Euclidean approach [9, 10, 11] provides a natural way to derive black hole thermodynamical properties. Doing a Wick's rotation $t \rightarrow -i\tau$ in the Schwarzschild metric one gets the metric with the Euclidean signature. The corresponding manifold with the Euclidean metric is regular if and only if the imagine time τ is periodic and the period is $2\pi/\kappa$, where κ is the surface gravity of the black hole. The corresponding

*E-mail address: frolov@phys.ualberta.ca

regular Euclidean metric is known as the Gibbons-Hawking instanton. The period $2\pi/\kappa$ of the imagine time τ is naturally identified with the inverse temperature of the black hole, while the Euclidean action, calculated for the Gibbons-Hawking instanton, is directly related with its free energy. York and collaborators [12, 13] made an important observation that the formal derivation of the black hole entropy using the Euclidean approach requires certain modifications for its consistency. Namely, to get a well defined canonical ensemble, one needs to consider a black hole in a box of finite size and fix the corresponding thermodynamical quantities (temperature) at the boundary. Box filled with thermal radiation and a black hole in it in thermal equilibrium with radiation is thermodynamically stable if the radius of a spherical box r is less than $3M$. If one changes the temperature of the box, the mass of a black hole inside the box also changes, as required by the conditions of the thermal equilibrium. As the result of this process the free energy F of the system is changed. One can use the relation (1.2) and single out the contribution to the entropy due to the black hole. In the classical approximation the so defined thermodynamical entropy coincides with the Bekenstein-Hawking entropy S^{BH}.

The simple relation between the thermodynamical entropy of a black hole and its surface area is characteristic for Einstein theory. The definition of the black hole entropy can be generalized to non-Einstein versions of gravitational theory, provided they allow existence of black holes. Wald [14] showed that in the general case the entropy of a black hole is defined by Nöther charge related with the Killing vector. (For application and developing this idea, see [15, 16, 17].) Recently the interest to the problem of black hole entropy was increased by observation that the entropy of extremal black hole might vanish [18, 19].

The success of the thermodynamical analogy in black hole physics allows one to hope that this analogy may be even deeper and it is possible to develop statistical-mechanical foundation of black hole thermodynamics. It is worthwile to remind that the thermodynamical and statistical-mechanical definitions of the entropy are logically different. *Thermodynamical entropy* S^{TD} is defined by the response of the free energy F of a system to the change of its temperature:

$$dF = -S^{TD} dT. \tag{1.3}$$

(This definition applied to a black hole determines its Bekenstein-Hawking entropy.)

Statistical-mechanical entropy S^{SM} is defined as

$$S^{SM} = -\mathrm{Tr}(\hat{\rho}\ln\hat{\rho}), \tag{1.4}$$

where $\hat{\rho}$ is the density matrix describing the internal state of the system under consideration. It is also possible to introduce the *informational entropy* S^I by counting different possibilities to prepare a system in a final state with given macroscopical parameters from different initial states

$$S^I = -\sum_n p_n \ln p_n, \tag{1.5}$$

with p_n being the probabilities of different initial states. In standard case all three definitions give the same answer.

Is the analogy between black holes thermodynamics and the 'standard' thermodynamics complete? Are there internal degrees of freedom of a black hole responsible for its entropy? Is it possible to apply the statistical-mechanical and informational definitions of the entropy to black holes and how are they related with the Bekenstein-Hawking entropy? These are the questions that are to be answered.

Historically first attempts of the statistical-mechanical foundation of the entropy of a black hole were connected with the informational approach [2, 20]. According to this approach the black hole entropy is interpreted as "the logarithm of the number of quantum mechanically distinct ways that the hole could have been made" [5, 20]. The so defined informational entropy of a black hole is simply related to the amount of information lost by stretching the horizon, and as was shown by Thorne and Zurek it is equal to the Bekenstein-Hawking entropy [5, 20]. Quite interesting results relating informational and statistical-mechanical entropies can be obtained in a special model proposed by Bekenstein and Mukhanov [21, 22]. According to this model a black hole is identified with a system having discrete internal states, so that an absorption or emission of a particle by a black hole is accompanied by the transition from one state to another. In such model the mass of a black hole is quantized[1]. Unfortunately the physical origin of internal degrees of freedom of a black hole in the model and their discreteness is not derived but postulated.

The dynamical origin of the entropy of a black hole and the relation between the statistical-mechanical and Bekenstein-Hawking entropy have remained unclear. In the present talk I describe some new results obtained in this direction.

2 Dynamical Degrees of Freedom of a Black Hole

The problem of the dynamical origin of the black hole entropy was intensively discussed recently. The proposed basic idea is to relate the dynamical degrees of freedom of a black hole with its quantum excitations. This idea has different realizations[2].

In the framework of the *membrane paradigm* the dynamical degrees of freedom are identified with different possible states of thermal atmosphere of a black hole, while the entropy of a black hole is identified with the amount of information about the state of thermal atmosphere, which is lost by stretching the horizon[5, 20].

In his *brick wall model* 't Hooft [25] proposed to consider a mirror-like boundary, located outside a black hole at the close distance to its horizon. He assumed that outside the boundary there exist thermal radiation with temperature equal to the Hawking temperature. He has shown that the entropy of such thermal radiation is

[1] Another approach to the quantization of the mass of a black hole can be found in [23].

[2] For recent review of the problem of the dynamical origin of the entropy of a black hole, see also [24].

of the same order of magnitude at the Bekenstein-Hawking entropy, provided the distance of the mirror-like boundary from the horizon is chosen to be of order of Planck's length.

Bombelli *et al* [26] attracted attention to the fact that even in a flat spacetime in a Minkowski vacuum state one can obtain a non-vanishing entropy if one restricts his observations to the spatially bounded part of the space. The corresponding *entanglement entropy* arises due to the presence of correlations of those modes of zero-point fluctuations which are propagating in the vicinity of the boundary of chosen statial region. It was proposed to relate black hole entropy with the entanglement entropy related with the presence of the horizon [26, 27]

We (with Igor Novikov) arrived to the similiar idea independently by analyzing the gedanken experiment proposed in our earlier paper [28]. Namely we assumed that there exist a traversible wormhole, and its mouths are freely falling into a black hole. If one of the mouths crosses the gravitational radius earlier than the other, then rays passing through the first mouth can escape from the region lying inside the gravitational radius. Such rays would go through the wormhole and enter the outside region though the second mouth. As the result during the period of time when the first of the mouth is inside and the other mouth is outside the gravitational radius the surface area of the horizon decreases. If we assume that the black hole entropy is related with the surface area of a black hole, then the only possibility to escape contradictions with the second law is to assume that during such process the decrease of entropy is related with the possibility to get access to some new information concerning black hole internal states. At first sight it looks like a puzzle. We know that (at least in the classical General Relativity) a black hole at late time is completely specified by finite number of parameters. For a non-rotating unchaged black hole one need to know only one parameter (mass M) to describe all its properties. It is true not only for the exterior where this property is the consequence of no-hair theorems, but also for the black hole interior [29]. The reason why it is impossible for an isolated black hole at late time to have non-trivial classical states in its interior in the vicinity of the horizon is basically the same as for the exterior regions. These states might be excited only if a collaping body emits the pulse of fields or particles immideately after it crosses the event horizon. Due to red-shift effect the energy of the emitted pulse must be exponentially large in order to reach the late time region with any reasonable energy. After short time (say $100M$) after the formation of a black hole it is virtually impossible because it requires the energy of emission much greater than the black hole mass M.

In quantum physics the situation is completely different. The above mentioned puzzle can be solved if we remember that any quantum field has zero-point fluctuations. To analyze states of a quantum field it is convenient to use its decomposition into modes. Besides the positive-frequency modes which have positive energy, there exist also positive-frequency modes with negative energy. In a non-rotating uncharged black holes such modes can propagate only inside the horizon where the Killing vector used to define the energy is spacelike. It is possible to show that at late time for any regular initial state of the field these states are thermally excited and the corresponding temperature coincides with the black hole temperature $T_H = (8\pi M)^{-1}$. (For more details see, e.g.[30].) In principle by using a traversable wormhole described above one can get information concerning these internal modes propagating near the horizon and change their states.

In the framework of this *dynamical-black-hole-interior* model we proposed to identify the dynamical degrees of freedom of a black hole with the internal modes of all physical fields. The set of the fields must include the gravitational one. It can be shown that for any chosen field the number of the modes infinitely grows as one considers the regions located closer and closer to the horizon. For this reason the contribution of a field to the statistical-mechanical entropy of a black hole calculated by counting the internal modes of a black hole is formally divergent. In order to make it finite one might restrict himself by considering only those modes which are located at the proper distance from the horizon greater than some chosen value l. For this choice of the cut-off the contribution of a field to the statistical-mechanical entropy of a black hole is

$$S^{SM} = \alpha\frac{A}{l^2}, \tag{2.1}$$

where A is the surface area of a black hole, and a dimensionless parameter α depends on the type of the field.

The calculation of the black hole entropy in the dynamical-black-hole-interior model is made by counting the number of thermally excited internal modes existing at given moment of time (more accurately on the chosen spacelike surface, crossing the horizon). This calculations can be simplified by using the following trick. Consider an eternal version of a black hole, i.e. an eternal black hole with the same mass M as the original black hole formed as the result of collapse. At late time the geometry of both holes are identical. One can trace back in time all the perturbations, propagating at late time by using the geometry of the eternal version of a black hole . As the result one can relate perturbations at late time in a spacetime of a real black hole with initial data on the Einstein-Rosen bridge (spatial slice of t =const) of the eternal black hole geometry.

By using this idea in [31] it was proposed to introduce a wave function of a black hole as the functional over the space of deformations of the Einstein-Rosen bridge. In this approach the wave function of a black hole depends on data located at the outside part Σ_+ (external degrees of freedom) as well as on the data located on the inner part Σ_- (internal degrees of freedom) of the Einstein-Rosen bridge. The density matrix describing the internal state of a black hole can be obtained from this wave function by its averaging over external degrees of freedom. It was shown that the 'no-boundary proposal' analogous to the 'no-boundary ansatz' in quantum cosmology [36] singles out a special wave function of a black hole. This no-boundary wave function describes the ground state of a black hole. It depends on the mass M of a black hole and on the values of the fields $\varphi_\pm$ on the upper (+) and lower (−) parts of the Einstein-Rosen bridge. In the the semiclassical approximation it is of the form

$$\Psi(M,\varphi_+,\varphi_-) = P\,\mathrm{e}^{-4\pi M^2 - \boldsymbol{I}_2(\varphi_+,\varphi_-)}, \tag{2.2}$$

where

$$\boldsymbol{I}_2(\varphi_+,\varphi_-) = \frac{1}{2}\sum_\lambda \left\{ \frac{\omega_\lambda \cosh(\beta\,\omega_\lambda/2)}{\sinh(\beta\,\omega_\lambda/2)}(\varphi^2_{\lambda,+} + \varphi^2_{\lambda,-}) - \frac{2\,\omega_\lambda}{\sinh(\beta\,\omega_\lambda/2)}\varphi_{\lambda,+}\varphi_{\lambda,-}\right\}. \tag{2.3}$$

Here $\varphi_{\lambda\pm}$ are coefficients of the decomposition of the field φ over the complete set of spatial harmonics of the field on the $\boldsymbol{\Sigma}_\pm$

$$\varphi_\pm(\boldsymbol{x}) = \sum_\lambda \varphi_{\lambda,\pm} R_{\lambda,\pm}(\boldsymbol{x}). \tag{2.4}$$

For fixed M the density matrix for internal variables φ_- of a black hole is defined as

$$\hat{\rho}_\beta(\varphi_-,\varphi'_-) = \int D\varphi_+ \Psi_\beta(\varphi_+,\varphi_-)\Psi_\beta(\varphi_+,\varphi'_-), \tag{2.5}$$

and it is of the form

$$\hat{\rho}(\varphi_-,\varphi'_-) = P'\,\mathrm{e}^{-\tilde{\boldsymbol{I}}_2(\varphi_-,\varphi'_-)}, \tag{2.6}$$

where $\tilde{\boldsymbol{I}}_2$ is given by the expression (2.3) with β changed by 2β. It is easy to show that

$$\hat{\rho}(\varphi_-,\varphi'_-) = N\langle\varphi_-|\mathrm{e}^{-\beta\hat{\boldsymbol{H}}}|\varphi'_-\rangle, \tag{2.7}$$

where N is a normalization constant and $\hat{\boldsymbol{H}}$ is the Hamiltonian of free fields φ propagating on the Schwarzschild background. The statistical-mechanical entropy S^{SM} of a black hole obtained by using this density matrix coincides with the expression (2.1).

The statistical-mechanical entropy S^{SM} in this as well as other 'dynamical' approaches possesses the following main properties: (1) $S^{SM} \sim A$, where A is the surface area of a black hole; (2) S^{SM} is divergent and requires regularization $S^{SM} \sim A/l^2$, where l is the cut-off parameter; (3) S^{SM} depends on the number of fields, which exist in Nature; (4) For $l \sim l_{Pl}$ $S^{SM} \sim S^{BH}$.

The following two problems are of importance: (1) What is the relation between the statistical-mechanical entropy S^{SM} introduced by counting the internal degrees of freedom of a black hole and its thermodynamical entropy S^{TD}? In particular how to explain the universality of the Bekenstein-Hawking entropy S^{BH}, while S^{SM} is not universal and depends on the number of fields? (2) The formal expression for the Bekenstein-Hawking entropy S^{BH} contains the Planck length. Does it mean that by studying the thermodynamical properties of black holes we can obtain certain conclusions concerning physics at Planckian scales?

In what follows we shall try to clarify this questions.

3 Renormalized Effective Action and Free Energy

The complete information concerning the canonical ensemble of black holes with a given inverse temperature β at the boundary is contained in the partition function $Z(\beta)$ given by the Euclidean path integral [11]

$$Z(\beta) = \int D[\phi] \exp(-I_E[\phi]). \tag{3.1}$$

Here the integration is taken over all fields including the gravitational one that are real on the Euclidean section and are periodic in the imaginary time coordinate τ with period β. The quantity ϕ is understood as the collective variable describing the fields. In particular it contains the gravitational field. Here $D[\varphi]$ is the measure of the space of fields ϕ and I_E is the Euclidean action of the field configuration. The action I_E includes the Euclidean Einstein action. The state of the system is determined by the choice of the boundary conditions on the metrics that one integrates over. For the canonical ensemble for the gravitational fields inside a sperical box of radius r_0 at temperature T one must integrate over all the metrics inside r_0 which are periodically identified in the imaginary time direction with period $\beta = T^{-1}$. Such a partition function must describe in particular the thermal ensemble of black holes. The partition function Z is related with the effective action $\Gamma = -\ln Z$ and with the free energy $F = \beta^{-1}\Gamma = -\beta^{-1}\ln Z$.

By using the stationary-phase approximation one gets

$$\beta F \equiv -\ln Z = I_E[\phi_0] - \ln Z_1 + \dots . \tag{3.2}$$

Here ϕ_0 is the (generally speaking, complex) solution of classical field equations for action $I[\phi]$ obeying the required periodicity and boundary conditions. Besides the tree-level contribution $I_E[\phi_0]$, the expression (3.2) includes also one-loop corrections $\ln Z_1$, connected with the contributions of the fields perturbations on the background ϕ_0, as well as higher order terms in loops expansion, denoted by $(\dots)$. The one-loop contribution of a field ϕ can be written as follows $\ln Z_1 = -\frac{1}{2}\mathrm{Tr}\ln(-D)$, where D is the field operator for the field ϕ inside the box r_0. The one-loop contribution contains divergences and required the renormalization. In order to be able to absorb these divergences in the renormalization of the coefficients of the initial classical action we chose the latter in the form

$$I_{cl} = \int d^4x\sqrt{g}L, \tag{3.3}$$

$$L = \left[-\frac{\Lambda_B}{8\pi G_B} - \frac{R}{16\pi G_B} + c_B^1 R^2 + c_B^2 R_{\mu\nu}^2 + c_B^3 R_{\alpha\beta\mu\nu}^2\right]. \tag{3.4}$$

By using heat-kernel representation for $\ln Z_1$ one can write

$$-\frac{1}{2}\ln\det(-D) = \frac{1}{2}\int_{\delta^2}^{\infty}\frac{ds}{s}\mathrm{Tr}K(s), \tag{3.5}$$

where $K(s)$ is the heat-kernel of the operator D which has the following Schwinger-DeWitt expansion

$$K(s) = e^{-sD} = \frac{1}{16\pi^2 s^2} \sum a_n s^n, s \to 0. \tag{3.6}$$

For the particular case of a scalar massless field

$$a_0 = 1, \qquad a_1 = (1/6 - \xi)R, \tag{3.7}$$

$$a_2 = \frac{1}{180} R^2_{\alpha\beta\mu\nu} - \frac{1}{180} R^2_{\mu\nu} - \frac{1}{6}(\frac{1}{5} - \xi)\Box R + \frac{1}{2}(\frac{1}{6} - \xi)^2 R^2 \tag{3.8}$$

By substituting this expansion into (3.5) one can conclude that the one-loop contribution Γ_1 to the effective action can be written in the form

$$\Gamma_1 = \Gamma_1^{div} + \Gamma_1^{fin}. \tag{3.9}$$

where

$$\Gamma_1^{div} = -\frac{1}{32\pi^2} \int d^4x \sqrt{g} \left[\frac{a_0}{2\delta^4} + \frac{a_1}{\delta^2} - 2a_2 \ln(\delta) \right]. \tag{3.10}$$

The divergent part of the one-loop effective action has the same structure as the initial classical action (3.3) and hence one can write

$$\Gamma = \Gamma_{cl}^{ren} + \Gamma_1^{ren}, \tag{3.11}$$

$$\Gamma_1^{ren} = \Gamma_1 - \Gamma_1^{div} = \Gamma_1^{fin}. \tag{3.12}$$

Here Γ_{cl}^{ren} is identical to the initial classical action with the only change that all the bare coefficients Λ_B, G_B, and c_B^i are substituted by their renormalized versions Λ_{ren}, G_{ren}, and c_{ren}^i

$$\frac{\Lambda_{ren}}{G_{ren}} = \frac{\Lambda_B}{G_B} + \frac{1}{8\pi\delta^4}, \tag{3.13}$$

$$\frac{1}{G_{ren}} = \frac{1}{G_B} + \frac{1}{2\pi\delta^2}(\frac{1}{6} - \xi), \tag{3.14}$$

$$c_{ren}^i = c_B^i + \alpha^i \ln \delta. \tag{3.15}$$

We shall refer to (3.11) as to the loop expansion of the renormalized effective action. After multiplying the the renormalized effective action by β^{-1} we get the expansion for the renormalized free energy.

The effective action Γ contains the complete information about the system under consideration. In particular the variation of Γ with respect to the metric provides one with the equations for the quantum average metric $\bar{g} = \langle g \rangle$:

$$\frac{\delta\Gamma}{\delta\bar{g}} = 0. \tag{3.16}$$

One usually assumes that quantum corrections are small and solves this equation perturbatively:

$$\bar{g} = g_{cl} + \delta g, \tag{3.17}$$

where g_{cl} is a solution of the classical equations. At this point we need to make an important remark. In principle, there exist two possibilities: either to begin with the solution of the classical equations for the action (3.3), or its renormalized version Γ_{cl}^{ren}, which is written in terms of the renormalized constants. One usually assumes that the renormalized values of Λ_{ren} and c_{ren}^i vanish $\Lambda_{ren} = c_{ren}^i = 0$. It means that in general case their initial values were not vanishing unless one is dealing with some special type of theory (e.g. assuming sypersimmetry). In other words the global properties of the solutions for I_{cl} and Γ_{cl}^{ren} are generally different. So to provide the condition that δg is small, one is to begin with the metric g_{cl} that is an extremum of Γ_{cl}^{ren}

$$\frac{\delta \Gamma_{cl}^{ren}}{\delta g_{cl}} = 0. \tag{3.18}$$

We assume that the renormalization of the coupling constants in the classical action is made from the very beginning and we shall assume that the 'classical' field g_{cl} is a solution of the equation (3.18). In our case g_{cl} is the Euclidean black hole metric, while the metric $\bar{g}$ describes the Gibbons-Hawking instanton deformed due to the presence of quantum corrections to the metric. The quantity $\Gamma[\bar{g}]$ being expressed as the function of boundary conditions (β and r_0) specifies the thermodynamical properties of a black hole.

4 Thermodynamical Entropy

For the above described canonical ensemble of gravitational fields the leading tree-level contribution to the renormalized effective action is given by the Euclidean gravitational action for the Euclidean black hole solution (the so-called Gibbons-Hawking instanton).

Under the assumption that $\Lambda_{ren} = 0$ and $c_{ren}^i = 0$ the tree-level contribution to the renormalized free energy of the black hole is [12, 13]

$$F_0^{ren} \equiv \beta^{-1}\Gamma_{cl}^{ren}[g_{cl}] = r_0\left(1 - \sqrt{1 - r_+/r_0}\right) - \pi r_+^2 \beta^{-1}. \tag{4.1}$$

Here $r_+ = 2G_{ren}M$ is the gravitational radius of a black hole of mass M, which for a given temperature β^{-1} at the boundary r_0 is defined by the relation $\beta = 4\pi r_+(1 - r_+/r_0)^{1/2}$. According to definition the thermodynamical entropy of a black hole S_0^{TD} is determined by the response of the free energy of a system including a black hole to the change of the temperature. One can easily verify that

$$S_0^{TD} = -\frac{dF_0^{ren}}{dT} \equiv \beta^2 \frac{dF_0^{ren}}{d\beta} = \frac{A_H}{4l_{\rm P}^2}, \tag{4.2}$$

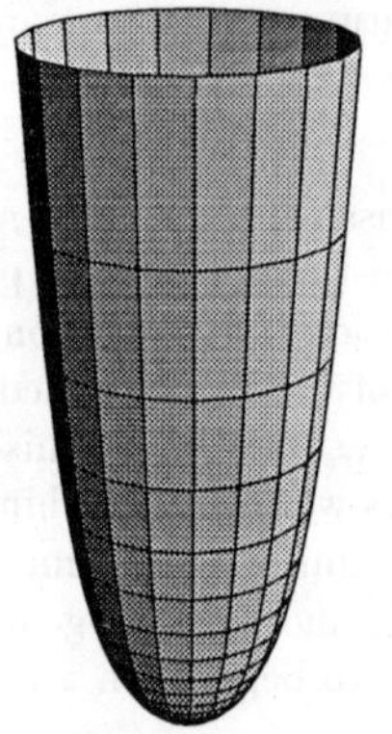

Figure 1: Embedding diagram for a two-dimentional (τ, r) section of the Gibbons-Hawking instanton. Regularity condition at the Euclidean horizon $r = r_+$ requires $\beta = \beta_H \equiv 8\pi G_{ren} M$.

and hence it coincides with the Bekenstein-Hawking expression S^{BH}. (It is assumed that r_+ in F_0^{ren} is expessed in terms of β and r_0 before differentiation with respect to β.) One-loop contribution in Eq.(3.2) describes quantum correction to the entropy of a black hole as well as the entropy of thermal radiation in its exterior. The latter evidently depends on the radius r_0 of the boundary. Since the Euclidean black hole background is regular the corresponding contribution F_1^{ren} is finite. For this reason the quantum corrections to the Bekenstein-Hawking entropy $4\pi M^2$ are small unless the mass of a black hole M is comparable with the Planckian mass [37]. Due to the presence of the conformal anomalies one might expect that the leading corrections are of the order $\ln M$ (see, e.g. [38]).

5 Statistical-Mechanical Entropy

The derivation of the thermodynamical entropy of a black hole requires the *on-shell* calculations. It means that one uses only a regular Euclidean metrics that are the solutions of the field equations (3.16). The discussion of the relation of the thermodynamical and statistical-mechanical entropy of a black hole requires *off-shell* calculations. The reason for this is quite simple and can be explained, for example, by using the approach based on the no-boundary wavefunction of BFZ [31]. The matrix elements in the $|\varphi_-\rangle$ basis of the operator $\hat{\rho} \ln \hat{\rho}$ which enters the definition of the

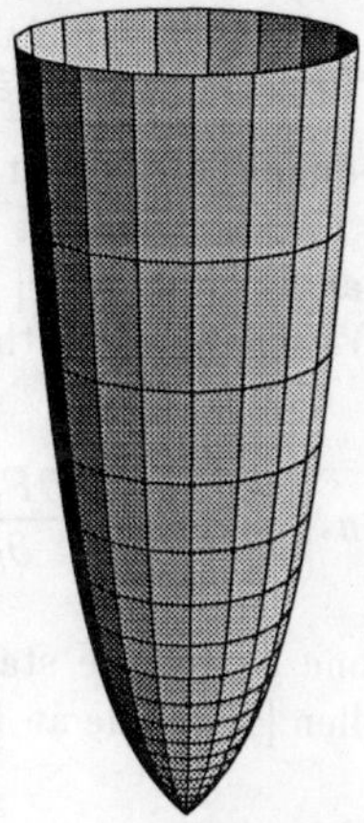

Figure 2: Embedding diagram for a two-dimentional (τ, r) section of the Euclidean black hole with cone-like singularity ($\beta \neq \beta_H \equiv 8\pi G_{ren} M$).

statistical-mechanical entropy (1.4) can be obtained by partially differentiating (2.7) with respect to β. On the other hand the Hamiltonian $\hat{H}$ depends on the black-hole geometry, and hence on the mass M of a black hole. That is why to obtain the expression for the statistical-mechanical entropy one needs to be able to use β and M as independent parameters. Strictly speaking for $\beta \neq \beta_H \equiv 8\pi M$ there are no regular Euclidean solutions with the Euclidean black-hole topology $R^2 \times S^2$. Such solutions can be obtained only if one exclude a horizon sphere S^2. In other words, there exist non-vacuum solutions with the cone-like singularity. For $\beta = \beta_H$ the singularity dissapears. In order to be able to discuss the statistical-mechanical entropy and its relation to the thermodynamical entropy one must generalize the calculation of the one-loop contribution to the renormalized free energy to the case of spaces with cone-like singularity. The new feature which arises is that the corresponding renormalized one-loop corrections would contain new type of divergence, which is directly connected with the presence of cone singularity. In order to make the answer finite one might introduce spatial cut-off in the volume integrals near the cone-singularity. It is convenient to restrict the integration by some proper distance l from singularity. This cut-off was present in the 't Hooft's 'brick wall model' [25]. In the model with 'dynamical black-hole-interior' the presence of such a cut-off was connected with the quantum fluctuations of the horizon [30]. Similar cut-off naturally arises in the string theory [33]. The renormalized one-loop contribution F_1^{ren} to the free energy is of the form

$$F_1^{ren} = F_1^{ren}[\beta, \beta_H, \varepsilon], \tag{5.1}$$

where $\varepsilon = (l/2G_{ren}M)^2$ is the dimensionless cut-off parameter. For $\varepsilon \to 0$ and $\beta \neq \beta_H$ the one-loop free energy F_1^{ren} is divergent $F_1^{ren} \sim \varepsilon^{-1} f(\beta, \beta_H)$. For $\beta = \beta_H$ the divergence dissappears, so that $F_1^{ren}[\beta_H, \beta_H, 0]$ is finite. This quantity is directly related with quantum (one-loop) corrections to the thermodynamical entropy of a black hole

$$S_1^{TD} = \beta_H^2 \frac{d}{d\beta_H} F_1^{ren}[\beta_H, \beta_H, 0] \equiv \left[\beta^2 \frac{\partial F_1^{ren}}{\partial \beta} + \beta_H^2 \frac{\partial F_1^{ren}}{\partial \beta_H}\right]_{\beta=\beta_H}. \tag{5.2}$$

The expression (5.1) allows one to get the statistical-mechanical entropy S^{SM}. Dowker and Kennedy [41] and Allen [42] made an important observation that

$$F_1^{ren} = F_{vac}^{ren} + F_{therm}^{ren}, \qquad \frac{\partial F_{vac}^{ren}}{\partial \beta} = 0, \tag{5.3}$$

$$F_{therm}^{ren} = -\beta^{-1} \ln \mathrm{Tr}\left[\mathrm{e}^{-\beta \hat{H}}\right] = \ln[\sum \exp(-\beta E_n)], \tag{5.4}$$

where E_n is the energy (eigenvalue of the Hamiltonian $\hat{H}$ of the field φ). By using the expansion in eigenfunctions one can obtain

$$F_{therm}^{ren} = \sum_\lambda f(\beta\omega_\lambda) = \int d\omega N(\omega|\beta_H, \varepsilon) f(\beta\omega). \tag{5.5}$$

$f(\beta\omega) = \beta^{-1} \ln[1 - \exp(-\beta\omega)]$ is free energy of an oscillator of frequency ω at inverse temperature β, and $N(\omega|\beta_H)$ is the density of number of states at the given energy ω in a spacetime of a black hole of mass M. This density of number of states diverges. In order to make it finite we introduced the cut-off ε. We include ε as the argument of N in order to remind about this. The expression (5.5) is usually a starting point for 'brick wall' model.

The statistical-mechanical entropy S^{SM} is

$$S^{SM} = \left[\frac{\partial F_1^{ren}}{\partial \beta}\right]_{\beta_H} = \left[\frac{\partial F_{therm}^{ren}}{\partial \beta}\right]_{\beta_H} \tag{5.6}$$

$$= \int d\omega N(\omega|\beta_H, \varepsilon) s(\beta\omega) \tag{5.7}$$

Here $s(\beta\omega) = \beta\omega/(e^{\beta\omega} - 1) - \ln(1 - e^{-\beta\omega})$ is the entropy of a quantum oscillator of frequency ω with inverse temperature β. S^{SM} is divergent in the limit $\varepsilon \to 0$. The divergence is directly related with the divergency of the density of number of states located in the narrow region in the vicinity of the horizon.

By comparing the expressions (5.2) and (5.6) we can conclude that S^{TD} and S^{SM} differs from one another. It happens for the following two reasons: (1) Vacuum polarization (F_{vac}^{ren}) depends on M and hence on β_H; (2) $d/d\beta$ does not commute with Tr-operation. In general case one gets

$$S_1^{TD} = S^{SM} + \Delta S. \tag{5.8}$$

In the limit $\varepsilon \to 0$ $\Delta S \neq 0$ is also divergent, but S_1^{TD} remains finite and (for $M \gg \mu_{Pl}$) small. The relation (5.8) provides explanation of the entropy renormalization procedure by Thorne and Zurek [20].

Fursaev and Solodukhin [45, 46, 47] recently proposed another approach for off-shell calculations of thermodynamical characteristics of a black hole. Namely, instead of cutting-off the vicinity of a cone singularity, they proposed to calculate the effective action directly on a spacetime with cone-like singularity. In order to make this mathematically well defined one might at first consider a manifold with the topology $R^2 \times S^2$ which is smooth at the fixed-point sphere (horizon) and differs from the cone metric only in the very narrow region of size l near the horizon. The effective action must be considered as the function of l and the parameter l must be finally put to zero. The one-loop correction to the free energy F_1^{cone} is divergent in this limit, but it is possible to show that the divergence is proportional to $(\beta - \beta_H)^2$. That is why the corresponding contribution $S_1^{cone} = [\beta^2 \partial F_1^{cone}/\partial\beta]_{\beta=\beta_H}$ to the entropy is finite. This approach gives the same expression for the thermodynamical entropy and might be considered as useful tools for such calculations. The relation of this off-shell entropy S_1^{cone} to the statistical-mechanical entropy S^{SM} is not clear. Among other approaches to the calculation of the entropy we mention also the approach based on the Pauli-Villars regularization [43].

6 Black Hole Thermodynamics and Physics at Planckian Scales

The expression for the Bekenstein-Hawking entropy (1.1) contains the Planckian length in the denominator. Does it mean that this and other thermodynamical characteristics of a black hole for their understanding require the knowledge of physics at Planckian scales?

When we are discussing black hole solutions and their properties we use gravitational equations. The coupling constants in these equations are assumed to coincide with 'observable' values. Due to the existence of ultraviolet divergencies the observable coupling constants differ from their initial bare values. Any procedure which gives sence to this renormalization procedure finally must deal with the problems of physics at Planckian scale. But in this sence black hole physics does not differ from the usual Newtonian theory. Calculations in the Newtonian theory also use the renormalized ('observable') gravitational constant, and hence in order to derive the same results in the framework of quantum gravity beginning from some initial background theory one must pass through all the complications connected with the remormalization procedure and redefining the coupling constant. One can do it from very beginning or develope more complicated scheme and made all the renormalizations only at the end. The same is true also for black holes. Besides this in the case of black hole there are situations when quantum gravity becomes really important. It is well known that the final stage of a black hole evoparation as well as the structure near singularity inside a black hole for their consideration require quantum gravity. Quantum gravity might be also useful for study of small quantum corrections to black hole characteristics. But these corrections can essentially change parameters of a black hole when the curvature at the horizon becomes comparable with Planckian curvature, i.e. for black holes of Planckian mass. These remarks are of course trivial. But if we exclude these evident cases do we still need quantum gravity to explain properties of macroscopic black holes?

This question is not new. The standard derivation of Hawking quantum radiation of a black hole formally requires the integration over all (including much higher than Planckian) frequencies of the initial zero-point-fluctuations of a quantum field. There is a belief that one can escape the formal usage of super-Planckian energies in the calculations. This point of view was supported by recent result by Unruh [44]. He considered a model in which due to the presence of dispersion the frequencies of zero-point-fluctuations are restricted. Unruh has shown that nevertheless the Hawking radiation at late time remains constant and with high accuracy thermal.

A similar problem arises in connection of a black-hole entropy. We saw, for example, that the statistical-mechanical entropy S^{SM} is dependent on the cut-off parameter. One might argue that such a cut-off must be provided by quantum gravity. For this reason S^{SM} is (at least potentially) the quantity which for its knowledge requires Planckian scale physics. Does it mean that the study of the thermodynamical properties might give us information about these scales? The above discussion indicates that it is impossible. In the stardard gedanken experiments the observable quantity is S^{TD}. S^{SM} (at least in the leading order) does not contribute to S^{TD} and hence one cannot measure it. The statistical-mechanical entropy might be useful for description of exitations in the close vicinity of the horizon (for example of their damping). The main contribution to S^{SM} is given by very high frequency modes inside the gravitational barrier propagating very close to the horizon. It looks like that the only reasonable way to measure S^{SM} is to excite these modes. For example, one can do it by colliding particles of superhigh energy near the horizon. This experiment requires very high (super-Planckian) energies. But having these energies available one can use them for study Planckian physics in usual Minkowski spacetime without any black holes.

To conclude, quantum gravity is required for understanding very fundamental problems of black holes, such as the problem of final state, but it also looks like that the thermodynamics of macroscopical black hole does not provide us with any new powerful tools for verifying the theory of quantum gravity.

7 Acknowledgements

This work was supported by the Natural Sciences and Engineering Research Council of Canada.

References

[1] J. D.Bekenstein, Nuov.Cim.Lett. **4** (1972) 737.

[2] J. D. Bekenstein, Phys.Rev. **D7** (1973) 2333.

[3] J. M. Bardeen, B. Carter, and S. W. Hawking, Comm.Math.Phys. **31**, (1973) 181.

[4] J. D. Bekenstein, Phys.Rev. **D9** (1974) 3292.

[5] K. S. Thorne, W. H. Zurek, and R. H. Price, in *Black Holes: The Membrane Paradigm* , edited by K. S. Thorne, R. H. Price, and D. A. MacDonald (Yale University Press, New Haven, 1986), p.280.

[6] I. Novikov and V. Frolov, *Physics of Black Holes.* (Kluwer Academic Publ., Dordrech-Boston-London), 1989.

[7] R. W. Wald, In:*Black Hole Physics* (Eds.V. DeSabbata and Z. Zhang), (Kluwer Academic Publ., Dordrech-Boston-London), 1992).

[8] V. Frolov and D. N. Page, Phys.Rev.Lett. **71** (1993) 3902.

[9] J.B.Hartle and S.W.Hawking, Phys.Rev. **D13**, 2188 (1976).

[10] G. W. Gibbons and S. W. Hawking, Phys.Rev. **D15** (1976) 2752.

[11] S. W. Hawking, In: *General Relativity: An Einstein Centenary Survey.* (eds. S. W. Hawking and W. Israel), Cambridge Univ.Press, Cambridge, 1979.

[12] J. W. York, Phys.Rev. **D33** (1986) 2092.

[13] H. W. Braden, J. D. Brown, B. F. Whiting, and J. W. Jork, Phys.Rev. **D42** (1990) 3376.

[14] R. M. Wald, Phys.Rev. **D48** (1993) 3427.

[15] T. Jacobson and R. C. Myers, Phys.Rev. Lett. **70** (1993) 3684.

[16] V. Iyer and R. M. Wald, Phys.Rev. **D50** (1994) 846.

[17] T. Jacobson, G. Kang, and R. C. Myers, Phys.Rev. **D49** (1994) 6587.

[18] G. W. Gibbons and R. E. Kallosh, Phys.Rev. **D51** (1995) 2839.

[19] S. W. Hawking, G. T. Horowitz, and S. F. Ross, Phys.Rev. **D51** (1995) 4302.

[20] W. H. Zurek and K. S. Thorne, Phys.Rev.Lett. **54** (1985) 2171.

[21] V. F. Mukhanov, JETP Lett. **44** (1986) 63.

[22] J. D. Bekenstein and V. F. Mukhanov. 'Spectroscopy of quantum black holes' Preprint gr-qc/9505012 (1995).

[23] V. A. Berezin. Phys.Lett., **B241** (1990) 194.

[24] J. D. Bekenstein, preprint gr-qc/9409015 (1994).

[25] G. 't Hooft, Nucl.Phys. **B256** (1985) 727.

[26] L.Bombelli, R.K.Koul, J.Lee, and R.Sorkin, Phys.Rev. **D34** 373 (1986).
[27] M.Srednicki, Phys.Rev.Lett. **71**, 666 (1993).
[28] V. Frolov and I. Novikov, Phys.Rev. **D48** (1993) 1607.
[29] A. G. Doroshkevich and I. D. Novikov, Soviet Phys. JETP. **63** (1993) 1538.
[30] V. Frolov and I. Novikov, Phys.Rev. **D48** (1993) 4545.
[31] A. I. Barvinsky, V. P. Frolov, and A. I. Zelnikov, Phys.Rev.**D51** (1995) 1741.
[32] S. Carlip and C. Teitelboim, Preprint gr-qc/9312002 (1993).
[33] H. J. de Vega and N. Sánchez, Nucl.Phys. **B299** (1988) 818.
[34] L. Susskind and J. Uglum, Phys.Rev. **D50** (1994) 2700.
[35] D. Garfinkle, S. B. Giddings, and A. Strominger, Phys.Rev. **D49** (1994) 958.
[36] J.B.Hartle and S.W.Hawking, Phys.Rev. **D28**, 2960 (1983).
[37] P. R. Anderson, W. A. Hiscock, J. Whitesell, and J. W. York, Phys.Rev. **D50** (1994) 6427.
[38] D. V. Fursaev, Phys.Rev.**D51** (1995) 5352.
[39] S. W. Hawking, Comm.Math.Phys. **43**, (1975) 199.
[40] V. Frolov, Phys.Rev.Lett. **74** (1995) 3319.
[41] J.S.Dowker and G.Kennedy, J.Phys. **A 11** (1978) 895.
[42] B. Allen, Phys.Rev. **D33** (1986) 3640.
[43] J.-G. Demers, R. Lafrance, and R. C. Myers. Phys.Rev.**D52** (1995) 2245.
[44] W. G. Unruh, Phys.Rev.**D51** (1995) 2827.
[45] D. V. Fursaev and S. N. Solodukhin, On one-loop renormalization of black hole entropy, Preprint E2-94-462, hep-th/9412020 (1994).
[46] S. N. Solodukhin, Phys.Rev.**D51** (1995) 609.
[47] D. V. Fursaev and S. N. Solodukhin, Phys.Rev.**D52** (1995) 2133.

Null Hypersurface Canonical Formalism

Joshua Goldberg

Department of Physics, Syracuse University, Syracuse, NY 13244-1130

Abstract

We use the null surface canonical formalism of Goldberg, Robinson, and Soteriou to examine the behavior of the gravitational field in the vicinity of null infinity. After a brief review of the Hamiltonian and the resulting constraints, the radial equations are integrated. This is first done omitting logarithmic behavior and then how and where the logarithms appear is discussed. One finds that the logarithmic behavior does not interfere with the definition of energy-momentum and the radiation of gravitational energy at future null infinity. On the other hand, the logarithms occur before angular momentum terms can be defined at infinity.

1. Introduction.

From the analysis of the Einstein equations at null infinity by Bondi[1] and Sachs[2] and by Newman and Penrose[3,4] , we learned that the constraint equations of general relativity can be integrated along the null generators of an outgoing null surface. Bondi and Sachs use the metric variables and as initial data they give the conformal metric on a foliation of the null surface by two surfaces. Newman and Penrose give a component of the cȯnformal tensor, Ψ_0 in their notation, which then allows them to integrate the equations using the Bianchi identities as field equations. In both cases, additional information must be given at null infinity in order to complete the propagation. Although the Newman-Penrose approach is simpler, in that it involves solving a system of linear equations for a larger number of variables, it has not been used in quantum gravity because an action integral does not exist in terms of their variables. The introduction of a phase space for general relativity consisting of an SL(2,C) connection and a densitized triad has allowed us to introduce a canonical formalism which lies between the metric formalism of Bondi-Sachs and that of Newman-Penrose[5]. Whether this canonical formalism on a null surface will aid in the construction of quantum gravity remains to be seen.

In these lectures, which is based on work done with Chrysis Soteriou[6], I shall outline the Hamiltonian, discuss problems which remain in trying to quantize the theory, and then apply the formalism to repeate the analysis at null infinity. This is interesting for two reasons. The Ashtekar variables are complex and reality conditions must be applied. It is interesting to see how those conditions are treated in the classical domain. Secondly, it turns out that the variables used strongly suggest studying the possibility of logarithmic behavior in the neighborhood of null infinity. In order to avoid the appearance of terms in $\log r$, the data for the connection must skip two orders in $1/r$. Going back to insert these terms, one finds logarithmic behavior which preserves null infinity, the definition of mass, and the radiation of gravitational energy, but does not permit the description of angular momentum or the regular fall off of the conformal tensor or peeling[7,3].

I shall begin with the construction of the Hamiltonian, then introduce sufficient coordinate condtions so that the integration of the constraint equations can be carried out easily, then solve the equations without log terms, and finally discuss the introduction of the log terms.

2. The Hamiltonian.

We assume that we are working with a space-time (M,g) which is asymptotically flat and which, in the vicinity of future null infinity, can be foliated by a family of null surfaces. Therefore, we may introduce a null basis of one-forms and its dual tetrad basis.

$$\begin{aligned}
\theta^{\mathbf{0}} &= (N + \alpha\nu^{\mathbf{1}}{}_i N^i)dt + \alpha\nu^{\mathbf{1}}{}_i dx^i, \\
\theta^{\mathbf{i}} &= \nu^{\mathbf{i}}{}_i(N^i dt + dx^i) \\
e_{\mathbf{0}} &= N^{-1}(\partial_t - N^i\partial_i), \\
e_{\mathbf{i}} &= (v_{\mathbf{i}}{}^i + \alpha\delta_{\mathbf{i}}{}^{\mathbf{1}} N^{-1} N^i)\partial_i - \alpha\delta_{\mathbf{i}}{}^{\mathbf{1}} N^{-1}\partial_t,
\end{aligned} \tag{2.1}$$

The indices all have the range 1-3 and $\nu^{\mathbf{i}}{}_i v_{\mathbf{j}}{}^i = \delta^{\mathbf{i}}{}_{\mathbf{j}}$. Bold face indices refer to one forms and tetrads. In terms of the one forms, the metric is

$$\begin{aligned}
ds^2 &= \eta_{\alpha\beta}\theta^\alpha \otimes \theta^\beta, \\
\eta_{01} &= -\eta_{23} = 1,
\end{aligned}$$

with the orientation

$$V = -i\theta^{\mathbf{0}} \wedge \theta^{\mathbf{1}} \wedge \theta^{\mathbf{2}} \wedge \theta^{\mathbf{3}}. \tag{2.2}$$

Thus, $-iN\nu$ is positive, where ν is the determinent of $\nu^{\mathbf{i}}{}_i$. (Greek letters range and sum from zero to three.)

The connection coefficients are defined by

$$d\theta^\alpha = \theta^\beta \wedge \omega^\alpha{}_\beta \tag{2.3}$$

and the Riemann tensor by

$$d\omega^\alpha{}_\beta + \omega^\alpha{}_\gamma \wedge \omega^\gamma{}_\beta = -\frac{1}{2}R^\alpha{}_{\beta\gamma\delta}\theta^\gamma \wedge \theta^\delta =: R^\alpha{}_\beta. \tag{2.4}$$

The self-dual components of the connection are

$$^+\omega^{\alpha\beta} = \frac{1}{2}(\omega^{\alpha\beta} - \frac{i}{2}\eta^{\alpha\beta}{}_{\mu\nu}\omega^{\mu\nu})$$

and are represented by $\Gamma^A{}_,$:

$$\Gamma^1 := \frac{1}{2}(\omega^{01} + \omega^{23}), \quad \Gamma^2 := \omega^{21}, \quad \Gamma^3 := \omega^{03}. \tag{2.5}$$

From these we obtain the self-dual components of the Riemann tensor as

$$\frac{1}{2}R^A = d\Gamma^a \; + \; \eta^A{}_{BC}\Gamma^B \wedge \Gamma^C. \tag{2.6}$$

In a $3+1$ decomposition, we have

$$\Gamma^A = A^A{}_i dx^i + B^A dt, \tag{2.7}$$

and

$$\begin{aligned}
R^A{}_{ij} &= 2A^A{}_{[i,j]} \; + \; 2\eta^A{}_{BC}A^B{}_{[j}A^C{}_{i]}, \\
R^A{}_{0i} &= D_i B^A \; - \; A^A{}_{i,0}.
\end{aligned} \tag{2.8}$$

The derivative operator D acts on the index A as

$$D_i f^A := \partial_i f^A + 2\eta^A{}_{BC} A^B{}_i f^C. \tag{2.9}$$

We take as a basis of self-dual two-forms

$$\begin{aligned} S^1 &= \frac{1}{2}(\theta^1 \wedge \theta^0 + \theta^3 \wedge \theta^2), \\ S^2 &= \theta^1 \wedge \theta^2, \\ S^3 &= \theta^3 \wedge \theta^0. \end{aligned} \tag{2.10}$$

These two-forms satisfy the equations $(A, B = 1, 2, 3)$

$$\frac{1}{2}\eta_{\iota\kappa}{}^{\rho\sigma} S^A{}_{\rho\sigma} dx^\iota \wedge dx^\kappa \;=\; {}^*S^A \;=\; iS^A, \tag{2.11}$$

$$S^A \wedge S^B \;=\; ig^{AB} V. \tag{2.12}$$

The $SO(3)$ invariant metric

$$g_{AB} = \begin{pmatrix} 2 & 0 & 0 \\ 0 & 0 & -1 \\ 0 & -1 & 0 \end{pmatrix} \tag{2.13a}$$

and its inverse,

$$g^{AB} = \begin{pmatrix} \frac{1}{2} & 0 & 0 \\ 0 & 0 & -1 \\ 0 & -1 & 0 \end{pmatrix} \tag{2.13b}$$

are used to raise and lower the uppercase Latin self-dual, triad indices.

We take the complex action to be[8,9]

$$S = \int R^A \wedge S^B g_{AB}. \tag{2.14}$$

which with the given 3+1 decomposition becomes

$$\begin{aligned} \mathcal{L} \;=\; &(A^A{}_{i,0}\Sigma_A{}^i + B^A D_i \Sigma_A{}^i + R^A{}_{ij} N^i \Sigma_A{}^j - \\ &- \underset{\sim}{N} v^i (R^1{}_{ij}\Sigma_3{}^j + R^2{}_{ij}\Sigma_1{}^j)) + \mu_i(\Sigma_2{}^i - \alpha v^i) + \rho(\alpha)^2. \end{aligned} \tag{2.15}$$

In the above we have introduced $v^i := \tilde{v}_2{}^i$ and

$$\Sigma_1{}^i := -\tilde{v}_1{}^i, \quad \Sigma_2{}^i := -\alpha\tilde{v}_2{}^i, \quad \Sigma_3{}^i := -\tilde{v}_3{}^i. \tag{2.16}$$

The over (under) tilde indicates a density of weight plus (minus) one. Thus, v^i and $\Sigma_A{}^i$ are densities of weight one while the other variables are not densitized except as indicated by the tilde. We have introduced the last two terms with Lagrange multipliers ρ and μ_i to enforce $\alpha = 0$ and, hence,

$\Sigma_2 = 0$ after variation of the action. The variation of α is needed to obtain one of the field equations which is otherwise missing.

The quantities we have introduced are complex and we shall have to introduce reality conditions to recover the Einstein equations. This will be discussed at an appropriate time.

Clearly, $A^A{}_i$ and $\Sigma_A{}^i$ define the phase space and have the Poisson bracket relations

$$\{A^A{}_i(x),\ \Sigma_B{}^i(x')\} = \delta^A{}_B\delta^3(x-x'). \tag{2.17}$$

The remaining variables are effectively Lagrange multipliers whose variation lead to hypersurface equations. N^i and B^A are arbitrary in the absence of coordinate and gauge conditions. The strange thing is that $\underset{\sim}{N}v^i$ appear together. We will see that only when we apply the reality conditions can we identify $\underset{\sim}{N}$.

The Hamiltonian clearly takes the form

$$H = \int d^3x\{\underset{\sim}{N}\mathcal{H}_0 + N^i\mathcal{H}_i - B^A\mathcal{G}_A - \mu_i C^i - \rho(\alpha)^2\} \tag{2.18}$$

where

$$\begin{aligned}
\mathcal{H}_0 &:= v^i\left(R^1{}_{ij}\Sigma_3{}^j + R^2{}_{ij}\Sigma_1{}^j\right) = 0,\\
\mathcal{H}_i &:= -R^A{}_{ij}\Sigma_A{}^j = 0,\\
\mathcal{G}_A &:= D_i\Sigma_A{}^i = 0,\\
\phi_i &:= R^1{}_{ij}\Sigma_3{}^j + R^2{}_{ij}\Sigma_1{}^j = 0\\
C^i &:= \Sigma_2{}^i + \alpha v^i = 0.
\end{aligned} \tag{2.19a}$$

are constraints which arise from varying the action with respect to $\underset{\sim}{N}, N^i, B^A, v^i$ and μ_i. respectively. Note that $v^i\phi_i \equiv \mathcal{H}_0$ and $\Sigma_1{}^i\phi_i \equiv \Sigma_3{}^i\mathcal{H}_i$. Other constraints come from varying ρ, and α :

$$\alpha = 0, \qquad v^i\mu_i = 0. \tag{2.19b}$$

Propagation of the constraint $C^i = 0$ leads to

$$\chi^i := 2\delta^B{}_2 D_j\left(\underset{\sim}{N}v^{[i}\Sigma_A{}^{j]}Q^A{}_B\right) - 2A^3{}_j N^{[i}\Sigma_1{}^{j]} - B^3\Sigma_1{}^i = 0, \tag{2.19c}$$

$$Q^A{}_B := \delta_3^A\delta_B^1 + \delta_1^A\delta_B^2.$$

The equations $\chi^i = 0$ are not constraints, but defining equations for v^i. Propagation of the constraint $\mathcal{G}_3 = 0$ leads to a condition on μ_i,

$$\mu_i\Sigma_1{}^i = R^1{}_{i0}\Sigma_3{}^i - R^1{}_{ij}\Sigma_3{}^i N^j, \tag{2.20}$$

which later gives rise to a trivial equation.

Of the 11 constraints, five are first class, , and the remaining six, including $\mathcal{H}_0$, are second class. Thus there are 16 conditions on the 18 phase space variables, leaving two functions as free data or dynamical variables. However, until gauge and coordinate conditions are imposed, we are left with five dynamical equations:

$$A^1{}_{i,0} = \delta^1{}_A D_i B^A + N^j R^1{}_{ij} - \underset{\sim}{N} v^j R^2{}_{ij}, \tag{2.21a}$$

$$A^2{}_{i,0} = \delta^2{}_A D_i B^A + N^j R^2{}_{ij} - \mu_i, \tag{2.21b}$$

$$A^3{}_{i,0} = \delta^3{}_A D_i B^A + N^j R^3{}_{ij} - \underset{\sim}{N} v^j R^1{}_{ij}, \tag{2.21c}$$

$$\Sigma_1{}^i{}_{,0} = 2\delta^B{}_1 D_j\left(\underset{\sim}{N} v^{[i}\Sigma_A{}^{j]}Q^A{}_B\right) - 2\delta^B{}_1 D_j(N^{[i}\Sigma_B{}^{j]}) - 2B^3\Sigma_3{}^i, \tag{2.21d}$$

$$\begin{aligned}\Sigma_3{}^i{}_{,0} = {} & 2\delta^B{}_3 D_j\left(\underset{\sim}{N} v^{[i}\Sigma_A{}^{j]}Q^A{}_B\right) - 2\delta^B{}_3 D_j(N^{[i}\Sigma_B{}^{j]}) \\ & + 2B^1\Sigma_3{}^i + B^2\Sigma_1{}^i.\end{aligned} \tag{2.21e}$$

3. Complex Solutions.

In order to solve the equations, we must impose coordinate and gauge conditions. The surfaces t =constant form a foliation of null surfaces near null infinity and give us one coordinate. Space-like cuts of these surfaces are assumed to have the topology S^2. Hence, the generators can be labeled by the angular coordinates (θ, ϕ). We follow the convention of Bondi and Sachs [B-S] in choosing the coordinate r along the null generators to be the luminosity distance, so that

$$-i\bar{r}^2 \sin\theta\, \eta_{1jk}\, v^j \Sigma_3{}^k = \nu^2. \tag{3.1a}$$

and we can set

$$\Sigma_1{}^i = -ir^2 \sin\theta \delta^i{}_1. \tag{3.1b}$$

Then the null rotations generated by $\mathcal{G}_1$ and $\mathcal{G}_2$ allow us to fix $\Sigma_3{}^i$ tangent to the surfaces r =constant, that is, $\Sigma_3{}^1 = 0$, and then to set $A^1{}_1 = 0$. The latter is equivalent to setting $\epsilon = 0$ in the Newman-Penrose formalism[3,4].

The Dirac brackets[10,11] of $(A^3{}_2,\ \Sigma_3{}^2)$ are equal to their Poisson brackets and, therefore, are appropriate initial data on which to build the solution. We shall assume that these variables take their Minkowski space values at null infinity and that the solution can be expanded in $1/r^n$. The equations are listed below in the order in which they are to be solved. Because of the coordinate conditions, some of the equations which are formally propagation equations also need be satisfied only on the initial surface. Thus, the first group of equations are solved algebraically or by radial integrations: (The indices $a, b, ..$ range over 2 and 3.)

$$\mathcal{G}_1 = 0 \Rightarrow \Sigma_1{}^i{}_{,1} + 2A^3{}_i\Sigma_3{}^i = 0 \tag{3.2a}$$

$$\mathcal{G}_2 = 0 \Rightarrow A^3{}_i\Sigma_1{}^i = 0 \tag{3.2b}$$

$$\mathcal{H}_1 = 0 \Rightarrow A^3{}_{i,1}\Sigma_3{}^i = 0 \tag{3.2c}$$

$$\mathcal{G}_3 = 0 \Rightarrow \Sigma_3{}^i{}_{,i} - 2A^1{}_i\Sigma_3{}^i - A^2{}_i\Sigma_1{}^i = 0 \tag{3.2d1}$$

$$\mathcal{H}_a = 0 \Rightarrow (A^1{}_a\Sigma_1{}^j)_{,j} + (A^3{}_j\Sigma_3{}^j)A^1{}_bZ^b{}_a + G^1{}_a = 0, \tag{3.2d2}$$

$$G^1a := A^3{}_a\Sigma_3{}^i{}_{,i} + (A^3{}_{a,j} - A^3{}_{j,a})\Sigma_3{}^j,$$

$$Z^b{}_a := \delta^b{}_a - \frac{\Sigma_3{}^bA^3{}_a}{\Sigma_3{}^cA^3{}_c},$$

$$\phi_a = 0 \Rightarrow A^2{}_{a,j}\Sigma_1{}^j - \Sigma_3{}^cA^3{}_cA^2{}_bZ^b{}_a + G^2{}_a = 0 \tag{3.2e}$$

$$G^2{}_a := (A^1{}_{a,b} - A^1{}_{b,a})\Sigma_3{}^b - (A^2{}_{j,a} + 2A^2{}_jA^1{}_a)\Sigma_1{}^j$$

$$\chi^a = 0 \Rightarrow (V^a\Sigma_1{}^j)_{,j} + (\Sigma_3{}^cA^3{}_c)Z^a{}_bV^b = 0, \tag{3.2f}$$

$$\underset{\sim}{N}v^a \equiv V^a$$

$$\chi^1 = 0 \Rightarrow B^3\Sigma_1{}^1 + (V^b\Sigma_1{}^1)_{,b} + 2A^1{}_bV^b\Sigma_1{}^1 - A^3{}_bN^b\Sigma_1{}^1 = 0 \tag{3.2g1}$$

$$\dot{\Sigma}_1{}^a = 0 \Rightarrow N^a{}_{,j}\Sigma_1{}^j - (V^a\Sigma_3{}^b - V^b\Sigma_3{}^a)_{,b} + 2A^2{}_j\Sigma_1{}^jV^a + 2B^3\Sigma_3{}^a = 0 \tag{3.2g2}$$

$$\dot{\Sigma}_1{}^1 = 0 \Rightarrow N^1\Sigma_1{}^j{}_{,j} + 2A^2{}_jV^j\Sigma_1{}^1 + (N^a\Sigma_1{}^1)_{,a} = 0 \tag{3.2h}$$

$$\dot{A}^1{}_1 = 0 \Rightarrow B^1{}_{,1} + A^2{}_1B^3 - N^a(A^1{}_{a,1} + A^3{}_aA^2{}_1) - V^a(A^2{}_{1,a} - A^2{}_{a,1} - 2A^1{}_aA^2{}_1) = 0 \tag{3.2i}$$

$$\dot{\Sigma}_3{}^1 = 0 \Rightarrow B^2\Sigma_1{}^1 - N^1{}_{,a}\Sigma_3{}^a - N^iA^2{}_i\Sigma_1{}^1 = 0 \tag{3.2j}$$

The second group are propagation equations which require separate discussion:

$$\dot{A}^1{}_a = B^1{}_{,a} + A^2{}_aB^3 - A^3{}_aB^2 + N^j[A^1{}_{a,j} - A^1{}_{j,a} + A^2{}_jA^3{}_a - A^3{}_jA^2{}_a] - V^b[A^2{}_{a,b} - A^2{}_{b,a} + A^1{}_bA^2{}_a - A^2{}_bA^1{}_a] \tag{3.3a}$$

$$\dot{A}^2{}_i = B^2{}_{,i} - 2A^1{}_iB^2 + 2A^2{}_iB^1 + N^j[A^2{}_{i,j} - A^2{}_{j,i} - 2A^1{}_jA^2{}_i + 2A^2{}_jA^1{}_i] - \mu_i \tag{3.3b}$$

$$\dot{A}^3{}_a = B^3{}_{,a} - 2A^3{}_aB^1 + 2A^1{}_aB^3 + N^j[A^3{}_{a,j} - A^3{}_{j,a} + 2A^1{}_jA^3{}_a - 2A^3{}_jA^1{}_a] - V^b[A^1{}_{b,a} - A^1{}_{a,b} + A^2{}_aA^3{}_b - A^3{}_aA^2{}_b] \tag{3.3c}$$

$$\dot{\Sigma}_3{}^a = -2\partial_j(N^{[a}\Sigma_3{}^{j]}) + 4A^1{}_jN^{[a}\Sigma_3{}^{j]} + 2A^2{}_jN^{[a}\Sigma_1{}^{j]} - 2A^2{}_bV^{[a}\Sigma_3{}^{b]} - 2B^1\Sigma_3{}^a + B^2\Sigma_1{}^a \tag{3.3d}$$

Because the Poisson brackets of the constraints with the Hamiltonian vanish modulo the constraints themselves, the propagation of the variables is determined by that of $(A^3{}_2,\ \Sigma_3{}^2)$ and data which must be given on null infinity.[1-4] We shall see that these additional functions are associated with mass, angular momentum, and the news or radiation.

To proceed to the solutions, we write

$$\begin{aligned} A^3{}_2 &= -\frac{1}{\sqrt{2}}\{1+\frac{{}^3A}{r^3}+\ldots\} \\ \Sigma_3{}^2 &= -\frac{ir\sin\theta}{\sqrt{2}}\{1+\frac{{}^1\Sigma}{r}+\frac{{}^3\Sigma}{r^3}+\ldots\} \end{aligned} \tag{3.4}$$

Recall from (3.1b) that $\Sigma_1{}^1 = -ir^2\ \sin\theta$. Then, $\mathcal{G}_1$ and $\mathcal{G}_2$ tell us that

$$\Sigma_3{}^a A^3{}_a = ir\sin\theta, \tag{3.5a}$$

$$A^3{}_1 = 0. \tag{3.5b}$$

We introduce the eigen-forms and eigenvectors of the flat space part of the matrix $Z^a{}_b$, (a_a, b_a) and (s^a, t^a). They satisfy the following algebraic relations:

$$a_a s^a = b_a t^a = 1, \qquad a_a t^a = b_a s^a = 0.$$

Their properties as spin weighted objects are discussed in an Appendix.

Now, continuing in the order of the equations as given above, we find

$$A^3{}_a = -a_a + -\frac{{}^3A}{r^3}b_a \tag{3.6a}$$

$$\Sigma_3{}^a = -ir\sin\theta\{s^a+\frac{{}^1\Sigma}{r}t^a+\frac{{}^3\Sigma}{r^3}t^a\}. \tag{3.6b}$$

$$\begin{aligned} A^1{}_a =& [\frac{\cot\theta}{2\sqrt{2}}-\frac{1}{r}\eth\,{}^1\Sigma+\frac{1}{r^2}C^1{}_s-\frac{1}{r^3}\eth\,{}^3\Sigma]a_a+ \\ & [-\frac{\cot\theta}{2\sqrt{2}}+\frac{1}{r^2}C^1-\frac{1}{r^3}\bar\eth\,{}^3A]b_a \end{aligned} \tag{3.6c}$$

$$A^2{}_1 = \frac{1}{r^2}[\eth\,{}^1\Sigma-\frac{1}{r}C^1{}_s+\frac{1}{r^2}(\eth\,{}^3\Sigma-C^1{}_t)] \tag{3.6d}$$

$$\begin{aligned} A^2{}_a =& [\mathcal{A}+\frac{1}{r}\bar\eth\eth\,{}^1\Sigma-\frac{1}{2r^2}(\mathcal{M}\,{}^1\Sigma+\bar\eth C^1{}_s+2(\eth\,{}^1\Sigma)^2+{}^1\Sigma\eth^2\,{}^1\Sigma]a_a \\ & +[-\frac{1}{2}+\frac{1}{r}\mathcal{M}-\frac{1}{r^2}\bar\eth C^1{}_t]b_a \end{aligned} \tag{3.6e}$$

$$V^a = -\frac{1}{r}\Big[(\frac{1}{r}{}^1\mathcal{V}+\frac{1}{2r^3}{}^0\mathcal{V})s^a+({}^0\mathcal{V}{}^3A+\frac{1}{2r^2}{}^1\mathcal{V}{}^1\Sigma)t^a\Big] \tag{3.6f}$$

$$B^3 = O(\frac{1}{r^3}) \tag{3.6g}$$

$$N^a = -\frac{1}{r^2}[\bar\eth\,{}^1\mathcal{V}s^a+{}^0\mathcal{V}\eth\,{}^1\Sigma t^a] \tag{3.6h}$$

$$N^1 = -\frac{1}{2}\{{}^0\mathcal{V}-\frac{1}{r}[2{}^0\mathcal{V}\mathcal{M}+\bar\eth^2\,{}^1\mathcal{V}+\eth({}^0\mathcal{V}\eth\,{}^1\Sigma+2\mathcal{A}{}^1\mathcal{V})]\} \tag{3.6i}$$

$$B^1 = \frac{1}{2r^2}{}^0\mathcal{V}(\eth^2\,{}^1\Sigma - \mathcal{M}) \tag{3.6j}$$

$$B^2 = \frac{1}{r^2}\bar{\eth}\{({}^0\mathcal{V}\mathcal{M}) + \frac{1}{2}[\bar{\eth}^2\,{}^1\mathcal{V} + \eth({}^0\mathcal{V}\eth\,{}^1\Sigma)] - {}^1\mathcal{V}\bar{\eth}\mathcal{A}\} \tag{3.6k}$$

Note that each succeeding function depends only on those previously defined except for the six r–independent integration functions $C^1{}_s = C^1{}_a s^a$, $C^1{}_t = C^1{}_a t^a$, $\mathcal{M}$, $\mathcal{A}$, ${}^0\mathcal{V}$, and ${}^1\mathcal{V}$. The first four of these functions are related to the angular momentum, mass, and radiation. The remaining two will be fixed by the reality conditions.

The integration for N^a also introduces a function independent of r. However, such a term can be removed by a coordinate transformation[1,2]. Furthermore, N^1 will grow like r at null infinity unless we require that

$$\eth\,{}^0\mathcal{V} = \bar{\eth}\,{}^0\mathcal{V} = 0.$$

The above solutions have been written with this requirement so that at most, ${}^0\mathcal{V}$ can be a function of t alone.

The propagation equations determine the evolution of the r-independent functions. Because the constraints form a closed system, no further conditions on the general solution arise. From the equation for $\dot{A}^1{}_i$ we get

$${}^0\mathcal{V}\mathcal{A} = {}^1\dot{\Sigma} \tag{3.7a}$$

and

$$\dot{C}_s = -\bar{\eth}\,{}^2B^1 + {}^2B^2 + {}^0\mathcal{V}[-\eth(\bar{\eth}\eth\,{}^1\Sigma - \frac{1}{2}{}^1\Sigma) + \bar{\eth}\mathcal{M}] \tag{3.7b}$$

$$\dot{C}_t = -\eth\,{}^2B^1 + {}^1\mathcal{V}\bar{\eth}\mathcal{A} \tag{3.7c}$$

C_s and C_t are identified with the dipole aspect and, hence, the above equations are connected with the conservation of angular momentum. The equation for $\dot{A}^2{}_i\Sigma_1{}^i$ defines μ_1 which leads to an identity with Eq. (2.19). $\mu_i\Sigma_3{}^i$ is the null part of the conformal tensor, Ψ_4 in the Newman-Penrose notation[3]. Recall that variation of α gives $\mu_i v^i = 0$, so that the equation for $\dot{A}^2{}_a V^a$ gives a condition which with the reality conditions leads to conservation of energy. Thus,

$$-\frac{\partial}{\partial t}({}^0\mathcal{V}\mathcal{M} - {}^1\mathcal{V}{}^1\Sigma) = -{}^1\dot{\mathcal{V}}{}^1\dot{\Sigma} \tag{3.8}$$

The remaining equations propagate $A^3{}_a$ and $\Sigma_3{}^a$. This completes our discussion of the complex equations.

4. Real Solutions.

Solving the complex equations expresses the connection in terms of the tetrad. Therefore, one need apply the reality conditions only to the triad and then make the further identification that the complex conjugate of the connection is the anti-self-dual part of the real connection. For the tetrad, the reality conditions are

$$N = \bar{N}, \qquad N^i = \bar{N}^i, \qquad -i\nu = i\bar{\nu} > 0,$$

$$\Sigma_1{}^i = -\bar{\Sigma}_1{}^i, \qquad v^i = \bar{\Sigma}_3{}^i. \tag{4.1}$$

$\Sigma_1{}^1$ already satisfies the reality condition and we find

$$\begin{aligned} \Sigma_3 &= -ir\sin\theta\{s^a + \frac{1}{r}{}^1\Sigma t^a + \frac{1}{r^3}{}^3\Sigma t^a\}, \\ v^a &= ir\sin\theta\{t^a + \frac{1}{r}{}^1\bar{\Sigma}s^a + \frac{1}{r^3}{}^3\bar{\Sigma}s^a\}. \end{aligned} \tag{4.2}$$

From this we find the determinent ν,

$$\nu^2 = \eta_{ijk}\Sigma_1{}^i v^j \Sigma_3{}^k = -r^4\sin^2\theta[1 - \frac{1}{r^2}{}^1\Sigma{}^1\bar{\Sigma}].$$

The requirement $-i\nu > 0$ gives us

$$\nu = ir^2\sin\theta\sqrt{1 - \frac{1}{r^2}{}^1\Sigma{}^1\bar{\Sigma}}. \tag{4.3}$$

From the boundary condition that N should be 1 at null infinity and from $\underset{\sim}{N}v^a = V^a$, we find

$$\begin{aligned} &\frac{N}{r}(1 + \frac{1}{2r^2}{}^1\Sigma{}^1\bar{\Sigma})[t^a + \frac{1}{r}{}^1\bar{\Sigma}s^a + \frac{1}{r^3}{}^3\bar{\Sigma}s^a] \\ &= -\frac{1}{r}[({}^0\mathcal{V} + \frac{1}{2r^2}{}^1\mathcal{V}{}^1\Sigma)t^a + (\frac{1}{r}{}^1\mathcal{V} + \frac{1}{2r^3}{}^0\mathcal{V}\,{}^3A)s^a, \end{aligned}$$

which implies $N = 1$ and

$${}^0\mathcal{V} = -1, \qquad {}^1\mathcal{V} = -{}^1\bar{\Sigma}, \qquad \frac{1}{2}{}^3A = {}^3\bar{\Sigma} + \frac{1}{2}{}^1\Sigma{}^1\bar{\Sigma}{}^1\bar{\Sigma}. \tag{4.4}$$

Comparison with the Bondi-Sachs form of the metric [1,2]

$$ds^2 = \frac{Ve^{2b}}{r}du^2 + 2e^{2b}dudr - r^2h_{ij}(dx^i - U^i du)(dx^j - U^j du), \tag{4.5}$$

give us

$$\nu = -ie^{2b}r^2\sin\theta, \qquad N = 1, \qquad N^1 = \frac{V}{r}, \tag{4.6}$$

and for the principal spin coefficients [3,4]

$$\begin{aligned} \rho &= -\frac{1}{\nu}A^3{}_a\Sigma_3{}^a = -\frac{1}{r} - \frac{{}^1\Sigma{}^1\bar{\Sigma}}{r^3}, \\ \sigma &= \frac{1}{\nu}A^3{}_a v^a = -\frac{{}^1\bar{\Sigma}}{r^2} + \frac{1}{r^4}({}^3A - \frac{1}{2}{}^1\Sigma{}^1\bar{\Sigma}{}^1\bar{\Sigma} - {}^3\bar{\Sigma}), \\ \mu &= \frac{1}{\nu}A^2{}_a v^a = -\frac{1}{2r} + \frac{1}{r^2}(\mathcal{M} - {}^1\dot{\Sigma}{}^1\bar{\Sigma}), \\ \bar{\lambda} &= -\frac{1}{\nu}A^2{}_a\Sigma_3{}^a = -\frac{{}^1\dot{\bar{\Sigma}}}{r} + \frac{1}{r^2}(\bar{\eth}\eth{}^1\Sigma - \frac{1}{2}{}^1\Sigma). \end{aligned} \tag{4.7}$$

This shows that the Newman-Penrose asymptotic shear $\sigma^0 = -{}^1\bar{\Sigma}$ and

$$-\mathcal{M} + {}^1\dot{\Sigma}{}^1\bar{\Sigma} = \frac{1}{2}[\psi_2^0 + \bar{\psi}_2^0 + \bar{\eth}^2\sigma^0 + \eth^2\bar{\sigma}^0 + {}^1\dot{\Sigma}{}^1\bar{\Sigma} + {}^1\dot{\bar{\Sigma}}{}^1\Sigma]. \tag{4.8}$$

Note that the right hand side of (4.8) is real and comparison with (4.7) shows that it is just the negative of the Bondi-Sachs mass aspect.

From the equation for $\mathcal{M}$ above, we obtain the real mass loss expicitly by writing the integral over a sphere at null infinity as (dS is the area element of the unit sphere)

$$\frac{dM}{dt} = -\frac{1}{16\pi}\oint {}^1\dot{\Sigma}\,{}^1\dot{\bar{\Sigma}}dS, \tag{4.9}$$

$$\begin{aligned} M &:= \frac{1}{16\pi}\oint [\mathcal{M} + \bar{\mathcal{M}} - {}^1\bar{\Sigma}{}^1\dot{\Sigma} - {}^1\Sigma{}^1\dot{\bar{\Sigma}}]dS \\ 0 &= \oint [\mathcal{M} - \bar{\mathcal{M}} - {}^1\bar{\Sigma}{}^1\dot{\Sigma} + {}^1\Sigma{}^1\dot{\bar{\Sigma}}] \end{aligned} \tag{4.10}$$

This definition of the mass aspect agrees with that of Bondi-Sachs and Newman-Penrose. Thus, (4.9) describes the mass loss from gravitational radiation.

5. Logarithmic terms.

In obtaining the solutions in the previous section, we omitted the terms in $1/r$ and $1/r^2$ in ${A^3}_2$ and in $1/r^2$ in ${\Sigma_3}^2$. This was done in order to avoid the appearance of logarithmic terms in the expansion. In this section we reintroduce these terms and examine their effect on the solutions. We note at what level in $1/r \log r$ terms are induced in ${\mathcal{A}^3}_2$ and ${\Sigma_3}^2$ by the propagation equations. Then with this behavior we go through the equations again and find that the iterated solutions have no lower powers in $1/r$ appearing with logarithms. Then we ask whether energy and and angular momentum still have a meaning. Previous work [12,13] has shown that the conformal boundary introduced by Penrose still exists, but that the fall-off of the conformal tensor changes. In agreement with that work, our explicit calculations show that energy-momentum is still well defined, but not angular momentum.

The calculations follow the previous pattern so only the important differences will be shown explicitly. We find that $\mathcal{G}_1, \mathcal{G}_2$, and $\mathcal{H}_1$ give us

$$\begin{aligned} {\Sigma_3}^a &= -ir\sin\theta\{s^a + \frac{1}{r}{}^1\Sigma t^a + \frac{1}{r^2}({}^2\Sigma t^a - \tfrac{1}{2}{}^1A{}^1\Sigma s^a\} \\ {A^3}_a &= -\{a_a + \frac{1}{r}{}^1\mathcal{A}b^a + \frac{1}{r^2}({}^2\mathcal{A}b^a - \tfrac{1}{2}{}^1A{}^1\Sigma a^a\} \end{aligned} \tag{5.1}$$

At this point one can show that the solution of Eqs.(3.2f) for V^a requires that ${}^1A = 0$. To the order considered, it is otherwise the same as the previous result. Omitting this term, the integration for the connection gives us

$$\begin{aligned} {A^1}_a = &\Big[\frac{\cot\theta}{2\sqrt{2}} - \frac{1}{r}\eth\,{}^1\Sigma + \eth\,{}^2\Sigma\frac{\ln r}{r^2} + \frac{1}{r^2}{C^1}_s\Big]a_a + \\ &\Big[-\frac{\cot\theta}{2\sqrt{2}} + \bar{\eth}\,{}^2A\frac{\ln r}{r^2} + \frac{1}{r^2}{C^1}_t\Big]b_a \end{aligned} \tag{5.2a}$$

$${A^2}_1 = \frac{1}{r^2}\Big[\eth\,{}^1\Sigma + 2\eth\,{}^2\Sigma\,\frac{\ln r}{r}\Big] \tag{5.2b}$$

$$\begin{aligned} {A^2}_a = &\Big[-{}^1\dot{\Sigma} + \frac{1}{r}\bar{\eth}\eth\,{}^1\Sigma + {}^1\Sigma\bar{\eth}\eth\,{}^2\Sigma\,\frac{\ln r}{r^2} + \frac{1}{r^2}{}^2A\Big]a_a \\ &+ \Big[-\frac{1}{2} + \frac{1}{r}\mathcal{M} - \frac{\ln r}{r^2}(\bar{\eth}^2\,{}^2A - \eth^2\,{}^2\Sigma)\Big]b_a \end{aligned} \tag{5.2c}$$

We see that the logarithms appear below the leading $1/r$ behavior. In paricular, note that the logarithms in $A^1{}_a$ come in before the terms for angular momentum whereas in $A^2{}_a$ they come in after the mass aspect This basic pattern is repeated in all the remaining hypersurface equations. From the propagation equations we find that $A^3{}_a$ develops a term in $\ln r/r^3$ and $\Sigma_3{}^a$ a term in $\ln r/r^2$ which follows the term in ${}^1\Sigma$. It is the appearance of the logarithms at this low order that guarantees that the equations can be interated consistently. The equation for $\dot{A}^2{}_a v^a$ leads to the same expression for mass and mass loss as before. However, because the log terms come in before the angular momentum terms in $A^1{}_a$, one is no longer sure that angular momentum can be defined.

Below we exhibit the components of the conformal tensor using the notation of Newman and Penrose[3]:

$$\begin{aligned}
\Psi_4 &= R^2{}_{\rho\sigma} e_0{}^\rho e_3{}^\sigma \rightarrow \frac{{}^1\ddot{\Sigma}}{r}, \\
\Psi_3 &= R^1{}_{\rho\sigma} e_3{}^\rho e_0{}^\sigma \rightarrow \frac{-\eth\, {}^1\Sigma}{r^2}, \\
\Psi_2 &= R^2{}_{\rho\sigma} e_1{}^\rho e_2{}^\sigma \rightarrow \frac{\eth^2\, {}^1\Sigma - \mathcal{M}}{r^3}, \\
\Psi_1 &= R^1{}_{\rho\sigma} e_1{}^\rho e_2{}^\sigma \rightarrow 2\bar{\eth}\, {}^2A \frac{\ln r}{r^4} + \frac{2C^1{}_t - \bar{\eth}\, {}^2A}{r^4}, \\
\Psi_0 &= R^1{}_{\rho\sigma} e_1{}^\rho e_2{}^\sigma \rightarrow -\frac{{}^2A}{r^4} - 3\alpha\frac{\ln r}{r^5} - \frac{{}^3A - \alpha}{r^5}.
\end{aligned} \tag{5.3}$$

${}^1\Sigma = -\bar{\sigma}^\circ$ the asymptotic shear of Newman-Penrose, and α comes from the lowest order logarithm in $A^3{}_a$.

Note that in the absence of logarithmic terms, there is a regular increase in the fall-off from Ψ_4 to Ψ_0. This behavior is referred to as peeling. The presence of the logarithms breaks the peeling at Ψ_2. This result is in full agreement with the earlier analysis by Nowak and Goldberg[12,13].

Based on work by Andersson and Chrusciel[14], the results described here have also been obtained by Chrusciel, MacCallum, and Singleton[15] by an analysis following the work of Bondi and Sachs.

Appendix: The Minkowski Space Limit.

The metric takes the form

$$ds^2 = dt(dt + 2dr) - r^2(d\theta^2 + \sin^2\theta\, d\phi^2)$$

and the four one-forms and tetrad vectors are

$$\begin{aligned}
\theta^0 &= dt, & e_0 &= \partial_t - \tfrac{1}{2}\partial_r, \\
\theta^1 &= (dr + \tfrac{1}{2}dt), & e_1 &= \partial_r \\
\theta^2 &= \tfrac{1}{\sqrt{2}} r(d\theta - i\sin\theta\, d\phi), & e_2 &= \tfrac{1}{\sqrt{2}}(\partial_\theta + \tfrac{i}{\sin\theta}\partial_\phi) \\
\theta^3 &= \tfrac{1}{\sqrt{2}} r(d\theta + i\sin\theta\, d\phi), & e_3 &= \tfrac{1}{\sqrt{2}}(\partial_\theta - \tfrac{i}{\sin\theta}\partial_\phi)
\end{aligned} \tag{A1}$$

The triad densities and the self-dual connection one-forms are

$$\begin{aligned}
\Sigma_1{}^i &= -ir^2\sin\theta\, \delta^i{}_1; & A^1{}_i &= \tfrac{i}{2}\cos\theta\delta^i{}_3; \\
v^a &= ir\sin\theta\, t^a; & A^2{}_a &= -\tfrac{1}{2}b_a; \\
\Sigma_3{}^a &= -ir\sin\theta\, s^a; & A^3{}_a &= -a_a;
\end{aligned} \tag{A2}$$

$$s^a = \frac{1}{\sqrt{2}}\begin{pmatrix} 1 \\ -i/\sin\theta \end{pmatrix}; \qquad t^a = \frac{1}{\sqrt{2}}\begin{pmatrix} 1 \\ i/\sin\theta \end{pmatrix}; \tag{A3a}$$

$$a_a = \frac{1}{\sqrt{2}}(1,\ i\sin\theta);$$
$$b_a = \frac{1}{\sqrt{2}}(1,\ -i\sin\theta). \tag{A3b}$$

(s^a, a_a), and (t^a, b_a) are dual eigenvectors and eigenforms of

$$^0Z = \frac{1}{2}\begin{pmatrix} 1 & -i\sin\theta \\ i/\sin\theta & 1 \end{pmatrix} \tag{A4}$$

where ^0Z is the Minkowski space limit of

$$Z^a{}_b := \delta^a{}_b - \frac{\Sigma_3{}^a A^3{}_b}{\Sigma_3{}^c A^3{}_c}.$$

Thus

$$^0Z^a{}_b s^b = a_a {}^0Z^a{}_b = 0$$
$$^0Z^a{}_b t^b = t^a, \quad b_a {}^0Z^a{}_b = b_b \tag{A5}$$

We assign spin weight[16–18] +1 to (t^a, a_a) and $-1 to (s^a, b_a)$. The action of the spin weight raising and lowering operators on a function of spin weight s is

$$\eth\eta := -\frac{1}{\sqrt{2}}\sin^s\theta\Big(\frac{\partial}{\partial\theta} + \frac{i}{\sin\theta}\frac{\partial}{\partial\phi}\Big)(\sin^{-s}\theta\,\eta) \tag{A1.4a}$$

and

$$\bar{\eth}\eta := -\frac{1}{\sqrt{2}}\sin^{-s}\theta\Big(\frac{\partial}{\partial\theta} + \frac{i}{\sin\theta}\frac{\partial}{\partial\phi}\Big)(\sin^s\theta\,\eta). \tag{A6}$$

The action of these operators on the spin-weighted spherical harmonics can be found in Appendix A of reference 16 and further details can be found in references 17 and 18.

References.

1. H. Bondi, M. Van Der Burg, and A. Metzner, *Proc. Roy. Soc.* **A269**(1962) 21.
2. R. Sachs, *Proc. Roy. Soc.* **A270**(1962) 103.
3. E.T. Newman and R. Penrose, *J. Math. Phys.*3(1962) 566.
4. E.T. Newman and T. Unti, *J. Math. Phys.*3(1962) 891.
5. J.N. Goldberg, D.C. Robinson, and S. Soteriou, *Class. Quantum Gravity*9 (1992) 1309.
6. J.N. Goldberg and C. Soteriou, *Class. Quantum Gravity* to appear.
7. R. Sachs, *Proc. Roy. Soc.* **A264**1961 309.
8. P.A.M. Dirac, *Phys. Rev.*114 (1959) 924.
9. P.G. Bergmann and A. Komar, Phys. Rev. Lett.4 (1960) 432.
10. J. Samuel, *Pramana J. Phys.*28 (1987) L429.
11. T. Jacobson and L. Smolin, *Phys. Lett.*196B (1987) 39.
12. S. Novak and J.N. Goldberg, *Gen. Rel. Grav.*13(1981) 79.
13. S. Novak and J.N. Goldberg, *Gen. Rel. Grav.*14(1982) 655.
14. L. Andersson and P. Chrusciel, *Commun. Math. Phys.*161(1994) 533.

15. P.T. Chrusciel, M.A.H. MacCallum, and D.B. Singleton, *Phil. Trans. R. Soc. London A***350** (1995) 113.
16. E.N. Glass and J.N. Goldberg, *J. Math. Phys.***11**(1970) 3400.
17. E.T. Newman and R. Penrose, *J. Math. Phys.***7**(1966) 863.
18. R. Penrose and W. Rindler *Spinors and Space-time, vol 1* (Cambridge University Press, Cambridge, 1984).

On the Gauge Aspects of Gravity†

Frank Gronwald

and

Friedrich W. Hehl

Institute for Theoretical Physics, University of Cologne
D-50923 Köln, Germany
E-mail: fg@thp.uni-koeln.de, hehl@thp.uni-koeln.de

ABSTRACT

We give a short outline, in Sec. 2, of the historical development of the gauge idea as applied to internal ($U(1)$, $SU(2)$, ...) and external (R^4, $SO(1,3)$, ...) symmetries and stress the fundamental importance of the corresponding conserved currents. In Sec. 3, experimental results with neutron interferometers in the gravitational field of the earth, as interpreted by means of the equivalence principle, can be predicted by means of the Dirac equation in an accelerated and rotating reference frame. Using the Dirac equation in such a non-inertial frame, we describe how in a gauge-theoretical approach (see Table 1) the Einstein-Cartan theory, residing in a Riemann-Cartan spacetime encompassing torsion and curvature, arises as the simplest gravitational theory. This is set in contrast to the Einsteinian approach yielding general relativity in a Riemannian spacetime. In Secs. 4 and 5 we consider the conserved energy-momentum current of matter and gauge the associated translation subgroup. The Einsteinian teleparallelism theory which emerges is shown to be equivalent, for spinless matter and for electromagnetism, to general relativity. Having successfully gauged the translations, it is straightforward to gauge the four-dimensional affine group $R^4 \rtimes GL(4,R)$ or its Poincaré subgroup $R^4 \rtimes SO(1,3)$. We briefly report on these results in Sec. 6 (metric-affine geometry) and in Sec. 7 (metric-affine field equations (111, 112, 113)). Finally, in Sec. 8, we collect some models, currently under discussion, which bring life into the metric-affine gauge framework developed.

† Proc. of the 14th Course of the School of Cosmology and Gravitation on *Quantum Gravity*, held at Erice, Italy, May 1995, P.G. Bergmann, V. de Sabbata, and H.-J. Treder, eds. (World Scientific, Singapore 1996) to be published.

Contents

From a letter of A. Einstein to F. Klein of 1917 March 4 (translation)[70]:

"...Newton's theory ... represents the gravitational field in a seemingly complete way by means of the potential Φ. This description proves to be wanting; the functions $g_{\mu\nu}$ take its place. But I do not doubt that the day will come when that description, too, will have to yield to another one, for reasons which at present we do not yet surmise. I believe that this process of deepening the theory has no limits ..."

1. Introduction

• What can we learn if we look at gravity and, more specifically, at general relativity theory (GR) from the point of view of classical gauge field theory? This is the question underlying our present considerations. The answer

• leads to a better understanding of the interrelationship between the *metric* and *affine* properties of spacetime and of the *group* structure related to gravity. Furthermore, it

• suggests certain classical field-theoretical generalizations of Einstein's theory, such as Einstein–Cartan theory, Einsteinian teleparallelism theory, Poincaré gauge theory, Metric-Affine Gravity, that is, it leads to a deepening of the insight won by GR.

We recently published a fairly technical review article on our results[29]. These lectures can be regarded as a down-to-earth introduction into that subject. We refrain from citing too many articles since we gave an overview[a] of the existing literature in ref.([29]).

2. Remarks on the history of the gauge idea

2.1. General relativity and Weyl's $U(1)$-gauge theory

Soon after Einstein in 1915/16 had proposed his gravitational theory, namely general relativity (GR), Weyl extended it in 1918 in order to include – besides gravitation – electromagnetism in a unified way. Weyl's theoretical concept was that of recalibration or *gauge invariance* of length. In Weyl's opinion, the integrability of length in GR is a remnant of an era dominated by action-at-a-distance theories which should be abandoned. In other words, if in GR we displace a meter stick from one point of spacetime to another one, it keeps its length, i.e., it can be used as a standard

[a]In the meantime we became aware that in our Physics Reports we should have cited additionally the work of Mistura[58] on the physical interpretation of torsion, of Pascual-Sánchez[71,72,73] on teleparallelism, inter alia, of Perlick[76] on observers in Weyl spacetimes, and that of Ponomariov and Obukhov[78] on metric-affine spacetimes and gravitational theories with quadratic Lagrangians.

of length throughout spacetime; an analogous argument is valid for a clock. In contrast, Weyl's unified theory of gravitation and electromagnetism of 1918 is set up in such a way that the unified Lagrangian is invariant under recalibration or re-gauging.

For that purpose, Weyl extended the geometry of spacetime from the (pseudo-) Riemannian geometry with its Levi-Civita connection $\Gamma^{\{\}}_{\alpha\beta}$ to a Weyl space with an additional (Weyl) covector field $Q = Q_\alpha \vartheta^\alpha$, where ϑ^α denotes the field of coframes of the underlying four-dimensional differentiable manifold. The Weyl connection one-form reads

$$\Gamma^{\mathrm{W}}_{\alpha\beta} = \Gamma^{\{\}}_{\alpha\beta} + \frac{1}{2}(g_{\alpha\beta}\, Q - \vartheta_\alpha\, Q_\beta + \vartheta_\beta\, Q_\alpha)\,. \tag{1}$$

The additional freedom of having a new one-form (or covector) field Q at one's disposal was used by Weyl in order to accommodate Maxwell's field. He identified the electromagnetic potential A with Q.

Weyl's theory turned out to be non-viable, at least in the sense and on the level of ordinary length and time measurement. However, his concept of gauge invariance survived in the following way: When quantum mechanics was developed, it became clear that (in the Schrödinger representation) the wave function Ψ of an electron, for example, is only determined up to an arbitrary phase ϕ:

$$\Psi \longrightarrow e^{i\phi}\Psi\,, \qquad \phi = const\,. \tag{2}$$

The set of all phase transformations builds up the one-dimensional Abelian Lie group $U(1)$ of unitary transformations. If one substitutes, according to (2), the wave function in the Dirac equation by a phase transformed wave function, no observables will change; they are invariant under 'rigid', i.e., constant phase transformations. This is an elementary fact of quantum mechanics.

In 1929, Weyl revitalized his gauge idea: Isn't it against the spirit of field theory to implement a rigid phase transformation (2) 'at once' all over spacetime, he asked. Shouldn't we postulate a $U(1)$ invariance under a spacetime dependent change of the phase instead:

$$\Psi \longrightarrow e^{i\phi(x)}\Psi\,, \qquad \phi = \phi(x)\,? \tag{3}$$

If one does it, the original invariance of the observables is lost under the new 'soft' transformations. In order to kill the invariance violating terms, one has to introduce a compensating potential one-form A with values in the Lie algebra of $U(1)$, which transforms under the soft transformations in a suitable form. This couples A in a well determined way to the wave function of the electron, and, if one insists that the $U(1)$-potential A has its own physical degrees of freedom, then the field strength $F := dA$ is non-vanishing and the coupled Dirac Lagrangian has to be amended with a kinetic term quadratic in F. In this way one can reconstruct the whole classical Dirac-Maxwell theory from the naked Dirac equation together with the postulate of soft phase invariance. Because of Weyl's original terminology, one still talks about $U(1)$-gauge invariance – and not about $U(1)$-phase invariance, what it really is. Thus

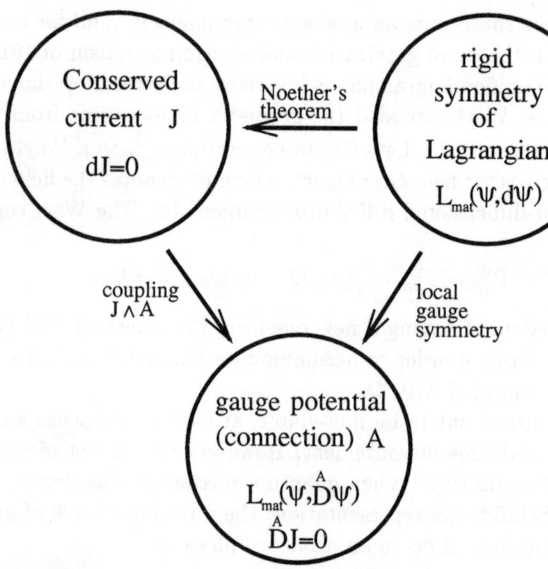

Fig. 1. The structure of a gauge theory à la Yang-Mills is depicted in this diagram, which is adapted from Mills[53]. Let us quote some of his statements on gauge theories: 'The gauge principle, which might also be described as a principle of *local symmetry*, is a statement about the invariance properties of physical laws. It requires that *every continuous symmetry be a local symmetry...*' 'The idea at the core of gauge theory...is the local symmetry principle: *Every continuous symmetry of nature is a local symmetry.*' The history of gauge theory has been traced back to its beginnings by O'Raifeartaigh[69], who also gave a compact review of its formalism[68].

the electromagnetic potential is an appendage to the Dirac field and not related to length recalibration as Weyl originally thought.

2.2. Yang-Mills and the structure of a gauge theory

Yang and Mills, in 1954, generalized the Abelian $U(1)$-gauge invariance to non-Abelian $SU(2)$-gauge invariance, taking the (approximately) *conserved* isotopic spin *current* as their starting point, and, in 1956, Utiyama set up a formalism for the gauging of any semi-simple Lie group, including the Lorentz group $SO(1,3)$. The latter group he considered as essential in GR. We will come back to this topic below.

In any case, the gauge principle historically originated from GR as a concept for removing as many action-at-a-distance concept as possible – as long as the group under consideration is linked to a conserved current. This existence of a conserved current of some matter field Ψ is absolutely vital for the setting-up of a gauge theory. In Fig.1 we sketched the structure underlying a gauge theory: A rigid symmetry of a Lagrangian induces, via Noether's theorem, a conserved current J, $dJ = 0$. It can happen, however, as it did in the electromagnetic and the $SU(2)$-case, that a conserved current is discovered first and then the symmetry deduced by a kind of a reciprocal Noether theorem (which is not strictly valid). Generalizing from the gauge approach to the Dirac-Maxwell theory, we continue with the following gauge procedure:

Extending the rigid symmetry to a soft symmetry amounts to turn the constant group parameters ε of the symmetry transformation on the fields Ψ to functions of spacetime, $\varepsilon \to \varepsilon(x)$. This affects the transformation behavior of the matter Lagrangian which usually contains derivatives $d\Psi$ of the field Ψ: The soft symmetry transformations on $d\Psi$ generate terms containing derivatives $d\varepsilon(x)$ of the spacetime-dependent group parameters which spoil the former rigid invariance. In order to counterbalance these terms, one is forced to introduce a compensating field $A = A_i{}^a \tau_a dx^i$ (a=Lie-algebra index, τ_a=generators of the symmetry group) – nowadays called *gauge potential* – into the theory. The one-form A turns out to have the mathematical meaning of a Lie-algebra valued *connection*. It acts on the components of the fields Ψ with respect to some reference frame, indicating that it can be properly represented as the connection of a frame bundle which is associated to the symmetry group. Thereby it is possible to replace in the matter Lagrangian the exterior derivative of the matter field by a gauge-covariant exterior derivative,

$$d \longrightarrow \overset{A}{D} := d + A\,, \quad L_{\text{mat}}(\Psi, d\Psi) \longrightarrow L_{\text{mat}}(\Psi, \overset{A}{D}\Psi)\,. \tag{4}$$

This is called *minimal coupling* of the matter field to the new gauge interaction. The connection A is made to a true dynamical variable by adding a corresponding kinematic term V to the minimally coupled matter Lagrangian. This supplementary term has to be gauge invariant such that the gauge invariance of the action is kept. Gauge invariance of V is obtained by constructing it from the field strength $F = \overset{A}{D}A$, $V = V(F)$. To derive field equations of order no more than two, one has to restrict V to be no more than quadratic in F. This leads to the general ansatz $V = F \wedge H$, where the field momentum or "excitation" H is implicitly defined by $H = -\partial V / \partial F$. The explicit form of H has to be specified from case to case.

By construction, the gauge potential in the Lagrangians couples to the conserved current one started with – and the original conservation law, in case of a non-Abelian symmetry, gets modified and is only gauge covariantly conserved,

$$dJ = 0 \quad \longrightarrow \quad \overset{A}{D}J = 0\,, \quad J = \partial L_{\text{mat}} / \partial A\,. \tag{5}$$

The physical reason for this modification is that the gauge potential itself contributes a piece to the current, that is, the gauge field (in the non-Abelian case) is charged. For instance, the Yang-Mills gauge potential B^a carries isotopic spin, since the $SU(2)$-

group is non-Abelian, whereas the electromagnetic potential, being $U(1)$-valued and Abelian, is electrically uncharged.

2.3. Gravity and the Utiyama-Sciama-Kibble approach

Let us come back to Utiyama (1956). He gauged the Lorentz group $SO(1,3)$, inter alia. Using some ad hoc assumptions, like the postulate of the symmetry of the connection, he was able to recover GR. This procedure is not completely satisfactory, as is also obvious from the fact that the conserved current, linked to the Lorentz group, is the angular momentum current. And this current alone cannot represent the source of gravity. Accordingly, it was soon pointed out by Sciama and Kibble (1961) that it is really the *Poincaré group* $R^4 \rtimes SO(1,3)$, the semi-direct product of the translation and the Lorentz group, which underlies gravity. They found a slight generalization of GR, the so-called Einstein-Cartan theory (EC), which relates – in a Einsteinian manner – the mass-energy of matter to the curvature and – in a novel way – the material *spin* to the torsion of spacetime. In contrast to the Weyl connection (1), the spacetime in EC is still metric compatible, i.e. governed by a Riemann-Cartan[b] (RC) geometry. Torsion is admitted according to

$$\Gamma^{\rm RC}_{\alpha\beta} = \Gamma^{\{\}}_{\alpha\beta} - \frac{1}{2}\left[e_\alpha \rfloor T_\beta - e_\beta \rfloor T_\alpha - (e_\alpha \rfloor e_\beta \rfloor T_\gamma)\vartheta^\gamma\right] . \tag{6}$$

Incidentally, approaches to the gauging of the Poincaré group on a *fixed* Minkowski background yield effectively similar structures, see Wiesendanger[96].

In order to fix the notation, let us shortly recapitulate the structures emerging in spacetime geometry. A four-dimensional differential manifold has at each point a tangent space, spanned by the four basis vectors $e_\alpha = e^i{}_\alpha \partial_i$, with ∂_i as the vectors tangent to the coordinate lines and $\alpha, \beta = 0, 1, 2, 3$, and the cotangent space, spanned by the four one-forms $\vartheta^\beta = e_j{}^\beta dx^j$, with dx^j as coordinate one-forms. The two types of bases are dual to each other $e_\alpha \rfloor \vartheta^\beta = \delta^\beta_\alpha$, where $\rfloor$ denotes the interior product, i.e., in components, $e^k{}_\alpha \, e_k{}^\beta = \delta^\beta_\alpha$. This is the underlying manifold.

On top of it we specify a linear connection. For Einstein[14] "... the essential achievement of general relativity, namely to overcome 'rigid' space (ie the inertial frame), is *only indirectly* connected with the introduction of a Riemannian metric. The directly relevant conceptual element is the 'displacement field' (Γ^l_{ik}), which expresses the infinitesimal displacement of vectors. It is this which replaces the parallelism of spatially arbitrarily separated vectors fixed by the inertial frame (ie the equality of corresponding components) by an infinitesimal operation. This makes it possible to construct tensors by differentiation and hence to dispense with the introduction of 'rigid' space (the inertial frame). In the face of this, it seems to be of

[b] The terminology is not quite uniform. Borzeskowski and Treder[9], in their critical evaluation of different gravitational variational principles, call such a geometry a Weyl-Cartan gemetry.

secondary importance in some sense that some particular Γ field can be deduced from a Riemannian metric..." In this vein, we introduce a linear connection

$$\Gamma_\alpha{}^\beta = \Gamma_{i\alpha}{}^\beta dx^i \,, \tag{7}$$

with values in the Lie-algebra of the linear group $GL(4, R)$. These 64 components $\Gamma_{i\alpha}{}^\beta(x)$ of the 'displacement' field enable us, as pointed out in the quotation by Einstein, to get rid of the rigid spacetime structure of special relativity (SR).

In order to be able to recover SR in some limit, the primary structure of a connection of spacetime has to be enriched by the secondary structure of a metric

$$g = g_{\alpha\beta}\, \vartheta^\alpha \otimes \vartheta^\beta \,, \tag{8}$$

with its 10 component fields $g_{\alpha\beta}(x)$. At least at the present stage of our knowledge, this additional postulate of the existence of a metric seems to lead to the only practicable way to set up a theory of gravity. In some future time one may be able to 'deduce' the metric from the connection and some extremal property of the action function – and some people have tried to develop such type of models, but without success so far.

2.4. E. Cartan's analysis of general relativity and its consequences

Besides the gauge theoretical line of development which, with respect to gravity, culminated in the Sciame-Kibble approach, there was a second line dominated by E. Cartan's (1923) geometrical analysis of GR. The concept of a linear connection as an independent and primary structure of spacetime, see (7), developed gradually around 1920 from the work of Hessenberg, Levi-Civita, Weyl, Schouten, Eddington, and others. In its full generality it can be found in Cartan's work. In particular, he introduced the notion of a so-called torsion – in holonomic coordinates this is the antisymmetric and therefore tensorial part of the components of the connection – and discussed Weyl's unified field theory from a geometrical point of view.

For this purpose, let us tentatively call

$$\left(g_{\alpha\beta},\, \vartheta^\alpha,\, \Gamma_\alpha{}^\beta\right) \tag{9}$$

the *potentials* in a gauge approach to gravity and

$$\left(Q_{\alpha\beta},\, T^\alpha,\, R_\alpha{}^\beta\right) \tag{10}$$

the corresponding *field strengths*. Later, in Sec. 6, inter alia, we will see why this choice of language is appropriate. Here we defined

$$\text{nonmetricity} \quad Q_{\alpha\beta} \;:=\; -\overset{\Gamma}{D}\, g_{\alpha\beta} \,, \tag{11}$$

$$\text{torsion} \quad T^\alpha \;:=\; \overset{\Gamma}{D}\, \vartheta^\alpha = d\vartheta^\alpha + \Gamma_\beta{}^\alpha \wedge \vartheta^\beta \,, \tag{12}$$

$$\text{curvature} \quad R_\alpha{}^\beta \;:=\; {}''\,\overset{\Gamma}{D}\, \Gamma_\alpha{}^\beta\,'' = d\Gamma_\alpha{}^\beta - \Gamma_\alpha{}^\gamma \wedge \Gamma_\gamma{}^\beta \,. \tag{13}$$

Then symbolically we have

$$\left(Q_{\alpha\beta},\, T^{\alpha},\, R_{\alpha}{}^{\beta}\right) \sim \overset{\Gamma}{D}\left(g_{\alpha\beta},\, \vartheta^{\alpha},\, \Gamma_{\alpha}{}^{\beta}\right). \tag{14}$$

By means of the field strengths it is straightforward of how to classify the spacetime manifolds of the different theories discussed so far:

$$\text{GR (1915)}: \quad Q_{\alpha\beta} = 0, \quad T^{\alpha} = 0, \quad R_{\alpha}{}^{\beta} \neq 0. \tag{15}$$
$$\text{Weyl (1918)}: \quad Q_{\gamma}{}^{\gamma} \neq 0, \quad T^{\alpha} = 0, \quad R_{\alpha}{}^{\beta} \neq 0. \tag{16}$$
$$\text{EC (1923/61)}: \quad Q_{\alpha\beta} = 0, \quad T^{\alpha} \neq 0, \quad R_{\alpha}{}^{\beta} \neq 0. \tag{17}$$

Note that Weyl's theory of 1918 requires only a nonvanishing *trace* of the nonmetricity, the *Weyl covector* $Q := Q_{\gamma}{}^{\gamma}/4$. For later use we amend this table with the Einsteinian teleparallelism ($\text{GR}_{||}$), which was discussed between Einstein and Cartan in considerable detail (see Debever[12]) and with metric-affine gravity[29] (MAG), which presupposes the existence of a connection *and* a (symmetric) metric that are completely independent from each other (as long as the field equations are not solved):

$$\text{GR}_{||}\ \text{(1928)}: \quad Q_{\alpha\beta} = 0, \quad T^{\alpha} \neq 0, \quad R_{\alpha}{}^{\beta} = 0. \tag{18}$$
$$\text{MAG (1976)}: \quad Q_{\alpha\beta} \neq 0, \quad T^{\alpha} \neq 0, \quad R_{\alpha}{}^{\beta} \neq 0. \tag{19}$$

Both theories, $\text{GR}_{||}$ and MAG, were originally devised as unified field theories with no sources on the right hand sides of their field equations. Today, however, we understand them[10,29] as gauge type theories with well-defined sources.

Cartan gave a beautiful geometrical interpretation of the notions of torsion and curvature. Consider a vector at some point of a manifold, that is equipped with a connection, and displace it around an infinitesimal (closed) loop by means of the connection such that the (flat) tangent space, where the vector 'lives' in, rolls without gliding around the loop. At the end of the journey[29] the loop, mapped into the tangent space, has a small closure failure, i.e. a translational misfit. Moreover, in the case of vanishing nonmetricity $Q_{\alpha\beta} = 0$, the vector underwent a small rotation or – if no metric exists – a small linear transformation. The torsion of the underlying manifold is a measure for the emerging translation and the curvature for the rotation (or linear transformation):

$$\text{translation} \longrightarrow \text{torsion } T^{\alpha} \tag{20}$$
$$\text{rotation (lin. transf.)} \longrightarrow \text{curvature } R_{\alpha}{}^{\beta}. \tag{21}$$

Hence, if your friend tells you that he discovered that torsion is closely related to electromagnetism or to some other nongravitational field – and there are many such 'friends' around, as we can tell you as referees – then you say: 'No, torsion is related to translations, as had been already found by Cartan in 1923.' And translations – we

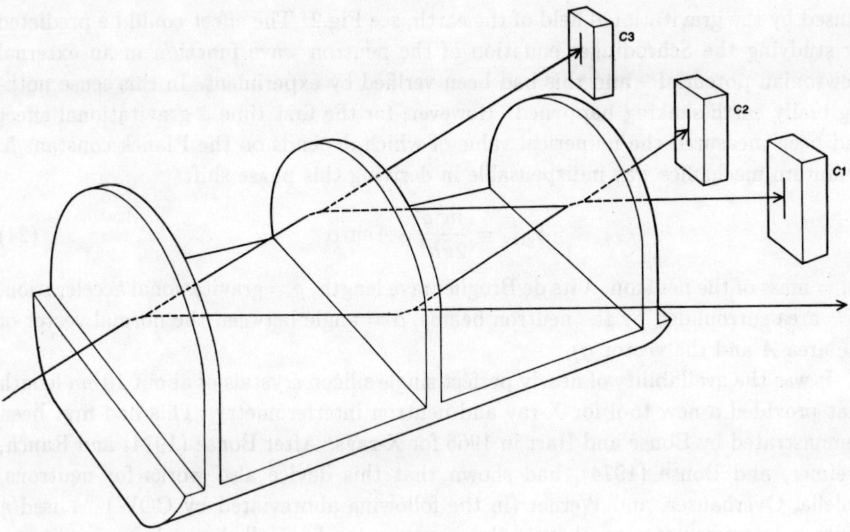

Fig. 2. The neutron interferometer of the COW-experiment[11,18]: A neutron beam is split into two beams which travel in different gravitational potentials. Eventually the two beams are reunited and their relative phase shift is measured.

hope that we don't tell you a secret – are, via Noether's theorem, related to energy-momentum,[c] i.e. to the source of gravity, and to nothing else. We will come back to this discussion in Sec.4.

For the rest of these lectures, unless stated otherwise, we will choose the frame e_α, and hence also the coframe ϑ^β, to be *orthonormal*, that is,

$$g(e_\alpha, e_\beta) \stackrel{*}{=} o_{\alpha\beta} := \mathrm{diag}(-+++)\,. \tag{22}$$

Then, in a Riemann-Cartan space, we have the convenient antisymmetries

$$\Gamma^{RC}_{\alpha\beta} \stackrel{*}{=} -\Gamma^{RC}_{\beta\alpha} \qquad \text{and} \qquad R^{RC}_{\alpha\beta} \stackrel{*}{=} -R^{RC}_{\beta\alpha}. \tag{23}$$

3. Einstein's and the gauge approach to gravity

3.1. Neutron matter waves in the gravitational field

Twenty years ago a new epoch began in gravity: Colella-Overhauser-Werner measured by interferometric methods a phase shift of the wave function of a neutron

[c]Not long ago, one of the editors of Physical Review D, Lowell S. Brown, decreed that all papers with (Cartan's) torsion in the title or the abstract are automatically to be rejected. You may guess how the authors of such papers reacted. I was always wondering what this clever physicist would do, if 'energy-momentum' appeared in a title. My guess being that he would not even recognize the need for becoming active again.

caused by the gravitational field of the earth, see Fig.2. The effect could be predicted by studying the Schrödinger equation of the neutron wave function in an external Newtonian potential – and this had been verified by experiment. In this sense nothing really earth-shaking happened. However, for the first time a gravitational effect had been measured the numerical value of which depends on the Planck constant $\hbar$. Quantum mechanics was indispensable in deriving this phase shift

$$\theta_{\text{grav}} = \frac{m^2 g}{2\pi\hbar^2}\,\lambda\,A\sin\alpha \tag{24}$$

(m = mass of the neutron, λ its de Broglie wave length, g = gravitational acceleration, A = area surrounded by the neutron beams, α = angle between the normal vector of the area A and the vector $\boldsymbol{g}$).

It was the availability of nearly perfect single silicon crystals of about 10 cm length that provided a new tool for X-ray and neutron interferometry. This had first been demonstrated by Bonse and Hart in 1965 for X-rays. After Bonse (1974) and Rauch, Treimer, and Bonse (1974), had shown that this device also works for neutrons, Colella, Overhauser, and Werner (in the following abbreviated by COW) "...used a neutron interferometer to observe the quantum-mechanical phase shift of neutrons caused by their interaction with the Earth's gravitational field"[11]. Their experiment is sketched in Fig.3.

They used neutrons cooled to room temperature such that their resulting mean velocity $v_n \simeq 10^{-5}\,c$ is non-relativistic. Their mass is $m_n = 1.67 \times 10^{-21}\,kg$, and the de Broglie wave length $\lambda_n := 2\pi\hbar/p \approx 0.2\,nm$. A beam of $1\,cm$ width enters the first 'ear' of the interferometer at a Bragg angle in the range of 20° to 30°. It is *coherently* scattered by planes of atoms perpendicular to the surface of the crystal. This Laue scattering gives rise to a transmitted and a diffracted beam, with opposite Bragg angles. Due to the Borrman effect, the beam travels through the crystal at first along the planes and the splitting occurs actually only after it emerges from the ear again.

When the interferometer gets rotated in the gravitational field of the earth, the upper and lower beams travel at a vertical distance of about $2\,cm$ and encounter a potential difference of $\Delta\varphi/E_{\text{kin}} = (m_n g \Delta h/(1/2)m_n v^2) \approx 10^3/(3\times10^5)^2 \approx 10^{-8}$, which is only a tiny fraction of the kinetic energy. Nevertheless, this leads to a measurable effect on the *phase* of the neutron's coherent wave which oscillates about $10\,cm/\lambda_n \approx 10^9$ times during the horizontal flight. Although the oscillation rate of the upper beam is 'redshifted' merely by a factor of 10^{-7}, the upper beam manages to make $\theta_{\text{grav}} \approx 10^9/10^7 = 100$ oscillations more than the lower beam. This phase shift can be observed by the interference pattern of the recombined beams.

In the actual experiment, side effects have to be taken care of: Gravity produces distortions in the single crystal. Contributions from this can be eliminated by comparing X-ray and neutron interference patterns in the same interferometer. Moreover, the neutron beam itself is bent into a parabolic path with $4\times10^{-7}cm$ loss in altitude. This yields, however, no significant influence on the phase.

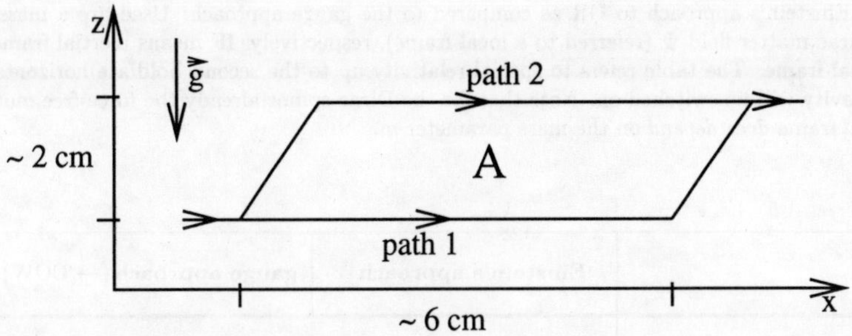

Fig. 3. COW experiment schematically.

In the COW experiment, the single-crystal interferometer is at rest with respect to the laboratory, whereas the neutrons are subject to the gravitational potential. In order to compare this with the effect of acceleration relative to the laboratory frame, Bonse and Wroblewski[8] let the interferometer oscillate horizontally by driving it via a pair of standard loudspeaker magnets. Thus these experiments of BW and COW test the effect of local acceleration and local gravity on matter waves and prove its *equivalence* up to an accuracy of about 4%.

3.2. Accelerated and rotating reference frame

In order to be able to describe the interferometer in an accelerated frame, we first have to construct a non-inertial frame of reference. If we consider only *mass points*, then a non-inertial frame in the Minkowski space of SR is represented by a *curvilinear* coordinate system, as recognized by Einstein[13]. Einstein even uses the names 'curvilinear co-ordinate system' and 'non-inertial system' interchangeably.

According to the standard gauge model of electro-weak and strong interactions, a neutron is not a fundamental particle, but consists of *one* up and *two* down quarks which are kept together via the virtual exchange of gluons, the vector bosons of quantum chromodynamics, in a permanent 'confinement phase'. For studying the properties of the neutron in a non-inertial frame and in low-energy gravity, we may disregard its extension of about 0.7 fm, its form factors, etc. In fact, for our purpose, it is sufficient to treat it as a Dirac particle which carries spin 1/2 but is *structureless otherwise.*

A Dirac particle has to be described by means of a four-component Dirac *spinor.* And this spinor is a half-integer representation of the (covering group $SL(2, C)$ of

Table 1. Einstein's approach to GR as compared to the gauge approach: Used are a mass point m or a Dirac matter field Ψ (referred to a local frame), respectively. IF means inertial frame, NIF non-inertial frame. The table refers to special relativity up to the second boldface horizontal line. Below, gravity will be switched on. Note that for the Dirac spinor already the force-free motion in an *inertial* frame *does depend* on the mass parameter m.

	Einstein's approach	**gauge approach** ($\rightarrow$ COW)
elementary object in SR	mass point m	Dirac spinor $\Psi(x)$
inertial frame	Cartesian coord. system x^i $ds^2 \stackrel{*}{=} o_{ij}\, dx^i dx^j$	holonomic orthon. frame $e_\alpha = \delta^i_\alpha\, \partial_i, \quad e_\alpha \cdot e_\beta = o_{\alpha\beta}$
force-free motion in IF	$\dot{u}^i \stackrel{*}{=} 0$	$(i\gamma^i \partial_i - m)\Psi \stackrel{*}{=} 0$
non-inertial frame	arbitrary curvilinear coord. system $x^{i'}$	anholonomic orthon. frame $e_\alpha = e^i{}_\alpha\, \partial_i$ coframe $\vartheta^\alpha = e_i{}^\alpha dx^i$
force-free motion in NIF	$\dot{u}^{i'} + u^{j'} u^{k'} \left\{ {}^{\;i'}_{j'k'} \right\} = 0$	$\left[i\gamma^\alpha e^i{}_\alpha (\partial_i + \Gamma_i) - m \right] \Psi = 0$ $\Gamma_i := \frac{1}{2}\, \Gamma_i{}^{\beta\gamma} \rho_{\beta\gamma}$ Lorentz
non-inertial objects	$\left\{ {}^{\;i'}_{j'k'} \right\}$ 40	$\vartheta^\alpha, \quad \Gamma^{\alpha\beta} = -\Gamma^{\beta\alpha}$ 16 + 24
constraints in SR	$\tilde{R}(\partial\{\}, \{\}) = 0$ 20	$T(\partial e, e, \Gamma) = 0, \; R(\partial\Gamma, \Gamma) = 0$ 24 + 36
global IF	$g_{ij} \stackrel{*}{=} o_{ij}\, , \; \left\{ {}^{\,i}_{jk} \right\} \stackrel{*}{=} 0$	$(e_i{}^\alpha,\, \Gamma_i{}^{\alpha\beta}) \stackrel{*}{=} (\delta_i^\alpha, 0)$
switch on gravity	$\tilde{R} \neq 0$ *Riemann*	$T \neq 0, \quad R \neq 0$ *Riemann − Cartan*
local IF ('Einstein elevator')	$g_{ij}\vert_P \stackrel{*}{=} o_{ij}\, , \; \left\{ {}^{\,i}_{jk} \right\}\vert_P \stackrel{*}{=} 0$	$(e_i{}^\alpha,\, \Gamma_i^{\alpha\beta})\vert_P \stackrel{*}{=} (\delta_i^\alpha, 0)$
field equations	$\tilde{Ric} - \frac{1}{2} tr(\tilde{Ric}) \sim mass$ **GR**	$Ric - \frac{1}{2} tr(Ric) \sim mass$ $Tor + 2\; tr(Tor) \sim spin$ **EC**

the) Lorentz group $SO(1,3)$. Therefore at any one point of spacetime we need an *orthonormal reference frame* in order to be able to describe the spinor. Thus, as soon as matter fields are to be represented in spacetime, the notion of a reference system has to be generalized from Einstein's curvilinear coordinate frame ∂_i to an arbitrary, in general anholonomic, orthonormal frame e_α, with $e_\alpha \cdot e_\beta = o_{\alpha\beta}$.

It is possible, of course, to introduce in the Riemannian spacetime of GR arbitrary orthonormal frames, too. However, in the heuristic process of setting up the *fundamental structure* of GR, Einstein and his followers (for a recent example, see the excellent text of d'Inverno[36], Secs.9 and 10) restricted themselves to the discussion of mass points and holonomic (natural) frames. Matter waves and arbitrary frames are taboo in this discussion. In Table 1, in the middle column, we displayed the Einsteinian method. Conventionally, *after* the Riemannian spacetime has been found and the dust settled, then electrons and neutron and what not, and their corresponding wave equations, are allowed to enter the scene. But before, they are ignored. This goes so far that the well-documented experiments of COW (1975) and BL (1983) – in contrast to the folkloric Galileo experiments from the leaning tower – seemingly are not even mentioned in d'Inverno[36] (1992).

Prugovečki[79], one of the lecturers here in Erice at our school, in his discussion of the classical equivalence principle, recognizes the decisive importance of orthonormal frames (see his page 52). However, in the end, within his 'quantum general relativity' framework, the good old Levi-Civita connection is singled out again (see his page 125). This is perhaps not surprising, since he considers only zero spin states in this context.

We hope that you are convinced by now that we should introduce arbitrary orthonormal frames in SR in order to represent non-inertial reference systems for matter waves – and that this is important for the setting up of a gravitational gauge theory[2,42]. The introduction of accelerated observers and thus of non-inertial frames is somewhat standard, even if during the Erice school one of the lecturers argued that those frames are inadmissible. Take the text of Misner, Thorne, and Wheeler[57]. In their Sec.6, you will find an appropriate discussion. Together with Ni[30] and in our Honnef lectures[27] we tailored it for our needs.

Suppose in SR a non-inertial observer locally measures, by means of the instruments available to him, a three-acceleration $\boldsymbol{a}$ and a three-angular velocity $\boldsymbol{\omega}$. If the *laboratory coordinates* of the observer are denoted by $x^{\bar{i}}$, with $\overline{\boldsymbol{x}}$ as the corresponding three-radius vector, then the non-inertial frame can be written in the succinct form[30,27]

$$\begin{aligned} e_{\hat{0}} &= \frac{1}{1+\boldsymbol{a}\cdot\overline{\boldsymbol{x}}/c^2}\left[\partial_{\overline{0}} - \left(\frac{\boldsymbol{\omega}}{c}\times\overline{\boldsymbol{x}}\right)^{\overline{B}}\partial_{\overline{B}}\right], \\ e_A &= \partial_{\overline{A}}\,. \end{aligned} \tag{25}$$

Here 'naked' capital Latin letters, $A,\ldots = \hat{1},\hat{2},\hat{3}$, denote spatial anholonomic components. For completeness we also display the coframe, that is, the one-form basis, which one finds by inverting the frame (25):

$$\begin{aligned} \vartheta^{\hat{0}} &= \left(1+\frac{\boldsymbol{a}\cdot\overline{\boldsymbol{x}}}{c^2}\right) dx^{\overline{0}} = N\, dx^{\overline{0}}\,, \\ \vartheta^{A} &= dx^{\overline{A}} + \left(\frac{\boldsymbol{\omega}}{c}\times\overline{\boldsymbol{x}}\right)^{\overline{A}} dx^{\overline{0}} = dx^{\overline{A}} + N^{\overline{A}}\, dx^{\overline{0}}\,. \end{aligned} \tag{26}$$

In the (3 + 1)-decomposition of spacetime, N and $N^{\overline{A}}$ are known as *lapse function* and *shift vector*, respectively.

Starting with the coframe, we can read off the connection coefficients (since torsion vanishes in SR) by using Cartan's first structure equation $d\vartheta^\alpha = -\Gamma_\beta{}^\alpha \wedge \vartheta^\beta$ with $\Gamma_\beta{}^\alpha = \Gamma_{\overline{i}\beta}{}^\alpha\, dx^{\overline{i}}$. We find:

$$\begin{aligned} \Gamma_{\overline{0}\hat{0}A} &= -\Gamma_{\overline{0}A\hat{0}} = \frac{a_A}{c^2}\,, \\ \Gamma_{\overline{0}AB} &= -\Gamma_{\overline{0}BA} = \epsilon_{ABC}\,\frac{\omega^C}{c}\,. \end{aligned} \tag{27}$$

It is important to note that the first index is holonomic, whereas the second and the third indices are anholonomic. If we transform the first index, by means of the frame coefficients $e^{\overline{i}}{}_\alpha$, with $e_\alpha = e^{\overline{i}}{}_\alpha\, \partial_{\overline{i}}$, into an anholonomic one, then we find the totally anholonomic connection coefficients as follows:

$$\begin{aligned} \Gamma_{\hat{0}\hat{0}A} &= -\Gamma_{\hat{0}A\hat{0}} = \frac{a_A}{c^2} \Big/ \left(1+\boldsymbol{a}\cdot\overline{\boldsymbol{x}}/c^2\right)\,, \\ \Gamma_{\hat{0}AB} &= -\Gamma_{\hat{0}BA} = \epsilon_{ABC}\,\frac{\omega^C}{c} \Big/ \left(1+\boldsymbol{a}\cdot\overline{\boldsymbol{x}}/c^2\right)\,. \end{aligned} \tag{28}$$

These connection coefficients (28) will enter the Dirac equation referred to a non-inertial frame.

In order to assure ourselves that we didn't make mistakes in computing the 'non-inertial' connection (27, 28) by hand, we used for checking its correctness the EXCALC package on exterior differential forms of the computer algebra system REDUCE, see Puntigam et al.[80] and the literature given there.

3.3. Dirac matter waves in a non-inertial frame of reference

The phase shift (24) can be derived from the Schrödinger equation with a Hamilton operator for a point particle in an external Newton potential. For setting up a gravitational theory, however, one better starts more generally in the special relativistic domain. Thus we have to begin with the Dirac equation in an external gravitational field or, if we expect the equivalence principle to be valid, with the Dirac equation in an accelerated and rotating, that is, in a non-inertial frame of reference.

Take the Minkowski spacetime of SR. Specify Cartesian coordinates. Then the field equation for a massive fermion of spin 1/2 is represented by the Dirac equation

$$i\hbar\gamma^i\,\partial_i\psi \stackrel{*}{=} mc\psi\,, \tag{29}$$

where the Dirac matrices γ^i fulfill the relation

$$\gamma^i\gamma^j + \gamma^j\gamma^i = 2\,o^{ij}\,. \tag{30}$$

For the conventions and the representation of the γ's, we essentially follow Bjorken-Drell[7].

Now we straightforwardly transform this equation from an inertial to an accelerated and rotating frame. By analogy with the equation of motion in an arbitrary frame as well as from gauge theory, we can infer the result of this transformation: In the non-inertial frame, the partial derivative in the Dirac equation is simply replaced by the *covariant derivative*

$$\partial_i \quad\Rightarrow\quad D_\alpha := \partial_\alpha + \frac{i}{4}\sigma^{\beta\gamma}\,\Gamma_{\alpha\beta\gamma}\,, \qquad \partial_\alpha := e^i{}_\alpha\partial_i \equiv e_\alpha\,, \tag{31}$$

where $\Gamma_{\alpha\beta\gamma}$ are the anholonomic components of the connection, see (28), and x^i the Cartesian coordinates of the lab system (which we called $x^{\bar{i}}$ previously; we drop the bar for convenience). The anholonomic Dirac matrices are defined by

$$\gamma^\alpha := e_i{}^\alpha\,\gamma^i \quad\Rightarrow\quad \gamma^\alpha\gamma^\beta + \gamma^\beta\gamma^\alpha = 2\,o^{\alpha\beta}\,. \tag{32}$$

The six matrices $\sigma^{\beta\gamma}$ are the infinitesimal generators of the Lorentz group and fulfill the commutation relation

$$[\gamma^\alpha, \sigma^{\beta\gamma}] = 2i(o^{\alpha\beta}\gamma^\gamma - o^{\alpha\gamma}\gamma^\beta)\,. \tag{33}$$

For Dirac spinors, the Lorentz generators can be represented by

$$\sigma^{\beta\gamma} := (i/2)(\gamma^\beta\gamma^\gamma - \gamma^\gamma\gamma^\beta)\,, \tag{34}$$

furthermore,

$$\boldsymbol{\alpha} := \gamma^{\hat{0}}\boldsymbol{\gamma} \qquad \text{with} \qquad \boldsymbol{\gamma} = \{\gamma^{\Xi}\}\,. \tag{35}$$

Then, the Dirac equation, formulated in the orthonormal frame of the accelerated and rotating observer, reads

$$i\hbar\gamma^\alpha D_\alpha\psi = mc\psi\,. \tag{36}$$

Although there appears now a 'minimal coupling' to the connection, which is caused by the change of frame, there is no new physical concept involved in this equation. Only for the measuring devices in the non-inertial frame we have to assume hypotheses similar to the clock hypothesis. This proviso can always be met by a suitable construction and selection of the devices. Since we are still in SR, torsion and curvature of spacetime both remain zero. Thus (36) is just a reformulation of the 'Cartesian' Dirac equation (29).

The rewriting in terms of the covariant derivative provides us with a rather elegant way of explicitly calculating the Dirac equation in the non-inertial frame of an accelerated, rotating observer: Using the anholonomic connection components of (28) as well as $\boldsymbol{\alpha} = -i\{\sigma^{\hat{0}\Xi}\}$, we find for the covariant derivative:

$$\begin{aligned} D_{\hat{0}} &= \frac{1}{1+\boldsymbol{a}\cdot\boldsymbol{x}/c^2}\left(\partial_0 + \frac{1}{2c^2}\,\boldsymbol{a}\cdot\boldsymbol{\alpha} - \frac{i}{c\hbar}\,\boldsymbol{\omega}\cdot\boldsymbol{J}\right), \\ D_\Xi &= \partial_\Xi\,. \end{aligned} \tag{37}$$

The total three-angular momentum operator

$$\boldsymbol{J} := \boldsymbol{L} + \boldsymbol{S} = \boldsymbol{x}\times\frac{\hbar}{i}\frac{\partial}{\partial\boldsymbol{x}} + \frac{1}{2}\,\hbar\boldsymbol{\sigma} = \boldsymbol{x}\times\boldsymbol{p} + \frac{1}{2}\,\hbar\boldsymbol{\sigma} \tag{38}$$

is composed, in the canonical manner, from the orbital piece $\boldsymbol{L}$ and the spin piece $\boldsymbol{S}$.

The physical effects in our lab frame can be most easily understood by going over to the Hamiltonian. After multiplying the Dirac equation by $\beta := \gamma^{\hat{0}}$ and $c(1+\boldsymbol{a}\cdot\boldsymbol{x}/c^2)$, we get

$$i\hbar\frac{\partial\psi}{\partial t} = H\psi \quad \text{with } H = \beta mc^2 + \mathcal{O} + \mathcal{E}\,. \tag{39}$$

After substituting the covariant derivatives, the operators $\mathcal{O}$ and $\mathcal{E}$, which are odd and even with respect to β, read, respectively[30]:

$$\mathcal{O} := c\boldsymbol{\alpha}\cdot\boldsymbol{p} + \frac{1}{2c}\{(\boldsymbol{a}\cdot\boldsymbol{x})(\boldsymbol{p}\cdot\boldsymbol{\alpha}) + (\boldsymbol{p}\cdot\boldsymbol{\alpha})(\boldsymbol{a}\cdot\boldsymbol{x})\}\,, \tag{40}$$

$$\mathcal{E} := \beta m(\boldsymbol{a}\cdot\boldsymbol{x}) - \boldsymbol{\omega}\cdot(\boldsymbol{L}+\boldsymbol{S})\,. \tag{41}$$

Up to now, these are mathematically *exact* results. For later purposes we introduce $\mathcal{O}_1 = c\boldsymbol{\alpha}\cdot\boldsymbol{p}$ and $\mathcal{O}_2 = \mathcal{O} - \mathcal{O}_1$.

Similarly as in quantum electrodynamics[7], a non-relativistic approximation can be obtained by applying successive Foldy-Wouthuysen transformations. After three such steps we find[30], up to the order of c^{-2},

$$\begin{aligned} H' &= \beta mc^2 + \frac{\beta}{2m}\boldsymbol{p}^2 - \frac{\beta}{8m^3c^2}\boldsymbol{p}^4 + \beta m(\boldsymbol{a}\cdot\boldsymbol{x}) - \boldsymbol{\omega}\cdot(\boldsymbol{L}+\boldsymbol{S}) \\ &+ \frac{\beta}{2m}\boldsymbol{p}\cdot\frac{\boldsymbol{a}\cdot\boldsymbol{x}}{c^2}\boldsymbol{p} + \frac{\beta\hbar}{4mc^2}\boldsymbol{\sigma}\cdot\boldsymbol{a}\times\boldsymbol{p} + O(\frac{1}{c^3})\,. \end{aligned} \tag{42}$$

The different non-inertial effects of a fermion are displayed in Table 2. Besides the rest mass and the usual kinetic term, we obtain terms which account for the *redshift effect* due to acceleration, thereby verifying the BW and, if the acceleration $\boldsymbol{a}$ is substituted by the gravitational acceleration $\boldsymbol{g}$, the COW experiments. Moreover the

Table 2. Inertial effects for a massive fermion of spin 1/2 in non-relativistic approximation.

$\beta m\,(\boldsymbol{a}\cdot\boldsymbol{x})$	Redshift (Bonse-Wroblewski → COW)
$-\boldsymbol{\omega}\cdot\boldsymbol{L}$	Sagnac type effect (Page-Werner et al.)
$-\boldsymbol{\omega}\cdot\boldsymbol{S}$	Spin-rotation effect (Mashhoon)
$\beta\,\boldsymbol{p}\cdot(\boldsymbol{a}\cdot\boldsymbol{x})\,\boldsymbol{p}\,/\,(2mc^2)$	Redshift effect of kinetic energy
$\beta\,\hbar\,\boldsymbol{\sigma}\cdot(\boldsymbol{a}\times\boldsymbol{p})\,/\,(4mc^2)$	New inertial spin-orbit coupling

'Sagnac type' effect occurs in the same manner as in the non-relativistic Schrödinger equation, and a spin-rotation effect is found which, for the neutron interferometer, has first been proposed by Mashhoon[46]. This term could not have been obtained by using the Schrödinger equation.

Thus we demonstrated that the Dirac equation in a Minkwoski spacetime[d] referred to a non-inertial frame, yields the BW phase shift and, if the equivalence principle is assumed, also that of the COW experiment. The claim of our colleague in Erice, that in SR such non-inertial frames are illegitimate, is thus disproved. Rather we have shown the usefulness of non-inertial frames ϑ^α, see (26), with $d\vartheta^\alpha \neq 0$.

3.4. 'Deriving' a theory of gravity: Einstein's method as opposed to the gauge procedure

Now we turn back to Table 1: Einstein's approach, in the middle column, is compared with the gauge approach represented in the right column. The basic idea is that a mechanical system is considered in an inertial and then in a non-inertial frame. In Einstein's approach the object under investigation is a point particle and the non-inertial frame a *natural* (or coordinate) coframe dx^i, in the gauge approach the object is a Dirac wave function and the non-inertial frame an *arbitrary* (or anholonomic) coframe ϑ^α, with $d\vartheta^\alpha \neq 0$. Einstein's 'trick' was to search for the fields emerging in the non-inertial frame and to identify them as pseudo-gravitational. 'Pseudo', since the spacetime is a flat and uncontorted Minkowski space without gravitational fields present. This implies that the pseudo-gravitational fields have to obey some constraints, otherwise it wouldn't be guaranteed that the Minkowski background prevails.

[d]These considerations can be generalized to a Riemannian spacetime, see Huang[34] and the literature quoted there.

The gauge approach had been developed by Utiyama et al. long before BW and COW set up their experiments. These experiments confirmed the appropriateness of the earlier theoretical development. Clearly, in some WKB-limit, compare the Foldy-Wouthuysen technique in Sec. 3.3, the Einsteinian approach can be recovered as a limiting case of the more general gauge approach.

In the Einstein case the pseudo-gravitational field is represented by the Christoffel symbol (of the second kind), in the gauge case by the coframe and the Lorentz-connection. The transition from SR to a spacetime with 'real' gravity consists in the *relaxation* of the constraints formulated earlier. We are finding a Riemannian and a Riemann-Cartan space,[e] respectively.

But beware! Locally we have to recover the Minkowskian behavior of spacetime: In the Einsteinian free-fall elevator gravity does not show up. In the Riemann space this is technically achieved by means of Riemannian normal coordinates, in the case of the Riemann-Cartan space there exist *normal frames* which realize an 'anholonomic' elevator, as has been first pointed out by von der Heyde[31] and developed by one of us[25], by Modanese & Toller[59], Iliev[35], and Hartley[23]. In Hartley's formulation[23], we have the following proposition:

Let M be a manifold with metric g and (arbitrary) metric–compatible connection ∇ (Riemann–Cartan space). For any single point $P \in M$, there exist coordinates $\{x^i\}$ and an orthonormal frame $\{e_\alpha\}$ in a neighborhood of P such that

$$\left\{ \begin{array}{lcl} e_\alpha & = & \delta^i_\alpha \, \partial_{x^i} \\ \Gamma_\alpha{}^\beta & = & 0 \end{array} \right\} \quad \text{at } P\,, \tag{43}$$

where $\Gamma_\alpha{}^\beta$ are the connection 1–forms referred to the frame $\{e_\alpha\}$.

Not too many people know of this theorem, even though it had been proposed by von der Heyde already twenty years ago. Usually, if we talk to relativists about the normal frames in a Riemann-Cartan space, they state that those cannot exists, since torsion, as a tensor, cannot be transformed to zero. In this context it is tacitly assumed that the starting point are Riemannian normal coordinates and the torsion is 'superimposed'. However, since only a *natural* frame is attached to Riemannian normal coordinates, one is too restrictive in the discussion right from the beginning. And, of course, the curvature is also of tensorial nature – and still Riemannian normal coordinates do exist.

In Table 1 there are two subtleties involved which we want to mention:

In the gauge column we formulated two constraints in SR, namely the vanishing of torsion and that of curvature. One could well wonder whether we have to relax

[e]Recently Hammond[22] gave a very pronounced and interesting plea in favor of the existence of torsion in nature. Only his attempts to derive torsion from some new potential, we don't find convincing, since it looks ad hoc to us. Moreover, the coframe is some sort of potential already, see (12) and Table 5 – and why should we multiply the number of potentials beyond necessity?

both constraints at the same time. We could allow for torsion only, thereby ending up with a spacetime carrying a teleparallelism, or, alternatively, we could only admit the emergence of curvature, thereby recovering the Riemannian spacetime of GR. It is perhaps surprising that also in the teleparallelism case, see Secs.4 and 5 below, by a suitable choice of a torsion square Lagrangian, one can arrive at a theory which is equivalent to GR. Nevertheless, there are good reasons for relaxing both constraints: Firstly, by mimicking the Einsteinian procedure of the middle column of Table 1, we cannot see any reason why we should lift only one constraint; secondly, by admitting a Riemann-Cartan space, we can still 'trivialize' the gravitational potentials at a point P and its neighborhood, as found in proposition (43), in spite of the presence of torsion and curvature. This implies that a Riemann-Cartan space, if described in terms of suitable coframes, looks Minkowskian locally. Eqs.(43) supply the strongest reason for taking the four-dimensional Riemann-Cartan space seriously as a model for spacetime.

A second subtlety is related to the fact that the minimally coupled Dirac equation (36), in a Riemann-Cartan space, slightly differs from the Euler-Lagrange equation of the minimally coupled Dirac *Lagrangian*. However, this shouldn't cause headaches: The gauge theoretical set up, see Fig. 1, is so closely linked to the Lagrangian formalism – not to speak of the fundamental importance of the Feynman path integral or of Schwinger's variational principle – that the *minimal coupling* procedure should be implemented on the *Lagrangian level*.

Eventually concentrating our attention to the last row of Table 1, we read off the field equations of GR and of the EC-theory: They result from the simplest conceivable Lagrangian, namely from the curvature scalar of the Riemann or the Riemann-Cartan spacetime respectively. The EC-theory differs from GR in a very weak spin-contact interaction which is unmeasurable at the present time. In this sense, the EC-theory is a *viable* theory of gravitation.

4. Conserved momentum current, the heuristics of the translation gauge

4.1. Motivation

We have in mind to derive gravity from a symmetry principle. But what is the right symmetry to derive gravity from? As already indicated in Sec.2, we think that gravity stems from translation symmetry. Our motivation is the following:

We start from SR. The invariance of the action $W = \int L_{\text{mat}}(\Psi, d\Psi)$ of an isolated material system under rigid spacetime translations yields, by the application of the Noether theorem, a *conserved energy-momentum current* three-form $\mathcal{T}_j$ via

$$\mathcal{T}_j := \frac{\delta L_{\text{mat}}}{\delta dx^j} = \frac{1}{3!}\, \mathcal{T}_{klmj}\, dx^k \wedge dx^l \wedge dx^m\,, \qquad d\,\mathcal{T}_j = 0\,. \tag{44}$$

(One obtains the "usual" energy-momentum tensor T_{ij} from $\mathcal{T}_j$ by means of $T_{ij} = \epsilon_i{}^{klm}\mathcal{T}_{klmj}$.) The corresponding charge $M := \int d^3x \mathcal{T}_0$ is conserved in time. In other words: Rigid translational invariance is attributed to the classical field-theoretical equivalent to mass(-energy density), which is the source of Newton-Einstein gravity.

The analogy to electrodynamics guides us of how to actually generate the gravitational interaction. In electrodynamics one finds from rigid $U(1)$-invariance of an action $W = \int L_{\text{mat}}(\Psi, d\Psi)$ a conserved electromagnetic current with corresponding electric charge Q. As we discussed in length in Sec.2, it is possible to generate the electromagnetic interaction by gauging the rigid $U(1)$-symmetry. Following this example, we expect the gravitational interaction to emerge from gauging the rigid translational symmetry. To quote Feynman[15]: "...gravity is that field which corresponds to a gauge invariance with respect to displacement transformations." But before we gauge the translation group we should think about...

4.2. Active and passive translations

Whenever we describe quantities with respect to some reference system (a certain basis, a certain coordinate system) we do encounter the possibility of active or passive transformations. Actively transforming a quantity or passively transforming its reference system will usually change the labels of the quantity which are related to the reference system. How can we decide whether an active, a passive, or even a mixture of both kinds of transformation took place when those labels of a quantity changed? Is it necessary to distinguish between active and passive transformations? For the case of active and passive translations, we will answer this question to the extent that we show the mathematical equivalence of actively or passively translating a matter field Ψ in Minkowski spacetime. From this we will later proceed in order to gauge the translation group.

The field Ψ is thought to be some p-form, possibly Lie-algebra valued. We discard any internal structure of Ψ since we are only interested in its spacetime properties. For the Minkowski spacetime we assume for convenience a gauge in which the connection components are identically zero, $\Gamma \stackrel{*}{=} 0$. This can be achieved by choosing pseudo-Cartesian coordinates x^i and using the corresponding holonomic (co-)frame. The total variation $\delta_t\Psi$ of Ψ under translations is the sum of both active and passive variations:

$$\delta_t\Psi = \delta_a\Psi + \delta_p\Psi\,. \tag{45}$$

An active translation of Ψ is generated by the flow of a vector field $\xi = \xi^i\partial_i$, that is, $\Psi \to \xi^*\Psi$. In this case the active variation is given by the Lie-derivative l_ξ,

$$\delta_a\Psi = l_\xi\Psi\,, \qquad l_\xi = d(\xi\rfloor) + (\xi\rfloor)d\,. \tag{46}$$

Despite possible exterior indices of Ψ, we don't have to use a 'covariant Lie derivative' since $\Gamma \stackrel{*}{=} 0$. We will be interested only in infinitesimal variations. Then the Lie derivative describes the difference between the actively translated $\xi^*\Psi(x^i)$ and $\Psi(x^i + \xi^i)$.

Passive translations are generated by coordinate transformations. To find the passive transformation which corresponds to a given active transformation, we have to know the explicit form of the operator δ_p and the actual coordinate transformation which corresponds to a specific $\delta_a = l_\xi$. For the actual coordinate transformation, we look at the total variation of the coordinate functions x^i,

$$\delta_t x^i = \delta_a x^i + \delta_p x^i = l_\xi x^i + \delta_p x^i = \xi^i + \delta_p x^i \,. \tag{47}$$

The coordinate transformation $\delta x^i = \delta_p x^i = -\xi^i$ makes $\delta_t x^i$ to vanish. Thus the passive translation $\delta_p x = \xi$ is equivalent to $\delta_a x = l_\xi x$ (what we intuitively expected without any calculation). Having this candidate for the appropriate coordinate transformation, we look for an operator δ_p such that $\delta_t \Psi = l_\xi \Psi + (\delta_p \Psi)(-\xi) = 0$. We set

$$(\delta_p \Psi)(-\xi) = \Psi(x^i - \xi^i) - \Psi(x^i) \,, \tag{48}$$

i.e., we take the value of Ψ at a point in the coordinate system $x^i - \xi^i$ and subtract the value of Ψ at the same point while using the coordinate system x^i. Writing

$$(\delta_p \Psi)(-\xi) = \Psi_{i_1 \dots i_p}(x^i - \xi^i) d(x^{i_1} - \xi^{i_1}) \wedge \dots \wedge d(x^{i_p} - \xi^{i_p}) \,, \tag{49}$$

Taylor expanding $\Psi_{i_1 \dots i_p}(x^i - \xi^i)$, and keeping terms up to order ξ, we can show that

$$(\delta_p \Psi)(-\xi) = -l_\xi \Psi \,. \tag{50}$$

Thereby we just rederived the definition of the Lie derivative from a passive point of view, and the coordinate transformation $\delta x = \xi$ becomes clearly equivalent to an active translation generated by l_ξ.

4.3. Heuristic scheme of translational gauging

In order to gauge the translation group we will follow the general gauge scheme which we set up in Sec.2. Therefore we will begin with a rigidly translation invariant action $W = \int L_{\mathrm{mat}}(\Psi, d\Psi)$. We expect to have to introduce four gauge potentials A^α in order to extend from rigid to soft translation invariance of S. Each A^α will compensate one of the four independent soft translations l_{ξ_α}.

We note that soft translations do not commute any longer: Expanding the flow-generating vectors ξ_α according to $\xi_\alpha = e^i{}_\alpha \, \partial_i$, with spacetime dependent functions $e^i{}_\alpha = e^i{}_\alpha(x)$, we can use the formula

$$[l_{\xi_\alpha}, l_{\xi_\beta}] = l_{[\xi_\alpha, \xi_\beta]} \tag{51}$$

to show that (the equality $\stackrel{*}{=}$ requires the gauge $\Gamma \stackrel{*}{=} 0$)

$$[l_{\xi_\alpha}, l_{\xi_\beta}] = -e^j{}_{[\alpha} \, e^i{}_{\beta]} \, e_i{}^\gamma{}_{,j} \, \xi_\gamma \stackrel{*}{=} -T_{\alpha\beta}{}^\gamma \, \xi_\gamma \,. \tag{52}$$

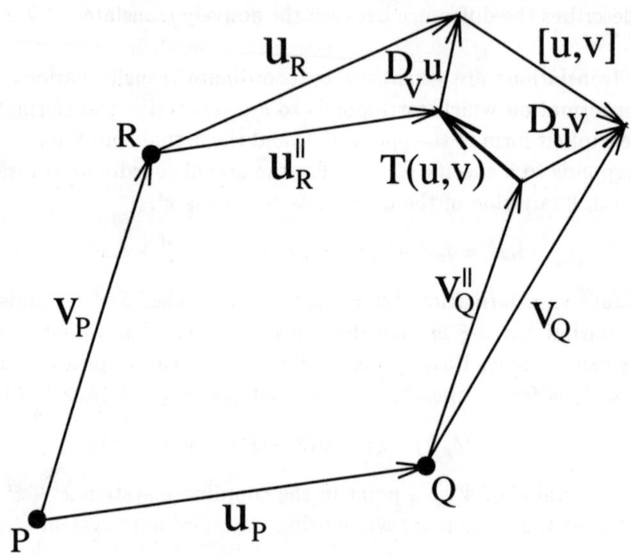

Fig. 4. On the geometrical interpretation of torsion: It represents a closure failure of infinitesimal displacements.

The $T_{\alpha\beta}{}^{\gamma}$ are the components of the torsion two-form $T^{\gamma} = \frac{1}{2} T_{\alpha\beta}{}^{\gamma} \vartheta^{\alpha} \wedge \vartheta^{\beta}$ expanded in the anholonomic coframe $\vartheta^{\alpha} = e_i{}^{\alpha} dx^i$. This softening of the Lie algebra of the translation group should be seen in correspondence to the structure

$$\text{commutator of local symmetry transformations} = \text{field strength}\,, \tag{53}$$

which is well known from Yang-Mills theory. Therefore it is tempting to view the torsion tensor as the translational field strength.

The torsion tensor represents a translational misfit, as already indicated in Sec.2.4. A more detailed discussion is given by de Sabbata and Sivaram[82] who also collected numerous other facts and results on torsion. Torsion measures the noncommutativity of displacements of points in analogy to the curvature tensor which measures the noncommutativity of displacements of vectors. This is explained in Fig. 4: Let P be a point of the (spacetime) manifold M and v_P, u_P two linearly independent vectors of $T_P M$. We regard v_P, u_P as infinitesimally small. Then they define two points, R and Q, on M. (We can make this mathematically precise by defining R, Q via the exponential map $\exp_P : T_P M \to M$, $\exp_P(v) = R$, $\exp_P(u) = Q$.) In other words: The vector v_P displaces P infinitesimally to R, the vector u_P displaces P infinitesimally to Q. The prescription to perform two successive displacements v, u is to first displace P by means of v_P to R and second parallel transport u_P to $u_R^{\parallel}$

and displace R by means of $u_R^{\|}$. Fig. 4 shows that the commutator of two successive displacements won't vanish in general – the gap between the two resulting points is, by definition, a measure for the torsion. It is also shown in Fig.4 that this definition matches the usual textbook definition

$$T(u,v) := D_u v - D_v u - [u,v]\,. \tag{54}$$

To make contact with our present notation, we evaluate the components of T with respect to an arbitrary frame e_α:

$$\begin{aligned} T(e_\beta, e_\gamma) &= D_{e_\beta} e_\gamma - D_{e_\gamma} e_\beta - [e_\beta, e_\gamma] \\ &= \Gamma_{\beta\gamma}{}^\alpha e_\alpha - \Gamma_{\gamma\beta}{}^\alpha e_\alpha + C_{\beta\gamma}{}^\alpha e_\alpha \\ &= T_{\beta\gamma}{}^\alpha e_\alpha\,. \end{aligned} \tag{55}$$

Hence

$$T_{\beta\gamma}{}^\alpha = C_{\beta\gamma}{}^\alpha + \Gamma_{\beta\gamma}{}^\alpha - \Gamma_{\gamma\beta}{}^\alpha\,, \tag{56}$$

or, by using the object of anholonomity

$$C^\alpha := d\vartheta^\alpha = \frac{1}{2} C_{\beta\gamma}{}^\alpha\, \vartheta^\beta \wedge \vartheta^\gamma\,, \tag{57}$$

we find:

$$T^\alpha = d\vartheta^\alpha + \Gamma_\beta{}^\alpha \wedge \vartheta^\beta\,. \tag{58}$$

The three irreducible pieces of the torsion under $SO(1,3)$-decomposition are displayed for later use in Tab.3.

Table 3. Irreducible decomposition of the torsion $T^\alpha = {}^{(1)}T^\alpha + {}^{(2)}T^\alpha + {}^{(3)}T^\alpha$ under the Lorentz group $SO(1,3)$

	explicit expression	number of components	name
${}^{(2)}T^\alpha$	$\frac{1}{3}\vartheta^\alpha \wedge (e_\beta \rfloor T^\beta)$	4	TRATOR
${}^{(3)}T^\alpha$	$-\frac{1}{3}{}^*(\vartheta^\alpha \wedge {}^*(T^\beta \wedge \vartheta_\beta))$	4	AXITOR
${}^{(1)}T^\alpha$	$T^\alpha - {}^{(2)}T^\alpha - {}^{(3)}T^\alpha$	16	TENTOR

From the translational field strength T^α, see (52) and (53), we come back to the translational gauge potential A^α. According to the general gauge scheme, it will couple to the energy-momentum current according to

$$\mathcal{T}_\alpha = \frac{\delta L}{\delta A^\alpha} \,. \tag{59}$$

The left hand side of (44) suggests that the gauge process will replace the holonomic coframe dx^i by the potentials A^α. Do the A^α have something to do with an anholonomic coframe ϑ^α? A consistent relation between $T^\alpha, \vartheta^\alpha$, and A^α is established if we assume that

$$\vartheta^\alpha = \delta^\alpha_i dx^i + A^\alpha \,. \tag{60}$$

Then we obtain

$$dA^\alpha = d\vartheta^\alpha \stackrel{*}{=} T^\alpha \,. \tag{61}$$

From (61) we get the 'Bianchi' identity $dT^\alpha \stackrel{*}{=} 0$. Comparison to the general Bianchi identity $DT^\alpha = R_\alpha{}^\beta \wedge \vartheta^\beta$ indicates that the structures found so far are part of a more general framework. This is, indeed, the case. In order to derive GR from as little input as possible, we refrain at this point from introducing additional fields, as for example an independent linear connection Γ. We collected the structures relevant for the translational gauge scheme in Tab.4.

Table 4. The relevant structures in a gauge approach of the four parameter translation group

		number of components
conserved momentum current	$\mathcal{T}_\alpha$	4×4
translational gauge potential	$\vartheta^\alpha \simeq A^\alpha$ (coframe)	4×4
translational field strength	T^α (torsion)	4×6
1st Bianchi identity ($\Gamma \stackrel{*}{=} 0$)	$dT^\alpha \stackrel{*}{=} 0$	4×4

5. Theory of the translation gauge: From Einsteinian teleparallelism to GR

As we argued in the last section, GR should be derivable from gauge ideas in a fairly straightforward manner. The key to arrive at GR is to start with the conserved momentum current and to gauge the translation group that is connected with it.

5.1. Translation gauge potential

We commence with the very basics: Consider again a field theory in Minkowski spacetime (pseudo-Cartesian coordinates) defined by the Lagrangian $L_{\rm mat} = L_{\rm mat}(\Psi, d\Psi)$.

An explicit dependence of L on the coordinates x^i is already forbidden by rigid translational invariance, which we started from. The coordinates enter more implicitly: The field Ψ has to be expressed in terms of differential forms in order to build the Lagrangian 4-form L as the appropriate integrand of the action. Therefore one needs the differentials dx^i as natural (or holonomic) basis for the physical field. They are invariant under rigid but not under local translations:

$$\delta x^i = \varepsilon^i(x) \quad \Longrightarrow \quad \delta dx^i = d\varepsilon^i \,. \tag{62}$$

Referring to the equivalence between active and passive transformations, which is valid at this stage (see Sec.4.2), we view in this approach the translations as passive transformations. The differentials dx^i are no longer sufficient to build up an invariant Lagrangian. According to the gauge principle, we have to introduce a gauge potential A^α with transformation behavior $\delta A^\alpha = -\delta_i^\alpha d\varepsilon^i$ such that

$$\vartheta^\alpha := \delta_i^\alpha \hat{D} x^i := \delta_i^\alpha dx^i + A^\alpha \tag{63}$$

transforms like

$$\delta\vartheta^\alpha = 0 \,. \tag{64}$$

The anholonomic one-form basis ϑ^α serves as an appropriate form basis for Lagrangians L_{mat} since it automatically incorporates soft translational invariance. One usually refers to ϑ^α, instead of A^α, as the translational potential. Normally, a distinction between ϑ^α and A^α is not necessary, since their field components differ just by a constant (1 or 0).

5.2. Lagrangian

The corresponding field strength $T^\alpha \stackrel{*}{=} d\vartheta^\alpha$ can be used to construct a kinematic supplementary term for ϑ^α to the Lagrangian. The double role of ϑ^α as both, a dynamical gauge potential and an orthonormal frame (defining a new metric via $g = o_{\alpha\beta}\, \vartheta^\alpha \otimes \vartheta^b$), explains the transition from Minkowski space to a dynamical spacetime, which is due to translational invariance. For the kinematic term we make the quadratic ansatz $V = d\vartheta^\alpha \wedge H_\alpha$. What would be a good choice for H_α? Eyeing at Yang-Mills theory, we are tempted to put $H_\alpha = \frac{1}{2\ell^2}\, {}^*d\vartheta_\alpha$, with ℓ = Planck length. But we would like to end up with a softly Lorentz invariant theory. The Lagrangian $V = \frac{1}{2\ell^2}\, {}^*d\vartheta^\alpha \wedge d\vartheta_\alpha$ is rigidly but not softly Lorentz invariant, though. Note that this postulate of local Lorentz symmetry is not equivalent to a gauging of the Lorentz group! We won't cure the lack of Lorentz invariance by the introduction of some new gauge field Γ for the Lorentz field, but will use it just as a criterion of choosing a good Lagrangian.

The most general term V quadratic in $d\vartheta^\alpha$ is obtained by choosing H_α as

$$H_\alpha = \frac{1}{2\ell^2}\, {}^*\left(a_1\, {}^{(1)}d\vartheta_\alpha + a_2\, {}^{(2)}d\vartheta_\alpha + a_3\, {}^{(3)}d\vartheta_\alpha\right) , \tag{65}$$

see also Mielke[50,49]. The pieces ${}^{(I)}d\vartheta^\alpha$ correspond to the irreducible pieces ${}^{(I)}T^\alpha$ of the torsion, compare Table 3:

$$\begin{aligned}
{}^{(2)}d\vartheta^\alpha &:= \frac{1}{3}\vartheta^\alpha \wedge (e_\beta \rfloor d\vartheta^\beta)\,, \\
{}^{(3)}d\vartheta^\alpha &:= -\frac{1}{3}{}^*\{\vartheta^\alpha \wedge {}^*(d\vartheta^\beta \wedge \vartheta_\beta)\}\,, \\
{}^{(1)}d\vartheta^\alpha &:= d\vartheta^\alpha - {}^{(2)}d\vartheta^\alpha - {}^{(3)}d\vartheta^\alpha\,.
\end{aligned} \tag{66}$$

The postulate of local Lorentz invariance leads to a solution for the constant and real parameters a_I in the following way:

Infinitesimal Lorentz rotations are expressed by $\delta\vartheta^\alpha = \varepsilon^\alpha{}_\beta\,\vartheta^\beta$ where $\varepsilon_{\alpha\beta} = -\varepsilon_{\beta\alpha}$ are the antisymmetric Lorentz group parameters. It is easy to check that the gauge Lagrangian $V = d\vartheta^\alpha \wedge H_\alpha$, with H_α given by (65), is invariant under *rigid* Lorentz rotations, $\delta V = 0$. The general expression for δV reads

$$\delta V = \left(\frac{\partial V}{\partial \vartheta^\alpha} - d\frac{\partial V}{\partial d\vartheta^\alpha}\right) \wedge \delta\vartheta^\alpha + d\left(\frac{\partial V}{\partial d\vartheta^\alpha} \wedge \delta\vartheta^\alpha\right)\,. \tag{67}$$

Hence we have $\delta V = 0$ for rigid Lorentz rotations. However, for *local* Lorentz rotations with spacetime-dependent group parameters $\varepsilon_{\alpha\beta} = \varepsilon_{\alpha\beta}(x)$, we get from (67) the offending term

$$\delta_{(\text{local})} V = d\varepsilon^\alpha{}_\beta \wedge \frac{\partial V}{\partial d\vartheta^\alpha} \wedge \vartheta^\beta\,. \tag{68}$$

In order to preserve Lorentz invariance, this term has to be canceled, modulo an exact form. Using the Leibniz rule, we obtain

$$d\varepsilon^\alpha{}_\beta \wedge \frac{\partial V}{\partial d\vartheta^\alpha} \wedge \vartheta^\beta = \varepsilon^\alpha{}_\beta\, d\left(\frac{\partial V}{\partial \vartheta^\alpha} \wedge \vartheta^\beta\right) - d\left(\varepsilon^\alpha{}_\beta \frac{\partial V}{\partial \vartheta^\alpha} \wedge \vartheta^\beta\right)\,. \tag{69}$$

The second term on the r.h.s. is already exact. From the first term we get the condition

$$\frac{\partial V}{\partial d\vartheta^{[\alpha}} \wedge \vartheta_{\beta]} = \text{exact form} \tag{70}$$

for local Lorentz invariance of V. We plug in the explicit expression for V and obtain, after some algebra,

$$\begin{aligned}
\frac{\partial V}{\partial d\vartheta^{[\alpha}} \wedge \vartheta_{\beta]} &= \left(\frac{1}{3}a_1 - \frac{1}{3}a_3\right) d\eta_{\alpha\beta} - \left(\frac{2}{3}a_3 + \frac{1}{3}a_1\right) d\vartheta_{[\alpha} \wedge \vartheta_{\beta]} \\
&+ \left(\frac{1}{6}a_1 + \frac{1}{6}a_2 - \frac{1}{3}a_3\right)(e_\gamma \rfloor d\vartheta^\gamma) \wedge \eta_{\alpha\beta}\,.
\end{aligned} \tag{71}$$

The last two terms can be made vanishing by choosing

$$a_2 = -2a_1\,, \qquad a_3 = -\frac{1}{2}\,a_1\,. \tag{72}$$

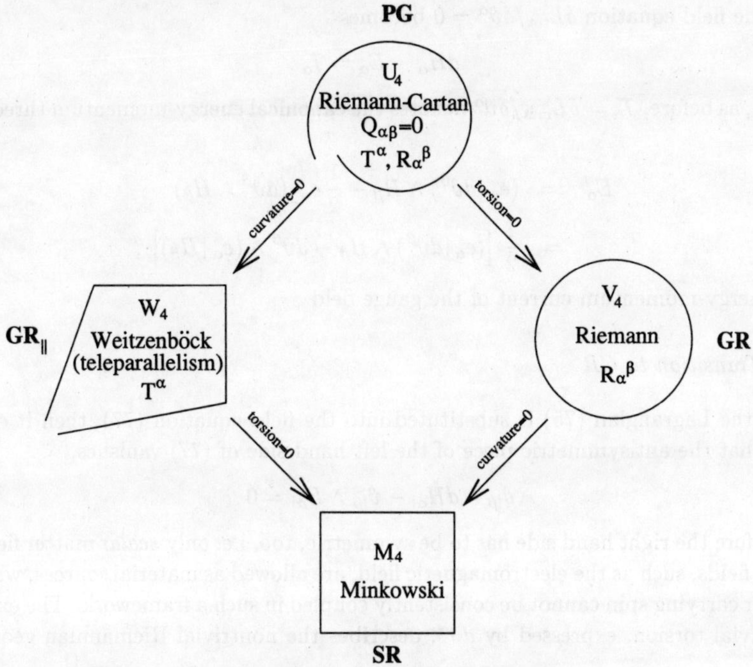

Fig. 5. A spacetime with a metric and a metric compatible connection (nonmetricity=0) is called a Riemann-Cartan space U_4. It can either become a Weitzenböck space W_4, if its curvature vanishes, or a Riemann space V_4, if the torsion happens to vanish. These different models of spacetime are the arenas for different gravitational theories.

Then we obtain

$$\frac{\partial V}{\partial d\vartheta^{[\alpha}} \wedge \vartheta_{\beta]} = \frac{a_1}{2}\, d\eta_{\alpha\beta}\,. \tag{73}$$

The constant a_1 can be absorbed by a suitable choice of the coupling constant ℓ in V, see (65). According to the usual conventions, we put $a_1 = -1$, i.e. V is locally Lorentz invariant for the choice of parameters

$$a_1 = -1\,, \qquad a_2 = 2\,, \qquad a_3 = \frac{1}{2}\,. \tag{74}$$

Hence

$$V_{||} = \frac{1}{2\ell^2}\, d\vartheta^\alpha \wedge {}^*\left(-{}^{(1)}d\vartheta_\alpha + 2\,{}^{(2)}d\vartheta_\alpha + \frac{1}{2}\,{}^{(3)}d\vartheta_\alpha\right)\,. \tag{75}$$

The total Lagrangian reads

$$L_{\rm tot} = V_{||} + L_{\rm mat}(\Psi, d\Psi, \vartheta^\alpha)\,, \tag{76}$$

and the field equation $\delta L_{\rm tot}/\delta\vartheta^\alpha = 0$ becomes

$$dH_\alpha - E_\alpha = \mathcal{T}_\alpha\,, \tag{77}$$

where, as before, $\mathcal{T}_\alpha = \delta L_{\rm mat}/\delta\vartheta^\alpha$ denotes the canonical energy-momentum three-form and

$$\begin{aligned} E_\alpha &:= (e_\alpha \rfloor d\vartheta^\beta)\wedge H_\beta - \frac{1}{2}\, e_\alpha\rfloor(d\vartheta^\beta\wedge H_\beta) \\ &= \frac{1}{2}\left[(e_\alpha\rfloor d\vartheta^\beta)\wedge H_\beta - d\vartheta^\beta\wedge(e_\alpha\rfloor H_\beta)\right], \end{aligned} \tag{78}$$

the energy-momentum current of the gauge field.

5.3. Transition to GR

If the Lagrangian (75) is substituted into the field equation (77), then it can be seen that the antisymmetric piece of the left hand side of (77) vanishes,

$$\vartheta_{[\beta}\wedge dH_{\alpha]} - \vartheta_{[\beta}\wedge E_{\alpha]} = 0\,. \tag{79}$$

Therefore the right hand side has to be symmetric, too, i.e. only *scalar* matter fields or *gauge* fields, such as the electromagnetic field, are allowed as material sources, whereas matter carrying spin cannot be consistently coupled in such a framework. The existing nontrivial torsion, expressed by $d\vartheta^\alpha$, describes the nontrivial Riemannian geometry of spacetime. This is because we have tied the Christoffel connection $\overset{\{\}}{\Gamma}_{\alpha\beta}$, which is determined by the metric, to the contortion tensor $K_{\alpha\beta}$ by means of the teleparallel condition $\Gamma_{\alpha\beta} \stackrel{*}{=} 0$:

$$\overset{\{\}}{\Gamma}_{\alpha\beta} - K_{\alpha\beta} = \Gamma_{\alpha\beta} \stackrel{*}{=} 0 \qquad \Longrightarrow \qquad \overset{\{\}}{\Gamma}_{\alpha\beta} \stackrel{*}{=} K_{\alpha\beta}\,. \tag{80}$$

In other words: Meaningful teleparallel theories do *not* presuppose spinning matter as a source for nontrivial torsion, in contrast to what is sometimes stated in literature[39].

Doesn't all this look like general relativity? The Levi-Civita (or Christoffel) connection, corresponding to the metric $g = o_{\alpha\beta}\,\vartheta^\alpha\otimes\vartheta^\beta$, is given by

$$\overset{\{\}}{\Gamma}_{\alpha\beta} = \frac{1}{2}\left(e_\alpha\rfloor d\vartheta_\beta - e_\beta\rfloor d\vartheta_\alpha - (e_\alpha\rfloor e_\beta\rfloor d\vartheta_\gamma)\wedge\vartheta^\gamma\right). \tag{81}$$

The corresponding Riemannian curvature reads

$$\overset{\{\}}{R}_{\alpha\beta} = d\overset{\{\}}{\Gamma}_{\alpha\beta} - \overset{\{\}}{\Gamma}_{\alpha\gamma}\wedge\overset{\{\}}{\Gamma}{}^\gamma{}_\beta\,. \tag{82}$$

We use the last two equations to replace on the Lagrangian level the variable $d\vartheta^\alpha$ by $\overset{\{\}}{\Gamma}_{\alpha\beta}$. Using (81) and (82), one can prove the remarkable identity

$$\frac{1}{2}\overset{\{\}}{R}{}^{\alpha\beta}\wedge\eta_{\alpha\beta} - \ell^2\,V = d(\vartheta^\alpha\wedge{}^*d\vartheta_\alpha)\,, \tag{83}$$

with V given by (75). Therefore one finds that the kinematic term V, with the above choice of parameters a_I, is equivalent to the Hilbert-Einstein action modulo an exact term. Replacing V in the action S by means of (83) leads, via $\delta L_{\text{tot}}/\delta\vartheta^\alpha = 0$, to Einstein's equation

$$\overset{\{\}}{G}_\alpha := \frac{1}{2}\,\eta_{\alpha\beta\gamma}\wedge \overset{\{\}}{R}{}^{\beta\gamma} = \ell^2\,\mathcal{T}_\alpha\,. \tag{84}$$

In such a way, we arrive at GR in its original form. Shifting back and forth from the variable pair $(\vartheta^\alpha, \overset{\{\}}{\Gamma}{}_\alpha{}^\beta)$ to $(\vartheta^\alpha, d\vartheta^\alpha)$ means shifting back and forth from original GR to its teleparallel equivalent $\text{GR}_{||}$. This is displayed in Fig.5: A general PG induces a Riemann-Cartan space U_4 with nonvanishing torsion and curvature. Such a U_4 can be reduced to either a Weitzenböck space W_4 (curvature $= 0$) or a Riemann space V_4 (torsion $= 0$), i.e. to the geometries induced by $\text{GR}_{||}$ or GR, respectively.

We wish to point out that both theories can be obtained as special cases within the framework of the Poincaré gauge theory PG, see our Physics Reports[29], the literature given there, and the work of Pascual-Sánchez[72,73].

6. Gauging of the affine group $R^4 \rtimes GL(4,R)$

We recognized that the gauging of the translations yields a theory which, for spinless matter and for electromagnetism, that is, for symmetric energy-momentum currents, is equivalent to GR. Thus we have a new understanding of Einstein's theory. Why should we generalize $\text{GR}_{||}$ if it is consistent with experiment? Three somewhat interrelated arguments come to mind:

- The translations represent a subgroup of the Poincaré group. Only to gauge the translations and to leave the Lorentz subgroup of the Poincaré group untouched, would seem unnatural. This argument is all the more convincing, since the *semi-direct* product structure of the Poincaré group interrelates its two mentioned subgroups stronger than it were the case for a direct product.

- A Weitzenböck spacetime is a degenerate Riemann-Cartan space. The gauge arguments against the rigidity of a teleparallelism were already advanced by Weyl in the twenties against Einstein's corresponding theory.

- The translational gauge procedure, as it is obvious from the field equation (77) with (75), works only for spinless matter and for electromagnetism, since the field equation is symmetric, see (79), and supplies only 10 independent components.

We conclude that a gauging of the whole $4+6$ parameter Poincaré group is mandatory. The theory which emerges, the Poincaré gauge theory (PG), is formulated in a Riemann-Cartan spacetime, the Einstein-Cartan theory (EC) being a degenerate subcase of it. And the EC is a viable gravitational theory!

In order to see the built-up of the different structures of spacetime more clearly, we prefer to gauge[48] immediately the $4+16$ *affine* group[48] $A(4,R) = R^4 \otimes GL(4,R)$ — which lacks a metric structure altogether — and to introduce the metric *subsequently.* Thus we arrive at the metric-affine gauge theory (MAG) which encompasses the PG as a subcase. Symbolically, we may write

$$\text{MAG}/_{\text{nonmetricity}=0} \;=\; \text{PG}\,. \tag{85}$$

The actual gauging of the affine group was done in our Physics Reports[29], and we follow the presentation given there. A short outline of these results, see Secs. 3.1, 3.2, 3.3, loc. cit., will be given here. We start then in the flat n–dimensional affine space R^n. The rigid affine group $A(n,R) := R^n \otimes \; GL(n,R)$ is the semidirect product of the group of n-dimensional *translations* and n-dimensional *general linear transformations.* This transformation group, cf. ref.([40]) p. 27, acts on an affine n–vector $x = \{x^\alpha\}$ according to

$$x \qquad \to \qquad x' = \Lambda\, x + \tau\,, \tag{86}$$

where $\Lambda = \{\Lambda^\alpha{}_\beta\} \in GL(n,R)$ and $\tau = \{\tau^\alpha\} \in R^n$. Thus it is a generalization of the Poincaré group $P := R^4 \otimes \; SO(1,3)$, with the pseudo–orthogonal group $SO(1,n-1)$ being replaced by the general linear group $GL(n,R)$. It is convenient to work with a Möbius type representation for which we take the same symbol $A(n,R)$: It is that subgroup of $GL(n+1,R)$ which leaves the n-dimensional hyperplane $\overset{=}{R^n} := \left\{\overset{=}{x} = \begin{pmatrix} x \\ 1 \end{pmatrix} \in R^{n+1}\right\}$ invariant:

$$A(n,R) = \left\{ \begin{pmatrix} \Lambda & \tau \\ 0 & 1 \end{pmatrix} \in GL(n+1,R) \mid \Lambda \in GL(n,R),\ \tau \in R^n \right\}. \tag{87}$$

Thus, by an affine transformation, we obtain

$$\overset{=}{x}{}' = \begin{pmatrix} \Lambda & \tau \\ 0 & 1 \end{pmatrix} \begin{pmatrix} x \\ 1 \end{pmatrix} = \begin{pmatrix} \Lambda x + \tau \\ 1 \end{pmatrix}, \tag{88}$$

as is required for the action of the affine group on the flat affine space. The Lie algebra $a(n,R)$ consists of the generators P_γ, representing n-dimensional translations, and the $L^\alpha{}_\beta$, which span the Lie algebra $gl(n,R)$ of n-dimensional linear transformations.

In a matrix representation we can write the affine *gauge* group as

$$\mathcal{A}(n,R) = \left\{ \begin{pmatrix} \Lambda(x) & \tau(x) \\ 0 & 1 \end{pmatrix} \mid \Lambda(x) \in \mathcal{GL}(n,R), \quad \tau(x) \in \mathcal{T}(n,R) \right\}. \tag{89}$$

Having already had some experience with the Yang-Mills type gauge approach, we are aware of the need of introducing a gauge potential in order to step from rigid to soft group invariance. Accordingly, by gauging the affine group, the softening of the affine group transformations is accompanied by the introduction of the *generalized affine* connection (→ potential)

$$\stackrel{\approx}{\Gamma} = \begin{pmatrix} \Gamma^{(L)} & \Gamma^{(T)} \\ 0 & 0 \end{pmatrix} = \begin{pmatrix} \Gamma^{(L)\beta}_{\alpha} L^{\alpha}{}_{\beta} & \Gamma^{(T)\alpha} P_{\alpha} \\ 0 & 0 \end{pmatrix}. \tag{90}$$

It is a one–form $\stackrel{\approx}{\Gamma} = \stackrel{\approx}{\Gamma}_i dx^i$ and transforms inhomogeneously under an affine gauge transformation,

$$\stackrel{\approx}{\Gamma} \stackrel{A^{-1}(x)}{\longrightarrow} \stackrel{\approx}{\Gamma}{}' = A^{-1}(x)\stackrel{\approx}{\Gamma} A(x) + A^{-1}(x)dA(x)\,, \qquad A(x) \in \mathcal{A}(n,\, R), \tag{91}$$

where the transformation is formed with respect to the group element

$$A^{-1}(x) = \begin{pmatrix} \Lambda^{-1}(x) & -\Lambda^{-1}(x)\tau(x) \\ 0 & 1 \end{pmatrix}. \tag{92}$$

The corresponding affine curvature is given by

$$\stackrel{\approx}{R} := d\stackrel{\approx}{\Gamma} + \stackrel{\approx}{\Gamma} \wedge \stackrel{\approx}{\Gamma} = \begin{pmatrix} R^{(L)} & R^{(T)} \\ 0 & 0 \end{pmatrix} = \begin{pmatrix} d\Gamma^{(L)} + \Gamma^{(L)} \wedge \Gamma^{(L)} & d\Gamma^{(T)} + \Gamma^{(L)} \wedge \Gamma^{(T)} \\ 0 & 0 \end{pmatrix}. \tag{93}$$

It transforms covariantly under the affine gauge group:

$$\stackrel{\approx}{R} \stackrel{A^{-1}(x)}{\longrightarrow} \stackrel{\approx}{R}{}' = A^{-1}(x)\stackrel{\approx}{R} A(x)\,. \tag{94}$$

The exterior covariant derivative $\stackrel{\approx}{D} := d + \stackrel{\approx}{\Gamma}\wedge$ acts on an affine p–form $\stackrel{\approx}{\Psi} = \begin{pmatrix} \Psi \\ 1 \end{pmatrix}$ as follows:

$$\stackrel{\approx}{D}\stackrel{\approx}{\Psi} = \begin{pmatrix} d\Psi + \Gamma^{(L)} \wedge \Psi + \Gamma^{(T)} \\ 0 \end{pmatrix} = \begin{pmatrix} D\Psi + \Gamma^{(T)} \\ 0 \end{pmatrix}. \tag{95}$$

After substitution of (90) and (92), the inhomogeneous transformation law (91) splits into

$$\Gamma^{(L)} \stackrel{A^{-1}(x)}{\longrightarrow} \Gamma^{(L)\prime} = \Lambda^{-1}(x)\,\Gamma^{(L)}\,\Lambda(x) + \Lambda^{-1}(x)d\Lambda(x)\,, \tag{96}$$

and

$$\Gamma^{(T)} \stackrel{A^{-1}(x)}{\longrightarrow} \Gamma^{(T)\prime} = \Lambda^{-1}(x)\,\Gamma^{(T)} + \Lambda^{-1}(x)D\tau(x)\,. \tag{97}$$

The local translations $\tau(x)$ automatically drop out in (96) due to the one–form structure of $\Gamma^{(T)}$. Thereby (96) acquires the conventional transformation rule (with the exterior derivative d) of a Yang–Mills–type connection for $\mathcal{GL}(n, R)$. Thus we can identify $\Gamma^{(L)} = \Gamma = \Gamma_{\alpha}{}^{\beta} L^{\alpha}{}_{\beta}$ with the linear connection and are basically done with the gauging of the linear part of the affine group.

Things are more involved for the translational part, though. From what we have learned by gauging the translation group we expect $\Gamma^{(T)}$ to be related to the coframe $\vartheta := \vartheta^{\alpha} P_{\alpha}$, i.e. to a one–form with values in the Lie algebra of R^n. But due to the

Table 5. Gauge fields in metric-affine gauge theory MAG

Potential	Field strength	Bianchi identity
metric $g_{\alpha\beta}$	$Q_{\alpha\beta} = -Dg_{\alpha\beta}$	$DQ_{\alpha\beta} = 2R_{(\alpha}{}^{\mu}\, g_{\beta)\mu}$
coframe ϑ^α	$T^\alpha = D\vartheta^\alpha$	$DT^\alpha = R_\mu{}^\alpha \wedge \vartheta^\mu$
connection $\Gamma_\alpha{}^\beta$	$R_\alpha{}^\beta = d\Gamma_\alpha{}^\beta - \Gamma_\alpha{}^\mu \wedge \Gamma_\mu{}^\beta$	$DR_\alpha{}^\beta = 0$

covariant exterior derivative term $D\tau(x) := d\tau(x) + \Gamma^{(L)}\,\tau(x)$ in (97), the translational part $\Gamma^{(T)}$ does not transform as a covector, as is required for the coframe ϑ. To get a correspondence between $\Gamma^{(T)}$ and ϑ, we introduce a vector (vector–valued zero–form) $\overset{\approx}{\xi} = \begin{pmatrix} \xi \\ 1 \end{pmatrix} = \begin{pmatrix} \xi^\alpha P_\alpha \\ 1 \end{pmatrix}$ which transforms as $\overset{\approx}{\xi}' = A^{-1}(x)\,\overset{\approx}{\xi}$, i.e. as

$$\xi \overset{A^{-1}(x)}{\longrightarrow} \xi' = \Lambda^{-1}(x)\;(\xi - \tau(x)) \tag{98}$$

under an active affine gauge transformation. Then we define

$$\vartheta := \Gamma^{(T)} + D\xi \tag{99}$$

which transforms as a vector–valued one–form under the $\mathcal{A}(n,R)$, as required:

$$\vartheta \overset{A^{-1}(x)}{\longrightarrow} \vartheta' = \Lambda^{-1}(x)\,\vartheta\,. \tag{100}$$

The geometric and physical meaning of the relation (99), especially the role of the field ξ, is perhaps not completely satisfactorily clarified, yet. For an expanded discussion of this point we refer the reader again to ref.([29]), see also Mistura[58]. The translational piece of the affine connection, namely $\Gamma^{(T)}$, could play a decisive role in the gravitationally induced phase factor of a matter wave, as was pointed out by Morales-Técotl et al.[60]. Here, we confine ourselves to require the condition

$$D\xi = 0 \tag{101}$$

to hold. Then the generalized affine connection $\overset{\approx}{\Gamma}$ on the affine bundle $A(M)$ reduces to the *Cartan connection*

$$\overset{=}{\Gamma} = \begin{pmatrix} \Gamma^{(L)} & \vartheta \\ 0 & 0 \end{pmatrix} \tag{102}$$

on the bundle $L(M)$ of linear frames. Due to (100), this is not anymore a connection in the usual sense. We are thus left with the potentials $(\vartheta^\alpha, \Gamma_\alpha{}^\beta)$ and their corresponding field strengths $(T^\alpha, R_\alpha{}^\beta)$.

For this gauging of the affine group no metric was necessary. If additionally a *metric* is given, we recover the metric–affine geometrical arena of Sec. 2.4, as is summarized in Table 5.

We close this geometric section by introducing a further geometric ingredient, the "η-bases", which span the graded algebra of dual exterior forms on each cotangent space and which turn out to be quite useful in practical calculations. First, owing to the existence of a metric, we can define the scalar density $\sqrt{|\det g_{\mu\nu}|}$ and the familiar Hodge star operator *. The Hodge star operator maps a p–form Ψ into an $(n-p)$–form $^*\Psi$ by means of the explicit formula

$$^*\Psi := \frac{1}{(n-p)!\,p!}\sqrt{|\det g_{\mu\nu}|}\, g^{\alpha_1\gamma_1}\cdots g^{\alpha_p\gamma_p}\,\epsilon_{\alpha_1\cdots\alpha_p\beta_1\cdots\beta_{n-p}}\,\Psi_{\gamma_1\cdots\gamma_p}\,\vartheta^{\beta_1}\wedge\cdots\wedge\vartheta^{\beta_{n-p}}\,. \tag{103}$$

Now we define the g-volume element n–form

$$\eta := \sqrt{|\det g_{\mu\nu}|}\,\vartheta^{\hat{1}}\wedge\cdots\wedge\vartheta^{\hat{n}} = \frac{1}{n!}\sqrt{|\det g_{\mu\nu}|}\,\epsilon_{\alpha_1\cdots\alpha_n}\vartheta^{\alpha_1}\wedge\cdots\wedge\vartheta^{\alpha_n} = {}^*1\,, \tag{104}$$

dual to the unit zero–form. Picking a (pseudo-)orthonormal positively oriented coframe $\overset{\circ}{\vartheta}{}^{\alpha}$, the g-volume element simplifies to

$$\eta = \overset{\circ}{\vartheta}{}^{\hat{1}}\wedge\cdots\wedge\overset{\circ}{\vartheta}{}^{\hat{n}}\,. \tag{105}$$

Having this n–form at our disposal, we can successively contract it by means of the frame e_α, thereby arriving at an $(n-1)$–form, an $(n-2)$–form, etc., until we terminate the series with a zero–form:

$$\begin{aligned}
\eta_{\alpha_1} &:= e_{\alpha_1}\rfloor\eta = \frac{1}{(n-1)!}\,\eta_{\alpha_1\alpha_2\cdots\alpha_n}\vartheta^{\alpha_2}\wedge\cdots\wedge\vartheta^{\alpha_n} = {}^*\vartheta_{\alpha_1}\,,\\
\eta_{\alpha_1\alpha_2} &:= e_{\alpha_2}\rfloor\eta_{\alpha_1} = \frac{1}{(n-2)!}\,\eta_{\alpha_1\alpha_2\alpha_3\cdots\alpha_n}\vartheta^{\alpha_3}\wedge\cdots\wedge\vartheta^{\alpha_n} = {}^*(\vartheta_{\alpha_1}\wedge\vartheta_{\alpha_2})\,,\\
&\ \ \vdots\\
\eta_{\alpha_1\cdots\alpha_n} &:= e_{\alpha_n}\rfloor\cdots\rfloor e_{\alpha_1}\rfloor\eta = {}^*(\vartheta_{\alpha_1}\wedge\cdots\wedge\vartheta_{\alpha_n})\,.
\end{aligned} \tag{106}$$

Thus, in four dimensions, which we are concentrating on, we have an η-basis at our disposal (η, η_α, $\eta_{\alpha\beta}$, $\eta_{\alpha\beta\gamma}$, $\eta_{\alpha\beta\gamma\delta}$) that we will meet in numerous applications.

7. Field equations of metric–affine gauge theory (MAG)

The gauge procedure of the affine group led to the identification of the gauge potentials of the MAG, see Table 5. The material currents are then expected to couple to these potentials in the Yang-Mills type fashion, as is indicated in Fig. 1, and as we know already from the material energy-momentum current and the coframe according to the line after (77). If Ψ, as a p-form, represents a matter field (fundamentally a representation of the $SL(4,R)$ or of some of its subgroups[61,62]), its first order Lagrangian L will be embedded in metric-affine spacetime by the minimal coupling procedure, that is, exterior covariant derivatives feature in the kinetic terms of the Lagrangian instead of only exterior ones. Then the material currents are defined as follows:

$$\sigma^{\alpha\beta} := 2\,\frac{\delta L}{\delta g_{\alpha\beta}}\,, \qquad \Sigma_\alpha := \frac{\delta L}{\delta \vartheta^\alpha}\,, \qquad \Delta^\alpha{}_\beta := \frac{\delta L}{\delta \Gamma_\alpha{}^\beta}\,. \tag{107}$$

From GR we expect that $\sigma^{\alpha\beta}$ is the metric (and symmetric) energy-momentum current of matter ('Hilbert current'), whereas Secs. 4 and 5 lead us to the believe that Σ_α should be the corresponding canonical energy-momentum current ('Noether current'). The canonical hypermomentum current $\Delta^\alpha{}_\beta$, which couples to the linear connection, can be decomposed according to

$$\begin{array}{ccccccc} \Delta_{\alpha\beta} & = & \tau_{\alpha\beta} & + & \frac{1}{4}\,g_{\alpha\beta}\,\Delta & + & \widehat{\not\Delta}_{\alpha\beta} \\ & \sim & \text{spin current} & \oplus & \text{dilation current} & \oplus & \text{shear current}\,, \end{array} \tag{108}$$

with $\tau_{\alpha\beta} := \Delta_{[\alpha\beta]}$ as (dynamical) spin current, $\Delta := \Delta^\gamma{}_\gamma$, as dilation current, and $\widehat{\not\Delta}_{\alpha\beta}$ as symmetric and tracefree shear current.

If one applies the Noether theorem to the material Lagrangian, then, Σ_α, $\tau_{\alpha\beta}$, and Δ can be identified with the special-relativistic canonical Noether currents of energy-momentum, spin, and dilation, respectively. This leaves only the shear current $\widehat{\not\Delta}_{\alpha\beta}$, see our earlier work[21] for a detailed discussion, as a concept that is a bit more remote from direct observation than the other currents. In any case, the definitions (107) and their physical interpretations are completely justified by these considerations.

Adding to the matter Lagrangian a metric-affine first order gauge Lagrangian V,

$$W = \int \left[V(g_{\alpha\beta}, \vartheta^\alpha, Q_{\alpha\beta}, T^\alpha, R_\alpha{}^\beta) + L(g_{\alpha\beta}, \vartheta^\alpha, \Psi, D\Psi) \right], \tag{109}$$

and applying the action principle, we find[29] the matter and the gauge field equations (of Yang-Mills type):

$$D\left(\frac{\partial L}{\partial (D\Psi)}\right) - (-1)^p\,\frac{\partial L}{\partial \Psi} \;=\; 0\,, \qquad \text{(MATTER)} \tag{110}$$

$$D\left(2\,\frac{\partial V}{\partial Q_{\alpha\beta}}\right) + 2\,\frac{\partial V}{\partial g_{\alpha\beta}} \;=\; -\,\sigma^{\alpha\beta}\,, \qquad \text{(ZEROTH)} \tag{111}$$

$$D\left(\frac{\partial V}{\partial T^\alpha}\right) + \frac{\partial V}{\partial \vartheta^\alpha} \;=\; -\,\Sigma_\alpha\,, \qquad \text{(FIRST)} \tag{112}$$

$$D\left(\frac{\partial V}{\partial R_\alpha{}^\beta}\right) + \vartheta^\alpha \wedge \frac{\partial V}{\partial T^\beta} + 2\,g_{\beta\gamma}\,\frac{\partial V}{\partial Q_{\alpha\gamma}} \;=\; -\,\Delta^\alpha{}_\beta\,. \qquad \text{(SECOND)} \tag{113}$$

In SECOND a Noether identity for V has already been employed. Analogously, in FIRST, the canonical energy-momentum of the translational gauge potential ϑ^α can be expressed explicitly as

$$\frac{\partial V}{\partial \vartheta^\alpha} = e_\alpha \rfloor V - (e_\alpha \rfloor T^\beta) \wedge \frac{\partial V}{\partial T^\beta} - (e_\alpha \rfloor R_\beta{}^\gamma) \wedge \frac{\partial V}{\partial R_\beta{}^\gamma} - (e_\alpha \rfloor Q_{\beta\gamma}) \frac{\partial V}{\partial Q_{\beta\gamma}} \,. \tag{114}$$

This structure is known from Minkowski's energy-momentum tensor of the Maxwellian field. It is interesting to note that, provided SECOND is fulfilled, FIRST and ZEROTH are equivalent, i.e., one of them is redundant. It is for this reason that we abstain, for $\partial V/\partial g_{\alpha\beta}$, to display a formula similar to (114).

The field equation of a Yang-Mills theory reads $\overset{A}{D}\,(\partial V_{\rm YM}/\partial F) = -J$. The field equations ZEROTH, FIRST, and SECOND are of this type. However, because of the universality of 'external' interactions (gravitation), additional *tensor*-valued *gauge* currents ($\partial V/\partial\vartheta^\alpha$ etc.) surface in the field equations. This is the distinguishing feature of MAG as compared to gauge theories of internal groups ($U(1)$, $SU(2)$, ...). In Yang-Mills theory the non-linearity of the gauge field is hidden in the non-tensorial pieces of the gauge covariant exterior derivative $\overset{A}{D}$ occurring on the left hand side $\overset{A}{D}\,(\partial V_{\rm YM}/\partial F)$ of the Yang-Mills equation. In gravitational gauge theory, there are *additional* non-linearities, besides those in $\overset{\Gamma}{D}$, namely the ones represented by (114) etc.. This result is non-trivial as a simple argument will show:

Suppose we had a Hilbert-Einstein gauge Lagragian linear in the curvature $R_\alpha{}^\beta$. Then in FIRST the leading term in differential order on its left hand side will vanish and we are left with (114). The surviving terms are

$$\frac{\partial V_{\rm HE}}{\partial \vartheta^\alpha} \sim e_\alpha \rfloor V_{\rm HE} - (e_\alpha \rfloor R_\beta{}^\gamma) \wedge \frac{\partial V_{\rm HE}}{\partial R_\beta{}^\gamma} \sim -\Sigma_\alpha \,. \tag{115}$$

Since $\partial V_{\rm HE}/\partial R_\beta{}^\gamma$ will be a constant, we recover the Einstein three-form (corresponding to the Einstein tensor in Ricci calculus) from this equation, giving substance to Schrödinger's dictum[83] that the left hand side of Einstein's equation is, in some sense, the gravitational energy-momentum tensor. Consequently, Einstein's field equation of GR is encapsulated in $\partial V/\partial\vartheta^\alpha$ of FIRST and, as such, has a distinctive anti-Yang-Mills flavor. In contrast, the Einsteinian teleparallelism $\text{GR}_{||}$, with its torsion-square Lagrangian $V_{||}$, picks up an essential piece from the proper Yang-Mills term of FIRST,

$$D\frac{\partial V_{||}}{\partial T^\alpha} + e_\alpha \rfloor V_{||} - (e_\alpha \rfloor T^\beta) \wedge \frac{\partial V_{||}}{\partial T^\beta} \sim -\Sigma_\alpha \,, \tag{116}$$

compare (77). It has much more of the Yang-Mills spirit than GR has – and this is the reason why $\text{GR}_{||}$ turned up when we gauged the translations in Secs. 4 and 5.

If, within our formalism, one desires to correctly derive the field equation for GR, see (115), and for $\text{GR}_{||}$, see (116), then one has to put on *Lagrange multipliers.* In the case of (115) they have to kill nonmetricity and torsion and for (116) to remove nonmetricity and curvature. The details have been worked out in our earlier review[29], see also Kopczyński[38]. **MAG** (metric-affine gauge theory, the general framework), **PG** (Poincaré gauge theory, vanishing nonmetricity, hence in a Riemann-Cartan spacetime), **EC** (Einstein-Cartan theory, the PG with the curvature scalar as gravitational Lagrangian), $\mathbf{GR}_{||}$ (Einsteinian teleparallelism, vanishing nonmetricity, vanishing curvature, hence in a Weitzenböck spacetime, specific torsion square Lagrangian), and **GR** (general relativity, vanishing nonmetricity, vanishing torsion, hence in a Riemannian space, curvature scalar as Lagrangian) are different (sub-)cases of this general scheme.

A further remark is in order: Gauging the affine group yields the gauge potentials $(\vartheta^\alpha, \Gamma_\alpha{}^\beta)$, see Table 5. If a metric exists on top of that linearly connected manifold, then a further independent geometrical field variable is at hand. Following Trautman[88], we are taking $(g_{\alpha\beta}, \vartheta^\alpha, \Gamma_\alpha{}^\beta)$ as independent variables in the action (109). Because of the redundancy of ZEROTH or FIRST, *provided* SECOND is fulfilled, one could argue that one should drop, say, the coframe as independent variable, as done by Tucker[f] and Wang[95], for example. Because of different arguments, however, we feel more comfortable with our procedure: (i) Both energy-momentum currents, the Hilbertian $\sigma^{\alpha\beta}$ and the Noetherian Σ_α, have a good and direct physical interpretation in SR. Dropping a gauge variable means dropping one of these useful quantities as a fundamental current. (ii) *A posteriori*, we do find the identity causing the redundancy mentioned. This seems safer than to assume something to that extend *a priori*. (iii) If we dropped the metric as an independent variable, for example, then we would have to take *orthonormal* frames as field variables, giving away a piece of freedom which we had earlier in being able to work with whatever frame we liked, be it orthonormal or oblique.

Let us finally remind ourselves: As soon as the gauge Lagrangian V is specified explicitly, we can find the field equations ZEROTH, FIRST, and SECOND by sheer partial differentiation. Hence the using of the general form of the field equations may save a lot of work.

8. Model building: Einstein-Cartan theory and beyond

The missing piece within the framework that we finally established in Sec. 7 is the gauge field Lagrangian $V(g_{\alpha\beta}, \vartheta^\alpha, Q_{\alpha\beta}, T^\alpha, R_\alpha{}^\beta)$. The hope is that the model, with a suitably chosen V – perhaps combined with some symmetry breaking mechanism which, for example, reduces the linear group $GR(4, R)$ to the Lorentz group $SO(1,3)$ – can be consistently quantized. For 1+1 dimensional curvature square models[51,64,67] successful quantization methods are already available, see Kloesch and Strobl[37] and the literature given there.

In testing a new framework, one first wants, in some limit, to recover old ground

[f] We are grateful to Robin W. Tucker (Lancaster) for an interesting discussion on this question.

where one feels at home. In our case this is GR. There are at least two ways of how to achieve this: One can choose the curvature scalar à la Hilbert (in exterior calculus: the corresponding four-form $R^{\alpha\beta}\wedge\eta_{\alpha\beta}$ à la Trautman[88]) and take Lagrange multipliers for extinguishing nonmetricity and torsion:

$$V_{\rm GR} = -\frac{1}{2\ell^2}\, R^{\alpha\beta}\wedge\eta_{\alpha\beta} + \frac{1}{2}\, Q_{\alpha\beta}\wedge {}^{(1)}\lambda^{\alpha\beta} + T^\alpha\wedge {}^{(2)}\lambda_\alpha\,. \tag{117}$$

Note that $R_\alpha{}^\beta$ is the curvature tensor of the independent field variable $\Gamma_\alpha{}^\beta$, for the η's see (106).

8.1. *Einstein-Cartan theory EC*

A second way is to start with the Einstein-Cartan Lagrangian (here with cosmological constant)

$$V_{\rm EC} = -\frac{1}{2\ell^2}\, R^{\alpha\beta}\wedge\eta_{\alpha\beta} + \frac{1}{2}\,\Lambda\,\eta + \frac{1}{2}\, Q_{\alpha\beta}\wedge\lambda^{\alpha\beta}\,, \tag{118}$$

and to derive the corresponding field equations:

$$\frac{1}{2}\,\eta_{\alpha\beta\gamma}\wedge R^{\beta\gamma} + \Lambda\,\eta_\alpha \;=\; \ell^2\,\Sigma_\alpha\,, \tag{119}$$

$$\frac{1}{2}\,\eta_{\alpha\beta\gamma}\wedge T^\gamma \;=\; \ell^2\,\tau_{\alpha\beta}\,. \tag{120}$$

In components in a holonomic frame they read:

$$Ric_{ij} - \frac{1}{2}\, g_{ij} Ric_k{}^k + \Lambda\, g_{ij} \;=\; \ell^2\,\Sigma_{ij}\,, \tag{121}$$

$$T_{ij}{}^k + 2\,\delta^k_{[i}\, T_{j]l}{}^l \;=\; \ell^2\,\tau_{ij}{}^k\,. \tag{122}$$

We recover GR for vanishing spin $\tau_{\alpha\beta} = 0$. In this context only the vanishing of nonmetricity had to be assumed. The vanishing of torsion, for spinless matter, was the result of the second field equation (120). We will see below, in Sec. 8.2, that also the last Lagrange multiplier can be abandoned if one amends the gravitational part of the Lagrangian with a piece quadratic in the Weyl one-form.

Basically, the two EC field equations (119) and (120) are *first order* partial differential equations in $\Gamma_\alpha{}^\beta$ and ϑ^α, with a spin fluid as source, see ref.([65]). As we explained already in the last Sec. 7, this is the anti-Yang-Mills flavor of GR or EC caused by the absence of an explicit torsion piece in the Lagrangian. A physical consequence is that in EC we have the usual gravitational interaction of the Newton-Einstein type plus a very weak (non-propagating) spin contact interaction. Up to the fifties, *weak* interaction was also thought to be of contact type (in fact, of a vector-axial vector type). Later, following gauge ideas, the short-range intermediate W-boson (and the Z) were postulated that made the weak interaction propagate.

8.2. Poincaré gauge theory PG, the quadratic version

The simplest appearance of explicit torsion pieces is that in the $\mathrm{GR}_{||}$ Lagrangian[24,25] studied in Sec. 5:

$$V_{||} = -\frac{1}{2\ell^2} T^\alpha \wedge {}^* \left(-{}^{(1)}T_\alpha + 2\,{}^{(2)}T_\alpha + \frac{1}{2}{}^{(3)}T_\alpha \right) + \frac{1}{2}\, Q_{\alpha\beta} \wedge {}^{(1)}\lambda^{\alpha\beta} + R_\alpha{}^\beta \wedge {}^{(2)}\lambda^\alpha{}_\beta \,. \quad (123)$$

This can be considered as a starting point for turning to Lagrangians quadratic in the field strengths. Amongst the simplest model cases is the *purely* quadratic von der Heyde et al.[32,6] Lagrangian

$$V_{\mathrm{vdH}} = -\frac{1}{2\ell^2} \left(T^\alpha \wedge \vartheta^\beta \right) \wedge {}^* \left(T_\beta \wedge \vartheta_\alpha \right) - \frac{1}{2\kappa} R^{\alpha\beta} \wedge {}^* R_{\alpha\beta} + \frac{1}{2}\, Q_{\alpha\beta} \wedge \lambda^{\alpha\beta} \,. \quad (124)$$

This Lagrangian has been 'derived' by means of the Gordon decomposition argument[32,25], see also Rumpf[81]. The torsion square piece in this Lagrangian, in a Weitzenböck spacetime, has a classical Newtonian limit. It differs from the torsion pieces in the teleparallel Lagrangian (123) by a quadratic axial torsion piece (then in (123) we had $-{}^{(3)}T_\alpha$ instead).

A number of exact classical solutions has been found for the model (124), a Kerr-NUT solution with torsion[4] is amongst the most prominent ones. For illustrating the basic features of such solutions, we display the less complicated subcase, namely the Baekler-Lee[3,43] solution – this is the Reissner-Nordström solution with dynamic torsion – as a fairly transparent example. We choose Schwarzschild coordinates (t, r, θ, ϕ), M = mass, q = electric charge, and find the orthonormal coframe

$$\begin{aligned} \vartheta^{\hat t} &= \frac{1}{2}[(\Phi+1)dt + (1-\frac{1}{\Phi})dr]\,, \\ \vartheta^{\hat r} &= \frac{1}{2}[(\Phi-1)dt + (1+\frac{1}{\Phi})dr]\,, \\ \vartheta^{\hat\theta} &= r\,d\theta\,, \\ \vartheta^{\hat\varphi} &= r\sin\theta\, d\varphi\,. \end{aligned} \quad (125)$$

The corresponding metric reads:

$$ds^2 = -\Phi dt^2 + \frac{1}{\Phi} dr^2 + r^2 (d\theta^2 + \sin^2\theta\, d\varphi^2)\,. \quad (126)$$

In the Reissner-Nordström function

$$\Phi := 1 - \frac{2(Mr - q^2)}{r^2} - \frac{\kappa}{4\ell^2} r^2 \quad (127)$$

the 'cosmological term' is induced by the Yang-Mills type curvature square piece in the Lagrangian ('strong gravity', cf. ref.([63])). If we additionally had a 'naked' cosmological constant, we could shift its value by a suitable choice of the 'strong' coupling constant κ. Torsion and curvature are given by

$$\begin{aligned} T^{\hat{t}} = T^{\hat{r}} &= \frac{Mr - 2q^2}{r^3}\, \vartheta^{\hat{t}} \wedge \vartheta^{\hat{r}} , \\ T^{\hat{\theta}} &= \frac{Mr - q^2}{r^3} \left(\vartheta^{\hat{t}} \wedge \vartheta^{\hat{\theta}} - \vartheta^{\hat{r}} \wedge \vartheta^{\hat{\theta}} \right), \\ T^{\hat{\varphi}} &= \frac{Mr - q^2}{r^3} \left(\vartheta^{\hat{t}} \wedge \vartheta^{\hat{\varphi}} - \vartheta^{\hat{r}} \wedge \vartheta^{\hat{\varphi}} \right), \end{aligned} \tag{128}$$

and

$$R^{\alpha\beta} = \frac{\kappa}{4\ell^2}\, \vartheta^\alpha \wedge \vartheta^\beta + \frac{Mr - q^2}{r^2}\, {}^{(4)}\mathcal{R}^{\alpha\beta} , \tag{129}$$

respectively, where

$$\begin{aligned} {}^{(4)}\mathcal{R}^{\hat{t}\hat{\theta}} &= {}^{(4)}\mathcal{R}^{\hat{r}\hat{\theta}} := \frac{\kappa}{4\ell^2} \left(\vartheta^{\hat{t}} \wedge \vartheta^{\hat{\theta}} - \vartheta^{\hat{r}} \wedge \vartheta^{\hat{\theta}} \right), \\ {}^{(4)}\mathcal{R}^{\hat{t}\hat{\varphi}} &= {}^{(4)}\mathcal{R}^{\hat{r}\hat{\varphi}} := \frac{\kappa}{4\ell^2} \left(\vartheta^{\hat{t}} \wedge \vartheta^{\hat{\varphi}} - \vartheta^{\hat{r}} \wedge \vartheta^{\hat{\varphi}} \right), \end{aligned} \tag{130}$$

represent a tracefree symmetric Ricci piece of the curvature two-form. The Coulomb field shows up in the electromagnetic field strength:

$$F = \frac{2\, q}{\ell\, r^2}\, \vartheta^{\hat{t}} \wedge \vartheta^{\hat{r}} . \tag{131}$$

It is a spherically symmetric vacuum solution with Maxwell field, i.e. a Reissner-Nordström solution with dynamic torsion. We don't display the solution in its original frame but in a suitably rotated one[5] such that the torsion two-form (128) has a 'Coulombic' look without global extra factors in front of the corresponding expressions. Note the factor of *two* in the $T^{\hat{t}}$ component of (128) in the q^2-piece. One interesting feature of this solution is that the curvature square Lagrangian supplies a constant curvature background proportional to κ, see the first piece on the right hand side of (129). The *torsion* is – this is not an unexpected feature in the light of our teleparallelism 'philosophy' – induced by ordinary *Newton-Einstein gravity*, as can be read off from (128), a fact usually hard to swallow by colleagues who relate torsion with obscurity. In (128) torsion is visibly the translation field strength, a fact made possible by the purely quadratic torsion piece in the Lagrangian, without a Hilbert-Einstein type admixture.

The von der Heyde Lagrangian (124) is a subcase of the general quadratic PG Lagrangian[25,17,66]

$$\begin{aligned} V_{\rm QPG} = \frac{\Lambda}{\ell^2}\,\eta \; &+ \; \frac{a_0}{2\ell^2}\, R^{\alpha\beta} \wedge \eta_{\alpha\beta} + \frac{1}{2\ell^2}\, T^\alpha \wedge {}^*\!\left(\sum_{M=1}^{3} a_{(M)}\, {}^{(M)}T_\alpha \right) \\ &+ \; \frac{1}{2\kappa}\, R^{\alpha\beta} \wedge {}^*\!\left(\sum_{N=1}^{6} b_{(N)}\, {}^{(N)}R_{\alpha\beta} \right) + \frac{1}{2}\, Q_{\alpha\beta} \wedge \lambda^{\alpha\beta} . \end{aligned} \tag{132}$$

Each of the three irreducible torsion and six irreducible curvature pieces contributes to the Lagrangian with an individual weight. The propagating modes of this Lagrangian were investigated by Sezgin and van Nieuwenhuizen[85] and by Kuhfuss and Nitsch[41]. A subclass of Lagrangians survived their selection criteria motivated by quantum field theoretical considerations (ghost-freeness, positive energy). Minkevich[54,55], amongst others, studied Freidmann type cosmological models resulting from such a Lagrangian.

8.3. Coupling to a scalar field

It is near at hand to add a dilaton type massless scalar field to this Lagrangian:

$$V_{\Phi\text{grav}} = V_{\text{QPG}} + \frac{1}{2}\, d\Phi \wedge {}^{*}d\Phi \,. \tag{133}$$

We found for this model a remarkable exact solution, a *torsion kink*[6]. Attributing distinctive Higgs-like features to the scalar field, we arrive at the more general model of Floreanini and Percacci[16]:

$$\begin{aligned} V_{\text{FP}} \;=\; & c_1 d\Phi \wedge {}^{*}d\Phi + c_2 d\Phi \wedge {}^{*}(e_\mu \rfloor T^\mu) + U(\Phi) + \frac{a_0}{2\ell^2}\,\Phi^2 R^{\alpha\beta} \wedge \eta_{\alpha\beta} \\ & + \frac{1}{2\ell^2}\, T^\alpha \wedge {}^{*}\left(\sum_{M=1}^{3} a_M\, {}^{(M)}T_\alpha\right) + \frac{1}{2}\, R^{\alpha\beta} \wedge {}^{*}\left(\sum_{N=1}^{6} b_N\, {}^{(N)}R_{\alpha\beta}\right) + \frac{1}{2}\, Q_{\alpha\beta} \wedge \lambda^{\alpha\beta} . \end{aligned} \tag{134}$$

The scalar Φ couples to the EC term in a Jordan-Brans-Dicke type way. The explicit form of the 'Higgs' potential $U(\Phi)$ is left open. Note the direct coupling $d\Phi \wedge T$ in (134). Its authors, on a quantum field theoretical level, investigated the renormalizability properties of their model.

Being relativists, we would be ill-advised if we didn't try to give the scalar field Φ a geometrical meaning, perhaps in the context of the Weyl one-form Q which is of the type of a gauge potential for dilations anyways. Therefore we lift the last Lagrange multiplier and now turn to...

8.4. Metric-affine gauge theory MAG

Let us, however, first continue the discussion of above of how to arrive at GR in spite of relaxing the last constraint and liberating thereby the connection completely from its dominance by the metric. The naive way, namely just to take a term proportional to $R^{\alpha\beta} \wedge \eta_{\alpha\beta}$, doesn't work. The projective transformation

$$\Gamma_\alpha{}^\beta \longrightarrow \Gamma_\alpha{}^\beta + \delta_\alpha^\beta P \,, \tag{135}$$

with some one-form field P, leaves the Hilbert-Einstein type Lagrangian invariant. Consequently, in such a model, the connection would only be determined up to a one-form (with four components). This is unsatisfactory. Moreover, if one coupled the gauge fields to matter, then only projectively invariant matter Lagrangians would be allowed, an a priori constraint without physical justification. Therefore one has to remove the projective invariance from the gravitational Lagrangian.

Since a projective transformation changes the trace $\Gamma_\gamma{}^\gamma$ of a connection and this trace is closely related to the Weyl one-form,

$$\Gamma_\gamma{}^\gamma = 2\,Q + d\ln\sqrt{|\det g_{\alpha\beta}|}\,, \qquad d\Gamma_\gamma{}^\gamma = R_\gamma{}^\gamma = 2\,dQ\,, \tag{136}$$

see Eq.(3.10.13) of ref.([29]), an obvious way to remove the projective invariance is to add the square of the Weyl one-form[26,47]:

$$V_{\mathrm{GR}'} = \frac{1}{2\ell^2}\left(-R^{\alpha\beta}\wedge\eta_{\alpha\beta} + \beta\,Q\wedge{}^*Q\right)\,. \tag{137}$$

A moment's reflection will remind ourselves what we have achieved by this innocently looking Lagrangian (137): Varying metric, frame, and connection independently, in vacuum – that is, in the absence of matter – yields the *Einstein vacuum equation in a Riemannian spacetime.* Generically, if matter is present and supplies energy-momentum and hypermomentum currents, then the *hypermomentum*, via the second field equation (113), turns out to be proportional to the *post*-Riemannian pieces of the connection.

The problem with the Lagrangian (137) is that it has the same defect as the EC Lagrangian. We have Newton-Einstein gravity and no further propagating modes. The obvious remedy is to add kinetic terms and, since we called the Weyl one-form already in earlier, we may want to make these modes propagating[28,78]:

$$V_{\mathrm{GRQ}} = \frac{1}{2\ell^2}\left(-R^{\alpha\beta}\wedge\eta_{\alpha\beta} + \alpha\,dQ\wedge{}^*dQ + \beta\,Q\wedge{}^*Q\right)\,. \tag{138}$$

In our first order formalism the Lagrangian must be expressed in terms of the gauge potential and their first derivatives. Since Q is already on the level of a field strength, one may be tempted to forbid the kinetic Q-terms. However, by means of the trace of the zeroth Bianchi identity, $dQ = \frac{1}{2}R_\gamma{}^\gamma$, we can cast out the pseudo-second order terms and find the well-behaved first order Lagrangian

$$V_{\mathrm{GRQ}} = \frac{1}{2\ell^2}\left(-R^{\alpha\beta}\wedge\eta_{\alpha\beta} + \frac{\alpha}{4}\,R_\beta{}^\beta\wedge{}^*R_\gamma{}^\gamma + \beta\,Q\wedge{}^*Q\right)\,. \tag{139}$$

This might be the *simplest* reasonable metric-affine Lagrangian with propagating modes. Probably it deserves closer investigation.

The trace $R_\beta{}^\beta$ of the curvature in (139) represents an irreducible piece of the curvature, in fact the piece which we numbered as *ten* and called, in our computer algebra programs[84], DILCURV. Altogether, in a metric-affine space, the curvature has *eleven* irreducible pieces, see ref.([29]), Table 4. If we recall that the nonmetricity has *four* irreducible pieces, then the general quadratic Lagrangian in MAG reads:

$$\begin{aligned} V_{\rm QMA} &= \frac{\Lambda}{\ell^2}\,\eta + \frac{a_0}{2\ell^2}\,R^{\alpha\beta}\wedge\eta_{\alpha\beta} + \frac{1}{2\ell^2}\,T^\alpha\wedge{}^*\left(\sum_{M=1}^{3} a_{(M)}\,{}^{(M)}T_\alpha\right) \\ &+ \frac{1}{2\kappa}\,R^{\alpha\beta}\wedge{}^*\left(\sum_{N=1}^{11} b_{(N)}\,{}^{(N)}R_{\alpha\beta}\right) + \frac{1}{2\ell^2}\,Q_{\alpha\beta}\wedge{}^*\left(\sum_{K=1}^{4} c_{(K)}\,{}^{(K)}Q_{\alpha\beta}\right) \\ &+ \frac{1}{2\ell^2}\,\vartheta^\alpha\wedge{}^*T^\beta\wedge\left(\sum_{K=2,3,4} d_{(K)}\,{}^{(K)}Q_{\alpha\beta}\right). \end{aligned} \tag{140}$$

In spite of this affluence of generality, Tresguerres[91,g] was able to find two exact solutions of the corresponding vacuum field equations in a most remarkable piece of work. He also discussed[91] the reasons for introducing the mixed $\vartheta\wedge{}^*T\wedge Q$ term. Generically Tresguerres found a Baekler-Lee type solution with torsion, see (125) to (131) above, but the *dilaton charge takes the place of the electric charge.* Thus the Weyl one-form Q, the quasi-Maxwellian potential, has a $1/r$-behavior and is closely interwoven with the torsion vector.

Tucker and Wang[95] put $\beta=0$ in (138) or (139),

$$V_{\rm TW} = \frac{1}{2l^2}(-R^{\alpha\beta}\wedge\eta_{\alpha\beta} + \alpha\,dQ\wedge{}^*dQ) = \frac{1}{2\ell^2}\left(-R^{\alpha\beta}\wedge\eta_{\alpha\beta} + \frac{\alpha}{4}\,R_\beta{}^\beta\wedge{}^*R_\gamma{}^\gamma\right), \tag{141}$$

i.e. they excluded the massive term of the Weyl field from further consideration. Rather, they explored the analogy of the Weyl one-form Q to the Maxwell potential A under these circumstances. Their field equations can be read off from (112), (113), and (114) straightforwardly as,

$$-e_\alpha\rfloor V_{\rm TW} + (e_\alpha\rfloor R_\beta{}^\gamma)\wedge\frac{\partial V_{\rm TW}}{\partial R_\beta{}^\gamma} = \Sigma_\alpha\,, \tag{142}$$

$$-D\left(\frac{\partial V_{\rm TW}}{\partial R_\alpha{}^\beta}\right) = \Delta^\alpha{}_\beta\,, \tag{143}$$

or

$$\frac{1}{2}\,\eta_{\alpha\beta\gamma}\wedge R^{\beta\gamma} + \frac{\alpha}{8}\left[(e_\alpha\rfloor R_\beta{}^\beta)\wedge{}^*R_\gamma{}^\gamma - (e_\alpha\rfloor{}^*R_\beta{}^\beta)\wedge R_\gamma{}^\gamma\right] = \ell^2\,\Sigma_\alpha\,, \tag{144}$$

$$\frac{1}{2}\,D\eta^\alpha{}_\beta - \frac{\alpha}{4}\,\delta^\alpha_\beta\,d{}^*R_\gamma{}^\gamma = \ell^2\,\Delta^\alpha{}_\beta\,. \tag{145}$$

Tucker and Wang[95] found Baekler-Lee type vacuum solutions with dilation ('Weyl') charge, similar as Tresguerres, but, in addition, they presented a highly interesting solution with propagating massless spinor matter as source.

[g]See also Tresguerres[89,90,92,93], Macías et al.[44], and, for cosmological models, Minkevich[56].

Nevertheless, one should recognize that the Lagrangian (141) is not without problems. Because the massive piece $\sim Q \wedge {}^*Q$ is missing, V_{TW} is invariant under the special projective transformation

$$\Gamma_\alpha{}^\beta \longrightarrow \Gamma_\alpha{}^\beta + \delta_\alpha^\beta \, dp \,. \tag{146}$$

Whereas such a type of 'gauge' transformation is desirable for an internal $U(1)$-connection – like in Maxwell's theory – it is definitely dangerous in the context of a dilation transformation. The linear connection is only determined up to the transformation (146), that is, 4 of its 64 components are left undetermined in the TW-model, see also the analysis of Teyssandier and Tucker[86].

The case studies of Tresguerres and Tucker-Wang taught us that the Weyl one-form plays a particular role in representing a dilation tpye field that ought to be useful in the context of the breaking of the dilation symmetry. Therefore there were attempts by Mielke et al., see ref.([29]) Sec. 6, to superimpose on the Lagrangian (140) an additional conformal symmetry in order to have a massless theory, free of dimensionful coupling constants at the beginning. Then one couples to a hypothetical dilaton field σ,

$$V = \sigma^2 \, R^{\alpha\beta} \wedge \eta_{\alpha\beta} + \frac{1}{2} \sigma \, D \wedge {}^*D\sigma + \frac{\lambda}{4} \sigma^4 \, \eta \,, \tag{147}$$

and breaks the dilation symmetry, thereby arriving at a low energy massive Lagrangian. Analogous approaches were proposed by Gregorash and Papini[19,20], Hochberg and Plunien[33], and by Poberii[77]. Recently Pawłowski and Rączka[75,74,94] developed similar models in a Riemannian spacetime, though. Nevertheless, they required conformal invariance for the Lagrangian they started from, coupled to a dilation field etc.. There are close relationships between the gravitational sectors of these models which lead us to the belief that the P&R model should be redrafted in the framework of a Weyl-Cartan, if not of a metric-affine spacetime.

With all these developments from different quarters, we have here – perhaps for the first time – a consistent framework for gauge models carrying both, non-trivial torsion *and* nonmetricity.

9. Acknowledgments

One of the authors would like to thank Venzo de Sabbata, Peter Bergmann, and H.-J. Treder for the invitation to lecture at this Erice School and his coauthors[29] Dermott McCrea[†] (Dublin), Eckehard Mielke (Kiel/Mexico City), and Yuval Ne'eman (Tel Aviv/Austin) for great help. Moreover, he is very obliged to H.A. Kastrup (Aachen), G. Süssmann (Munich), Ryszard Rączka (Warsaw), and H. Petry (Bonn) for seminar invitations where different versions of these lectures were given. We are most grateful to our colleagues David Hartley (St.Augustin), Yuri Obukhov

(Moscow/Cologne), and Romualdo Tresguerres (Madrid) for help and many clarifying discussions.

10. References

1. A. Ali, J. Ellis, and S. Randjbar-Daemi (eds.): *Salamfestschrift.* A Collection of Talks from the Conference on Highlights of Particle and Condensed Matter Physics (World Scientific, Singapore, 1994).
2. J. Audretsch, F.W. Hehl, and C. Lämmerzahl: *Matter wave interferometry and why quantum objects are fundamental for establishing a gravitational theory.* In: 'Relativistic Gravity Research' Proc., Bad Honnef, Germany 1991, J. Ehlers and G. Schäfer, eds., Lecture Notes in Physics (Springer) **410** (1992) 368–407.
3. P. Baekler: *A spherically symmetric vacuum solution of the quadratic Poincaré gauge field theory of gravitation with Newtonian and confinement potentials.* Phys. Lett. **B99** (1981) 329–332.
4. P. Baekler, M. Gürses, F.W. Hehl, and J.D. McCrea: *The exterior gravitational field of a charged spinning source in the Poincaré gauge theory: A Kerr-Newman metric with dynamic torsion.* Phys. Lett. **A128** (1988) 245-250.
5. P. Baekler, R. Hecht, F.W. Hehl, and T. Shirafuji: *Mass and spin of exact solutions of the Poincaré gauge theory.* Progr. Theor. Phys. **78** (1987) 16–21.
6. P. Baekler, E.W. Mielke, R. Hecht, and F.W. Hehl: *Kinky torsion in a Poincaré gauge model of gravity coupled to a massless scalar field.* Nucl. Phys. **B288** (1987) 800-812.
7. J.D. Bjorken and S. Drell: *Relativistic Quantum Mechanics* (McGraw-Hill, New York, 1964).
8. U. Bonse and T. Wroblewski: *Measurement of neutron quantum interference.* Phys. Rev. Lett. **51** (1983) 1401-1404.
9. H.-H.v. Borzeszkowski and H.-J. Treder: *Mixed field theories and Weyl-Cartan geometry.* Lecture given at this Erice School (1995).
10. Y. M. Cho: *Einstein Lagrangian as the translational Yang-Mills Lagrangian.* Phys. Rev. **D14** (1976) 2521-2525.
11. R. Colella, A.W. Overhauser, and S.A. Werner: *Observation of Gravitationally Induced Quantum Interference.* Phys. Rev. Lett. **34** (1975) 1472-1474.
12. R. Debever (ed.): *Elie Cartan – Albert Einstein: Lettres sur le Parallélisme Absolu 1929–1932.* Original text with English translation. Académie Royale de Belgique and Princeton University Press 1979.
13. A. Einstein: *The Meaning of Relativity,* 5th ed. (Princeton University Press, Princeton, 1955).
14. A. Einstein: *Preface as of 1955 April 04* in 'Cinquant'anni di Relatività

1905–1955.' M. Pantaleo, ed. Edizioni Giuntine and Sansoni Editore, Firenze (1955). Translation from the German by D. Hartley and the authors. The German original of the piece we quoted reads as follows: "...die wesentliche Leistung der allgemeinen Relativitätstheorie, nämlich die Überwindung (des) 'starren' Raumes, d.h. des Inertialsytsems, ist *nur indirekt* mit der Einführung einer Riemann-Metrik verbunden. Das unmittelbar wesentliche begriffliche Element ist das die infinitesimale Verschiebung von Vektoren ausdrückende 'Verschiebungsfeld' (Γ^l_{ik}). Dieses nämlich ersetzt den durch das Inertialsystem gesetzten Parallelismus räumlich beliebig getrennter Vektoren (nämlich Gleichheit entsprechender Komponenten) durch eine infinitesimale Operation. Dadurch wird die Bildung von Tensoren durch Differentiation ermöglicht und so die Einführung des 'starren' Raumes (Inertialsystem) entbehrlich gemacht. Dem gegenüber erscheint es in gewissem Sinne von sekundärer Wichtigkeit, dass ein besonderes Γ-Feld sich aus der Existenz einer Riemann-Metrik deduzieren lässt..."

15. R. Feynman: *Lectures on Gravitation.* Lecture notes by F.B. Morinigo and W.G. Wagner (California Institute of Technology, Pasadena, California 1962/63).
16. R. Floreanini and R. Percacci: *Renormalization-group flow of a dilaton potential.* Phys. Rev. **D52** (1995) 896-911.
17. H. Goenner: *Alternative theories of gravity.* In: 'General Relativity and Gravitation, GR11'. M.A.H. MacCallum, ed. (Cambridge University Press, Cambridge, 1987) pp. 262–273.
18. D.M. Greenberger and A.W. Overhauser: *The role of gravity in quantum theory.* Scientific American **242** (May 1980) 54-64.
19. D. Gregorash and D. Papini: *Weyl–Dirac theory with torsion.* Nuovo Cimento **B55** (1980) 37–51.
20. D. Gregorash and D. Papini: *Weyl–Dirac theory with torsion. II. Foundations and conservation equations.* Nuovo Cimento **B56** (1980) 21–38.
21. F. Gronwald and F.W. Hehl: *Stress and hyperstress as fundamental concepts in continuum mechanics and in relativistic field theory.* In: 'Advances in Modern Continuum Dynamics', International Conference in Memory of Antonio Signorini, Isola d'Elba, June 1991. G. Ferrarese, ed. (Pitagora Editrice, Bologna, 1993) pp. 1–32.
22. R.T. Hammond: *New fields in general relativity.* Contemporary Physics **36** (1995) 103–114.
23. D. Hartley: *Normal frames for non-Riemannian connections.* Class. Quantum Grav. **12** (1995) L103–L105.
24. K. Hayashi and T. Shirafuji: *New general relativity.* Phys. Rev. **D19** (1979) 3524–3553.
25. F.W. Hehl: *Four lectures on Poincaré gauge theory.* In: Proc. of the 6th

Course of the School of Cosmology and Gravitation on 'Spin, Torsion, Rotation, and Supergravity', held at Erice, Italy, May 1979, P.G. Bergmann, V. de Sabbata, eds. (Plenum, New York. 1980) pp. 5–61.

26. F.W. Hehl, G.D. Kerlick, and P. von der Heyde: *On a new metric affine theory of gravitation.* Phys. Lett. **63B** (1976) 446–448.
27. F.W. Hehl, J. Lemke, and E.W. Mielke: *Two lectures on fermions and gravity.* In: 'Geometry and Theoretical Physics', J. Debrus and A.C. Hirshfeld (eds.) (Springer, Berlin, 1991) pp. 56–140.
28. F.W. Hehl, E.A. Lord, L.L. Smalley: *Metric–affine variational principles in general relativity. II. Relaxation of the Riemannian constraint.* Gen. Relat. Grav. **13** (1981) 1037–1056.
29. F.W. Hehl, J.D. McCrea, E.W. Mielke, and Y. Ne'eman: *Metric-Affine Gauge Theory of Gravity: Field Equations, Noether Identities, World Spinors, and Breaking of Dilation Invariance.* Phys. Rep. **258** (1995) 1–171.
30. F.W. Hehl and W.-T. Ni: *Inertial effects of a Dirac particle.* Phys. Rev. **D42** (1990) 2045–2048.
31. P. von der Heyde: *The equivalence principle in the U_4 theory of gravitation.* Lettre al Nuovo Cimento **14** (1975) 250–252.
32. P. von der Heyde: *Is gravitation mediated by the torsion of spacetime?* Z. Naturf. **31a** (1976) 1725-1726.
33. D. Hochberg and G. Plunien: *Theory of matter in Weyl spacetime.* Phys. Rev. **D43** (1991) 3358–3367.
34. J.C. Huang: *Dirac particle, gravity, and inertial effects.* Ann. Physik (Leipzig) **3** (1994) 53–67.
35. B.Z. Iliev: *Special bases for derivations of tensor algebras. I. Cases in the neighborhood and at a point. II. Case along paths. III. Case along smooth maps with separable points of selfintersection.* Preprints E5-92-507, E5-92-508, E5-92-543 (Joint Institute for Nuclear Research, Dubna, 1992).
36. R. d'Inverno: *Introducing Einstein's Relativity.* (Clarendon Press, Oxford, 1992).
37. T. Kloesch and T. Strobl: *Classical and quantum gravity in 1+1 dimensions, part I: A unifying approach; part II: The universal coverings.* Electronic archive Los Alamos, gr-qc/9508020 and gr-qc/9511081 (1995).
38. W. Kopczyński: "The Palatini principle with constraints", Bull. de l' Academie Pol. des Sci., Sér. math. astr. et phys. **23** (1975) 467.
39. W.Kopczyński: *Problems with metric-teleparallel theories of gravitation.* J. Phys. **A15** (1982) 493–506.
40. W. Kopczyński and A. Trautman: *Spacetime and Gravitation.* Translated from the Polish (Wiley, Chichester and PWN – Polish Scientific Publishers, Warszawa, 1992).
41. R. Kuhfuss and J. Nitsch: *Propagating modes in gauge field theories of grav-*

ity. Gen. Relat. Grav. **18** (1986) 1207–1227.

42. C. Lämmerzahl: *On the equivalence principle in quantum theory.* Preprint University of Konstanz (1995)
43. C.H. Lee: *A spherically symmetric electro-vacuum solution of the Poincaré gauge field theory of gravitation.* Phys. Lett. **B130** (1983) 257–259.
44. A. Macías, E.W. Mielke, and H.A. Morales-Técotl: *Torsion and Weyl convector in metric-affine models of gravity,* J. Math. Phys. **36** (1995) 5868–5976.
45. A. Macías, E.W. Mielke, and H.A. Morales–Técotl: *Chiral fermions coupled to chiral gravity.* In: 'Proc. of CAP 50 aniversary', E. Lesard et al., eds. (Springer Verlag, Berlin) in print.
46. B. Mashhoon: *Neutron interferometry in a rotating frame of reference.* Phys. Rev. Lett. **61** (1988) 2639–2642.
47. J.D. McCrea, F.W. Hehl, and E.W. Mielke: *Mapping Noether identities into Bianchi identities in general relativistic theories of gravity and in the field theory of static lattice defects.* Int. J. Theor. Phys. **29** (1990) 1185–1206.
48. E.W. Mielke: *Geometrodynamics of Gauge Fields* — On the geometry of Yang-Mills and gravitational gauge theories (Akademie–Verlag, Berlin, 1987).
49. E.W. Mielke: *Positive-gravitational–energy proof from complex variables?* Phys. Rev. **D42** (1990) 3388-3394.
50. E.W. Mielke: *Ashtekar's complex variables in general relativity and its teleparallelism equivalent.* Ann. Phys. (N.Y.) **219** (1992) 78-108.
51. E.W. Mielke, F. Gronwald, Yu.N. Obukhov, R. Tresguerres, and F.W. Hehl: *Towards complete integrability of two dimensional Poincaré gauge gravity.* Phys. Rev. **D48** (1993) 3648-3662.
52. E.W. Mielke, Yu.N. Obukhov, and F.W. Hehl: *Yang–Mills configurations from 3D Riemann–Cartan geometry.* Phys. Lett. **A192** (1994) 153-162.
53. R. Mills: *Gauge fields.* American J. Physics **57** (1989) 493-507.
54. A.V. Minkevich: *Generalized cosmological Friedmann equations without gravitational singularity.* Phys. Lett. **A80** (1980) 232–234.
55. A.V. Minkevich: *Generalized cosmological Friedman equation and de Sitter solution.* Phys. Lett. **A95** (1983) 422–424.
56. A.V. Minkevich and I.M. Nemenman: *On the influence of the gravitating vacuum on the dynamics of homogeneous isotropic models in gauge theories of gravity,* Class. Quantum Grav. **12** (1995) 1259-1265.
57. C.W. Misner, K.S. Thorne, and J.A. Wheeler: *Gravitation.* (Freeman, San Francisco, 1973).
58. L. Mistura: *Cartan connection and defects in Bravais lattices.* Int. J. Theor. Physics **29** (1990) 1207-1218.
59. G. Modanese and M. Toller: *Radial gauge in Poincaré gauge field theories.* J. Math. Phys. **31** (1990) 452–458.
60. H.A. Morales–Técotl, A. Macías, and E.W. Mielke (1995): *Geometrical*

phases and translations. In: '1st Mexican School on Gravitation and Mathematical Physics', A. Macías et al., eds. (World Scientific, Singapore) in print.
61. Y. Ne'eman: *Gravitational interaction of hadrons: Band–spinor representations of $GL(n,R)$.* Proc. Nat. Acad. Sci. (USA) **74** (1977) 4157–4159.
62. Y. Ne'eman: *Spinor–type fields with linear, affine and general co–ordinate transformations.* Ann. Inst. Henri Poincaré **A28** (1978) 369–378.
63. Y. Ne'eman and Dj. Šijački: *Chromogravity explains 'strong gravity'.* In ref. [1] pp. 540–551.
64. Yu.N. Obukhov: *Two-dimensional Poincaré gauge gravity with matter.* Phys. Rev. **D50** (1994) 5072-5086.
65. Yu.N. Obukhov and V.A. Korotky: *The Weyssenhoff fluid in Einstein-Cartan theory.* Class. Quant. Grav. **4** (1987) 1633-1657.
66. Yu.N. Obukhov, V.N. Ponomariev, and V.V. Zhytnikov: *Quadratic Poincaré gauge theory of gravity: a comparison with the general relativity theory.* Gen. Relat. Grav. **21** (1989) 1107-1142.
67. Yu.N. Obukhov, S.N. Solodukhin, and E.W. Mielke: *Coupling of lineal Poincaré gauge gravity to scalar fields.* Class. Quantum Grav. **11** (1994) 3069-3079.
68. L. O'Raifeartaigh: *Hidden gauge symmetry.* Rep. Progr. Phys. **42** (1979) 159-223.
69. L. O'Raifeartaigh: *The dawning of gauge theory.* In ref.[1] pp. 577–591.
70. A. Pais: *'Subtle is the Lord...' The Science and the Life of Albert Einstein* (Clarendon Press, Oxford, 1982).
71. J.-F. Pascual-Sánchez: *lh theory of hadrons and leptons.* Lett. Nuovo Cim. **28** (1980) 321-324.
72. J.-F. Pascual-Sánchez: *On the limits of Poincaré gauge theory.* Phys. Lett. **A108** (1985) 387-390.
73. J.-F. Pascual-Sánchez: *Constrained macroscopic limit of Poincaré gauge theory of gravity.* Lett. Nuovo Cim. **44** (1985) 326-328.
74. M. Pawłowski and R. Rączka: *A unified conformal model for fundamental interactions without dynamical Higgs field.* Found. Phys. **24** (1994) 1305–1327.
75. M. Pawłowski and R. Rączka: *A Higgs–free model for fundamental interactions. Part 1: Formulation of the model.* In: 'Modern Group Theoretical Methods in Physics', J. Bertrand et al. (eds.) (Kluwer Academic, Dordrecht, 1995) pp. 221–232.
76. V. Perlick: *Observer fields in Weylian spacetime models.* Class. Quantum Grav. **8** (1991) 1369–1385.
77. E.A. Poberii: *Metric-affine scale-covariant gravity.* Gen. Relat. Grav. **26** (1994) 1011–1054.
78. V.N. Ponomariov and Ju. Obukhov: *The generalized Einstein-Maxwell theory*

of gravitation. Gen. Relat. Grav. **14** (1982) 309-330.

79. E. Prugovečki: *Principles of Quantum General Relativity.* (World Scientific, Singapore, 1995).
80. R. Puntigam, E. Schrüfer, and F.W. Hehl: *The use of computer algebra in Maxwell's theory.* In: 'Computer Algebra in Science and Engineering', J. Fleischer et al. (eds.) (World Scientific, Singapore, 1995) pp. 195–211.
81. H. Rumpf: *On the translational part of the Lagrangian in the Poincaré gauge theory of gravitation.* Z. Naturf. **33a** (1978) 1224–1225.
82. V. de Sabbata and C. Sivaram: *Spin and Torsion in Gravitation* (World Scientific, Singapore, 1994).
83. E. Schrödinger: *Space–Time Structure.* Reprinted with corrections (Cambridge University Press, Cambridge, 1960).
84. E. Schrüfer, F.W. Hehl, and J.D. McCrea: *Exterior calculus on the computer: The REDUCE–package EXCALC applied to general relativity and the Poincaré gauge theory.* Gen. Rel. Grav. **19** (1987) 197–218.
85. E. Sezgin and P. van Nieuwenhuizen: *New ghost-free Lagrangians with propagating torsion.* Phys. Rev. **D21** (1980) 3269–3280.
86. P. Teyssandier and R.W. Tucker: *Gravity, gauges and clocks.* Class. Quantum Grav., to appear (1995/96).
87. W. Thirring: *Classical Field Theory.* A Course in Mathematical Physics 2, 2nd ed. (Springer, New York, 1986).
88. A. Trautman: *On the structure of the Einstein–Cartan equations.* In: 'Differential Geometry', Symposia Mathematica Vol. **12** (Academic Press, London 1973) p. 139–162.
89. R. Tresguerres: *An exact solution of* $(2+1)$*–dimensional topological gravity in metric–affine spacetime.* Phys. Lett. **A168** (1992) 174–178.
90. R. Tresguerres: *Topological gravity in a 3–dimensional metric–affine space.* J. Math. Phys. **33** (1992) 4231–4238.
91. R. Tresguerres: *Exact vacuum solutions of 4–dimensional metric–affine gauge theories of gravitation.* Z. Phys. **C65** (1995) 347-354.
92. R. Tresguerres: *Exact static vacuum solution of 4–dimensional metric–affine gravity with nontrivial torsion.* Phys. Lett. **A**, submitted (1994).
93. R. Tresguerres: *Weyl–Cartan model for the non–massive cosmological eras with a dilatedly polarized vacuum.* Phys. Lett. **A**, submitted (1994).
94. E. Tsantilis, R.A. Puntigam, and F.W. Hehl: *A quadratic curvature Lagrangian of Pawłowski and Rączka: A finger exercise with MathTensor.* In: 'Relativity and Scientific Computing – Computer Algebra, Numerics, Visualization –' F.W. Hehl, R.A. Puntigam, H. Ruder (eds.) (Springer, Berlin, 1996) to be published.
95. R.W. Tucker and C. Wang: *Black holes with Weyl charge and non-Riemannian waves.* Class. Quantum Grav. **12** (1995) 2587–2605.

96. C. Wiesendanger: *Poincaré gauge invariance and gravitation in Minkowski spacetime.* Electronic archive Los Alamos, gr-qc/9505049 (1995).

SUPERLUMINAL AND GRAVITATIONAL BEHAVIOR OF NONLOCAL PHOTONS

ARTHUR KOMAR

YESHIVA UNIVERSITY

500 W. 185 ST.

NEW YORK, N.Y. 10033

ABSTRACT

The gauge invariant photon states of Maxwell's theory of electromagnetism in Minkowski space-time are shown to be essentially nonlocal, in the sense that they have support which extends outside of the forward light cone of the sources that generate them. Upon modelling the effects of gravitation by means of the principle of equivalence (i.e. employing a uniformly accelerating reference frame) we find that the two orthogonal polarization states of the photon should be expected to respond measurably differently to a gravitational field. This consequence of combining the features of the nonlocal photon with the (decidedly local) principle of equivalence is of considerable interest in view of the fact that it is in principle experimentally verifiable.

INTRODUCTION

Whereas, in the classical Maxwell theory, it is quite easy to exhibit a complete set of local gauge invariant observables, this property is somewhat misleading when one considers quantum phenomena. The photon field, although a gauge invariant solution of Maxwell's equations, is highly nonlocal. This nonlocality becomes particularly conspicuous when we consider the manner in which the sources contribute to the photon field. Indeed, in this paper we shall show that the photon field grows a superluminal tail, despite the fact that the local observables remained confined by the forward light cone of the source. We shall also investigate the interaction of the nonlocal photon field with a constant gravitational field (modelled by a uniformly accelerating elevator.) We find that the gravitational field treats the two orthogonal polarization states of the photon measurably differently.

THE PHOTON FIELD

As is well known, although photons are vector particles, since they have vanishing rest mass, they have only two spin degrees freedom, both space-like and transverse to their direction of motion. The actual photon wave function, which is a positive energy eigenstate of the Maxwell Hamiltonian, is necessarily complex. However since the theory is linear, the real and imaginary parts of the wave function are separately gauge invariant real solutions. It will therefore prove more convenient to examine such real solutions.

Consider the standard Maxwell equations with sources (in rationalized natural units)

(1a) $$\vec{\nabla}\cdot\vec{H} = 0$$

(1b) $$\vec{\nabla}\times\vec{E} + \frac{\partial\vec{H}}{\partial t} = 0$$

(1c) $$\vec{\nabla}\cdot\vec{E} = \rho$$

(1d) $$\vec{\nabla}\times\vec{H} - \frac{\partial\vec{E}}{\partial t} = \vec{J}$$

Equation (1a) can easily be solved by introducing the vector potential, $\vec{A}$

(2) $$\vec{H} = \vec{\nabla} \times \vec{A}$$

As we know however, this solution is not unique, since the gradient of a scalar field may be added to $\vec{A}$ leaving $\vec{H}$ unaffected. Thus

(3) $$\vec{A}' = \vec{A} + \vec{\nabla}\psi$$

If we require that the scalar field, ψ, satisfy the Poisson equation

(4) $$\nabla^2 \psi = -\vec{\nabla} \cdot \vec{A}$$

the new vector potential will satisfy the relation (with prime dropped)

(5) $$\vec{\nabla} \cdot \vec{A} = 0$$

But equally important, due to the uniqueness of the solution of the Poisson equation with standard (nonlocal) boundary conditions, we know that no further gauge transformations are available to alter the resulting vector potential. In other words our resulting vector potential is gauge-invariant. It is consequently a nonlocally defined observable of the Maxwell theory, although it is not manifestly Lorentz covariant.

In order to understand the transformation properties of such observables, one should note that the full group of the Maxwell equations is the semidirect product of the Poincare group and the gauge group. Thus in order to obtain the Lorentz transform of our nonlocal observable it is necessary to coordinate a space-time frame transformation (i.e. an element of the Poincare group) with a corresponding gauge transformation, exactly as one handles spin transformations in the familiar manner employed in Dirac's electron theory. Such a transformation law is well defined, although somewhat intricate to express analytically.

Let us now substitute equation (2) into equation (1b). We obtain

$$\vec{\nabla} \times \left(\vec{E} + \frac{\partial \vec{A}}{\partial t} \right) = 0 \tag{6}$$

It follows immediately that there exists a scalar field, ϕ, such that

$$\vec{E} + \frac{\partial \vec{A}}{\partial t} = - \vec{\nabla} \phi \tag{7}$$

Substituting equation (7) into equation (1c) and employing the gauge condition, equation (5), we find that the scalar field must satisfy the Poisson equation

(8) $$\nabla^2 \phi = -\rho$$

Again we see that with standard (nonlocal) boundary conditions the scalar field, ϕ, is uniquely defined. Indeed its solution is

(9) $$\phi(\vec{r},t) = \frac{1}{4\pi}\int \frac{\rho(r',t)\,d_3 r'}{|\vec{r}-\vec{r}'|}$$

We note at this point that the standard local observables, $\vec{E}$ and $\vec{H}$, are, via equations (2) and (7), uniquely determined by the knowledge of the nonlocally defined potentials, $\vec{A}$ and ϕ .

It remains for us to solve equation (1d) in order to fully specify the solution. Substituting equations (2) and (7) into equation (1d), and employing the gauge condition, equation (5), the solution for the scalar potential, equation (9), and a standard vector calculus identity, we obtain the d'Alembert equation for the vector potential $\vec{A}$

(10) $$\Box \vec{A} = \frac{1}{4\pi}\int \frac{\frac{\partial \rho}{\partial t}(\vec{r}',t)}{|\vec{r}-\vec{r}'|}\,d_3 r' - \vec{J}$$

The consistency of the original Maxwell equations, (1), requires that the sources satisfy the conservation of charge equation

(11) $$\frac{\partial \rho}{\partial t} + \vec{\nabla} \cdot \vec{j} = 0$$

Employing this relation in equation (10) we finally can rewrite the propagation equation for $\vec{A}$ thus

(12) $$\Box \vec{A} = -\vec{j} - \frac{1}{4\pi} \vec{\nabla} \int \frac{\vec{\nabla}' \cdot \vec{j}(\vec{r}', t)}{|\vec{r} - \vec{r}'|} d_3 r'$$

Observe that the last term on the right hand side is exactly what is required to preserve the gauge condition, equation (5). But much more cogently, note that should we choose to solve equation (12) by the employment of retarded potentials, this last term acts as a smeared source for $\vec{A}$, extending its support beyond the forward light cone of the physical source, $\vec{j}$. It is in this sense that the propagation of the nonlocal field, $\vec{A}$, is superluminal. (The solution for the scalar potential, equation (9), is blatantly superluminal, reflecting the fact that the action-at-a-distance Coulomb law is incorporated into Maxwell's theory without modification. However it does not correspond to the field of the photon wave function in the manner that $\vec{A}$ does.)

The term in the source field which produces the superluminal effect is manifestly the gradient of a scalar field. Thus if we take the curl of equation (10) we find for the propagation of the local field, $\vec{H}$, the well defined local source, $-\vec{\nabla} \times \vec{j}$. Similarly, after some elaborate but straight forward computations, employing equation (7), we find that the support of the local field $\vec{E}$ is also confined to the forward light cone of the local source.

We can therefore conclude that while the local classical Maxwell fields are consistent with relativistic causality, the quantum photon fields strictly speaking are not. This clash between the nonlocal demands of quantum theory and the requirement of relativistic causality, illustrated in this simple linear field theory, can only be expected to become worse should we consider nonlinear general relativistic theories. For in such theories it does not even seem possible to construct local gauge invariant classical structures.

CORRESPONDENCE PRINCIPLE CONSIDERATIONS

Rather than proceeding to consider the problem of the photon field in a general Riemanian manifold, we wish to consider the complications which arise in Minkowski space when viewing the system in a uniformly accelerated frame of reference. By an appeal to the correspondence principle we may hope to learn at least qualitatively of new features which could occur when non-local photon structures interact with gravitation.

Let us therefore consider a uniformly accelerating elevator in Minkowski space. We shall require that, at each elevation in the elevator, acceleration be constant in the instantaneous rest frame and that the spatial metric be Euclidean at all times. For convenience we also require that in an initial frame the Minkowski coordinates, $\hat{x}^\mu$, agree with the elevator coordinates, x^μ . The coordinate axis are to be oriented so that the $x^3 = z$ direction is parallel to the motion of the elevator and z = 0 is chosen to be the equation of the elevator floor at whose elevation the metric remains permanently Minkowskian. With these assumptions (and defining g to be the constant acceleration of the elevator) we obtain for the coordinate transformation to the elevator frame

$$\begin{aligned} \hat{z} &= \left(z + \tfrac{1}{g}\right)\cosh gt - \tfrac{1}{g} \\ \hat{t} &= \left(z + \tfrac{1}{g}\right)\sinh gt \\ \hat{x} &= x \\ \hat{y} &= y \end{aligned} \tag{13}$$

One may easily check that in this new coordinate system the metric of Minkowski space becomes

$$ds^2 = \lambda^2 dt^2 - dx^2 - dy^2 - dz^2 \tag{14}$$

where

(15) $$\lambda = 1 + gz$$

In this frame the source-free Maxwell equations can be written (now employing the standard covariant notation)

(16) $$F_{[\alpha\beta,\gamma]} = 0$$
$$(\lambda g^{\alpha\mu} g^{\beta\nu} F_{\mu\nu})_{,\beta} = 0$$

The first set of these equations are the integrability conditions for the existence of a (four-)vector potential A_α such that

(17) $$F_{\alpha\beta} \equiv A_{\alpha,\beta} - A_{\beta,\alpha}$$

Although the vector potential is not given uniquely by this equation, physically equivalent solutions being obtained by means of a gauge transformation described by an arbitrary scalar field ψ via

(18) $$\bar{A}_\alpha = A_\alpha + \psi_{,\alpha}$$

we are particularly interested in the realization of the Maxwell field in terms of potentials since, as we have already noted, they are required for the representation of the photon states.

The nonlocality of the photon wave function arose, as we have seen above, by our requirement that the potential, which describes inter alia the spin degrees of freedom, be space-like and transverse to the direction of propagation. As we have retained the gauge freedom of the theory we are still free to impose gauge conditions such as equation (5). In the Minkowski frame we found that such a condition, together with the field equations, led to equation (9), which in the absence of sources implied the vanishing of the time-like component of the vector potential. In our present case it is easy to see that such is not the case. It is easy to confirm that if we employ equation (5) and the vanishing of A_c in the $\alpha = 0$ component of the second set of equations (16) (i.e. the constraint equation) we are forced to have

$$F_{03} = 0 \tag{19}$$

Were this correct it would imply that, in the presence of a gravitational field, photons could never be found polarized such that the electric vector is parallel to the gravitational field, an effect so conspicuous that it surely would have been observed decades ago. Nevertheless the number of degrees of freedom of the photon should not be expected to be altered by the presence of a gravitational field. We must therefore proceed slightly differently.

Let us examine the constraint relation of equations (16) in greater detail. Substituting equation (17) into equations (16), we find that we can group the terms and write the constraint equation thus

$$(20)\qquad \left(\frac{1}{\lambda}A_s\right)_{,s0} = \frac{1}{\lambda}\left(\nabla^2 A_0 - \frac{g}{\lambda}A_{0,3}\right)$$

It is therefore natural and consistent to employ the gauge freedom to impose the gauge condition

$$(21)\qquad \left(\frac{A_s}{\lambda}\right)_{,s} = 0$$

The time-like component of A_μ must consequently satisfy

$$(22)\qquad \nabla^2 A_0 - \frac{g}{\lambda}A_{0,3} = 0$$

A simple application of Gauss' theorem shows that if we require A_0 to vanish on the walls of the elevator, it must vanish everywhere in the interior. We demonstrate this result as follows.

Define the spatial vector field C_s thus

$$(23)\qquad C_s \equiv A_0 \nabla_s A_0 - \frac{1}{2}\frac{g}{\lambda}A_0^2\delta_{s3}$$

The divergence of this vector field is seen to be

$$(24)\qquad \vec{\nabla}\cdot\vec{C} = |\nabla A_0|^2 + \frac{g^2}{2\lambda^2}A_0^2$$

Integrating this expression over the entire volume of the elevator and employing the boundary condition for A_0 on its surface we obtain from Gauss' theorem

$$\int \vec{\nabla} \cdot \vec{C} \, d_3 x = \oint \vec{C} \cdot d_2 \vec{S} = 0 \tag{25}$$

However, since the integrand of equation (25) is a positive definite expression, the only way this equation can be satisfied is if we have

$$A_0 = 0 \tag{26}$$

The vector potential is thus found to be everywhere space-like. However the "transversality" condition is now somewhat more complicated. Qualitatively this is due to the fact that we should expect the trajectory of the photon to be deflected by the constant gravitational field which the elevator frame is modelling. Thus a horizontally moving photon polarized such that its electric vector lies parallel to the gravitational field should, upon deflection of its trajectory, be expected to grow a component of the electric field in the direction of motion (i.e. longitudinally) in order to remain instantaneously "transverse". Remarkably, such a rather ingenuous heuristic argument will mislead one if, instead, we take the magnetic vector to be polarized parallel to the gravitational field! We find that polar vectors and axial vectors behave quite differently in the presence of gravitation.

A more rigorous demonstration of the radical difference between the behaviors of the two orthogonal polarization states (TM and TE) with respect to the gravitational field will have to await the quantization of our equivalence principle model theory. The quantization of this model is seriously complicated by the modified "transversality" condition, equation (21). It is at this point that the nonlocality of the photon comes into play with a vengeance, since an unequivocal procedure for the Fourier analysis of the potentials is no longer available.

The development of the quantum field theoretic version of our model exceeds the purview of this paper and will be presented in a subsequent publication. We wish to emphasize however that the differing effect of gravitation upon the two transverse (nonlocal) polarization states of the photon is in principle an experimentally verifiable consequence of the (decidedly local) principle of equivalence, and therefore potentially very important.

Describing GR by Characteristic Surfaces

by

Simonetta Frittelli[a], Carlos Kozameh[b], Ezra T. Newman[a]

[a]Dept. of Physics and Astronomy, University of Pittsburgh
Pittsburgh, Pa. 15260, USA

[b]FAMAF, University of Cordoba, 5000 Cordoba, Argentina

Abstract

The vacuum Einstein equations are refomulated as differential equations for both a complex function and a real function on a manifold, $\mathcal{M} \times S^2$, with $\mathcal{M}$ eventually becoming the four-dimensional space-time and the S^2 becoming the sphere of null directions over $\mathcal{M}$. At the beginning there is no other other structure available: the structure arises from the two functions. The complex function, denoted by $\Lambda(\mathcal{M} \times S^2)$, encodes information about a sphere's worth of surfaces through each point of $\mathcal{M}$. From knowledge of Λ there is <u>defined</u> a second rank tensor on $\mathcal{M}$ which can be interpreted as a conformal metric, - (actually a particular member of the conformal class) - such that the "surfaces" are automatically characteristic surfaces of this conformal metric. The real function, $\Omega(\mathcal{M} \times S^2)$, which plays the role of a conformal factor, converts the conformal metric into a vacuum Einstein metric. Locally, all Einstein metrics can be obtained in this manner.

As the details of this work have recently been given elsewhere only an outline will be given here.

I. Introduction

The geometric character of general relativity, (GR), makes it essentially unique among physically interesting field theories. Nevertheless, from the fact that the geometry is usually described by the metric tensor $g_{ab}(x)$, it becomes simple to think of the metric as just another local classical field that is to be thought of and treated on the same level as, e.g., the Maxwell, Dirac or Yang-Mills fields, classically or quantum mechanically. There is however a clear distinction between GR and other physical theories: only in GR do the characteristic surfaces play a *dynamical or evolutionary role* in the theory. In other theories, although the characteristic surfaces are of fundamental importance, - determining the domains of influence - they are given once and for all by the "background structure" and have no independent dynamics; they play the role of fixed props on a "stage" for the other theories. In GR however, since the metric determines the characteristic surfaces, when one solves the Einstein equations one is simultaneously solving for both the metric and its characteristics - though, of course, one uses only the metric in the equations.

Recently, in a series of papers[1,2,3,4] we have described a new point of view towards GR, which we have refered to as "a null surface view of GR", where the characteristic surfaces themselves are the basic variables (with an auxiliary scalar function needed as a conformal factor). A metric *is not needed* ; it arises as a derived concept. This new view of GR begins with a four-manifold, $\mathcal{M}$, (eventually to be the space-time) and no other structure. The theory is built out of differential equations for *families of surfaces* ; a two-dimensional set of surfaces through each point of $\mathcal{M}$. (The equations are six-dimensional; four dimensions for $\mathcal{M}$ and two for the parametrized set of surfaces.) A non-degenerate two index tensor, which can be interpreted as a (conformal) metric, is constructed (by differentiation) from these surfaces. The surfaces then are (automatically) the characteristic surfaces for this conformal metric. The scalar function, which satisfies its own equations coupled to the surface variables, playing the role of a conformal factor, turns the conformal metric into an Einstein metric, yielding a description of space-time geometry completely equivalent to vacuum GR.

The main purpose of this note is to give a survey of *this view* of GR. We will start with a description of our basic variables (one complex and one real function of six independent variables), and our (six-dimensional) differential equations for them. We then show how metrics can be constructed from their solutions. A complete proof will not be given (though we do give a sketch) of our contention that the metrics constructed in this manner are the general (local) solution of the vacuum Einstein equations, as this has been done elsewhere.[1,2]. We give the geometric meaning of the variables, (as the scalar product of two connecting vectors associated with two null congruences) and then show how the the variables encode information about families of (characteristic) surfaces. (In an appendix, we discuss the details of this encoding procedure.) Along the way, we obtain all null geodesics and all null geodesic congruences on $\mathcal{M}$. In the last section, we review the basic ideas involved, what value it might have and what are some possible applications and generalizations.

II. A Summary of "the Null Surface View of GR"

To summarize and explain the "null surface version of GR", we begin with the notation and a description of our variables and the set of equations they satisfy.

• The two dependent variables, Λ (complex, spin-wt 2) and Ω (real, spin-wt 0) are functions on $R^4 x S^2$ with coordinates $\theta^i = (\theta^0, \theta^+, \theta^-, \theta^1) \equiv (u, R, \omega, \varpi)$ on the R^4 sector and complex stereographic coordinates $(\zeta, \bar{\zeta})$ on the S^2, i.e., we have

$$\Lambda = \Lambda(\theta^i, \zeta, \bar{\zeta}), \qquad \Omega = \Omega(\theta^i, \zeta, \bar{\zeta}).$$

Remark 1: Though at this point in the discussion, it is not clear why our manifold should be six dimensional and have the structure of $R^4 x S^2$, we will see later that this structure has its origin in two (related) ideas; (1), the R^4 arises from the space-time and the S^2 from the sphere of null directions or, (2), the space of null geodesics (five dimensional) and a parameter along null geodesics makes it six.

• With the notation

$$\Phi_{,1} = \frac{\partial}{\partial R}\Phi, \quad \Phi_{,0} = \frac{\partial}{\partial u}\Phi, \quad \Phi_{,+} = \frac{\partial}{\partial \omega}\Phi, \quad \Phi_{,-} = \frac{\partial}{\partial \varpi}\Phi$$

Λ and Ω satisfy the following set of three differential equations (two complex and one real):

$$\text{E)} \qquad \Omega_{,11} = Q(\Lambda_{,1})\Omega, \tag{1}$$

$$\text{m}_\text{I}) \qquad \eth\Omega = \tfrac{1}{2}W(\Lambda_{,1})\Omega,$$

$$\text{m}_\text{II}) \qquad \eth\Lambda_{,1} - 2\Lambda_{,-} = \{W + \eth(\ln q)\}\Lambda_{,1}$$

with Q and W defined by

$$q^2 Q = -\tfrac{1}{4}q\Lambda_{,11}\bar{\Lambda}_{,11} - \tfrac{3}{8}(q_{,1})^2 + \tfrac{1}{4}qq_{,11}, \tag{2}$$

$$W(1 - \tfrac{1}{4}\Lambda_{,1}\bar{\Lambda}_{,1}) = \Lambda_{,+} + \tfrac{1}{2}\bar{\eth}\Lambda_{,1} - \tfrac{1}{2}\eth(\ln q) - \tfrac{1}{4}\Lambda_{,1}\bar{\eth}\ln q + \tfrac{1}{2}\Lambda_{,1}\bar{\Lambda}_{,-} + \tfrac{1}{4}\Lambda_{,1}\eth\bar{\Lambda}_{,1},$$

$$q = 1 - \Lambda_{,1}\bar{\Lambda}_{,1}.$$

Remark 2: We claim that E, m_I and m_{II} are equivalent to the vacuum Einstein equations. Though, at this point, it is impossible to see why this should be so, m_I and m_{II} are the conditions that a metric should exist and E is the condition that the Einstein equations are satisfied by that metric. (Eventually it will be seen that the metric is expressed as a function of Λ and Ω: Though E is but *one* component of $R_{ab} = 0$ the component varies with $(\zeta, \bar{\zeta})$ yielding essentially the entire $R_{ab} = 0$.) The details are given in refs.[1,5].

Used in these equations are two partial differential operators, $\eth$ and $\bar{\eth}$, involving all six independent variables. They are quite complicated and have several subtle implications.

• They are *defined* by

$$\eth\Phi(\theta^i, \zeta, \bar{\zeta}) = \eth'\Phi(\theta^i, \zeta, \bar{\zeta}) + \omega\,\Phi_{,0} + \Lambda\,\Phi_{,+} + R\,\Phi_{,-} + \left(\frac{\Lambda_{,1}\bar{J} + J}{1 - \Lambda_{,1}\bar{\Lambda}_{,1}}\right)\Phi_{,1}\,, \quad (3)$$

(and the complex conjugate for $\bar{\eth}$) with

$$J = -2\omega + \bar{\omega}\,\Lambda_{,0} + \bar{\eth}'\Lambda + R\Lambda_{,+} + \bar{\Lambda}\,\Lambda_{,-}$$

and where $\eth'$ and $\bar{\eth}'$ are the "usual"[6] edh and $\overline{\text{edh}}$ operators, holding the θ^i fixed, (essentially the ζ and $\bar{\zeta}$ derivatives). Note that the operators are *dependent* on the θ^i's, both explicitly and implicitly via the dependent variable $\Lambda(\theta^i, \zeta, \bar{\zeta})$. This leads (see the discussion following (10)) to an integrability condition on $\Lambda(\theta^i, \zeta, \bar{\zeta})$ and $\bar{\Lambda}(\theta^i, \zeta, \bar{\zeta})$, namely

$$\text{(I)} \qquad \bar{\eth}^2\Lambda = \eth^2\bar{\Lambda}. \quad (4)$$

This condition *should* be thought of as an additional equation for Λ and $\bar{\Lambda}$, in addition to E, m_I, and m_{II}.

Remark 3: The partial derivatives, $\frac{\partial}{\partial\theta^i}$, do not commute with $\eth$. The detailed form of the commutators are given in ref[1].

• Assuming that the "field" equations (m_I, m_{II}, E, I) have been solved, i.e.,

$$\Lambda = \Lambda(\theta^i, \zeta, \bar{\zeta}), \qquad \Omega = \Omega(\theta^i, \zeta, \bar{\zeta})$$

are assumed to be known, then a spheres' worth of metrics, $g^{ij}(\theta^i, \zeta, \bar{\zeta})$, on R^4, parametrized by $(\zeta, \bar{\zeta})$ can be constructed by the following prescription:

The Ω plays the role of a conformal factor, the remaining terms being functions of Λ and its derivatives, i.e.,

$$g^{ij}(\theta^i, \zeta, \bar{\zeta}) = \Omega^2 \mathfrak{g}^{ij}(\Lambda), \tag{5}$$

with

$$\mathfrak{g}^{00} = \mathfrak{g}^{0+} = \mathfrak{g}^{0-} = 0, \tag{6}$$

$$\mathfrak{g}^{01} = 1, \qquad \mathfrak{g}^{+-} = -1$$

$$\mathfrak{g}^{++} = -\Lambda_{,1}, \qquad \mathfrak{g}^{--} = -\bar{\Lambda}_{,1},$$

$$\mathfrak{g}^{1+} = -\tfrac{1}{2}(\bar{\eth}\Lambda_{,1} + \overline{W}\Lambda_{,1}), \qquad \mathfrak{g}^{1-} = -\tfrac{1}{2}(\eth\bar{\Lambda}_{,1} + W\bar{\Lambda}_{,1}),$$

$$\mathfrak{g}^{11} = W\mathfrak{g}^{1-} + \eth\mathfrak{g}^{1-} - \mathfrak{g}^{i-}T_i{}^1.$$

The $T_i{}^1$ is defined as

$$qT_i{}^1 = \{\Lambda_{,i}\bar{\Lambda}_{,+} + \eth\bar{\Lambda}_{,i} + \bar{\Lambda}_{,-}\delta^1{}_i + \bar{\Lambda}_{,0}\delta^+{}_i - 2\delta^-{}_i\}\Lambda_{,1}$$
$$+ \bar{\eth}\Lambda_{,i} + \bar{\Lambda}_{,i}\Lambda_{,-} + \Lambda_{,+}\delta^1{}_i + \Lambda_{,0}\delta^-{}_i - 2\delta^+{}_i \quad . \tag{7}$$

• The metric thus has the form

$$g^{ij}(\theta^i, \zeta, \bar{\zeta}) = \Omega^2 \begin{bmatrix} 0 & 1 & 0 & 0 \\ 1 & x & x & x \\ 0 & x & x & -1 \\ 0 & x & -1 & x \end{bmatrix} . \tag{8}$$

It becomes immediately clear that for each fixed pair $(\zeta, \bar{\zeta})$, the coordinate $\theta^0 = u$ = constant are characteristic surfaces of the metric and the $R = \theta^1$ is a parameter (non-affine) along the null geodesics of the surface. (R *is* an affine parameter for the conformally equivalent metric (8) with $\Omega = 1$.)

Remark 4: As we mentioned earlier, this set of metrics satisfy the vacuum Einstein equations for each pair $(\zeta, \bar{\zeta})$. Furthermore, though again it is not at all obvious, the set of metrics, for all $(\zeta, \bar{\zeta})$, are equivalent in the sense that coordinate transformations exist so that they can be transformed into each other. This result follows from m_I and m_{II}. See ref[1].

- We return to the issue of the subtleties of the $\eth$ and $\bar{\eth}$ operators and the origin of the integrability conditions (I), i.e.,

$$\bar{\eth}^2\Lambda = \eth^2\bar{\Lambda}.$$

By applying $\eth$ and $\bar{\eth}$, from the definitions (3), to θ^i we find that

$$\eth\theta^0 = \omega = \theta^+, \qquad \eth\theta^- = R = \theta^1, \qquad \eth\theta^+ = \Lambda(\theta^i, \zeta, \bar{\zeta}) \tag{9}$$
$$\bar{\eth}\theta^0 = \bar{\omega} = \theta^-, \qquad \bar{\eth}\theta^+ = R = \theta^1, \qquad \bar{\eth}\theta^- = \bar{\Lambda}(\theta^i, \zeta, \bar{\zeta}).$$

Rewriting the third equation of each set, using the earlier ones, we have

$$\eth^2\theta^0 = \Lambda(\theta^0, \eth\theta^0, \bar{\eth}\theta^0, \bar{\eth}\eth\theta^0, \zeta, \bar{\zeta}), \tag{10}$$
$$\bar{\eth}^2\theta^0 = \bar{\Lambda}(\theta^0, \eth\theta^0, \bar{\eth}\theta^0, \bar{\eth}\eth\theta^0, \zeta, \bar{\zeta}).$$

A new idea is introduced here. The pair (10), *should be* thought of as two equations for the determination of θ^0 as a function of $(\zeta, \bar{\zeta})$. The $\eth$ and $\bar{\eth}$ in these equations act simply as the ordinary $\eth$ and $\bar{\eth}$ derivatives operating on spin-weight zero functions θ^0. The earlier, Eq.(3), complicated definition of $\eth$ and $\bar{\eth}$, arose as the total $(\zeta, \bar{\zeta})$ derivatives treating the θ^i as functions of $(\zeta, \bar{\zeta})$. With this idea, *equation (I) is just the integrability condition on the pair* (10), i.e., it is the condition that the solutions of the two equations (10) coincide. [$\eth$ and its complex conjugate $\bar{\eth}$ can be converted into two vector fields on the six-dimensional space $(\theta^i, \zeta, \bar{\zeta})$. (I) is the condition that *the two sets of integral curves*, of the vector fields obtained from (10), *are surface forming; i.e., their Lie bracket vanishes.* See the Appendix.] If the pair are then integrated, there will be four constants of integration, x^a, (interpreted as local space-time coordinates), so that the solutions will be represented by

$$\theta^0 = u = Z(x^a, \zeta, \bar{\zeta}). \tag{11}$$

Using equations (9), we see that all the θ^i are functions of $(x^a, \zeta, \bar{\zeta})$, i.e.,

$$\theta^i = \theta^i(x^a, \zeta, \bar{\zeta}), \tag{12}$$

with its inversion,

$$x^a = x^a(\theta^i, \zeta, \bar{\zeta}). \tag{13}$$

• The equations $x^a = x^a(\theta^i, \zeta, \bar{\zeta})$ and $\theta^i = \theta^i(x^a, \zeta, \bar{\zeta})$, can be thought of as a coordinate transformation between the coordinates $\theta^i \leftrightarrow x^a$ that is parametrized by $(\zeta, \bar{\zeta})$. This transformation leads to the metric $g_{ab}(x)$ via

$$g_{ab}(x^a) = \theta^i{}_{,a}\theta^j{}_{,b}\, g_{ij}(\theta^i, \zeta, \bar{\zeta}). \tag{14}$$

That the $g_{ab}(x)$ does not depend on $(\zeta, \bar{\zeta})$ is a consequence[1,3] of the two metricity conditions, m_I and m_{II}. In fact, m_I and m_{II} have been derived from this requirement.

We also see that $u = Z(x^a, \zeta, \bar{\zeta})$ can be interpreted as a spheres' worth of null surfaces thru each space-time point x^a. Associated with each choice of $(\zeta, \bar{\zeta})$ there will be a null geodesic congruence, the generators of the null surfaces.

Up to now the geometric meaning of the variable Λ has been obscure. With no details[4] we give a rough idea of the meaning. From Eq. (13) written as

$$x^a = x^a(R, u, \omega, \bar{\omega}, \zeta, \bar{\zeta}) \tag{15}$$

we have the five dimensional set of null geodesics, parametrized by $Y^\alpha = (u, \omega, \bar{\omega}, \zeta, \bar{\zeta})$, with R as the geodesic parameter. The set of connecting vectors, defined by the derivatives dx^a/dY^α, can be broken into two groups, derivatives with respect to $(u,\omega,\overline{\omega})$ and those with respect to $(\zeta,\overline{\zeta})$. It turns out[4] that Λ is the scalar product between connecting vectors, one from each group. Specifically we have

$$\Lambda = -g^{i+} g_{ab} \frac{\partial x^a}{\partial \theta^i} \bar{ð}' x^b \qquad (16)$$

III. Discussion

It is perhaps worthwhile reviewing our results. From the scalar functions, $\Omega(R,u,\omega,\overline{\omega},\zeta,\overline{\zeta})$ and $\Lambda(R,u,\omega,\overline{\omega},\zeta,\overline{\zeta})$, on a six-dimensional manifold, (satisfying Eqs. E, m_I, m_{II}, I), the geometry of vacuum GR can be described in a new way.

Understanding the structure of our formulation comes in two steps; if, first, we integrate the equations {a perturbative integration[2,4], is complicated but (apparently) is a straight-forward process} we obtain both $\Omega = \Omega(R,u,\omega,\overline{\omega},\zeta,\overline{\zeta})$ and $\Lambda = \Lambda(R,u,\omega,\overline{\omega},\zeta,\overline{\zeta})$. From Ω and Λ a spheres worth of metrics, *all diffeomorphically equivalent*, can be constructed by differentiation of Λ. The form of the metric, implies immediately that u = const. are null surfaces, that ω, $\overline{\omega}$ label the null geodesics and that R is a parameter along the geodesics. At this stage the Einstein equations have been solved, though the geometric meaning of the $\Lambda(R,u,\omega,\overline{\omega},\zeta,\overline{\zeta})$ is still rather obscure - it appears as some sort of potential for the metric. The second step, (which is not needed to obtain Einstein metrics - but is more analogous to integrating the geodesic equations), is to integrate the equations

$$ð^2\theta^0 = \Lambda(\theta^0,\ ð\theta^0,\ \bar{ð}\,\theta^0,\ \bar{ð}\ ð\theta^0, \zeta, \overline{\zeta}),$$

$$\bar{ð}^{\,2}\theta^0 = \overline{\Lambda}(\theta^0,\ ð\theta^0,\ \bar{ð}\,\theta^0,\ ð\ ð\theta^0, \zeta, \zeta)$$

in the form $\theta^0 \equiv u = Z(x^a, \zeta, \overline{\zeta})$ where the four x^a are constants of integration, interpreted now as local space-time coordinates. The surfaces, $u = Z(x^a, \zeta, \bar{\zeta})$ = constant, are (a spheres worth of) null surfaces of the already existing metric. From this, by applying $\eth$, $\bar{\eth}$ and $\eth\bar{\eth}$ to Z, we can define the full set of $\theta^i = \theta^i(x^a, \zeta, \bar{\zeta})$, (see Eq.(9)) and interpret them as the coordinate transformation from the (arbitrary) coordinates x^a to the $(\zeta, \bar{\zeta})$-dependent coordinates, θ^i, with the x^a and $(\zeta, \bar{\zeta})$ being coordinates of the bundle of null directions over space-time. Inverting these we obtain $x^a = x^a(R, u, \omega, \bar{\omega}, \zeta, \bar{\zeta})$, the parametric form for the full five dimensional set of null geodesics. One also obtains a geometric meaning for Λ; namely it is the scalar product of two connecting vectors obtained by varying ω and ζ separately.

A series of questions, conjectures and issues arise naturally if we adopt the point of view of this "null surface" form of GR.

1. One of the most important points, for us, is to note that "standard" GR is a classical field theory, of the gravitational *field* (in any one of its many versions), while from our point of view it is, in essence, a theory of surfaces; the metric appears as a derived concept. Only the conformal factor should be thought of as a "field". Two immediate observations follow from this.

a. In the conventional approach to quantum gravity one considers the fields, (*metric, connection, vierbein, holonomy, etc.*) on some manifold (three, four or higher dimensional) that, on quantization, become operator valued functions on the manifold. Here the basic objects are, instead, the characteristic surfaces, i.e., collections of points, the space-time points themselves. It becomes very difficult to see how this "null surface view" can be converted into any conventional relationship with quantum theory. In light of the difficulties with the construction of a "conventional" quantum theory of gravity, it might well be useful to think along alternative lines, perhaps suggested by the new view.

b. If we foliate the space-time by a fixed family of space-like three-surfaces then the intersection of (any) one of our null surfaces with one of the space-like surface is a two-surface, (a "membrane"). One can then follow the intersection of that null surface on to the succeeding members of the foliation and thus see that our "dynamics of characteristic surfaces" can be viewed as the evolution of two-surfaces. *In this manner we can think of GR as the evolution of membranes* - mod the conformal factor.

2. It seems possible that the view of GR present here could suggest or allow for variations, generalizations or modifications of standard GR.

a. One possible suggestion arises from the clear separation of the metric into the conformal metric and the conformal factor. Is it possible to formulate a quantum theory with the conformal factor being an operator while the conformal metric (e.g., our Λ) remains a c-number? This has the beauty of leaving the conformal structure of space-time classical, avoiding many unpleasant questions and just "quantizing the "field" Ω. It is not clear whether this is a feasible program. In any case it does not seem to have been investigated extensively.

b. Since null surfaces play such a dominant role in our version of GR, and, in fact, all null geodesics are contained in our solution, $u = Z(x^a, \zeta, \bar{\zeta})$, it is clear that there is a close connection with Penrose's twistor theory- or more precisely with ambi-twistor theory. The details remain to be investigated.

c. Our treatment of GR suggests a new version of loop variables. In the five-dimensional space of null geodesics, parametrized by $(u, \omega, \bar{\omega}, \zeta, \bar{\zeta}) \equiv Y^{\alpha}$ loops could be defined parametrically by $Y^{\alpha} = Y^{\alpha}(s)$. In the space-time, using our null geodesics, they become the two-surfaces,

$$x^{a} = x^{a}(R,s) = x^{a}(R, Y^{\alpha}(s))$$

but for fixed R they are loops. The attractive feature of this version is that their evolution (along R) is already completely given and does not depend on an arbitrary choice of lapse and/or shift function as is needed in the conventional approach. This also remains to be investigated.

Acknowledgments

We are deeply indebted to both Lionel Mason and Carlo Rovelli for many conversations and suggestions. We also thank Savi Iyer, whose Ph.D. Thesis played a crucial role in the developments that led to the present work. This research was supported in part by the National Science Foundation under Grants #PHY 89-04035 and PHY 92-05109. C.K. thanks CONICET for support.

Appendix

We discuss here a situation that is simpler but analogous to that in the main body of this work, where we show how to encode the properties of a three parameter family of surfaces into a four dimensional manifold $\mathcal{M}$ independent of the local coordinates x^{a}.

Starting from a four-dimensional manifold $\mathcal{M}$ with arbitrary local coordinates x^{a}, we consider a three-parameter family of *three-dimensional* surfaces in the manifold so that a two-parameter set (an R^{2} set) of surfaces pass through each point x^{a}. If we let the R^{2} be parametrized by s and s*, the surfaces $F(x^a,u,s,s^*) = 0$ can be expressed as

$$u = \mathfrak{z}(x^a, s, s^*) = \text{constant} \qquad (A.1)$$

We wish to encode the information about this family of surfaces in a manner intrinsic (in some sense) to the family itself and not be dependent on the choice of the original local coordinates, x^a.

We first define

$$w = \partial\mathfrak{z}(x^a, s, s^*)/\partial s \equiv \partial_s\mathfrak{z}, \qquad (A.2)$$
$$w^* = \partial\mathfrak{z}(x^a, s, s^*)/\partial s^* \equiv \partial_{s^*}\mathfrak{z},$$
$$r = \partial_s\partial_{s^*}\mathfrak{z}(x^a, s, s^*)$$

and introduce the notation, $(u, w, w^*, r) \equiv \theta^i \equiv (\theta^0, \theta^+, \theta^-, \theta^1)$, so that

$$\theta^i = \theta^i(x^a, s, s^*) \equiv (\mathfrak{z}, \partial_s\mathfrak{z}, \partial_{s^*}\mathfrak{z}, \partial_s\partial_{s^*}\mathfrak{z}).$$

We refer to the θ^i as an R^2's worth of coordinates defined by, (or "intrinsic" to) the surfaces (A.1). Inverting the transformation yields

$$x^a = x^a(\theta^i, s, s^*). \qquad (A.3)$$

Taking the second derivatives of (A.1) yields

$$\partial_s\partial_s\mathfrak{z} = \ell(x^a, s, s^*) \quad \text{and} \quad \partial_{s^*}\partial_{s^*}\mathfrak{z} = \ell^*(x^a, s, s^*),$$

which become, after eliminating the x^a's via (A.3),

$$\partial_s\partial_s\mathfrak{z} = \lambda(\theta^i, s, s^*) \quad \text{and} \quad \partial_{s^*}\partial_{s^*}\mathfrak{z} = \lambda^*(\theta^i, s, s^*) \qquad (A.4)$$

It is easy to see that the full information about the surfaces, i.e.,(1.1) is encoded into the functions, $\lambda(\theta^i, s, s^*)$ and $\lambda^*(\theta^i, s, s^*)$, of the intrinsic coordinates. To return to the original description of the surfaces,

$$u = \mathfrak{z}(x^a,s,s^*),$$

one writes (1.4) as the two second order differential equations

$$\partial_s\partial_s\mathfrak{z} = \lambda(\mathfrak{z}, \partial_s\mathfrak{z}, \partial_{s^*}\mathfrak{z}, \partial_s\partial_{s^*}\mathfrak{z}, s, s^*) \qquad (A.5)$$
$$\partial_{s^*}\partial_{s^*}\mathfrak{z} = \lambda^*(\mathfrak{z}, \partial_s\mathfrak{z}, \partial_{s^*}\mathfrak{z}, \partial_s\partial_{s^*}\mathfrak{z}, s, s^*)$$

and attempts to integrate them. From the construction, the solutions are the original $u = \mathfrak{z}(x^a, s, s^*)$, but now with the local coordinates, x^a arising as the constants of integration. [To obtain surfaces with special properties - not depending on the particular coordinates x^a - we *could have imposed* differential equations on the choice of $\lambda(\theta^i, s, s^*)$. This, in fact, is done in GR. See Sec.II.] Notice that if we had been given two *arbitrary* functions $\lambda(\theta^i, s, s^*)$ and $\lambda^*(\theta^i, s, s^*)$, (A.5) would not, in general, have a joint solution in the form $u = \mathfrak{z}(x^a, s, s^*)$.

It is important to understand when it is possible to work backwards from a given (λ, λ^*) pair, to the family of surfaces. Beginning with the six dimensional space of $y^A = (\theta^i, s, s^*)$, the following pair of vector fields can be constructed

$$\partial \equiv Q_A\partial/\partial y^A \equiv \partial/\partial s + \theta^+\partial/\partial u + \lambda\, \partial/\partial w + r\partial/\partial w^* + \kappa\, \partial/\partial r \qquad (A.6)$$

$$\partial^* \equiv Q^*_A\partial/\partial y^A \equiv \partial/\partial s^* + \theta^-\partial/\partial u + \lambda^*\, \partial/\partial w^* + r\partial/\partial w + \kappa^*\partial/\partial r.$$

with the functions κ and κ^* to be determined. Their integral curves satisfy, respectively the two sets of o.d.e.'s,

$$du/ds = w\,, \quad dw/ds = \lambda(\theta^i, s, s^*), \quad dw^*/ds = r, \qquad (A.7a)$$
$$dr/ds = \kappa\,, \quad ds/ds = 1, \quad ds^*/ds = 0\,,$$

and

$$du/ds^* = w^*\,, \quad dw^*/ds^* = \lambda^*(\theta^i, s, s^*), \quad dw/ds^* = r, \qquad (A.7b)$$
$$dr/ds^* = \kappa^*\,, \quad ds/ds^* = 0, \quad ds^*/ds^* = 1.$$

The two families of the integral curves of both ∂ and ∂^*, i.e., $y^A = y^A(s, \alpha^i)$ and $y^A = y^A(s^*, \beta^i)$, each depends on five parameters, the α^i and β^i. When ∂ and ∂^* commute, i.e., $\partial\partial^* - \partial^*\partial = 0$, the two vector fields are surface forming allowing a joint solution of the form

$$y^A = y^A(s, s^*, x^a)$$

or

$$u = \mathfrak{z}(x^a, s, s^*)\,, \qquad w = \partial_s \mathfrak{z}(x^a, s, s^*),$$

$$w^* = \partial_{s^*} \mathfrak{z}(x^a, s, s^*), \qquad r = \partial_s \partial_{s^*} \mathfrak{z}(x^a, s, s^*)$$

with the x^a, an arbitrary set of *four* parameters. This joint solution can be found by solving Eqs.(A.5), with $u = \mathfrak{z}$ since this is implied by the integral curves of (A.6):

$$d^2u/ds^2 = \lambda(u,\ du/ds,\ du/ds^*,\ d^2u/ds^*ds,\ s,\ s^*)$$

$$d^2u/ds^{*2} = \lambda^*(u,\ du/ds,\ du/ds^*,\ d^2u/ds^*ds,\ s,\ s^*). \qquad \text{(A.8)}$$

If the (x^a, s, s^*) are thought of as a new set of coordinates replacing the (θ^i, s, s^*), then both vector fields are simultaneously rectified in the new coordinates: i.e., they become

$$\partial = \partial/\partial s \quad \text{and} \quad \partial = \partial/\partial/s^* \ .$$

Requiring the vanishing of the commutator leads, after a brief calculation, to a unique form for the (κ, κ^*) in terms of the (λ, λ^*) as well as conditions on the (λ, λ^*) which are completely equivalent to the relations contained in the original construction, namely $\partial_s \partial_s \lambda^* = \partial_{s^*} \partial_{s^*} \lambda$.

In Sec.II we used the close analog of this structure where the R^2 is replaced by S^2 in four dimensional space-time.

References

1. *GR via Characteristic Surfaces,* by S. Frittelli, C. N. Kozameh, E. T. Newman, submitted to J. Math. Phys.

2. *Linearized Einstein Theory via Null Surfaces,* by S. Frittelli, C. N. Kozameh, E.T. Newman, submitted to J. Math. Phys.

3. *Lorentzian Metrics from Characteristic Surfaces,* by S. Frittelli, C. N. Kozameh, E.T. Newman, submitted to J. Math. Phys.

4. *On the Dynamics of Characteristic Surfaces,* by S. Frittelli, C. N. Kozameh, E.T. Newman, submitted to J. Math. Phys.

5. C. N. Kozameh and E. T. Newman, in *Asymptotic Behavior of Mass and Spacetime Geometry*, edited by F.J. Flaherty, Lecture Notes in Physics #202, Springer-Verlag , New York, (1984).

6. R. Penrose and W. Rindler, *Spinors and Space-Time,* Vol 1, Cambridge University Press (1984)

TWISTOR THEORY, THE EINSTEIN EQUATIONS, AND QUANTUM MECHANICS

Roger Penrose
Mathematical Institute
University of Oxford, England

1. The question of quantized space-time and quantum gravity

The normal procedures of quantum gravity involve leaving the principles of quantum mechanics (QT) intact, whilst attempting to modify general relativity (GR) - or some other classical space-time theory - in accordance with those principles. Yet GR is tested overall to an even greater accuracy (about 1 part in 10^{14} as implied by the precision of the Hulse Taylor observations of the binary pulsar PSR 1913+16 over a period of 20 years), than is QT (to 1 part in 10^{11} or 10^{12} as follows from the agreement between theory and experiment in quantum electrodynamics).Thus, my own viewpoint is to seek a new theory which modifies and units both GR and QT, and for which both emerge as limiting cases of this theory. *Twistor theory* represents an attempt in that direction.

Before embarking on a description of twistor theory, however, it will be appropriate to consider what I regard as the main problem of present-day quantum theory, namely how the quantum micro-world relates to the classical macro-world. There would appear to be one set of laws for the micro-world, governed by complex numbers (where the amplitudes have their moduli squared to become real-number probabilities). The fundamental difficulties is well illustrated by Schrödinger's famous cat.

Here I shall consider that a photon is emitted at a source and is put into a superposition of a transmitted and reflected state by a half-silvered mirror. The transmitted part of the photon's wave function causes a gun to fire which kills a cat, so that according to the standard unitarity procedures of quantum state evolution (Schrödinger's equation), the cat ought to be put into a superposition of death and life at the same time - with complex coefficients! (see Fig.1).

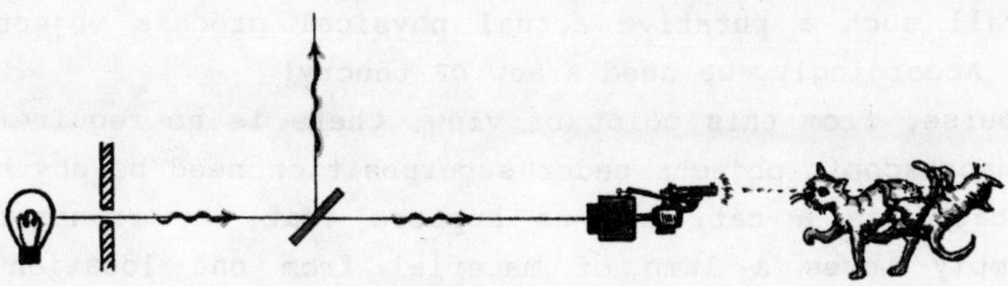

Fig.1

Schrödinger's cat
a half-silvered mirror splits a phothon's state into a superposition of two states, one of which kills the cat. Is the cat in a superposition of dead and alive?

But perhaps something goes wrong with superposition principle when it applies to significantly differing space-time geometries.

The live cat and the dead one will indeed be accompained by space-time geometries which slightly differ, so a superposition

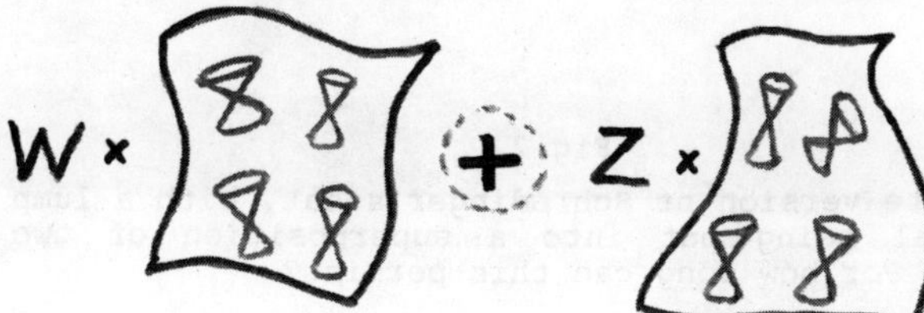

of two space-times might actually become

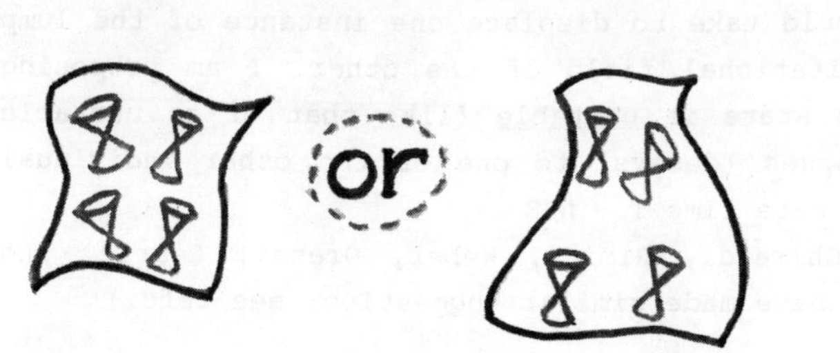

(with relative probabilities $|w|^2 : |z|^2$).

I shall call such a putative actual physical process Objective Reduction. Accordingly, we need a new **OR** theory!

Of course, from this point of view, there is no requirement that the macroscopic objects under superposition need be anything so complicated as a cat. Let us suppose that the transmitted photon simply moves a lump of material from one location to another (see Fig.2)

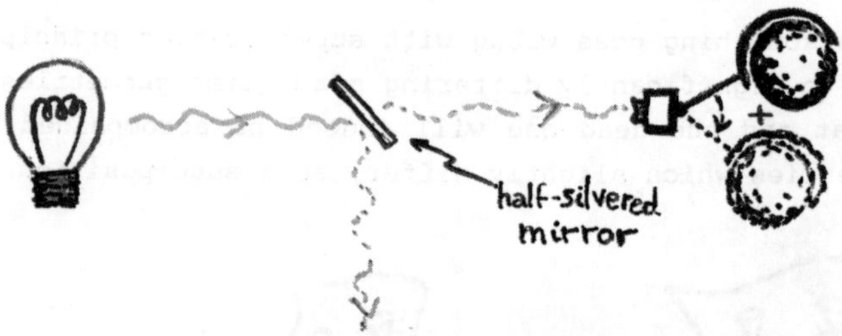

Fig.2

An inanimate version of Schrödinger's cat, with a lump of material being put into a superposition of two locations. For how long can this persist?

I am not going to attempt to provide such a theory here. But we can at least pose the following question: At what stage does a superpositiom (+) of the two lump positions become either one lump position or the other? My suggestion is as follows. Consider the energy E that it would take to displace one instance of the lump away from the gravitational field of the other. I am proposing that the superposed state is <u>unstable</u> (like that of an unstable nucleus), and collapses (decays) to one or the other individual state, in an approximate time $T \sim \hbar/E$
(Also Kávolyhásy, Ghirardi, Rimini, Weber, Grassi, Pearle, and particularly Diósi, have made similar suggestion; see refs.)

Stability of a superposition of a pair of lump locations

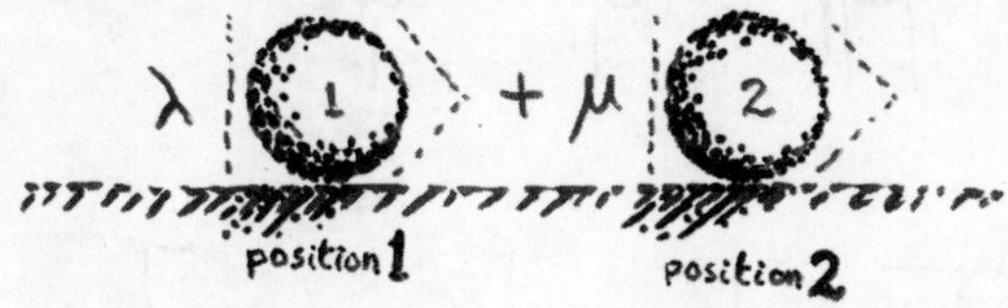

Fig.3

I want to give you a new argument supporting the idea that such superpositions of macroscopically displaced mass distributions are unstable. Let $|\psi_1\rangle$ and $|\psi_2\rangle$ be the quantum states for two alternative spatially displaced stationary lump locations. Ignoring the gravitational fields produced by the lump itself, in each location, we have a timelike Killing vector **T** for background static space-time, which plays the role of the "$\partial/\partial t$" in the Schrödinger equation. For each individual state, we have

$$i\hbar \mathbf{T}|\psi_1\rangle = F|\psi_1\rangle, \qquad i\hbar \mathbf{T}|\psi_2\rangle = F|\psi_2\rangle,$$

where F is its energy, whence for the superposed state,

$$i\hbar \mathbf{T}\left(\lambda|\psi_1\rangle + \mu|\psi_2\rangle\right) = F\left(\lambda|\psi_1\rangle + \mu|\psi_2\rangle\right).$$

We have a complete degeneracy: all superpositions of $|\psi_1\rangle$ and $|\psi_2\rangle$ are just as stationary as are $|\psi_1\rangle$ and $|\psi_2\rangle$ individually.

Now, bring in the gravitational field of the lump. Are all the superposed states still stationary, each with well-defined energy (F)?

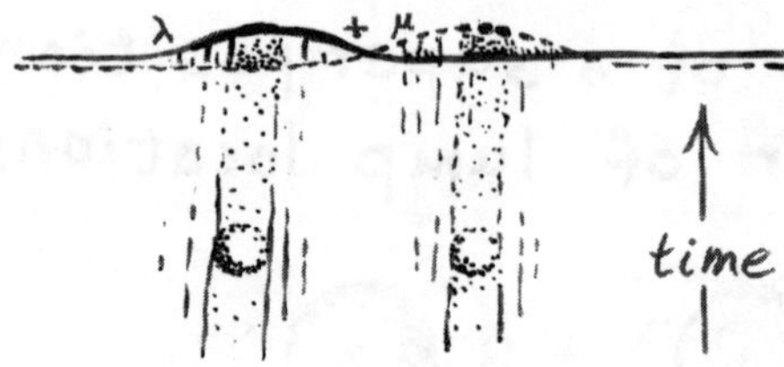

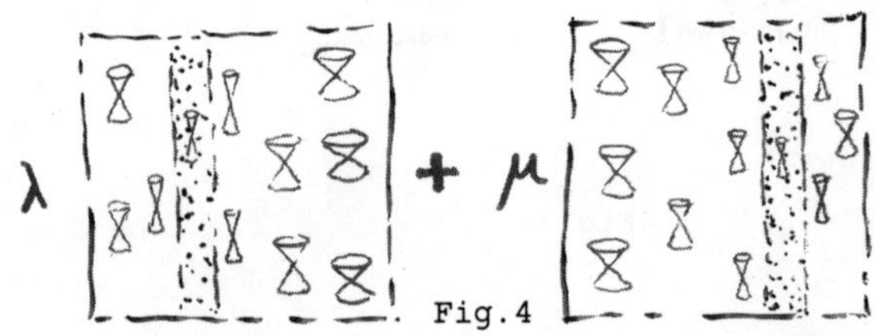

Fig.4

Superposition of differing space-times?

We have to consider superpositions of actual differing *space-time geometries*, since this is how the gravitational fields of the lump in its two alternative locations must be described. Quantum gravity usually treats such a superposition very formally, but we need to do physics *in* such a superposed space-time. It is hard to see how to make sense of such an idea - even to express the fact that the lump occupies a different location in the superposed space-times - without some kind of pointwise identification between the space-times. However the principle of general covariance is antagonistic to such a proposal and, moreover, such an identification would lead to a *superposed causality* (see Fig.5), which does not make sense. Accordingly, I

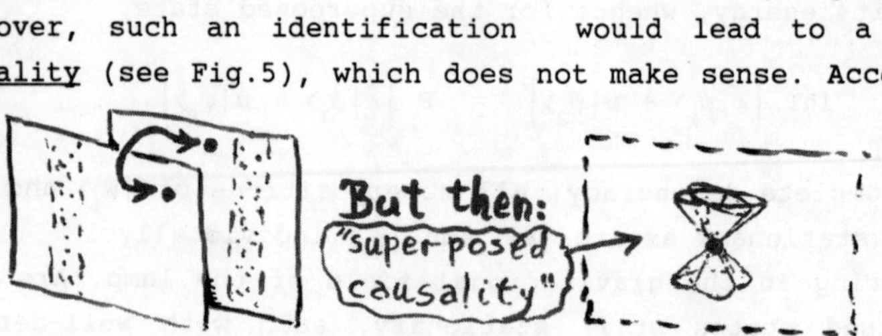

Fig.5

Try to identify points? - at least approximately

shall consider that there is some <u>approximate</u> notion of pointwise identification between the two space-times under superposition, where there is an imbuilt "uncertainty involved in the identification. This will lead to an ill-definedness in the notion of a Killing vector for the superposed space-times, and hence to an ill-definedness in the very notion of stationarity.

We can take as the "measure of ill - definedness" of the Killing vector, the local expression

$$\dot{g}_{00}/\dot{g}_{00}^{2} \quad \text{or} \quad \phi_1 - \phi_2$$

(where

$$\sqrt{g_{00}} \approx 1 + 2\phi, \text{ with } \phi = \text{Newtonian potential; } G=1)$$

or, rather, as its spatial variation. In the Newtonian limit, we can express this variation as

$$\vec{\nabla}(\phi_1 - \phi_2)$$

and then we can take as the total measure of ill-definedness, the expression

$$\int \vec{\nabla}(\phi_1 - \phi_2) \cdot \vec{\nabla}(\phi_1 - \phi_2)\, d^3x$$

$$= - \int (\phi_1 - \phi_2)\, \nabla^2(\phi_1 - \phi_2)\, d^3x$$

$$= 4\pi \int (\phi_1 - \phi_2)\,(\rho_1 - \rho_2)\, d^3x$$

$$= -4\pi \iint \frac{(\rho_1(x) - \rho_2(x))(\rho_1(y) - \rho_2(y))}{|x-y|}\, d^3x\, d^3y$$

where ρ is the mean density. This is just the gravitational self-energy of the difference mass distribution (i.e. of one mass distribution subtracted pointwise from the other).

In the case of a pair mass distributions in which one is just a rigid displacement of the other, this gives just the same energy

E as we had before, for the superposed displaced lump locations. (In that case E was considered to be the gravitational interaction energy, which was specifically Diósi's proposal. The two agree for mere rigid displacements, but it is not altogether clear which is a better measure, in the general case). The lack of definiteness in the notion of stationarity corresponds to this energy E, so we anticipate a lifetime of the superposed state of the order of $\hbar E^{-1}$ as asserted earlier.

For an isolated system, the lump movement itself is the only mass diaplacement which need be considered. In this case we can easily estimate $\hbar E^{-1}$ in simple cases. For a nucleon, considered as a solid sphere of about a fermi's radius in a superposition of two separate spatial locations, the state would jump to one location on the other in a timescale of a few million years. Of course, this need not be a very realistic model, but it tells us that we can ignore **OR** effects for a single nucleon which is just as well, since neutron interferometry is a confirmed fact). For a water speck of 10^{-3}cm radius, however, a decay time for the superposed state would be about a millionth of second. If of radius 10^{-4}cm (micron), we expect decay of the superposition in about 1/20 second; if of radius 10^{-5}cm, then it might take a few hours (Fig.6).

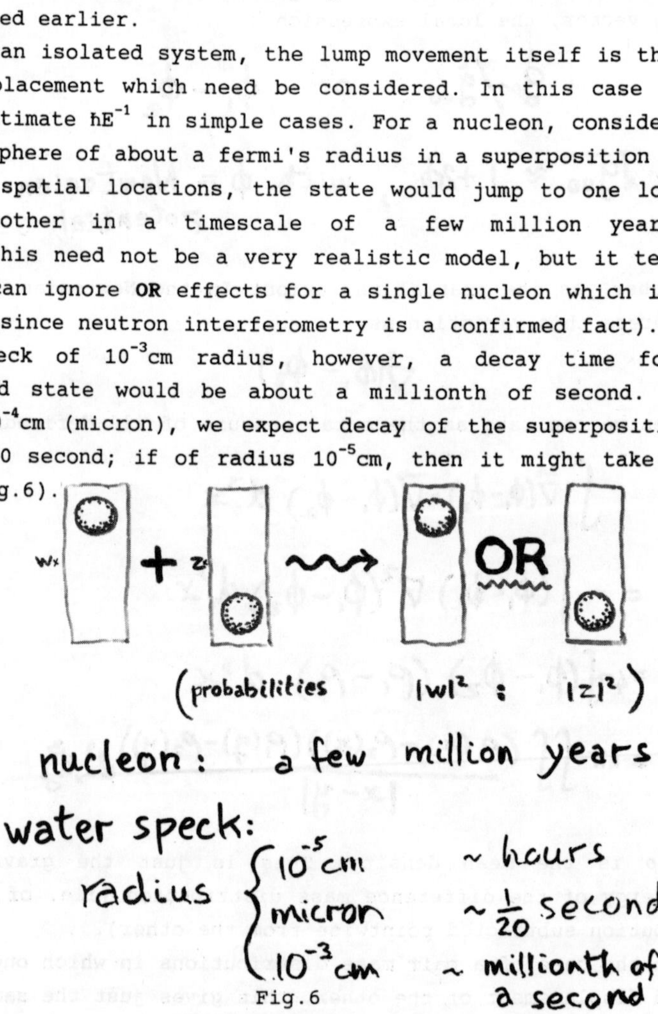

Fig.6

Decay times (isolated system)

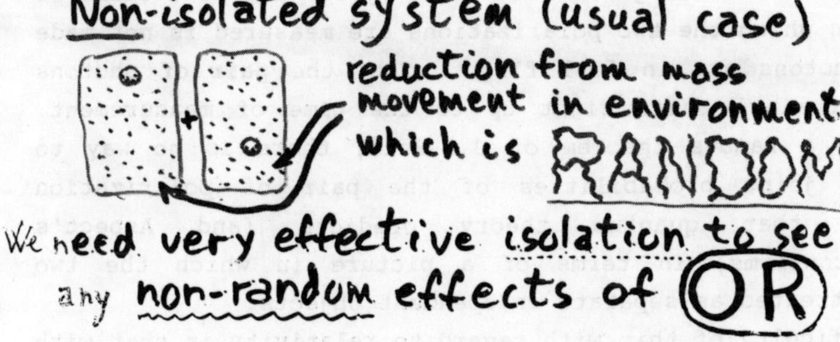

Fig.7

The effect of the environment on state-reduction

These times do not take into account the fact that, in most physical situations, the system under consideration is not isolated, so there may be a good deal of mass movement in the environment accompanying the movement of the lump. The quantum state will be an <u>entangled</u> one, and the major part of the contribution to E is then likely to come from the movement of the particles, in the environment, which are entangled with the state of the lump. Since the environment normally consists largely of randomly moving particles, their effect on the **OR** process would be an essentially random one, but where reduction would take place much more rapidly than would have been the case without the entangled environment. (see Fi.7). This leads, in effect, to nothing noticeably different from what is expected in the normal treatments of "reduction" due to environmental entanglement. It seems that the effects of an **OR** theory, consistent with the ideas being expressed here, would be very hard (though not impossible) to detect experimentally.

There would be many very difficult problems to be faced by any detailed **OR** theory. One of the most severe of these comes about because of the quantum entanglements that arise in EPR (Einstein-Podolsky-Rosen) -type situations. The most famous example of an experimentally verified EPR effect occurs with the experiment performed by Aspect and colleagues in 1982. Here there is a photon source which emits pairs of photons in opposite directions, and the polarization plane of each photon is measured, the two measurements being some 12 metres apart. The selections of directions in which the two polarizations are measured is not made until the photons are in full flight. Yet, the pair of photons behaves as a single unit right up to the time of measurement. According to a famous theorem of J.S.Bell, there is no way to explain the joint probabilities of the pair of polarization measurements that quantum theory predicts (and Aspect's experiment confirms) in terms of a picture in which the two photons are treated as separate independent objects.

The difficulty of this with regard to relativity is that with an entangled system (such as the pair of photons just described), the measurement of one member of the system appears instantaneously to affect the state of the other. Such a description would not be consistent with the principles of relativity (see Fig.8). This suggests to me that a radically different approach to space-time description will be needed if a fully objective picture of entangled quantum states is to be obtained. This is what twistor theory strives and achieves. In the following sections I shall try to indicate where this theory stands at present.

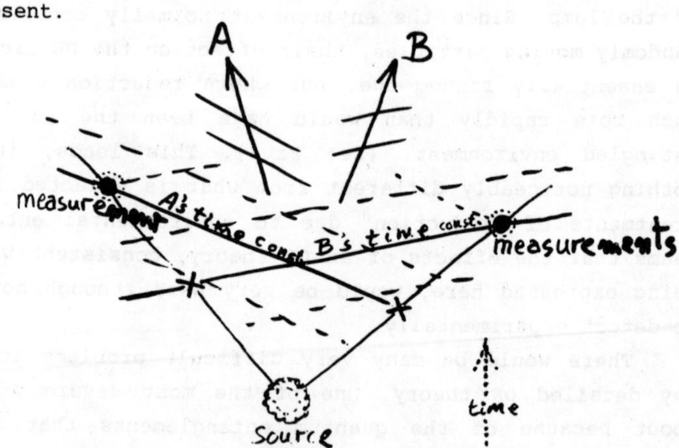

Fig.8

EPR effects and time

According to A, the left-hand measurement is first and "reduces" the right-hand photon

According to B, the right-hand measurement is first and "reduces" the left-hand photon

Do we need a radically space-time/quantum world view?

2. Motivations for twistor theory

The theory of twistors is partially motivated by a desire for a non-local approach to space-time physics, where space-time points (events) are secondary, the twistors themselves being primary in this picture. Other motivations come from the complex number structure of QM.

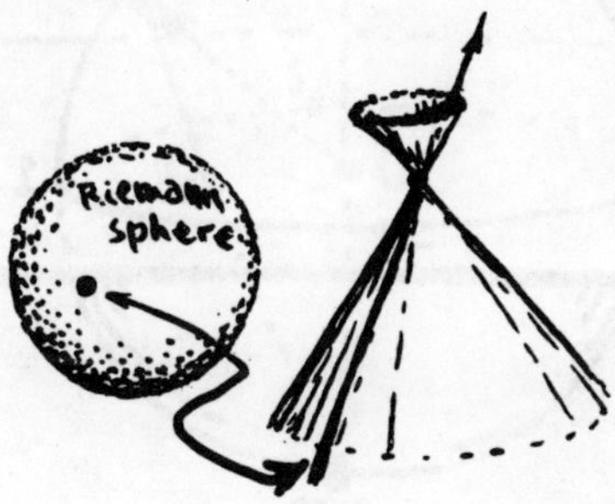

Fig.9
Twistors as light rays
The light rays through any space-time point constitute a Riemann sphere.

Although the complex numbers (amplitudes) that appear in quantum mechanics are usually treated rather abstractly, there is actually a fairly close relationship between these complex numbers and space-time geometry, even though this relationship is not always very manifest and is not very widely appreciated. The most obvious place where this relationship can be seen occurs with an ordinary spin 1/2·particle, such as electron, proton, or neutron. If we consider the two basic states to be "spin up" and "spin-down, described by the respective state vectors $|\uparrow\rangle$ and $|\downarrow\rangle$, the general state of spin

$$u\ |\uparrow\rangle\ +\ v\ |\downarrow\rangle$$

turns out to be eigenstate of a spin measurement in some other direction, the various possible spatial directions being representable as points on a Riemann sphere (complex plane together with a point at infinity). The complex number $z = v/u$ can be represented on this sphere by stereographic projection to the complex plane (see Fig.10). This shows up the close relationship between complex amplitudes and spatial geometry in the case of a spin 1/2 particle.

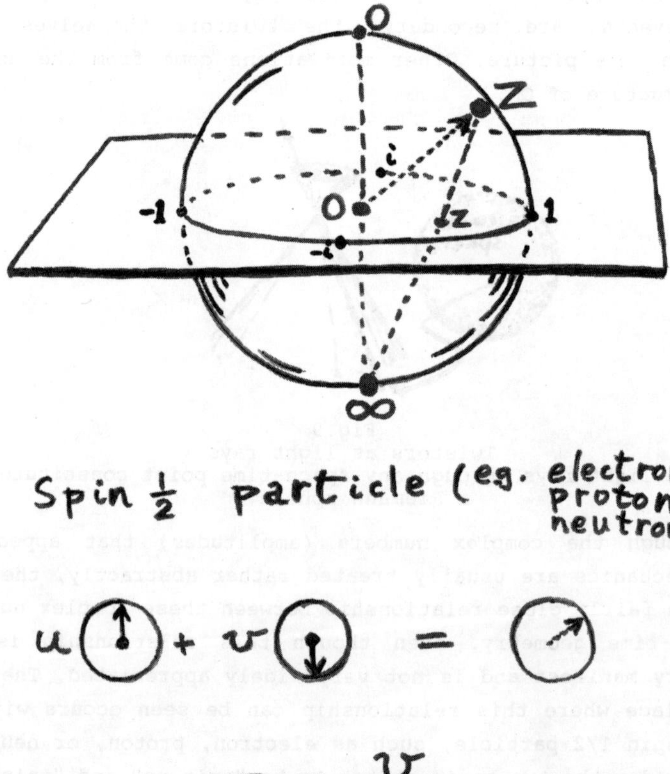

Fig.10
Spin 1/2: Stereographic projection of the Riemann sphere. It represents the complex ratio z = v/u, indicating a close relationship between spatial geometry and complex quantum amplitudes.

This might be thought to be a rather special property of spin 1/2. However, it turns out that there is a corresponding geometrical relationship for higher spin also. In Fig.11, the Majorana description for the general spin state, of spin n/2 is indicated. Here, there are n (unordered) points on the Riemann sphere representing the general state. They define the n rotes of a polynomial whose coefficients correspond to the various amplitudes for the eigenstates of spin in the upward direction. These eigenstates are the states with definite m-value - those states distinguished by a Stern-Gerlach apparatus with field oriented upwards. They have Majorana descriptions for which there are (1/2)n + m points all at the north pole and (1/2) - m points all at the south pole. See Fig.11.

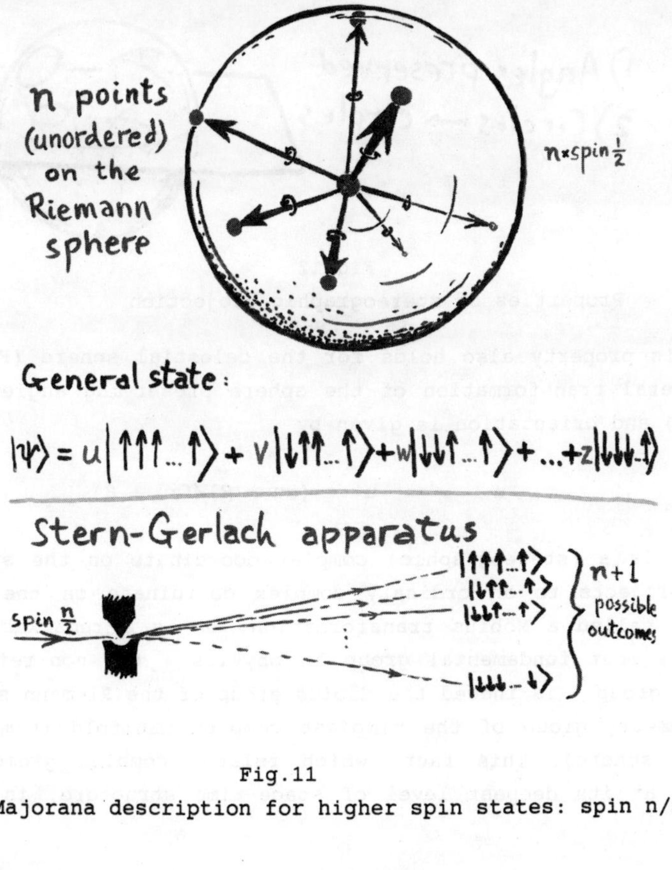

Fig.11
Majorana description for higher spin states: spin n/2

The Riemann sphere also has a role to play in relativity, as the <u>celestial sphere</u> of an observer. The reason that this is appropriate is that the transformations that take the celestial sphere of an observer who passes close by the first observer and looks out at the same sky, but transformed by abberation effects or by a rotation of frame of reference, are precisely the (holomorphic) transformations of the Riemann sphere. In other words, the action of the (restricted) Lorentz group on the celestial sphere is the same as the action of global complex-analytic maps of the Riemann sphere to itself.

The geometric nature of these transformation follows from the properties of stereographic projection: angles are preserved and circles map to circles (see Fig.12).

Fig.12

Properties of stereographic projection

This property also holds for the celestial sphere (Fig.13). The general transformation of the sphere preserving angles (also circles) and orientation is given by

$$u \Longrightarrow u' = (\alpha u + \beta)/(\gamma u + \beta),$$

where u is a (stereographic) complex coordinate on the sphere - which projects to an ordinary complex coordinate on the plane. This is called a Möbius transformation. It is a remarkable fact that the most fundamental group in physics - the non-reflective Lorentz group - is indeed the Möbius group of the Riemann sphere - the symmetry group of the simplest complex manifold (namely the Riemann sphere). This fact, which relates complex geometry to physics at its deepest level of space-time structure, is at the

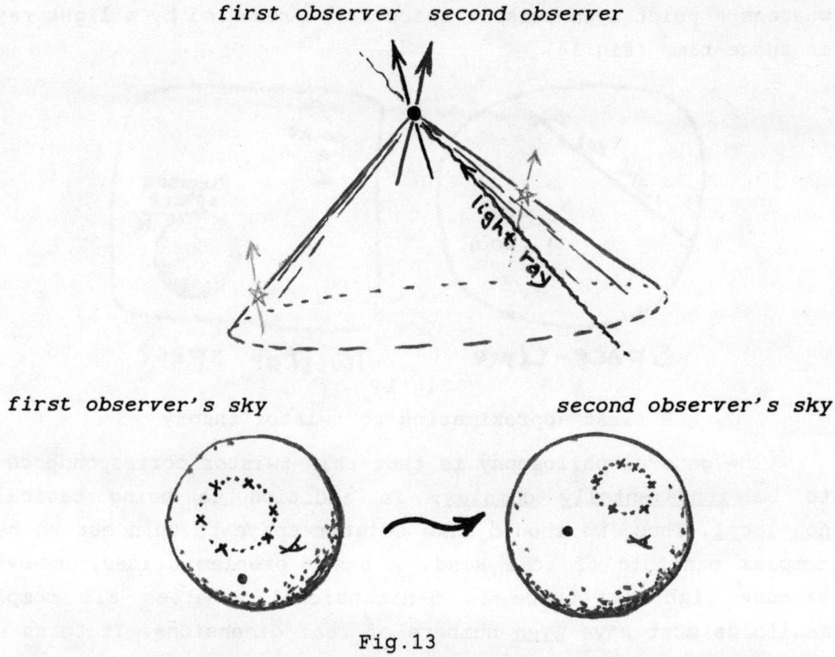

Fig.13
The conformal and circle-preserving property of Lorentz transformations

basis of the 2-spinor formalism. Carrying this philosophy further leads us into twistor theory.

3. Twistor geometry

Twistor geometry provides a different way of studying space-time structure from that provided by the usual formalism. Instead of space-time points (events) being the fundamental entities, we think of entire light rays as being more fundamental. Because of the Riemann-sphere structure of the celestial sphere, a point R of the space-time - as represented by the family of light rays through that point - corresponds to a Riemann sphere in light-ray space. As a first approximation to twistor theory, we may think of light-ray space as twistor space, so a point R in space-time is represented by a Riemann sphere R in twistor space whereas a point Z in twistor space is represented by a light ray Z in space-time (Fig.14).

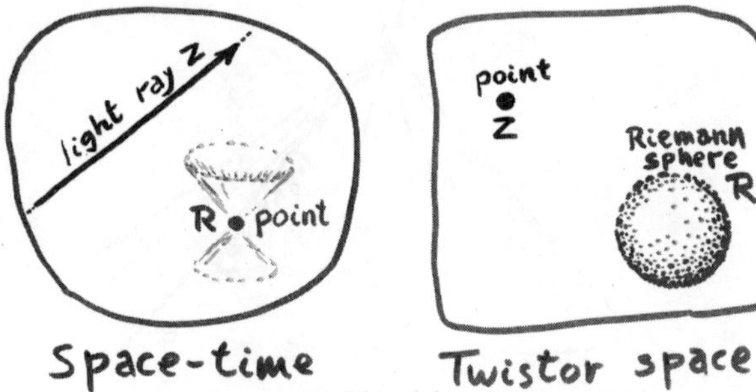

Fig.14

A first approximation to twistor theory

The general philosophy is that this twistor correspondence is to be <u>fundamentally</u> <u>complex</u>, in addition to being basically <u>non-local</u>. Thus, we should like twistor space to turn out to be a complex manifold of some kind. A basic problem arises, however, bacause light-ray space is 5-dimensional, whereas all complex manifolds must have <u>even</u> numbers of real dimensions. It turns out that for Minkowski space-time $\mathbb{M}$, there is an elegant solution to this problem. Light-ray space is 5-real-dimensional sub-manifold of a 6-real-dimensional manifold $\mathbb{PT}$ (called projective twistor space) which has a natural structure as a complex projective 3-space ($\mathbb{CP}^3$). The Riemann spheres corresponding to points in $\mathbb{M}$ are projective lines ($\mathbb{CP}^1$) in $\mathbb{PN}$ (Fig.15).

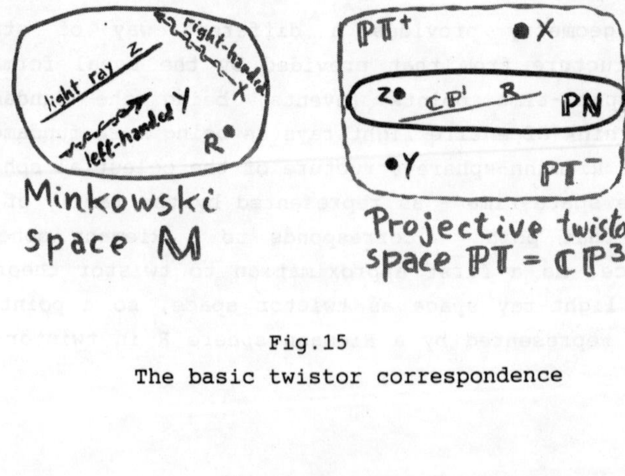

Fig.15

The basic twistor correspondence

To see how this correspondence comes about, take ordinary Minkowski coordinates for $\mathbb{M}$. A point R has space-time coordinates (r^0, r^1, r^2, r^3) while Z has twistor coordinates (Z^0, Z^1, Z^2, Z^3). The basic correspondence is expressed in the condition of incidence

$$\begin{pmatrix} Z^0 \\ Z^1 \end{pmatrix} = \frac{i}{\sqrt{2}} \begin{pmatrix} r^0+r^3 & r^1+ir^2 \\ r^1-ir^2 & r^0-r^3 \end{pmatrix} \begin{pmatrix} Z^2 \\ Z^3 \end{pmatrix}.$$

Geometrically, incidence expresses:

in $\mathbb{M}$	in $\mathbb{PT}$
point R lies on light ray Z	point Z lies on projective line R.

The space $\mathbb{PT}$ is defined by <u>projective</u> twistors, i.e. Z^α up to proportionality, i.e. by the ratios $Z^0:Z^1:Z^2:Z^3$. The twistors themselves define the non-projective vector space π.

In order to proceed further with twistor theory, it is convenient to introduce the 2-spinor notation. Here, each 4-tensor index a,b,c,.... is replaced by a pair of 2-spinor indices, one unprimed and one primed: a = AA', b = BB', c = CC',.... (using abstract indices). In standard coordinates, the vector

$$V^a = V^{AA'}$$

has components

$$\begin{pmatrix} V^{00'} & V^{01'} \\ V^{10'} & V^{11'} \end{pmatrix} = \frac{1}{\sqrt{2}} \begin{pmatrix} V^0+V^3 & V^1+iV^2 \\ V^1-iV^2 & V^0-V^3 \end{pmatrix}$$

whence

$$\det V^{AA'} = \frac{1}{2}\left[(V^0)^2-(V^1)^2-(V^2)^2-(V^3)^2\right] = \frac{1}{2} g_{ab} V^a V^b$$

$$= \frac{1}{2} \varepsilon_{AB}\, \varepsilon_{A'B'}\, V^{AA'} V^{BB'}$$

The quantities ϵ_{AB} , $\epsilon_{A'B'}$ are skew, having components which are standard Leci Civita symbols (0,1;-1,0). The spinor form of the metric tensor is then

$$g_{ab} = \mathcal{E}_{AB}\,\mathcal{E}_{A'B'}$$

the ϵ-quantities are used for lowering and raising spinor indices according to

$$\xi^A = \mathcal{E}^{AB}\xi_B \qquad \xi_A = -\mathcal{E}_{AB}\xi^B$$

Primed and unprimed indices are interchanged under complex conjugation. A single-index spinor ξ^A has (up to sign) a real-space-time representation as a <u>null flag</u>, i.e. as a null vector $x^a = \xi^A\bar{\xi}^{A'}$ together with an oriented null 2-plane element containing that null direction (see Fig.16).

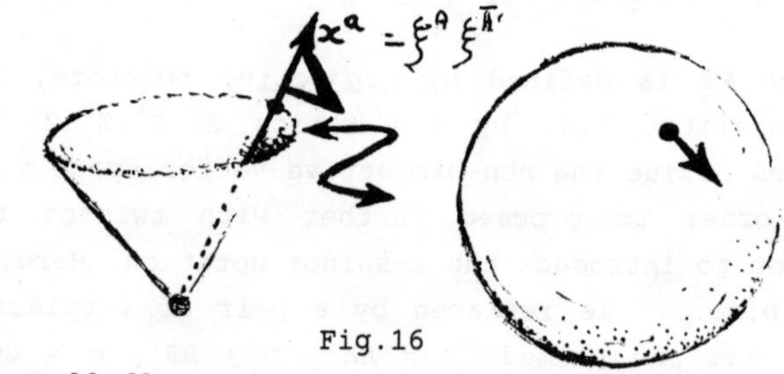

Fig.16

The null flag representation of a 2-spinor

Let O be an origin-point in $\mathbb{M}$. We can express the incidence relation as

$$\omega^A = i\Gamma^{AA'}\pi_{A'} ,$$

where $\Gamma^a = \Gamma^{AA'}$ is the position vector of some point R on a light ray Z, and where the four complex components of the twistor Z^α are the two components of each of the 2-spinors ω^A and $\pi_{A'}$, respectively. We write

$$Z^\alpha = (\omega^A, \pi_{A'})$$

and call ω^A and $\pi_{A'}$ the <u>spinor</u> <u>parts</u> of the twistor Z^α. The geometrical significance of ω^A and $\pi_{A'}$ (up to proportionality) is that the flagpole of $\pi_{A'}$ points along Z, whereas the flagpole of ω^A points at two light ray Z from the origin O (see Fig.17).

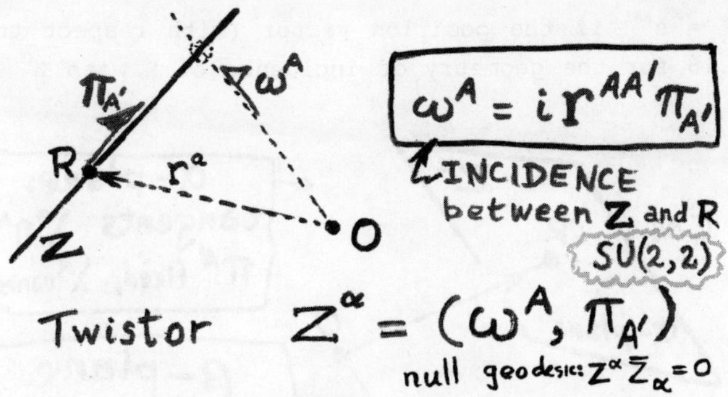

Fig.17
The spinor parts of a null twistor

The complex conjugate $\bar{Z}$ of a twistor Z^α is a dual twistor with spinor parts defined by

$$\bar{Z} = (\bar{\pi}_A \ , \ \bar{\omega}^{A'})$$

For a null ray, we must have

$$Z \bar{Z}_\alpha = 0, \text{ i.e. } \omega^A \bar{\pi}_A + \pi_{A'} \bar{\omega}^{A'} = 0 \ ,$$

this being the equation of ℙ𝕅, as depicted in Fig.15. When this equation does not hold, there are no real points R incident with the twistor Z^α. However, there will be complex points R (points of ℂ𝕄, the complexification of 𝕄) which are incident with Z^α, i.e. whose position vectors $\mathbf{r}^{AA'}$ (not required to be Hermitian) satisfy the incidence relation $\omega^A = i\mathbf{r}^{AA'}\pi_{A'}$ (so long as $\pi_{A'} \neq 0$, otherwise the points R would lie at infinity). For fixed Z^α, the complex points R incident with Z^α constitute a complex totally null plane (so of four real dimensions) called an α-plane. The tangent vectors to an α-plane are all null vectors, having the spinor form

$$\lambda^A \pi^{A'}$$

where $\pi^{A'}$ is fixed and λ^A varies. There is also another kind of totally null plane in ℂ𝕄, called a β-plane, whose tangent vectors have this same form but where now it is λ^A which is held fixed and $\pi^{A'}$ which varies. The points S of a β-plane are those incident with a dual twistor $W_\alpha = (\lambda_A \ , \ \mu^{A'})$, the incidence relation of S with W_α being expressed as

$$\mu^{A'} = -i\lambda_A s^{AA'}$$

where $s^a = s^{AA'}$ is the position vector (with respect to O) of S. See Fig.18 for the geometry of incidence of R with Z^α and S with W_α.

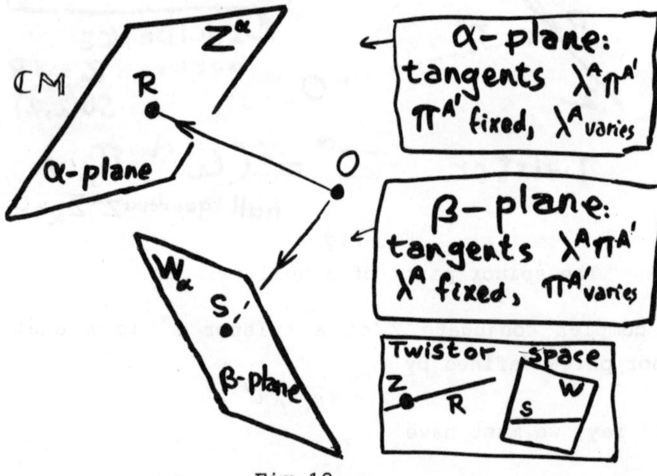

Fig.18
Complex picture

We can also express an incidence of Z^α with W_α as

$$Z^\alpha W_\alpha = 0$$

which means, geometrically, that the α-plane Z and the β-plane W have a point in common. When this happens we also have a whole <u>line</u> of intersection of the α-plane with the β-plane (Fig.19).

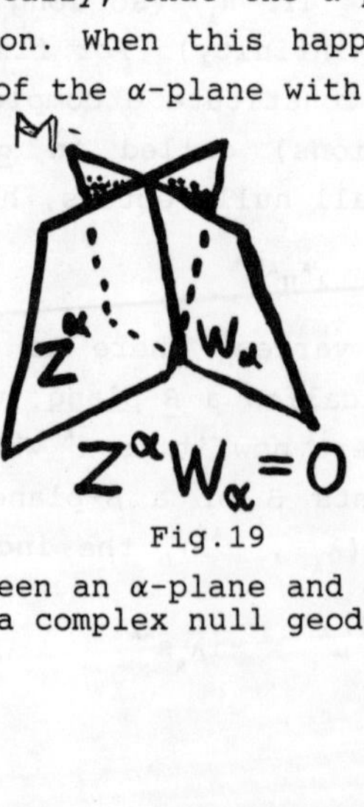

Fig.19
Incidence between an α-plane and a β-plane. They then meet in a complex null geodesic

Such a line is a complex light ray in $\mathbb{CM}$. Any complex light ray has a unique α-plane through it, and also a unique β-plane. (When the light ray is real, tha α-plane and the β-plane are complex conjugates of one another.)

Complexified Minkowski space, when compactified by the addition of appropriate elements at infinity can be thought of as a complex quadric in $\mathbb{CP}^5$. This is the Klein quadric representing the projective twistor space $\mathbb{PT}$. See Fig.20.

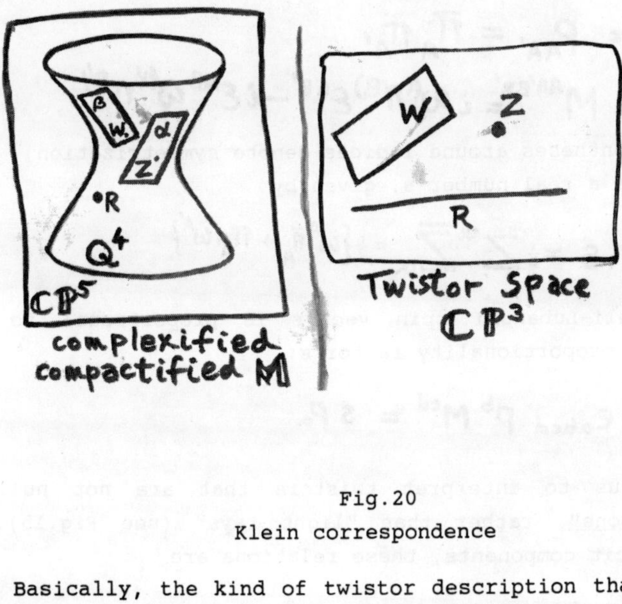

Fig.20
Klein correspondence

Basically, the kind of twistor description that I have been giving up to this point is what might be referred to as a Mk.1 twistor, given by a light ray (null geodesic) in Minkowski space $\mathbb{M}$, or by an α-plane in $\mathbb{CM}$. We shall shortly be coming to a Mk.2 twistor, which describes a spinning massless particle in terms of the 4-momentum and angular momentum structure for massless particle in $\mathbb{M}$. There is also the notion of a Mk.3 twistor, which is a helicity raising operator for massless fields in $\mathbb{M}$. There are partial generalizations of Mk.1 twistors to curved space-time $\mathcal{M}$: hypersurface twistors, asymptotic twistors, local twistors, 2-surface twistors, ambitwistors etc., which I shall not discuss here, but later on I shall say something about Mk.4 twistors, which may be thought of as charges for massless helicity 3/2 fields. These are consistent in Ricci-flat general curved space-times, but not fully understood, as yet.

4. Twistors, angular mementum, and helicity

A Mk.2 twistor represents the 4-momentum P_a and the 6-angular momentum M^{ab} structure of a massless particle:

$$P_a = P_{AA'} = \bar{\pi}_A \pi_{A'}$$

$$M^{ab} = M^{AA'BB'} = i\omega^{(A}\bar{\pi}^{B)}\varepsilon^{A'B'} - i\varepsilon^{AB}\bar{\omega}^{(A'}\pi^{B')}$$

(where the parentheses around indices denote symmetrization). The helicity is a real number s, given by

$$s = \tfrac{1}{2} Z^{\alpha}\bar{Z}_{\alpha} = \tfrac{1}{2}\{\omega^{A}\bar{\pi}_{A} + \pi_{A'}\bar{\omega}^{A'}\}$$

where the Pauli-Lubanski spin vector is proportional to the momentum, with proportionality factor s:

$$\tfrac{1}{2}\varepsilon_{abcd}\, p^{b} M^{cd} = s\, p_a$$

This enables us to interpret twistors that are not null as "spinning photons", rather than "light rays" (see Fig.15). In terms of explicit components, these relations are

$$p^0 = \tfrac{1}{\sqrt{2}}(Z^3\bar{Z}_1 + Z^2\bar{Z}_0)$$

$$p^1 = \tfrac{1}{\sqrt{2}}(Z^3\bar{Z}_1 - Z^2\bar{Z}_0)$$

$$p^2 = \tfrac{1}{\sqrt{2}}(-Z^2\bar{Z}_1 - Z^3\bar{Z}_0)$$

$$p^3 = \tfrac{i}{\sqrt{2}}(Z^2\bar{Z}_1 - Z^3\bar{Z}_0)$$

for the 4-momentum components

$$M^{01} = -M^{10} = \frac{i}{2}(Z^0\bar{Z}_0 - Z^1\bar{Z}_1 - Z^2\bar{Z}_2 + Z^3\bar{Z}_3)$$
$$M^{02} = -M^{20} = \frac{i}{2}(Z^0\bar{Z}_1 + Z^1\bar{Z}_0 - Z^2\bar{Z}_3 - Z^3\bar{Z}_2)$$
$$M^{03} = -M^{30} = \frac{1}{2}(Z^0\bar{Z}_1 - Z^1\bar{Z}_0 - Z^2\bar{Z}_3 + Z^3\bar{Z}_2)$$

$$M^{12} = -M^{21} = \frac{i}{2}(Z^0\bar{Z}_1 - Z^1\bar{Z}_0 + Z^2\bar{Z}_3 - Z^3\bar{Z}_2)$$
$$M^{32} = -M^{23} = \frac{1}{2}(Z^0\bar{Z}_0 - Z^1\bar{Z}_1 + Z^2\bar{Z}_2 - Z^3\bar{Z}_3)$$
$$M^{13} = -M^{31} = \frac{1}{2}(Z^0\bar{Z}_1 + Z^1\bar{Z}_0 + Z^2\bar{Z}_3 + Z^3\bar{Z}_2)$$
$$M^{00} = M^{11} = M^{22} = M^{33} = 0$$

for the 6-angular momentum components are

$$(\bar{Z}_0, \bar{Z}_1, \bar{Z}_2, \bar{Z}_3) = (\overline{Z^2}\ \overline{Z^3}, \overline{Z^0}, \overline{Z^1}).$$

For the helicity, we have

$$s = (1/2)(Z^0\bar{Z}_0 + Z^1\bar{Z}_1 + Z^2\bar{Z}_2 + Z^3\bar{Z}_3).$$

To understand the role of Mk.3 twistor we consider what happens when we move the origin O, in $\mathbb{M}$, to a new point X, and see how ω^A and $\pi_{A'}$ change, keeping the twistor Z^α fixed in its geometrical interpretation. We shall think of X as a variable field point, instead of using the fixed origin in O (see Fig.21.)

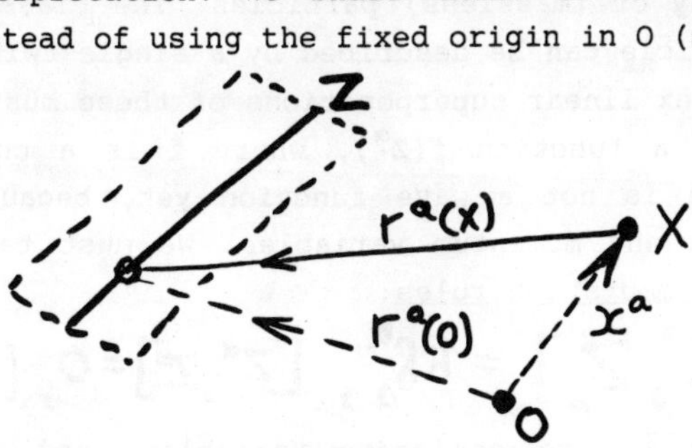

Fig.21
Making the origin a variable point

(ω^A, $\pi_{A'}$) become <u>fields</u> (keeping Z fixed) which satisfy

$$\nabla^{A'(A}\omega^{B)} = 0$$

which is called the twistor equation (which is conformally invariant). From this equation follows the existence of a spinor field $\pi_{A'}$ such that

$$\nabla_{AA'}\omega^B = -i\varepsilon_A{}^B\pi_{A'}$$

$$\nabla_{AA'}\pi_{B'} = 0$$

The twistor equation provides us with a twistor's role as a spin lovering operator. Take a massless field

$$\phi_{AB\ldots LM} = \phi_{(AB\ldots LM)}$$

subject to the field equation

$$\nabla^{AA'}\phi_{AB\ldots LM} = 0$$

then the spin-lovered field

$$\phi_{AB\ldots LM}\omega^M = \psi_{AB\ldots L}$$

satisfies same equation, but with one fewer indices.

5. Twistor quantum theory

Let us try to use twistor space as the framework for a quantum theory of (massless) particles. The classical state of a massless particle can be described by a single twistor; in quantum theory, complex linear superpositions of these must be considered, i.e. we have a function $f(Z^\alpha)$, where f is a complex function. However, this is not a wave function yet, because Z^α describes both position and momentum variables. We must take into account the <u>twistor</u> <u>commutation</u> <u>rules</u>:

$$[Z^\alpha, \bar{Z}_\beta] = \hbar\delta^\alpha_\beta, \quad [Z^\alpha, Z^\beta] = 0, \quad [\bar{Z}_\alpha, \bar{Z}_\beta] = 0.$$

Thus, Z^α and $\bar{Z}_\beta$ are conjugate variables, and a wave function should not involve both. A twistor wave function is a function of

Z^α independent of $\bar{Z}_\alpha$, which means it is holomorphik in Z^α. We can also consider <u>dual</u> twistor wave functions, which are anti-holomorphic in Z^α, i.e. holomorphic in a dual twistor W_α $(= \bar{Z}_\alpha)$.

The non-commutation of Z^α with $\bar{Z}_\beta$ could lead to possible factor ordering problems in the expressions for momentum and angular momentum given earlier. In fact, the expressions for p_a and M^{ab} are actually unaffected, but we find

$$s = (1/4)\left\{Z^\alpha \bar{Z}_\alpha + \bar{Z}_\alpha Z^\alpha\right\} = (\hbar/2)\left\{- 2 - Z^\alpha(\partial/\partial Z^\alpha)\right\} =$$

$$= (\hbar/2)\left\{ - 2 - \textit{homogeneity degree}\right\}$$

or, in terms of $W_\alpha = \bar{Z}_\alpha$,

$$s = (\hbar/2)\left\{2 + W_\alpha(\partial/dW_\alpha\right\} = (\hbar/2 + \textit{homogeneity degree})$$

The relationships between spin and helicity are illustrated in the table

TABLE

	$S/\hbar$	hom. deg. in Z^α	hom. deg in W_α
graviton	+2	−6	+2
photon	+1	−4	0
anti-neut.	+1/2	−3	−1
scalar	0	−2	−2
neutrino	−1/2	−1	−3
photon	−1	0	−4
graviton	−2	+2	−6

6. Twistor contour integrals

How do we get from a twistor wavefunction f to a space-time field ϕ.... ? This is done by means of a contour integral

$$\phi_{\cdots\cdots}(r^\cdot) = \oint \pi_\cdot\pi_\cdot\cdots\pi_\cdot \ \text{or} \ \frac{\partial}{\partial W_\cdot}\frac{\partial}{\partial W_\cdot}\cdots\frac{\partial}{\partial W_\cdot} f(Z^\cdot)\, \pi_\cdot d\pi^\cdot$$

where Z is constrained to be incident with **r**, i.e. where $\omega = \mathrm{i}\mathbf{r}\pi$, $Z = (\omega, \pi)$, the two components of ω^A being Z^0, Z^1, and the two components of $\pi_{A'}$ being Z^2, Z^3. The function f needs to be holomorphic in an appropriate region for the contour integration to give a non-trivial answer. Such a region is illustrated in Fog.22:

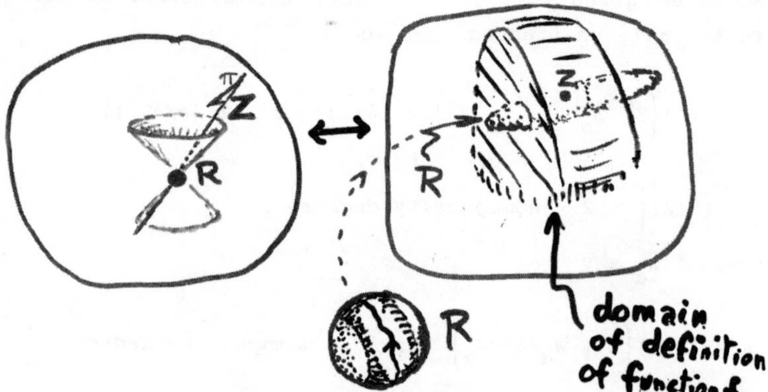

Fig.22

The geometry of the twistor contour integral

it intersects each Riemann sphere in $\mathbb{PT}$, representing a point of $\mathbb{M}$ (or $\mathbb{CM}$) where the field ϕ... is to be defined, in an <u>annular</u> region.

More correctly f should be thought of as a representative of an element of (first) <u>cohomology</u>. The region U of $\mathbb{CM}$, in which ϕ... is defined, corresponds to a region U of $\mathbb{PT}$, swept out by the lines (Riemann spheres) that represent the points of U. In the simplest situation, we cover U by a pair of open sets U_1, U_2 (with $U_1 \cup U_2 = U$), and f is defined on the intersection $U_1 \cap U_2$. More general, we might have several open sets in the covering $\{U_i\}$, and then we require a whole collection of holomorphic functions $\{f_{ij}\}$, where $f_{ij} = f_{ji}$ defined on the intersection $U_i \cap U_j$ of a <u>pair</u> of the sets. See Fig. 23.

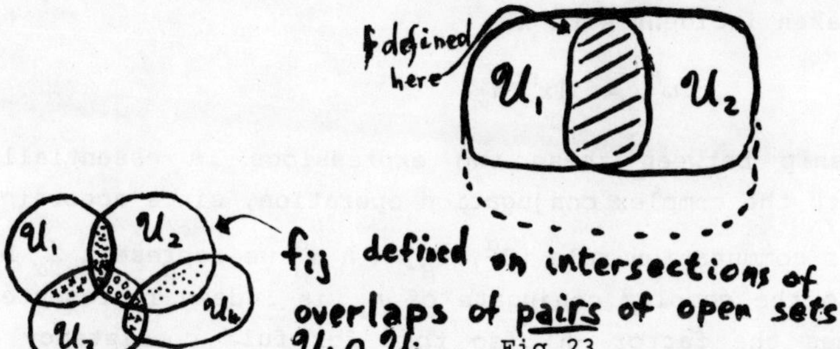

Fig.23

Strictly, f is a representative of (first) cohomology

The result of such a contour integral is always a solution of the massless free-field equations. These include the scalar wave equation (for spin zero), and I shall distinguish the positive helicity case from the negative helicity case by the use of primed rather than unprimed indices (where we assume that wavefunctions are given as complex fields of positive frequency). Thus we have (all field quantities being totally symmetric), for the massive free-field equations,

$$\nabla^{AA'}\phi_{AB\ldots L} = 0 \qquad \text{(helicity} < 0\text{)}$$

$$\Box\phi = 0 \qquad \text{(helicity} = 0\text{)}$$

$$\nabla^{AA'}\tilde{\phi}_{A'B'\ldots L'} = 0 \qquad \text{(helicity} > 0\text{)}$$

and we can distinguish the twistor contour integrals for negative helicity from those for positive helicity (zero helicity being a special case of each) by using the respective explicit forms

$$\text{helic.} \leq 0: \quad \phi_{AB\ldots L}(x) = \frac{1}{2\pi i}\oint \frac{\partial}{\partial\omega^A}\frac{\partial}{\partial\omega^B}\cdots\frac{\partial}{\partial\omega^L} f(\omega,\pi)\,\pi_{Q'}\,d\pi^{Q'}$$

$$\text{helic.} \geq 0: \quad \tilde{\phi}_{A'B'\ldots L'}(x) = \frac{1}{2\pi i}\oint \pi_{A'}\pi_{B'}\cdots\pi_{L'}\, f(\omega,\pi)\,\pi_{Q'}\,d\pi^{Q'}$$

where Z^α is taken incident with x^α

$$\omega^P = ix^{PP'}\pi_{P'}$$

The relationship between these two expressions is essentially consistent with the complex conjugation operation, since according to the twistor commutation rule $[Z^\alpha, \bar{Z}_\beta] = \hbar\, \delta^\alpha_\beta$ we represent $\bar{Z}_\alpha$ by $-\hbar\, \partial/\partial Z^\alpha$, so the complex conjugate of $\pi_{A'}$ is indeed represented by $\partial/\partial\omega^A$ (times the factor $-\hbar$), so that for full consistency a further constant factor may be included in the integrals, such as $(-\hbar)^{-n}$ in the first expression. Various familiar field equations, such as the Maxwell equations (for helicity ± 1) or the linearized Einstein equations (for helicity ± 2 written in the form of Bianchi identity applied to a weak-field curvature tensor K_{abcd}) are expressed by these massless free-field equations, where

$$F_{ab} = F_{AA'BB'} = \phi_{AB}\varepsilon_{A'B'} + \varepsilon_{AB}\tilde{\phi}_{A'B'}$$

$$K_{abcd} = \phi_{ABCD}\varepsilon_{A'B'}\varepsilon_{C'D'} + \varepsilon_{AB}\varepsilon_{CD}\tilde{\phi}_{A'B'C'D'}.$$

For complex wave functions the tilde quantities are independent of the non-tilde quantities, but for classical real fields, the tilde is replaced by a bar — the symbol for complex conjugation.

We can now see how the "Mk.3" interpretation of a twistor as a spin-lovering operator is more precisely interpreted as <u>helicity raising</u>. We represent the twistor Y^α by the operator $Y^\alpha\partial/\partial Z^\alpha$ acting on twistor functions f(Z) corresponding to

$$\phi_{AB\ldots LM} \Longrightarrow \phi_{AB\ldots LM}\eta^M$$

or

$$\tilde{\phi}_{A'\ldots L'} \Longrightarrow \eta^M\nabla_{MM'}\tilde{\phi}_{A'\ldots L'} - i(n+1)\zeta_{(M'}\tilde{\phi}_{A'\ldots L')}$$

where $Y^\alpha = (\eta^A, \zeta_{A'})$. Corresponding, a <u>dual</u> twistor plays a role as a helicity <u>lowering</u> <u>operator</u>.

7. Positive and negative frequency splitting

Quantum (field) theory requires a notion of splitting field quantities into <u>positive</u> and <u>negative</u> frequency parts, to express quantum fields propagating into the <u>future</u> and <u>past</u>, respectively. In <u>one</u> dimension, we understand this splitting, using the notion of Fourier transform by interpreting positive frequency as superpositions of e^{-ipx} for $p > 0$, and negative frequency as superpositions of $e^{-i\tilde{p}x}$ for $\tilde{p} < 0$. For space-time fields in <u>four</u>-dimensions, the <u>positive</u> frequency fields are superpositions of $\exp(-ip_a x^a)$ where the momentum 4-vector p^0 is future pointing and negative frequency fields are superpositions of $\exp(-i\tilde{p}_a x^a)$ where $\tilde{p}^a$ is past-pointing (see Fig.24).

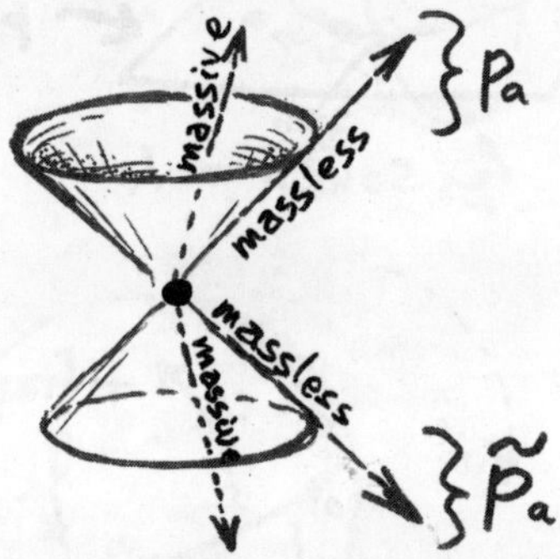

Fig.24
Momentum 4-vectors

There is also a more geometrical way of representing the notion of positive/negative frequency. In the 1-dimensional case, we can use the Riemann sphere - which, as we saw earlier (in Fig.10) can be related by stereographic projection, from the south pole, to the ordinary complex plane. In our previous picture, the equator on the sphere coincided with the unit circle on the plane. For convenience of description, however, it will be more appropriate to rotate the sphere so that the real axis (together with ∞) is represented as the equator (see Fig.25). The result that we shall use is that if a complex-valued function of a real variable is viewed as being defined on the equator of this Riemann sphere, then it has positive or negative frequency according to whether it can be extended holomorphically into the upper or lower hemispheres of the Riemann sphere.

Twistor theory provides us with a way of doing something remarkably similar, but now globally, for the whole of space-time, in the case of a massless field ϕ....(wave function) and its corresponding twistor function f.

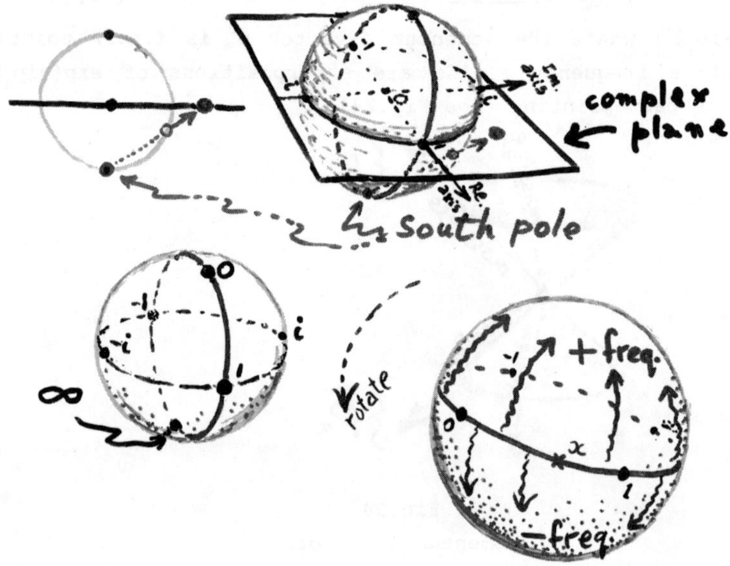

Fig.25
A complex function f(x) of a real variable x, defined on the equator of the Riemann sphere, is positive (negative) frequency if it extend as a complex holomorphic function for the upper (lower) hemisphere.

How do we ensure that f defines a wavefunction of positive or negative frequency? The answer is, regarding f as an element of first cohomology, that this cohomology element extends holomorphically from $\mathbb{PN}$ into the entire top-half [bottom-half] of twistor space ($\mathbb{PT}^+$ [or $\mathbb{PT}^-$]) if the wavefunction is of positive [negative] frequency. See Fig.26. This fact forms the basis of a good deal of the twistor approach to quantum field theory. Incomimg states, taken to have positive frequency, are represented as first cohomology elements on $\mathbb{PT}^+$ and outgoing states, taken to have negative frequency, as cohomology elements on $\mathbb{PT}^-$.

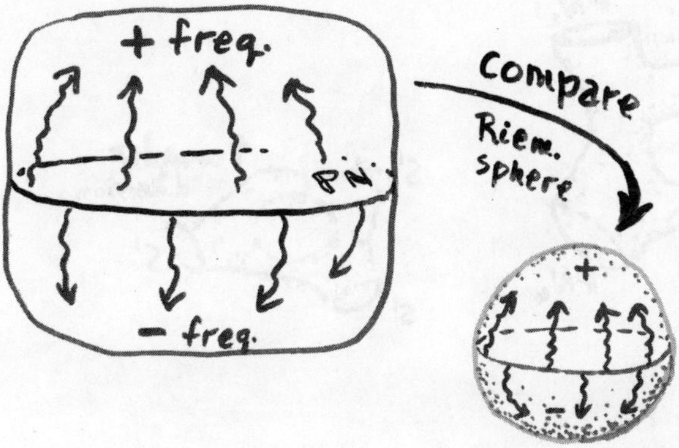

Fig.26
Positive-frequency twistor wavefunctions considered as elements of first cohomology, extend into the top half of twistor space if they have positive frequency and into the bottom half if negative frequency - similarly to the behaviour of ordinary functions of the Riemann sphere

Most of the detailed work on the twistor approach to quantum field theory has been following a perturbative approach using the twistor analogues of Feynman diagrams known as <u>twistor diagrams</u>. The subject hes been developed particularly by Andrew Hodges and his students, and some novel procedures for regularizing infinities have been discovered which have no direct analogues in conventional space-time, or momentum-space, descriptions. So far, the general approach lacks internally motivated driving principles and relies, instead, on translating directly from the conventional Feynman diagrams. An idea for possibly supplying such a driving principle was introduced by Michael Singer, and developed subsequently by Hodges and myself. In this approach, the (punctured) Riemann surfaces of conformal field theory are replaced by 3-complex-dimensional analogues obtained by glueing bits of projective twistor space together (see Fig.27).

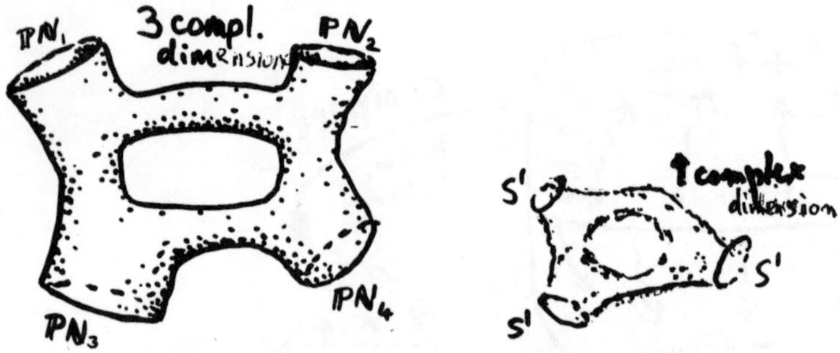

Fig.27

Generalization of conformal field theory for 4-dimensional space-time, where generalized (projective) twistor spaces take the place of Riemann surfaces

8. General relativity

There are numerous special applications of twistor theory to general relativity, but most of these will not be discussed here. Instead I shall try to indicate something of the general programme of bringing Einstein's general relativity into the scope of twistor theory. A major ingredient of this programme the "non-linear graviton construction" that I introduce some twenty years ago. This can be described roughly as follows.

Recall, first, the way in which (compactified) complex Minkowski space $\mathbb{CM}$ arises from $\mathbb{CP}^3$ (projective twistor space). Lines in $\mathbb{CP}^3$ represent points in $\mathbb{CM}$, where meeting lines represent null-separated points see (Fig.28). Now, let us try to represent a curved (complex) space-time by deforming the complex structure of $\mathbb{CP}^3$. In fact, by general theorems, there is <u>no</u> non-trivial complex deformation of $\mathbb{CP}^3$. However, there are non-trivial deformations of <u>portions</u> of $\mathbb{CP}^3$. We may consider a portion P of $\mathbb{CP}^3$ which is a tabular neighbourhood of a line in $\mathbb{CP}^3$ (representing an open neighbourhood of a point in $\mathbb{CM}$). We can deform P by cutting it (say) two (overlapping) pieces which are then re-glued, with the pieces somewhat displaced from their original attachment.

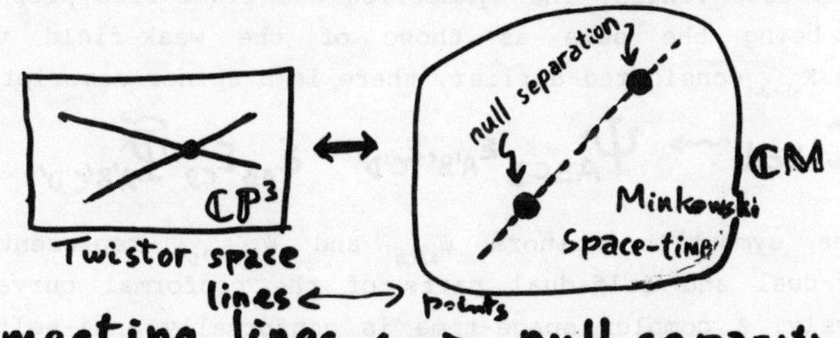

Fig.28

Twistor representation of (complex) flat space-time

It follows from general theorems by Kodaira (and Spencer) that, provided the deformation is within some limit, there is a 4-parameter family of holomorphic curves in this re-glued space (which I shall call $\mathbb{P}\mathfrak{J}$) - distinguished merely by the fact that they are holomorphic and belong to the right topological class - which represents the points of a 4-dimensional complex manifold $\mathcal{M}$. The manifold $\mathcal{M}$ obtains a complex conformal metric, where null separated points in $\mathcal{M}$ are represented by meeting lines in $\mathbb{P}\mathfrak{J}$, just as was the case for $\mathbb{CM}$ (see Fig.29)

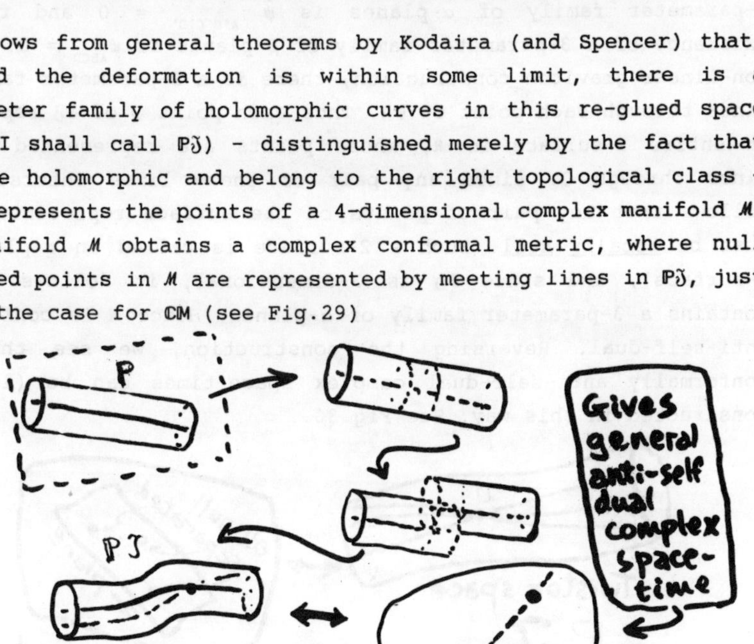

Fig.29

The non-linear graviton construction

In fact, this construction always provides a complex space-time $\mathcal{M}$ which has an (anti)-self-dual conformal curvature. We recall that the conformal curvature is described by the Weyl tensor $C^a{}_{bcd}$, which is the part of the Riemann tensor $R^a{}_{bcd}$ all of whose traces vanish. The symmetries and trace-free properties of C_{abcd} being the same as those of the weak-field vacuum curvature K_{abcd} considered earlier, there is a spinor description

$$C_{abcd} \rightsquigarrow \Psi_{ABCD}\varepsilon_{A'B'}\varepsilon_{C'D'} + \varepsilon_{AB}\varepsilon_{CD}\tilde{\Psi}_{A'B'C'D'},$$

where the symmetric spinors ψ_{ABCD} and $\tilde{\psi}_{A'B'C'D'}$ represent the anti-self-dual and self-dual parts of the conformal curvature, respectively. A complex space-time is conformally anti-self-dual if $\tilde{\psi}_{A'B'C'D'} = 0$ and conformally self-dual if $\psi_{ABCD} = 0$. It turns out that the integrability condition for the existence of a 3-parameter family of α-planes is $\tilde{\psi}_{A'B'C'D'} = 0$ and for the existence of a 3-parameter family of β-planes is $\psi_{ABCD} = 0$. In the non-linear graviton construction, there is a 2-parameter family of lines through each point of $\mathbb{P}\mathcal{T}$. Thus each point Q of $\mathbb{P}\mathcal{T}$ represents an entire 2-surface in $\mathcal{M}$, whose points are represented by the lines through Q. Since any pair of these lines intersect, it follows that any pair of points on the surface represented by Q must be <u>totally</u> <u>null</u>. Such a 2-surface is called an α-plane (or α-surface), and since $\mathbb{P}\mathcal{T}$ is 3-dimensional, it follows that $\mathcal{M}$ contains a 3-parameter family of α-planes. Hence $\mathcal{M}$ is conformally anti-self-dual. Reversing the construction, we see that all conformally anti-self-dual complex space-times can be (locally) constructed in this way. See Fig.30.

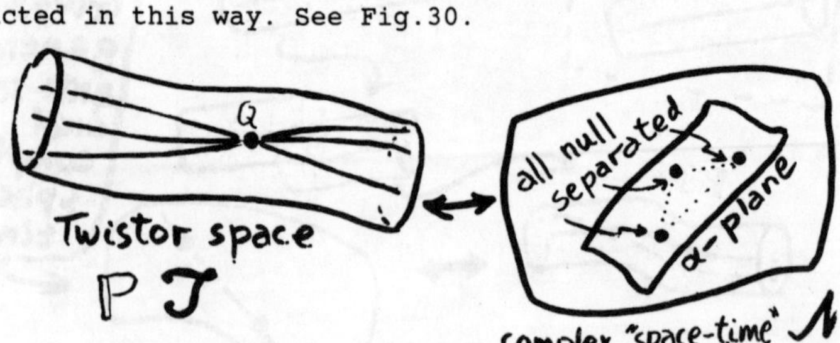

Fig.30

Each point of $\mathbb{P}\mathcal{T}$ represents an α-plane in $\mathcal{M}$

The condition that $\mathcal{M}$ also satisfy the Einstein vacuum equations, i.e. be Ricci-flat, is most easily expressed in terms of the non-projective curved twistor space $\mathcal{T}$. The space $\mathcal{T}$ is, in effect, a complex line bundled over $\mathbb{P}\mathcal{T}$. We can understand a point of $\mathcal{T}$ as representing an α-plane in $\mathcal{M}$ together with a spinor $\pi_{A'}$ which is parallelly propagated over the α-plane, each tangent vector to the α-plane having the form $\lambda^A\pi^{A'}$, for some λ^A. The consistency of this notion follows from the fact that $\tilde{\psi}_{A'B'C'D'} = 0$, this providing the necessary integrability condition for parallel $\pi_{A'}$-spinors over α-planes. Now the condition that there is a parallelism for $\pi_{A'}$-spinors (locally) over $\mathcal{M}$ (not just over α-planes) is that $\mathcal{M}$ be both anti-self-dual and Ricci flat. For we have

$$[\nabla_{AA'}, \nabla_{BB'}]\pi_{C'} = 0$$

as the integrability condition for this. Splitting this into various parts, we see

$$\text{Part } [AB]\,(A'B'C') = 0 \quad \text{by } \tilde{\Psi}_{A'B'C'D'} = 0\,;$$

$$\text{Part } [AB]\,(A'B')^{B'} = 0 \quad \text{by } R = 0\,;$$

$$\text{Part } (AB)\,[A'B']\,C' = 0 \quad \text{by } R_{ab} - \tfrac{1}{4}g_{ab} = 0.$$

The remaining curvature ψ_{ABCD} represents the non-existence of β-planes in $\mathcal{M}$, i.e. the non-existence of holomorphic "planes" in $\mathbb{P}\mathcal{T}$ (ruled by lines). We now have a notion of parallelism between α-planes (when their $\pi_{A'}$-spinors are parallel). This parallelism means that $\mathcal{T}$ is fibred over the space of (non-zero) $\pi_{A'}$-spinors; equivalently, that $\mathbb{P}\mathcal{T}$ is fibred over the Riemann sphere of projective $\pi_{A'}$-spinors (see Fig.31).

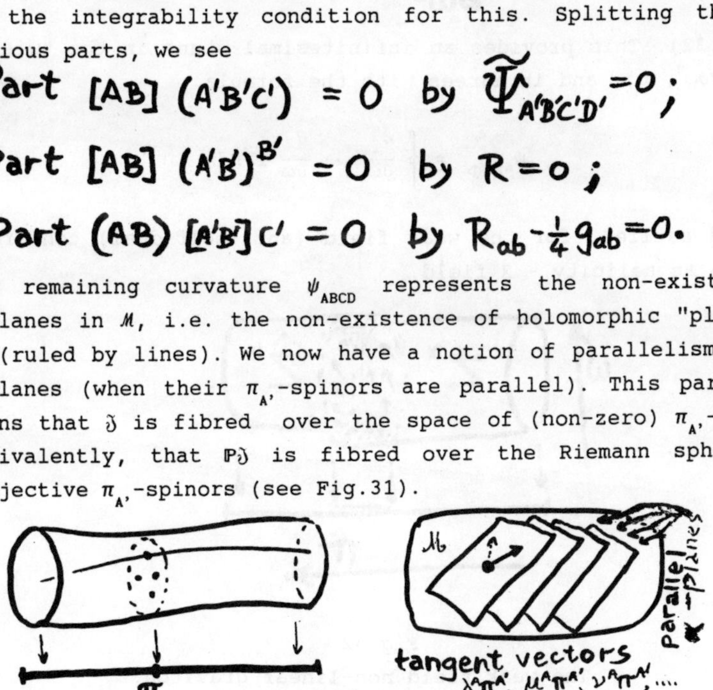

Fig.31

For $\mathcal{M}$ to be Ricci-flat the twistor space is fibred over the π-space, expressing a parallelism of α-planes in $\mathcal{M}$

The space $\mathfrak{Z}$ has a local structure given by a simple 2-form τ, a 4-form Δ, and a vector field Y (the Euler operator whose integral curves provide the spaces $\mathbb{P}\mathfrak{Z}$). This structure is locally the same as that of $\mathbb{T}$, where

$$\tau = (1/2)d\pi_{A'}\wedge d\pi^{A'}, \quad \Delta = (1/24)dZ^{\alpha}\wedge dZ^{\beta}\wedge dZ^{\gamma}\wedge dZ^{\delta}\epsilon_{\alpha\beta\gamma\delta}, \quad Y = Z^{\alpha}\partial/dZ^{\alpha}$$

The complex metric of $\mathcal{M}$ is determined by these forms. In the case of an infinitesimally deformed twistor space, there is a relation to a twistor function f(z) degree + 2, the deformation being given by the glueing together of two portions of twistor space, the twistor Z^{α} of one space being matched to the twistor $\hat{Z}^{\alpha}$ of the other by

$$\hat{\omega}^{A} = \omega^{A} + \epsilon\varepsilon^{AB}\frac{\partial f}{\partial\omega^{B}}, \qquad \hat{\pi}_{A'} = \pi_{A'}$$

(see Fig.32). This provides an infinitesimal shunt in the twistor by $(\epsilon^{AB}\partial f/\partial\omega^{B}, 0)$, and it agrees with the formula

$$\psi_{ABCD} = \oint\frac{\partial}{\partial\omega}\cdots\frac{\partial}{\partial\omega}f\,\pi d\pi,$$

described earlier, for the weak field (anti-self-dual) conformal curvature as helicity - 2 field.

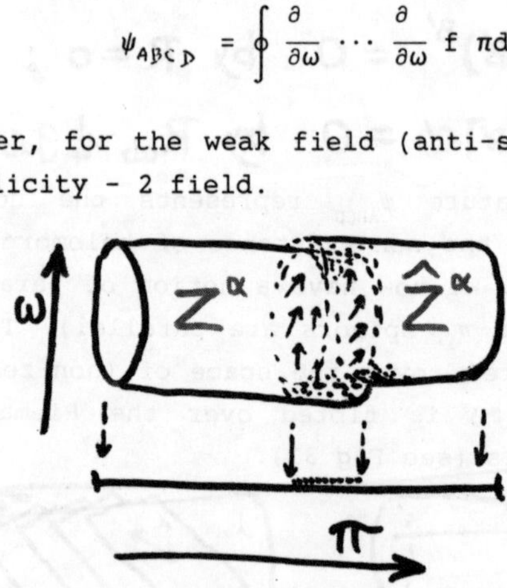

Fig.32
The weak field non-linear graviton.
Two patches of twistor space are displaced along a vector field given by $\epsilon^{AB}(\partial f/\partial\omega^{A})\partial/\partial\omega^{B}$, where f has homogeneity + 2.

9. Ward construction for (anti-)self-dual Yang-Mills fields

Closely analogous to the non-linear graviton construction for anti-self-dual Einstein fields is Richard Ward's construction for anti-self-dual Yang-Mills fields. A Yang-Mills connection on an (analytic) 4-space $\mathcal{M}$ with anti-self-dual conformal curvature is anti-self-dual if and omly if it is integrable on α-planes (in its complexification $\mathbb{C}\mathcal{M}$ if $\mathcal{M}$ is real - and therefore conformally flat). This integrability implies that the bundle over $(\mathbb{C})\mathcal{M}$, upon which this connection acts, can be carried over to the twistor space $\mathbb{P}\mathfrak{J}$ (or $\mathfrak{J}$). See Fig.33.

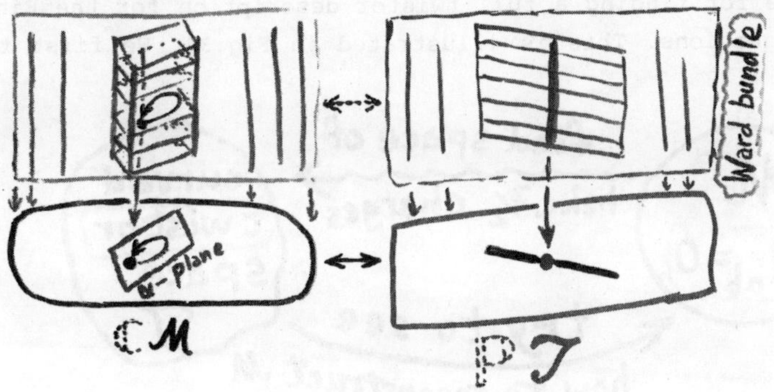

Fig.33

It turns out that the connection on the bundle over $(\mathbb{C})\mathcal{M}$ is completely determined by merely the holomorphic structure of the corresponding twistor bundle. moreover, this Yang-Mills connection is automatically anti-self-dual.

The Ward bundle can be constructed in terms of free transition functions and this fact leads to many applications to integrable systems (KdV, sine-Gordon non-linear Schrödinger, Toda lattice, Einstein with two commuting Killing vectors, etc.; work by Ward, Sparling, Mason, Woodhouse, and many others).

10. The general Einstein vacuum equations

The most promising idea for relating twistors to the Einstein (vacuum) equations $R_{ab} = 0$ is to continue together the following two facts.

(1) Twistors describe the *charges* for helicity 3/2 massless fields - in <u>flat</u> space $\mathbb{M}$

(2) The Einstein vacuum equations are *consistency conditions* for the equations describing massless helicity 3/2 fields in *curved* space-time $\mathcal{M}$.

Thus, we have a reasonably clear-cut (but difficult) programme for finding a full twistor description for the Einstein vacuum equations. This is illustrated in Fig.34. We first try to

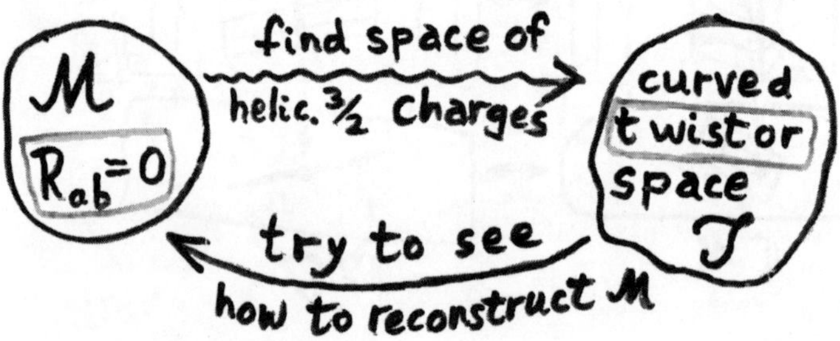

Fig.34
Programme for representing Ricci-flat space-times in terms of twistor space: highly non-local

find the appropriate space of "charges" for massless fields in $\mathcal{M}$ (assumed Ricci-flat) of helicity 3/2. This space is our desired curved twistor space $\mathfrak{T}$. We hope that $\mathfrak{T}$ can somehow be specified in a natural way in terms of free functions. We then try to see how to reconstruct $\mathcal{M}$ from $\mathfrak{T}$ in terms of some kind of global structure.

This is still very much "work in progress", but it will be of some interest to consider, here, certain ideas that are involved. To understand what is meant by the term "charges", as used here, consider the case of spin 1. We imagine a region of space-time in which a free Maxwell field F_{ab} is defined, and which surrounds some world-tube containing sources to the field. We can obtain the (conserved) total charge of the sources by performing a Gauss integral

$$e = \oint {}^{*}F$$

(ignoring factors of 4π, etc.) over some suitable topological sphere S^2, within the region of definition of the Maxwell field, surrounding (i.e. linking) the sorce region. Here ${}^{*}F$ is the 2-form defined by the dual Maxwell field:

$$\begin{aligned} {}^{*}\underset{\sim}{F} &= {}^{*}F_{ab}\, dx^a \wedge dx^b \\ {}^{*}F_{ab} &= \tfrac{1}{2} e_{abcd} F^{cd} \end{aligned}$$

We can also see a self-dual field

$${}^{+}F_{ab} \left(= \tfrac{1}{2} F_{ab} - \tfrac{i}{2} {}^{*}F_{ab}\right) = \varepsilon_{AB} \tilde{\varphi}_{A'B'},$$

which provides us with a <u>complex</u> charge whose real and imaginary parts are the magnetic charge μ and (minus) the electric charge e, where

$$\mu - ie = 2\oint {}^{+}\underset{\sim}{F} = \oint \underset{\sim}{\tilde{\varphi}},$$

with

$${}^{+}F = {}^{+}F_{ab} dx^a \wedge dx^b \quad \text{and} \quad \tilde{\varphi} = 2\tilde{\varphi}_{A'B'} dx_A^{A'} \wedge dx^{AB'}$$

In flat space-time $\mathbb{M}$, we can do something similar for the charges for spin 2, to obtain the mass, momentum, and angular momentum for a linearized gravitational field, given by a (self-dual) spinor $\tilde{\psi}_{A'B'C'D'}$. We can use the same Gauss integral $\oint\tilde{\varphi}$ as before, where we spin-lower twice:

$$\tilde{\varphi}_{A'B'} = \mu^{C'} \hat{\mu}^{D'} \tilde{\psi}_{A'B'C'D'},$$

where μ and $\tilde{\mu}$ are solutions of the (complex conjugated) twistor equation for dual twistors $W_\alpha = (\lambda_A , \mu^{A'})$ and $\hat{W}_\alpha$. Similarly, we can obtain the charges for spin 3/2 by

$$\tilde{\varphi}_{A'B'} = \mu^{C'}\psi_{A'B'C'} .$$

Then

$$\oint \tilde{\varphi} = e(W_\alpha) = Z^\alpha W_\alpha ,$$

since e is linear in W_α. This defines the *twistor charge* Z^α.

Let us examine the equations for helicity 3/2 more carefully. The field is given by a symmetric spinor $\psi_{A'B'C'}$ satisfying

$$\nabla^{AA'}\psi_{A'B'C'} = 0 .$$

We need potentials for this field. There are two wersion of this: the <u>Dirac</u> and <u>Rarita</u>-<u>Schwinger</u> forms of the potentials. For simplicity (and other reasons of convenience) I shall use the Dirac form here. In the following, the ticked (✓) equations can be consistently used in Ricci-flat curved space-time whereas the ones without a tick must be discarded when $\mathcal{M}$ is not flat. We have a first potential

$$\sigma_{A'B'}{}^{C} = \sigma_{(A'B')}{}^{C} \qquad (\checkmark)$$

subject to

$$\nabla^{AA'}\sigma_{A'B'}{}^{C} = 0 \qquad (\checkmark)$$

$$\nabla_{CC'}\sigma_{A'B'}{}^{C} = \psi_{A'B'C'}$$

and a second potential

$$\rho_{A'}{}^{BC} = \rho_{A'}{}^{(BC)} \qquad (\checkmark)$$

subject to

$$\nabla^{AA'}\rho_{A'}{}^{BC} = 0$$

$$\nabla_{BB'}\rho_{A'}{}^{BC} = 2i\,\sigma_{A'B'}{}^{C} . \qquad (\checkmark)$$

There is a gauge freedom, given by

$$\sigma_{A'B'}{}^{C} \Longrightarrow \sigma_{A'B'}{}^{C}\ \nabla_{B'}{}^{C}\pi_{B'} \qquad (\checkmark)$$

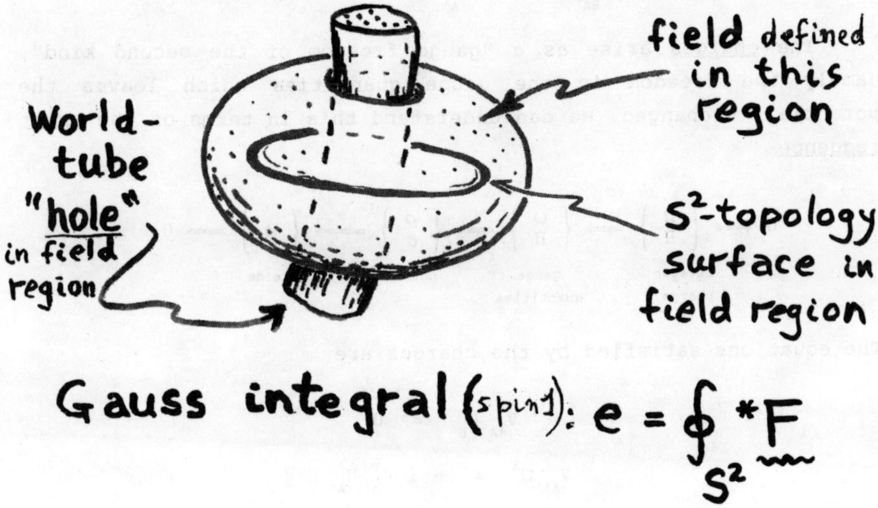

Fig.35
Charges: as properties of holes in regions of source free field

where

$$\nabla^{BB'}\pi_{B'} \;=\; 0 \qquad (\checkmark)$$

and

$$\rho_{A'}^{\;BC} \Longrightarrow \rho_{A'}^{\;BC} \;-\; i\,\varepsilon^{BC}\pi_{A'} \;+\; \nabla_{A'}^{\;C}\,\omega^{B} \qquad (\checkmark)$$

where

$$i\,\nabla_{BA'}\omega^{B} \;=\; 2\pi_{A'}\;. \qquad (\checkmark)$$

The <u>charges</u> arise as a "gauge freedom of the second kind", namely the freedom in the gauge quantities which leaves the potentials unchanged. We can understand this in terms of the <u>exact sequence</u>

$$0 \Longrightarrow \underset{\substack{\text{twistor}\\ \text{charge}}}{\begin{Bmatrix}\Omega\\ \Pi\end{Bmatrix}} \Longrightarrow \underset{\substack{\text{gauge}\\ \text{quantities}}}{\begin{Bmatrix}\omega\\ \Pi\end{Bmatrix}} \Longrightarrow \underset{\text{potentials}}{\begin{Bmatrix}\rho\\ \sigma\end{Bmatrix}} \Longrightarrow \underset{\text{fields}}{\begin{Bmatrix}\psi\end{Bmatrix}} \Longrightarrow 0$$

The equations satisfied by the charges are

$$\nabla_{AA'}\Pi_{B'} \;=\; 0$$

$$\nabla_{AA'}\Omega^{B} \;=\; -\,i\,\varepsilon_{A}^{\;B}\,\Pi_{A'}\;,$$

which are precisely the equations defining a twistor in $\mathbb{M}$.

Alternatively, we can use the Rarita-Schwinger form of the potentials. Here ρ, σ are <u>not</u> symmetric and the gauge is <u>free</u>. The equations for the potentials are a little more complicated than in the Dirac form, and I shall not use them here.

We cannot directly use the above exact sequence in curved (Ricci-flat) space-time because the inconsistency of the unticked equations. Nevertheless, it is useful to persue the topological implications of the existence of such a sequence. In $\mathbb{M}$, we can adopt a topological procedure for charges in flat space as "glitches" in the gauge quantities. See Fig.36. We take a region in $\mathbb{M}$ whose topology is that of a (tickered out) sphere, in which the field is globally defined. This is divided into two overlapping hemispherical regions in each of which there is a separate choice of potential for the field. The difference between the potentials in the two halves provides a zero-field potential on the annular overlap. This zero field potential is locally obtenaible from a pure gauge, but as one proceeds around the annulus this gauge jumps. The jump must be an element of the "gauge freedom of the second kind", i.e. an element of the charge space - a twistor.

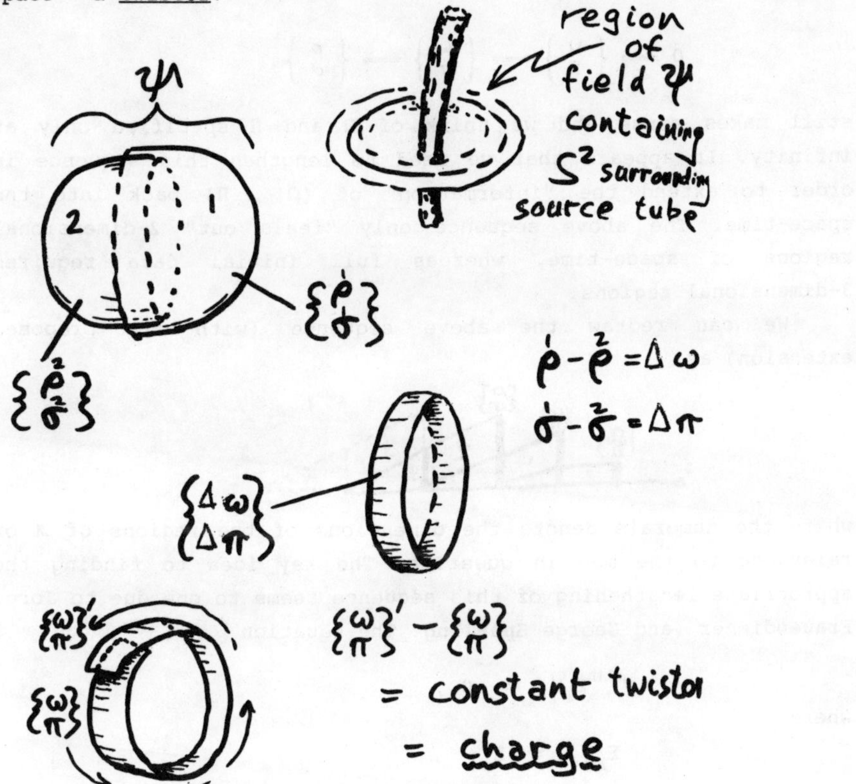

Fig.36

The procedure for obtaining the twistor charges for a helicicity 3/2 field in $\mathbb{M}$ as topological "glitches". Here, the notation $\Delta\omega + \nabla_{AA'}\omega^B + i\ \varepsilon_A^B\pi_{A'}$, $\Delta\pi = \nabla_{AA'}\pi_{B'}$ is being used.

In curved (Ricci-flat) space-time $\mathcal{M}$, neither the fields ψ nor the constant twistors $(\Omega\ ,\ \Pi)$ exist. The best idea seems to be to go to null infinity $\mathcal{J}$ (assuming $\mathcal{M}$ is asymptotically flat), to find

"(Ω,Π)" at $\mathcal{J}$ and to define "ψ" merely as potential modulo gauge. Since, in this procedure, different (Ω,Π)s come about in different null directions and in different retarded times it is a <u>non</u>-<u>linear</u> process. However, the sequence just considered seems inadequate in that it does not "feel out" enough of the space-time. In $\mathcal{M}$, we cannot use "ψ" directly, but the final map of the sequence

$$0 \Longrightarrow \left\{ \begin{matrix} \Omega \\ \Pi \end{matrix} \right\} \Longrightarrow \left\{ \begin{matrix} \omega \\ \pi \end{matrix} \right\} \Longrightarrow \left\{ \begin{matrix} \rho \\ \sigma \end{matrix} \right\}$$

still makes sense and we think of Ω and Π specified only at infinity. It appears that we need to lengthen this sequence in order to extend the "information" of $(\Omega\ ,\ \Pi)$ back into the space-time. The above sequence only "feels out" 2-dimensional regions of space-time, whereas full initial data requires 3-dimensional regions.

We can redraw the above sequence (with its proposed extension) as

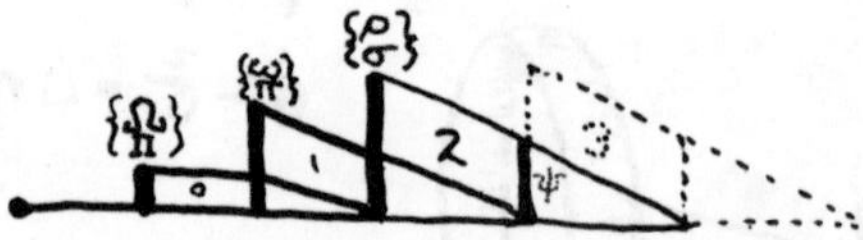

where the numerals denote the dimensions of the regions of $\mathcal{M}$ of relevance to the map in question. The key idea to finding the appropriate lengthening of this sequence seems to one due to Jorge Frauendiener (and George Sparling. The equation

$$\nabla^{B'(B}\Sigma_{A'B'}{}^{C)} = 0\ ,$$

where

$$\Sigma_{A'B'}{}^{C} = \Sigma_{B'A'}{}^{C}\ ,$$

is consistent (and conformally invariant) in arbitrary space-times.

In the Ricci-flat case, we have a (non-exact) sequence

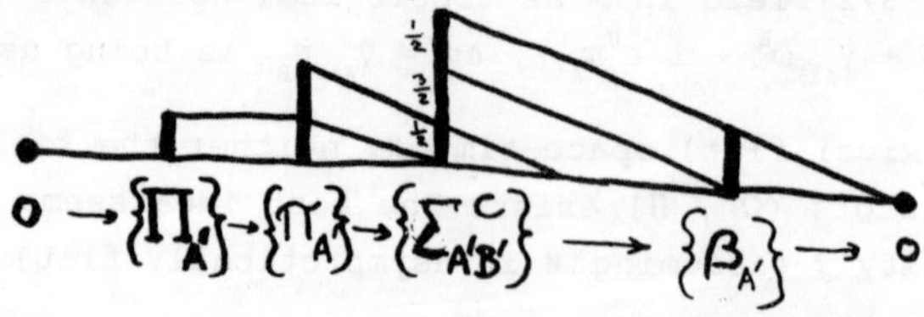

(the fractions indicating the spins of the fields relevant to the indicated portions of the spaces). The gauge freedom is given by

$$\Sigma_{A'B'}{}^{C} \Longrightarrow \Sigma_{A'B'}{}^{C} + \nabla_{B'}{}^{C}\pi_{A'}$$

where

$$\nabla^{A'C}\pi_{A'} = 0 .$$

Also, we have

$$\nabla^{A'C}\beta_{A'} = 0$$

where

$$\beta_{A'} = \nabla_{B}{}^{B'}\Sigma_{A'B'}{}^{B} .$$

We can rewrite the above sequence as an <u>exact</u> sequence

We can now introduce a second potential $P^{BC}_{A'}$ $(= P^{(BC)}_{A'})$, and we obtain a commutative diagram of exact sequences (which I shall not attempt to explain here in any detail):

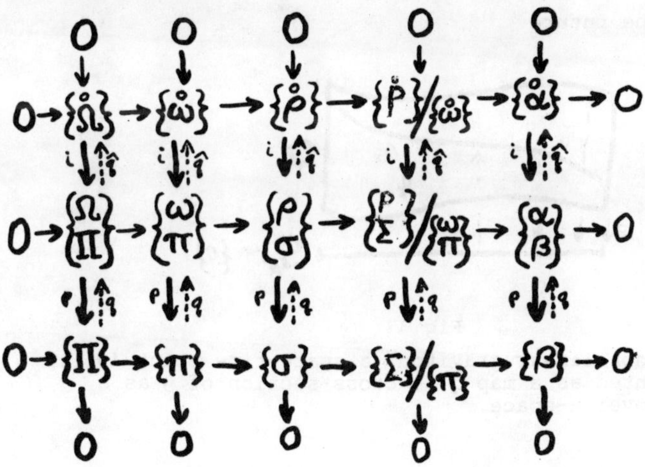

Here, i denotes injection and p, projection. The maps q and $\hat{q}$ derive from corresponding maps in $\mathbb{M}$,

$$\text{In } \mathbb{M}, \quad 0 \rightarrow \underset{\{\Omega^A\}}{\mathbb{S}^A} \underset{\hat{q}}{\overset{i}{\rightleftarrows}} \underset{\{\Omega^A \atop \Pi_{A'}\}}{\mathbb{T}^\alpha} \underset{q}{\overset{p}{\rightleftarrows}} \underset{\{\Pi_{A'}\}}{\mathbb{S}_{A'}} \rightarrow 0$$

which represents points q, $\hat{q}$ in space-time, defined by

$$\Pi_{A'} \overset{q}{\mapsto} (iq^{AA'}\Pi_{A'}, \Pi_{A'}) \,, \quad (\Omega^A, \Pi_{A'}) \overset{\hat{q}}{\mapsto} \Omega^A - iq^{AA'}\Pi_{A'}$$

The map q generalizes to anti-self-dual Ricci-flat $\mathcal{M}$. This is in accordance with the way in which space-time points are represented in the non-linear graviton construction (see Fig.37): a point q of $\mathcal{M}$, in that construction, is a cross-section of $\mathfrak{J}$, considered as a bundle over the (origin-free) primed spin space $\mathbf{S}_{A'} - \{0\}$ (π-space). This is a curved-space analogous of the map q considered above. This works when $\mathcal{M}$ is anti-self-dual. There should be a corresponding "googly" map $\hat{q}$ when $\mathcal{M}$ is self-dual. To see how space-time points are to be represented in twistors terms in the case of a general Ricci-flat space-time $\mathcal{M}$ is a major question for the future.

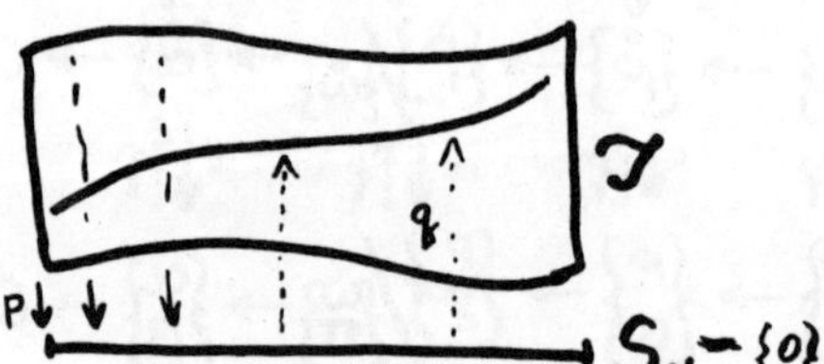

Fig.37

In the non-linear graviton, a space-time point is represented as a map q: a cross-section of $\mathfrak{J}$ as a bundle over π-space.

References

Aspect, A. , Grangier, P., and Roger, G. (1982). Experimental realization of Einstein-Podolsky-Rosen-Bohm *Gedankenexperiment*: a new violation of Bell's inequalities. *Phys Rev. Lett.* **48**, 91-4.

Bailey, T.N. and Baston, R.J., eds. (1990) *Twistors in Mathematics and Physics* London Mathematical Society Lecture Notes Series, 156 (Cambridge University Press)

Bell, J.S. (1966) On the problem of hidden variables in quantum theory, *Revs. Mod. Phys.* **38**, 447-52.

Buchdahl, H.A. (1958) On the compatibility of relativistic wave equations for particles of higher spin in the presence of a gravitational field *Nuovo Cim.* **10**, 96-103.

Deser, S. and Zumino, B. (1976) Consistent Supergravity *Phys. Lett. 62B*, 335-7.

Diósi, L. (1989) Models for universal reduction of macroscopic quantum fluctuations *Phys. Rev.* **A40**, 1165-74.

Dirac, P.A.M. (1936) Relativistic wave equations *Proc. Roy. Soc. London* **A155**, 447-59.

Eastwood M.G., Penrose, R., and Wells, R.O., Jr. (1981) Cohomology and massless fields, *Comm. Math. Phys.* **78**, 305-51.

Einstein, A., Podolsky, P. and Rosen, N. (1935) Can quantum-mechanical description of physical reality be considered complete? *Phys. Rev.* **47**, 777-80.

Fierz, M. (1940). Uber den Drehimpuls von Teichlen mit Ruhemasse null und beliebigem Spin. *Helv. Phys. Acta* **13**, 45-60.

Ghirardi, G.C., Rimini, A. and Weber, T. (1986) Unified dynamics for microscopic and macroscopic systems. *Phys. Rev.*, **D34**,470.

Ghirardi, G.C., Grassi, R., and Rimini, A. (1990) Continuous-spontaneous-reduction model involving gravity. *Phys. Rev.*, **A42**,1057-64.

Hodges, A.P. (1982) Twistor diagrams, *Physica* **114A**, 157-75.

Hodges, A.P. (1985) A twistor approach to the regularization of divergences, *Proc. Roy. Soc. London* **A397**, 341-74. Mass eigenstatates in twistor theory, *ibid*, 375-96.

Hodges, A.P., Penrose, R., and Singer, M.A. (1989) A twistor conformal field theory for four space-time dimensions, *Phys. Lett.* **B216**, 48-52.

Huggett, S.A. (Ed.) (1995) *Twistor Theory* (Marcel Dekker, New York)

Huggett, S.A. and Tod, K.P. (1985) *An Introduction to Twistor Theory* London Math. Soc. student texts (L.M.S. publ.)

Hughston, L.P. (1979) *Twistors and Particles*, Lecture Notes in Physics No. 97 (Springer-Verlag, Berlin).

Hughston, L.P. and Ward R.S. (eds.) (1979) *Advances in Twistor Theory* (Pitman, London).

Julia, B. (1982) Système linéare associé aux équations d'Einstein *Comptes. Rendus. Acad. Sci. Paris, Sér. II*, 113-6.

Károlyházy, F. (1974) Gravitation and quantum mechanics of macroscopic bodies, *Magyar Fizikai Polyoirat* 12, 24.

Károlyházy, F., Frenkel, A. and Lukács, B. (1986) On the possible role of gravity on the reduction of the wave function in *Quantum Concepts in Space and Time* eds. R.Penrose and C.J.Isham (Oxford University Press, Oxford).

Kibble, T.W.B. (1981) Is a semi-classical theory of gravity viable? In *Quantum Gravity 2: a Second Oxford Symposium*; eds. C.J. Isham, R. Penrose, and D.W. Sciama (Oxford Univ. Press, Oxford).

Kodaira, K. (1962) A theorem of completeness of characteristic systems for analytic submanifolds of a complex manifold. *Ann. Math.* (2) 75, 146-162.

Kodaira, K. and Spencer, D.C. (1958) On deformations of complex analytic structures I, II, *Ann. Math.* 67, 328-401, 403-466.

Majorana, E. (1932) Atomi orientati in campo magnetico variabile. *Nuovo Cimento* 9, 43-50.

Mason, L.J. and Hughston, L.P. (Eds.) (1990) *Further advances in twistor theory; Vol. I: The Penrose transform and its applications.* Pitman Research notes in Mathematics Series 231 (Longman Sci. & Tech., Harlow, Essex).

Mason, L.J., Hughston, L.P., and Kobak, P.Z. (Eds.) (1995) *Further advances in twistor theory; Vol. II: Integrable systems, conformal geometry and gravitation.* Pitman Research notes in Mathematics Series 232 (Longman Sci. & Tech., Harlow, Essex).

Mason, L.J. and Woodhouse, N.M.J. (1996) *Twistors, Self-Duality, and Integrable Systems* [approx. title] (Oxford Univ. Press).

Pearle, P. (1989) Combining stochastic dynamical state-vector reduction with spontaneous localization. Phys. Rev. **A39**, 2277-89.

Pearle, P. and Squires, E.J. (1995) Gravitaty, energy conservation and parameter values in collapse models. *Durham University preprint.* DTP/95/13.

Penrose, R. (1969) Solutions of the zero rest-mass equations, *J. Math. Phys.* 10, 38-9.

Penrose, R. (1976) Non-Linear gravitons and curved twistor theory *Gen. Rel. Grav.* 7, 31-52.

Penrose, R. (1982) Quasi-local mass and angular momentum in general relativity, *Proc. Roy. Soc. London* A381, 53-63.

Penrose, R. (1992) Twistors as spin 3/2 charges, in *Gravitation and Modern Cosmology* (P.G.Bergmann's 75th birthday vol.) eds. A. Zichichi, N. de Sabbata, and N.Sánchez (Plenum Press, New York).

Penrose, R. (1993) Gravity and quantum mechanics, in *General Relativity and Gravitation 1992. (Proc. 13th Int. Conf. on Gen. Rel. and Grav., Cordoba, Argentina. Part 1: Plenary Lectures.)* Eds. R.J.Gleiser, C.N.Kozameh, and O.M.Moreschi (Inst. of Phys. Bristol).

Penrose, R. (1994) *Shadows of the Mind; An approach to the missing science of consciousness* (Oxford Univ. Press, Oxford).

Penrose, R. (1996) On gravity's role in quantum state reduction *Gen. Rel. Grav.* (to appear).

Penrose, R. and Rindler, W. (1984) *Spinors and Space-Time, Vol. 1: Two-Spinor Calculus and Relativistic Fields* (Cambridge University Press, Cambridge).

Penrose, R. and Rindler, W. (1986) *Spinors and Space-Time, Vol. 2: Spinor and Twistor Methods in Space-Time Geometry* (Cambridge University Press, Cambridge.

Percival, I.C. (1995) Quantum spacetime fluctuations and primary state diffusion. *Proc. R. Soc. Lond.* A451, 503-13.

Ward, R.S. (1977) On self-dual gauge fields, *Phys. Lett.* 61A, 81-2.

Ward, R.S. and Wells, R.O. Jr (1989) *Twistor Geometry and Field Theory* (Cambridge University Press, Cambridge).

Woodhouse, N.M.J. and Mason, L.J. (1988) The Geroch group and non-Hausdorff twistor spaces *Nonlinearity* 1, 73-114.

Zeilinger, A., Gaehler,R., Shull, C.G., and Mampe, W. (1988) Single and double slit diffraction of neutrons. *Revs. Mod. Phys.* 60, 1067.

QUANTUM GEOMETRY AND GRAVITY

EDUARD PRUGOVEČKI
Department of Mathematics
University of Toronto
Toronto, Canada M5S 1A1

ABSTRACT

The geometro-stochastic method of quantization provides a framework for quantum general relativity, in which the principal frame bundles of local Lorentz frames that underlie the fibre-theoretical approach to classical general relativity are replaced by Poincaré-covariant quantum frame bundles. In the semiclassical regime for quantum field theory in curved spacetime, where the gravitational field is not quantized, the elements of these local quantum frames are generalized coherent states, which emerge naturally from phase space representations of the Poincaré group. Due to their informational completeness, these quantum frames are capable of taking over the role played by complete sets of observables in conventional quantum theory. The propagation of quantum-geometric fields proceeds by path integral methods, based on parallel transport along broken paths consisting of arcs of geodesics of the Levi-Civita connection. The formulation of quantum gravity within this framework necessitates the transition to quantum superframe bundles and a quantum gravitational supergroup capable of incorporating diffeomorphism invariance into the framework. This results in a geometric version of quantum gravity which shares some conceptual features with covariant as well as with canonical gravity, but which avoids the foundational and the mathematical difficulties encountered by these two approaches.

1. Introduction

During the preceding decade a new method of quantization was developed[1-4] in order to deal with some of the unresolved foundational problems[5,6] in relativistic quantum particle and field localization. This geometro-stochastic method eventually led[7,8] to a purely geometric framework for quantum general relativity, and proved capable of resolving foundational problems in quantum field theory in curved spacetime and in quantum gravity .[9,10]

The geometro-stochastic method of quantization shares[11] some features with the well-known geometric method of Kostant[12] and Souriau[13] in that it employs phase space representations of the Galilei and the Poincaré group, albeit of a different type[14] than the Kostant and Souriau method. In fact, the geometro-stochastic method is intrisically based on coherent states, and in this respect it displays common features (cf. Ref. 15, Sec. 16.2) with the method of Berezin.[16] However, in its scope,[9,10] it transcends both these methods, since it is applicable to quantum field theory (QFT) in the general relativistic context.

This is due to the fact that the geometro-stochastic quantization is based on a concept of quantum frame capable of taking over in the general relativistic regime, physically as well as mathematically, the role played in conventional quantum theory by complete sets of observables. In QFT in a given classical curved spacetime, principal bundles of such Poincaré quantum frame mediate the formulation of quantum geometries that are Hilbert or pseudo-Hilbert fibre bundles associated with principal frame bundles over curved spacetime. In turn, this enables a purely geometric type of formulation of propagation of quantum fields in such bundles by means of path integrals that conform to the strong equivalence principle of general relativity.[17] The corner stone on which this kind of propagation is based consists of quantum connections whose connection coefficients are provided by the Levi-Civita connection form in the given curved spacetime. Locally, such quantum geometric propagation is governed by energy density operators which arise from well-defined stress-energy tensor operators, which in the special relativistic regime give rise to infinitesimal generators of spacetime translations and Lorentz transformations.

The relation of this quantum-geometric framework for QFT to the conventional framework is arrived at by establishing from purely geometric considerations the existence of action integrals. A consistent physical interpretation of this framework is then obtained by proving the existence of Poincaré gauge invariant and locally conserved probability currents.

The extrapolation of these physical ideas and mathematical techniques to a geometric formulation of quantum gravity requires the use of geometric versions of BRST symmetries in superfibres over a quantum spacetime supermanifold.[9,10]

In Sec. 2 we describe those basic features of the geometro-stochastic method of quantization in the nonrelativistic regime which demonstrate its physical consistency with orthodox quantum mechanics. We then describe in Sec. 3 the special relativistic counterparts of these features. In Sec. 4 we extrapolate this framework into a geometric framework for quantum field theory in curved spacetime. In Sec. 5 we describe the basic features of the quantum-geometric mode of field propagation formulated within such a framework. We then outline in Sec. 6, in very general terms, how these physical ideas and techniques can be applied to quantum gravity; whereas, in the remaining Secs. 7-11 we highlight and explain the principal mathematical aspects of this application.

2. Nonrelativistic Geometro-Stochastic Quantization

It is well-known that the canonical quantization of a classical Hamiltonian system carried out in general canonical coordinate leads to results that are generically inconsistent with those obtained in the Cartesian coordinates related the adopted class of inertial reference frames. Therefore, the fundamental thesis underlying the geometro-stochastic method of quantization is that it is *the choice of reference frames that is of primary significance to the quantization procedure*, and not that of coordinates in general; moreover, that the choice of coordinates can be signicant only if they are operationally defined in relation to those frames. This suggests basing quantization on a group-theoretical foundation, in which the *operational* procedures required during the construction of *inertial* frames receives a natural interpretation that is directly related to the group of physical operations describing changes of such frames, and to the operational procedures carried out with them.

These fundamental physical ideas lead to the geometro-stochastic method of quantization, which, for the sake of simplicity, we shall describe for the spin-0 case. In the nonrelativistic regime the Galilei group plays the role of fundamental kinematical group. Moreover, the elements $(\mathbf{q},\mathbf{p})$ of the nonrelativistic phase space Γ are represented by the Cartesian coordinate triples $\mathbf{q},\mathbf{p}\in \mathbf{R}^3$ in relation to any given classical inertial frame of reference $\boldsymbol{u}$. The geometro-stochastic quantization scheme first introduced in Appendix C of Ref. 3 requires that representations of the canonical commutation relations with respect to $\boldsymbol{u}$ are to be derived from the irreducible subrepresentation of the unitary ray representation of the Galilei group in $L^2(\Gamma)$. Hence, let us denote by $(b,\mathbf{a},\mathbf{v},R)$ the generic element of that group, with b being the amount a time translation, the 3-vectors $\mathbf{a}$ and $\mathbf{v}$ representing a space translation and a velocity boost, respectively, and R a being a 3-rotation. Then the required representation is obtained by assigning to $(b,\mathbf{a},\mathbf{v},R)$ the unitary operator

$$(U(b,\mathbf{a},\mathbf{v},R)\psi)(\mathbf{q},\mathbf{p},t)$$
$$= \exp[i(-\tfrac{1}{2}m\mathbf{v}^2(t-b) + m\mathbf{v}\cdot(\mathbf{q}-\mathbf{a}))]\ \psi(R^{-1}[\mathbf{q}-\mathbf{a}-\mathbf{v}(t-b)], R^{-1}[\mathbf{p}-m\mathbf{v}], t-b)\ , \quad (2.1)$$

which acts on wave functions $\psi(\mathbf{q},\mathbf{p},t)$ that are, for each fixed value of $\mathbf{p} \in \mathbf{R}^3$, solutions of the free Schrödinger equation in $\mathbf{q}$ and t for the mass m (all expressed in Planck natural units), and which represent for each fixed value of t elements of the Hilbert space $L^2(\Gamma)$ of functions which are Lebesgue square-integrable[18] in $(\mathbf{q},\mathbf{p}) \in \mathbf{R}^6$.

The complete harmonic analysis of the above representation first carried out in Ref. 14 reveals an infinity of irreducible subrepresentations which are unitarily equivalent to the well-known irreducible ray representations of the Galilei group in configuration or momentum space. Each such irreducible phase space subrepresentation can be uniquely characterized by means of a unique resolution generator $\xi \in L^2(\Gamma)$ that gives rise, in conjunction with the elements $U(b,\mathbf{a},\mathbf{v},R)$ of that representation, to the generalized coherent states

$$\xi_{\mathbf{q},\mathbf{p}} = U(0,\mathbf{q},\mathbf{p}/m,I)\xi \in L^2(\Gamma)\ , \qquad (\mathbf{q},\mathbf{p}) \in \Gamma\ . \quad (2.2)$$

In turn, these generalized coherent states supply the orthogonal projector

$$\mathbf{P}_\xi = \int_{\mathbf{R}^6} \left|\xi_{\mathbf{q},\mathbf{p}}\right\rangle d\mathbf{q}d\mathbf{p} \left\langle \xi_{\mathbf{q},\mathbf{p}}\right|\ , \quad (2.3)$$

of $L^2(\Gamma)$ onto the subspace which carries that subrepresentation. The family of generalized coherent states in Eq. (2.2) is therefore uniquely assigned to the global classical frame of reference $\boldsymbol{u}$. Consequently, it can be envisaged[4,9] as constituting a quantum frame, operationally obtained by subjecting identical duplicates of a *quantum* test body, which has ξ as a *proper* state vector, to the basic kinematical procedures of spacetime translation, spatial rotation and velocity boost, in order to obtain an array of kinematically correlated microdetectors that can be then used for the spatio-temporal localization of quantum systems.

This interpretation is supported by the fact that the transition amplitudes

$$\psi(\mathbf{q},\mathbf{p}) = \langle U(0,\mathbf{q},\mathbf{p}/\mathbf{m},I)\xi \,|\, \psi\rangle\ , \qquad \psi \in \mathbf{P}_\xi L^2(\Gamma)\ , \quad (2.4)$$

between the normalized quantum state vector ψ of a quantum system and the constituents of such a quantum frame are directly related (cf. Ref. 9, Sec. 3.7) to the purely geometric concept of Fubini-Study distance between their representative rays in the projective space of unit Hilbert rays corresponding to elements in that subspace, and that the square of the absolute value in Eq. (2.4) can be interpreted as the probability density for detection in relation to these quantum frame elements. This interpretation follows from the fact that there is[2,3,14] a unitary map which assigns to each time-dependent wave function $\psi(\mathbf{q},\mathbf{p},t)$ in Eq. (2.1) a unique solution $\psi(\mathbf{x},t)$ of the free Schrödinger equation representing a wave function in configuration space. It can be then easily proved[4] that the probability density derivable from this phase-space wave function is related to the probability density derivable from the corresponding configuration space wave function as follows:

$$\rho_{(\xi)}(\mathbf{q},t) = \int_{\mathbf{R}^3} |\psi(\mathbf{q},\mathbf{p},t)|^2 d\mathbf{p} = (2\pi)^3 \int_{\mathbf{R}^3} |\psi(\mathbf{x},t)|^2 |\xi(\mathbf{x}-\mathbf{q})|^2 d\mathbf{x}\ . \quad (2.5)$$

Hence, in the *sharp-point limit*

$$(2\pi)^3 |\xi(\mathbf{x}-\mathbf{q})|^2 \quad \to \quad \delta^3(\mathbf{x}-\mathbf{q})\ , \quad (2.6)$$

physically corresponding to the limiting choice of *perfectly* pointlike quantum test bodies, the probability density in Eq. (2.5) converges to the probability density for infinitely precise position measurement outcomes originally postulated by Born,

$$\rho_{(\xi)}(\mathbf{q},t) \quad \to \quad |\psi(\mathbf{q},t)|^2 \ , \tag{2.7}$$

at all those points at which this configuration space wave function is continuous.

Of special significance for the extrapolation to the special relativistic regime in Sec. 3 are the resolution generators in Eq. (2.2) which, in the configuration space representation for the Galilei group, are supplied by the ground-state wave function,

$$\xi^{(\ell)}(\mathbf{x}) = (8\pi^3\ell^2)^{-3/4}\exp(-\mathbf{x}^2/4\ell^2) \ , \qquad \ell > 0 \ , \tag{2.8}$$

of a nonrelativistic harmonic oscillator. As a matter of fact, the special relativistic counterpart of each such resolution generator $\xi^{(\ell)}$ is the ground states of a corresponding relativistic harmonic oscillator that can be interpreted[4] as the ground state of the quantum metric operator introduced by Born[5], in which ℓ plays the role of fundamental length. Probability currents to which such a $\xi^{(\ell)}$ gives rise,

$$\mathbf{j}_\ell(\mathbf{q},t) = \int_{\mathbf{R}^3} \frac{\mathbf{p}}{m}\left|\psi_\ell(\mathbf{q},\mathbf{p},t)\right|^2 d\mathbf{p} \ , \qquad \psi_\ell(\mathbf{q},\mathbf{p}) = \left\langle U(0,\mathbf{q},\mathbf{p}/\mathbf{m},I)\xi^{(\ell)} \middle| \psi_\ell \right\rangle \ , \tag{2.9}$$

can be constructed[1,4] in complete analogy with their counterpart in classical statistical mechanics, and are Galilei-covariant and conserved. Moreover, in the case of a configuration space wave functions $\psi(\mathbf{x},t)$ with continuous first partial derivatives in the $\mathbf{x}$-variables, this new quantum probability current converges in the sharp-point limit to the conventional probability current in configuration space,

$$\mathbf{j}_\ell(\mathbf{x},t) \xrightarrow[\ell\to+0]{} (2im)^{-1}\psi^*(\mathbf{x},t)\overleftrightarrow{\nabla}\psi(\mathbf{x},t) \ . \tag{2.10}$$

This provides further support to the earlier described physical interpretation of the transition amplitude in Eq. (2.4).

On account of Eq. (2.3), the following family of transition probability amplitudes for free quantum propagation,

$$K_\ell(\mathbf{q}'',\mathbf{p}'',t'';\mathbf{q}',\mathbf{p}',t') \ = \ \left\langle U(t'',\mathbf{q}'',\mathbf{p}''/m,I)\xi^{(\ell)} \middle| U(t',\mathbf{q}',\mathbf{p}'/m,I)\xi^{(\ell)} \right\rangle \ , \tag{2.11}$$

displays the basic properties of a free propagator:

$$K_\ell(\mathbf{q}'',\mathbf{p}'',t'';\mathbf{q}',\mathbf{p}',t') = \int_{\mathbf{R}^6} K_\ell(\mathbf{q}'',\mathbf{p}'',t'';\mathbf{q},\mathbf{p},t)K_\ell(\mathbf{q},\mathbf{p},t;\mathbf{q}',\mathbf{p}',t')\,d\mathbf{q}d\mathbf{p} . \tag{2.12}$$

This phase-space free propagator can be used[4,19] in the formulation of path integrals for quantum propagators in the presence of interactions.

Upon suitable renormalization, these phase space free propagators converge in the sharp-point limit to the well-known free Feynman propagator for nonrelativistic quantum point particles:

$$(\pi/2\ell^2)^{3/2}K_\ell(\mathbf{q}'',\mathbf{p}'',t'';\mathbf{q}',\mathbf{p}',t') \xrightarrow[\ell\to+0]{} K(\mathbf{q}'',t'';\mathbf{q}',t') \ . \tag{2.13}$$

However, the above required renormalization constant obviously diverges in this sharp-point limit. Closer scrutiny reveals[4,19] that the basic formulae,

$$K_\ell(\mathbf{q}'',\mathbf{p}'',t'';\mathbf{q}',\mathbf{p}',t') = \lim_{\varepsilon\to+0}\int K_\ell(\mathbf{q}_N,\mathbf{p}_N,t_N;\mathbf{q}_{N-1},\mathbf{p}_{N-1},t_{N-1})$$

$$\times \prod_{n=N-1}^{1} K_\ell(\mathbf{q}_n,\mathbf{p}_n,t_n;\mathbf{q}_{n-1},\mathbf{p}_{n-1},t_{n-1})\,d\mathbf{q}_n d\mathbf{p}_n \ , \qquad \varepsilon = (t''-t')/N \ , \tag{2.14}$$

on which path integration in general relies, are mathematically well-defined in case of the phase space propagators in Eq. (2.11). On the other hand, in the case of the Feynman propagators which rely on pointlike localization in configuration space, the same type of formulae,

$$K(\mathbf{x}'',t'';\mathbf{x}',t') = \lim_{\varepsilon\to+0}\int K(\mathbf{x}(t_N);\mathbf{x}(t_{N-1}))\prod_{n=N-1}^{1}K(\mathbf{x}(t_n);\mathbf{x}(t_{n-1}))\,d\mathbf{x}(t_n), \quad (2.15)$$

are merely formal, since the above Lebesgue integrals exist if and only if their integrand are integrable in the absolute sense[18] – whereas that is not the case in Eq. (2.15). Hence, the mathematically rigorous treatment of Feynman path integrals require analytic continuations to imaginary time, leading to the well-known Feynman-Kac formula. On the other hand, the phase space path integrals resulting from Eq. (2.14) are mathematically well-defined in *real* time. This constitutes the basis of the usefulness of their special relativistic counterparts, described in the next section. Moreover, it presents decided advantages in the general relativistic regime, which will be our main concern in these lecture notes.

3. Special Relativistic Geometro-Stochastic Quantization

In the special relativistic regime the role played in geometro-stochastic quantization by the Galilei group is taken over by the Poincaré group. Similarly, that of nonrelativistic phase space is taken over by a relativistic phase space whose elements are labelled by the 4-vectors q and $p = mv$, with p restricted to the forward and backward mass hyperboloids in the case of particles and antiparticles, respectively. Hence, in the case of particles of rest mass m and zero spin, the Hilbert space $L^2(\Gamma)$ of Sec. 2 is replaced with the Hilbert space $L^2(\Sigma)$ of wave functions $\varphi(q,v)$ which are solutions in q of the free Klein-Gordon equation for rest mass m at each fixed value of v on the forward 4-velocity hyperboloid $\boldsymbol{V}^+$, and whose inner product is provided by the following integral over a hypersurface $\Sigma = \sigma\times\boldsymbol{V}^+$ in relativistic phase space

$$\langle\varphi|\varphi'\rangle = \int_\Sigma \varphi^*(q,v)\varphi'(q,v)\,d\Sigma(q,v)\,, \qquad d\Sigma(q,v) = 2v_\mu\delta(v^2-1)d\sigma^\mu(q)d^4v\ , \quad (3.1)$$

where the σ-component of Σ is a maximal spacelike hypersurface in Minkowski space. The above integration has to be carried out with respect to the unique (modulo a multiplicative constant) Lorentz covariant measure over Σ, so that the measure element $m^3d\Sigma$ reduces to $d\mathbf{q}d\mathbf{p}$ along any hypersurface Σ for which σ corresponds to a constant value of q^0 in the chosen global Lorentz frame of reference $\boldsymbol{u}$.

The nonrelativistic representations in (2.1) is now replaced by a representation which assigns the following operator,

$$(U(a,\Lambda)\varphi)(q,v) = \varphi(\Lambda^{-1}(q-a),\Lambda^{-1}v)\ . \quad (3.2)$$

to the generic element (a,Λ) of the Poincaré group, consisting of a spacetime translation a and Lorentz transformation Λ. As in the nonrelativistic case, the complete harmonic analysis of this representation reveals a host of irreducible subrepresentations that are unitarily equivalent to well-known momentum-space representations.[14] Each such subrepresentation is uniquely characterized by a resolution generator η providing the orthogonal projector of $L^2(\Sigma)$ onto the subspace which carries it, since it generates the generalized coherent states

$$\eta_\zeta = U(q,\Lambda_v)\eta\in L^2(\Sigma)\,, \qquad \zeta = (q,v)\,, \quad (3.3)$$

if Λ_v denotes the Lorentz boost to the 4-velocity v, and, moreover, that projector can be expressed in the form

$$\mathbf{P}_\eta = \int_\Sigma \left|\eta_\zeta\right\rangle d\Sigma(\zeta)\left\langle\eta_\zeta\right| . \tag{3.4}$$

Hence, by an extrapolation of the interpretation of (2.4) in the nonrelativistic case,

$$\varphi(q,v) = \langle U(q,\Lambda_v)\eta|\varphi\rangle , \qquad \varphi \in \mathbf{P}_\eta L^2(\Sigma) , \tag{3.5}$$

can be interpreted as a probability amplitude of detection of a quantum particle in a state described by the normalized vector φ in relation to the constituents of a special relativistic quantum frame associated to the classical Lorentz frame of reference $\boldsymbol{u}$.

We note that the above probability amplitude depends on the mean 4-velocity v of each micro-detector in the quantum frame relative to the corresponding classical frame. This is in complete accord with one of Born's key observations on an underlying reciprocity between spacetime and 4-momentum relationships in nature, as clearly indicated by the following quotation: "Ordinary relativity is based on the invariance of the 4-dimensional distance, or its square $R = x_k x^k$. Can one really define the distance of two particles in sub-atomic dimensions independently of their velocity? This seems to me not evident at all." (Ref. 5, p. 208.)

Born's idea of a quantum metric operator $D^2 = Q^2 + P^2$ containing a fundamental length ℓ that restricts the sharpness of spatio-temporal localizability leads (cf. Ref. 4, Sec. 4.5) to the *fundamental* quantum frames whose constituents are described by the generalized coherent states in (3.3), represented by the wave functions

$$\Phi^{\boldsymbol{u}}_{\ell,m;\zeta}(q',v') = \tilde{Z}^{-2}_{\ell,m} \int_{k^0>0} \exp\{[i(q-q') - \ell(v'+v)]\cdot k\}\, \delta(k^2 - m^2) d^4k \quad . \tag{3.6}$$

The normalization constant in Eq. (3.6) can be expressed in terms of the modified Bessel function K_2,

$$\tilde{Z}_{\ell,m} = 8\pi^4\, \mathrm{K}_2(2\ell m)/\ell m^2 = (4\pi^4/\ell^3 m^4) + O(\ell^{-1}) \xrightarrow[\ell\to+0]{} +\infty \, . \tag{3.7}$$

This constant diverges in the sharp-point limit $\ell \to 0$, but its value cannot be arbitrarily adjusted, since it is fixed by the fact that the operator in Eq. (3.4) is an orthogonal projection operator, so that

$$\mathbf{P}^{\boldsymbol{u}}_{\ell,m} = \int \left|\Phi^{\boldsymbol{u}}_{\ell,m;\zeta}\right\rangle d\Sigma(\zeta) \left\langle\Phi^{\boldsymbol{u}}_{\ell,m;\zeta}\right| = \left(\mathbf{P}^{\boldsymbol{u}}_{\ell,m}\right)^* = \left(\mathbf{P}^{\boldsymbol{u}}_{\ell,m}\right)^2 \, . \tag{3.8}$$

As a consequence, no counterparts of Eqs. (2.7) and (2.10) exist. This is in agreement with the fact that, as pointed out by Wigner in a lecture series on conventional nonrelativistic and special relativistic quantum theory, "every attempt to provide a precise definition of a position coordinate [for a quantum point particle] stands in direct contradiction with special relativity" (Ref. 6, p. 313).

Despite this fact, a consistent special relativistic quantum theory of spacetime localization is feasible if a *fixed* value $\ell > 0$ is adopted as a fundamental length, since then the following current, constructed in analogy with the one in Eq. (2.9),

$$j^\mu_\ell(q) = 2\int_{v^0>0} v^\mu \left|\varphi_\ell(q,v)\right|^2 \delta(v^2-1) d^4v \, , \qquad \varphi_\ell(q,v) = \left\langle \Phi^{\boldsymbol{u}}_{\ell,m;q,v} \middle| \varphi \right\rangle \, , \tag{3.9}$$

is Poincaré-covariant and conserved. Most importantly, it has a positive-definite timelike component, so that it provides a *bona fide* relativistic *probability* current.

This does not mean that the usual treatment of the Klein-Gordon equation has to be abandoned in the present framework, since the following counterpart,

$$J^\mu_\ell(q) = i\hat{Z}_{\ell,m} \int_{v^0>0} \varphi^*_\ell(q,v)\overleftrightarrow{\partial}^\mu \varphi_\ell(q,v)\, \delta(v^2-1) d^4v \, , \qquad \partial_\mu = \partial/\partial q^\mu \, , \tag{3.10}$$

of the familiar Klein-Gordon current also exists, and it is Poincaré-covariant as well as conserved. However, as it is the case in the conventional treatment of the Klein-Gordon equation, this latter current is *not* a probability current, since its timelike component is *not* positive definite even for wave functions corresponding to positive energies. Hence, as in the conventional approach, its use is restricted to the formulation of quantum field theoretical couplings of Klein-Gordon fields to other fields in the quantum-geometric field theory discussed in subsequent sections.

The normalization constant in Eq. (3.10) is uniquely determined by the following alternative form,[2,4]

$$\langle \varphi_\ell | \varphi'_\ell \rangle = i\hat{Z}_{\ell,m} \int_{v^0>0} \varphi^*_\ell(q,v)\overleftrightarrow{\partial}_\mu \varphi'_\ell(q,v)\, d\sigma^\mu(q)\, \delta(v^2-1) d^4v \quad , \qquad (3.11)$$

of the restriction of the inner product in (3.1) to subspace of $L^2(\Sigma)$ onto which the projector in Eq. (3.8) projects. Its value is finite for all $\ell > 0$, but it diverges in the sharp-point limit:

$$\hat{Z}_{\ell,m} = \mathrm{K}_2(2\ell m)/m\mathrm{K}_1(2\ell m) = (1/\ell m^2) + O(1) \xrightarrow[\ell\to+0]{} +\infty \quad . \qquad (3.12)$$

As shown in Refs. 9 and 10, the divergence in the sharp-point limit $\ell \to 0$ of the two normalization constants in Eqs. (3.7) and (3.12) provides an explanation of the root causes of the divergences that plague the conventional formulation of QFT. In essence, this explanation is that, as maintained already in the 1930s by Bohr, Born, Dirac, Heisenberg, and other founders of quantum mechanics, there *does* exist a fundamental length in nature. Hence, the appearance of all the divergences that plague conventional QFT is simply a manifestation of this underlying fact.

In the present context, as a consequence of (3.8), we have

$$K_\ell(q'',v'';q',v') = \langle \boldsymbol{\Phi}^{\boldsymbol{u}}_{q'',v''} | \boldsymbol{\Phi}^{\boldsymbol{u}}_{q',v'} \rangle = \int K_\ell(q'',v'';q,v) K_\ell(q,v;q',v') d\Sigma(q,v). \quad (3.13)$$

Hence, as was the case in Eqs. (2.11) and (2.12), the above transition probability amplitudes display all the basic properties of a propagator, and can be used[4,19] in the formulation of path integrals. On the other hand, in the sharp-point limit $\ell \to +0$ we can carry out their renormalization, so as to obtain,

$$2(2\pi)^{-3} Z_\ell^2 K_\ell(q'',p'';q',p') \xrightarrow[\ell\to+0]{} K_F(q''-q') \quad , \qquad q''^0 > q'^0 \quad , \qquad (3.14)$$

i.e., so that the forward time-ordered phase space propagator determined by (3.13) converges in that sharp-point limit to the corresponding forward part of the time-ordered Feynman propagator for Klein-Gordon particles, which is given by

$$K_F(x''-x') = 2(2\pi)^{-3} \int_{k^0>0} \exp[ik\cdot(x'-x'')]\, \delta(k^2-m^2) d^4k \,, \qquad x''^0 > x'^0 \,. \quad (3.15)$$

This basic feature contributes to the *formal* term-by-term convergence of formal perturbative expansions of transition amplitudes for collision processes in quantum-geometric field theory to their conventional counterparts in models involving Klein-Gordon quantum fields. Such term-by-term convergence can be achieved after other factors that diverge in the sharp-point limit have also been discarded – namely those divergent factors that appear in the formal sharp-point limit due to the independence of free Feynman propagators on the stochastic 4-velocity variable v .[9,10,17]

Totally analogous comparisons between the quantum-geometric and the conventional approach to QFT can be drawn[9,10] in the case of the Dirac equation, quantum electromagnetic fields, etc. However, in the presence of gravity, the retention of a fixed value $\ell > 0$ for the fundamental length becomes mandatory, since very basic measurement-theoretical considerations[20,21] show that Planck's length imposes an absolute lower bound on ℓ, so that no spacetime localization can exceed the precision stipulated by Planck's length.

4. Quantum Geometry and General Relativity

The basic physical idea in extending the geometro-stochastic method of quantization to the semiclassical general relativistic regime in which a Lorentzian manifold $(\mathbf{M}, g^{\mathrm{L}})$ is assumed to be *a priori* given, is based on the observation that, due to the strong equivalence principle, in classical general relativity (CGR) the role played by the Minkowski space $M^4 \equiv (\mathbf{R}^4, \boldsymbol{\eta})$ is taken over by the fibres $T_x\mathbf{M}$ of the tangent bundle $T\mathbf{M}$, whose typical fibre is $\mathbf{R}^4$; moreover, the role of the global Lorentz frames $\mathcal{L} = \{\boldsymbol{e}_i(x_0) | i = 0,1,2,3\}$, with their origins at points $x_0 \in M^4$, is taken over by the local Lorentz frames within the Lorentz frame bundle $L\mathbf{M}(g^{\mathrm{L}})$, which has the Lorentz group SO(3,1) as its structure group. Hence, all the tensor fields that appear in CGR represent sections of various tensor bundles associated with $L\mathbf{M}(g^{\mathrm{L}})$. The resulting Lorentz gauge invariance is underlined by the fact that the Levi-Civita connection employed in CGR is compatible with the metric g^{L}.

Mathematically, this Lorentz gauge invariance can be always extended into Poincaré gauge invariance based on the Poincaré frame bundle $P\mathbf{M}(g^{\mathrm{L}})$, that has the Poincaré group ISO(3,1) as its structure group. The Levi-Civita connection on $L\mathbf{M}(g^{\mathrm{L}})$ is then extended into a connection on $P\mathbf{M}(g^{\mathrm{L}})$, so that, as shown by Drechsler[22], the operator forms ∇ for covariant differentiation acquire four additional terms contained in the connection 1-forms

$$\tilde{\boldsymbol{\theta}}^i = \boldsymbol{\theta}^i + (\nabla \boldsymbol{a})^i \, , \qquad \boldsymbol{s} : x \mapsto (\boldsymbol{a}, \boldsymbol{e}_i) \, , \quad \boldsymbol{a} = a^i \boldsymbol{e}_i \, , \tag{4.1}$$

for any section $\boldsymbol{s}$ of $P\mathbf{M}(g^{\mathrm{L}})$ – where the canonical forms $\boldsymbol{\theta}^i$ are provided by the coframes $\{\boldsymbol{\theta}^i | \, i = 0,1,2,3\}$ dual to the Lorentz frames $\{\boldsymbol{e}_i \,|\, i = 0,1,2,3\}$ in $\boldsymbol{s}$; and where each Poincaré frame $\{(\boldsymbol{a}, \boldsymbol{e}_i) | \, i = 0,1,2,3\}$ is obtained by translating the corresponding local Lorentz frame by the 4-vector $\boldsymbol{a}$ within the tangent space within which that frame lies.

Such an extension is not esential to CGR itself, but it is essential from the quantum point of view. This is due to the fact that the kinematics and dynamics incorporated into such equations of motion as those of Klein-Gordon, Dirac, etc., which govern quantum propagation, gives rise to generators of infinitesimal spacetime translations for unitary representations of the Poincaré group.

Hence, in order to describe quantum-geometrically the propagation of massive spin-0 quantum fields, we shall construct the Fock-Klein-Gordon bundle

$$\mathcal{E}(\mathbf{M}, g^{\mathrm{L}}) = P\mathbf{M}(g^{\mathrm{L}}) \times_{\mathbf{G}} \mathcal{F} \, , \qquad \mathbf{G} = \mathrm{ISO}(3,1) \, , \tag{4.2}$$

associated with the principal bundle $P\mathbf{M}(g^{\mathrm{L}})$. The typical fibre of this quantum bundle is the Fock space

$$\mathcal{F} = \bigoplus\nolimits_{n=0}^{\infty} \mathcal{F}_n \, , \qquad \mathcal{F}_n = \mathbf{F} \underset{S}{\otimes} \cdots \underset{S}{\otimes} \mathbf{F} \, , \qquad \mathbf{F} = \mathbf{P}^{u_0}_{\ell,m} L^2(\Sigma) \, , \tag{4.3}$$

obtained from symmetrized tensor products of the Hilbert space determined for the standard frame $\boldsymbol{u}_0$ in $(\mathbf{R}^4, \boldsymbol{\eta})$ by Eqs. (3.5), (3.6) and (3.8). Hence, the representation provided by Eq (3.2) gives rise within this Fock space to the representation

$$\boldsymbol{U}(a, \Lambda) = \bigoplus_{n=0}^{\infty} U(a, \Lambda)^{\otimes n} \, , \qquad (a, \Lambda) \in \mathrm{ISO}(3,1) \, , \tag{4.4}$$

which can be used in the standard manner in the construction of the bundle $\mathbf{G}$-product[9,10] that appears in Eq. (4.2), so that the Poincaré group becomes the structure group of $\mathcal{E}(\mathbf{M}, g^{\mathrm{L}})$. We observe that the fibre $\mathcal{F}_x$ above $x \in \mathbf{M}$ contains the vacuum subfibre $\mathcal{F}_{0;x}$ spanned by a *local* Fock vacuum state vector $\boldsymbol{\Psi}_{0;x}$, which is left invariant by the counterpart of Eq. (4.4) in $\mathcal{F}_x$.

The operational description of measurements of spatio-temporal separations of events in a neighborhood of some base point $x \in \mathbf{M}$ where the relative curvature effects are small, so that classically they can be described exclusively in terms of Riemann normal coordinates for which $dx_k dx^k = \eta_{ik} dx^i dx^k$ at $x \in \mathbf{M}$, can be now supplemented,[10] in accordance with Born's[5] orginal physical and epistemological ideas, by fundamental uncertainties described by the local quantum fluctuation amplitudes

$$\Delta_x^{(+)}(\zeta';\zeta) = -i\, \boldsymbol{\Phi}_{\ell,m;\zeta}^{\boldsymbol{u}(x)}(\zeta'), \quad \zeta = (\boldsymbol{a} + q^i \boldsymbol{e}_i, v^i \boldsymbol{e}_i) \in T_x\mathbf{M} \times \mathbf{V}_x^+ , \quad \zeta = q - i\ell v \in \mathbf{C}^4 . \tag{4.5}$$

According to the interpretation of Eq. (3.5), these fluctuations result from the transition probability amplitudes of the geometrically[10] *local quantum frames* associated by the construction in Eq. (4.2) to each Poincaré frame $\boldsymbol{u}(x) = (\boldsymbol{a}(x), \boldsymbol{e}_i(x))$, that can be used to define those Riemann normal coordinates by means of the exponential map at x.

On account of Eqs. (3.8) and (3.11), we have

$$2\int \Delta_x^{(+)}(\zeta';\zeta)\Delta_x^{(+)}(\zeta;\zeta'')\,\delta(v^2-1)\,v_k\,d\sigma^k(q)d^4v = -i\Delta_x^{(+)}(\zeta';\zeta'') , \tag{4.6}$$

so that the 2-point functions in Eq. (4.5) can also play the role of geometrically local propagators. Hence, with their help we can define annihilation operators by their action upon n-exciton state vectors $\boldsymbol{\Psi}_{n;x} \in \mathcal{F}_{n;x}$ as follows:

$$\left(\varphi^{(-)}(x;\zeta)\boldsymbol{\Psi}_{n;x}\right)_{n-1}(\zeta_1,\ldots,\zeta_{n-1}) = in^{1/2}\int \Delta_x^{(+)}(\zeta;\zeta_n)\boldsymbol{\Psi}_{n;x}(\zeta_1,\ldots,\zeta_n)\,d\Sigma(\zeta_n) . \tag{4.7}$$

In conventional QFT the counterparts of these operators, and of their adjoints representing creation operators, have to be smeared with test functions. Indeed, in that context such operators are *proven* to exist only in the sense of operator-valued distributions (cf. Ref. 23, Sec. 10.4). On the other hand, the presence of the fundamental length ℓ in the 2-point functions in Eq. (4.5) renders them non-singular. This makes the annihilation operators in Eq. (4.7), as well as of their adjoints $\varphi^{(+)}(x;\zeta)$ representing creation operators, mathematically well-defined without any need for "smearing" with test functions. Hence, the Klein-Gordon *quantum frame field*

$$\varphi(x;\zeta) = \varphi^{(+)}(x;\zeta) + \varphi^{(-)}(x;\zeta) , \quad \zeta \in T_x\mathbf{M} \times \mathbf{V}_x^+ , \quad x \in \mathbf{M} , \tag{4.8}$$

is well-defined as it stands. As a consequence, it can be proved[8-10] that when the Poincaré group acts upon the elements $\boldsymbol{s}(x)$ of a section $\boldsymbol{s}$ of $P\mathbf{M}(\boldsymbol{g}^{\mathrm{L}})$, the infinitesimal generators of the counterparts

$$\boldsymbol{U}_{\boldsymbol{s}(x)}(a,\Lambda) = \bigoplus_{n=0}^{\infty} U_{\boldsymbol{s}(x)}(a,\Lambda)^{\otimes n} , \qquad \boldsymbol{s}(x) \in P\mathbf{M}(\boldsymbol{g}^{\mathrm{L}}) , \tag{4.9}$$

of the representation in Eq. (4.4) are given by the well-defined Bochner integrals (cf. Ref. 18, p. 480)

$$\boldsymbol{P}_{j;\boldsymbol{s}(x)} = \int :T_{jk}[\varphi(x;\zeta)]: d\sigma^k(q)\,\delta(v^2-1)\,d^4v , \quad Q^j_{\boldsymbol{s}(x)} = q^j - (i/m)\partial/\partial v_j , \tag{4.10}$$

$$\boldsymbol{M}^{ij}_{\boldsymbol{s}(x)} = \int :Q^i_{\boldsymbol{s}(x)}T^{jk}[\varphi(x;\zeta)] - Q^j_{\boldsymbol{s}(x)}T^{ik}[\varphi(x;\zeta)]: d\sigma_k(q)\,\delta(v^2-1)\,d^4v , \tag{4.11}$$

whose integrands are *bona fide* operator-valued functions, given by the normally-ordered values of the following *bona fide* operator-valued stress-energy tensor:

$$T_{jk}[\varphi] = \hat{Z}_{\ell,m}\left(\varphi_{,j}\varphi_{,k} + \tfrac{1}{2}\eta_{jk}(m^2\varphi^2 - \eta^{il}\varphi_{,i}\varphi_{,l})\right), \quad \varphi_{,j} = \partial\varphi/\partial q^j \ . \qquad (4.12)$$

However, as can be seen from Eq. (3.12), divergencies manifest themselves in the sharp-point limit $\ell \to +0$.

In each Fock fibre $\mathcal{F}_x$ we can introduce the Glauber-type coherent states

$$\boldsymbol{\Phi}_\mathbf{f} = \exp\left[-\tfrac{1}{2}\langle \mathbf{f}|\mathbf{f}\rangle + \varphi^{(+)}(\mathbf{f})\right]\Psi_{0;x}, \quad \varphi^{(+)}(\mathbf{f}) = \int \varphi^{(+)}(x;\zeta)\mathbf{f}(\zeta)d\Sigma(\zeta), \quad \mathbf{f} \in \mathcal{F}_{1;x}, \qquad (4.13)$$

defined by the strongly convergent power series for the above exponential. These coherent states constitute a family of second-quantized frames, in the sense that they give rise to the following continuous resolution of the identity operator $\mathbf{1}_x$ in each fibre $\mathcal{F}_x$,

$$\int_{\mathbf{F}_x} \left|\boldsymbol{\Phi}_\mathbf{f}\right\rangle \mathbf{df}\,\mathbf{df}^* \left\langle \boldsymbol{\Phi}_\mathbf{f}\right| = \mathbf{1}_x \ , \qquad \mathbf{F}_x = \mathcal{F}_{1;x} \ , \qquad (4.14)$$

where the above functional integral is mathematically rigorously defined by the method of Berezin.[25] These continuous resolutions of the identity will prove essential in Sec. 5) to the derivation of action integrals for quantum-geometric propagation, when taken in conjunction with the fact that the elements of these second-quantized frames are *bona fide* eigenvectors of the annihilation operators in Eq. (4.7):

$$\varphi^{(-)}(x;\zeta)\boldsymbol{\Phi}_\mathbf{f} = \mathbf{f}(\zeta)\boldsymbol{\Phi}_\mathbf{f} \ , \qquad \mathbf{f} \in \mathbf{F}_x \ . \qquad (4.15)$$

This quantum-geometric propagation emerges from the fact that the quantum bundle $\mathcal{E}(\mathbf{M},\boldsymbol{g}^\mathrm{L})$ is associated with $P\mathbf{M}(\boldsymbol{g}^\mathrm{L})$, and consequently each connection on $P\mathbf{M}(\boldsymbol{g}^\mathrm{L})$ gives rise to parallel transport within $\mathcal{E}(\mathbf{M},\boldsymbol{g}^\mathrm{L})$. In particular, this is the case with the earlier discussed extension to $P\mathbf{M}(\boldsymbol{g}^\mathrm{L})$ of the Levi-Civita connection on $L\mathbf{M}(\boldsymbol{g}^\mathrm{L})$. Therefore, for any choice of Poincaré gauge given by a section $\boldsymbol{s}$ of $P\mathbf{M}(\boldsymbol{g}^\mathrm{L})$, the corresponding parallel transport determines the quantum connection[7-10] (cf. Ref. 38 about the sign conventions)

$$\boldsymbol{\nabla} = \boldsymbol{d} + i\tilde{\boldsymbol{\theta}}^i\boldsymbol{P}_{i;\boldsymbol{s}} + \tfrac{i}{2}\tilde{\boldsymbol{\omega}}_{jk}\boldsymbol{M}_{\boldsymbol{s}}^{jk} \ , \qquad \boldsymbol{d} = \boldsymbol{\theta}^i\boldsymbol{\partial}_{e_i} \ , \qquad (4.16)$$

whose connection 1-forms are those derived from that extension, and whose infinitesimal generators for spacetime translations and Lorentz transformations are the ones in Eqs. (4.10) and (4.11). Thus, Poincaré gauge covariance is embedded into the parallel transport resulting from this above quantum connection.

5. Quantum-Geometric Propagation in Quantum Bundles

In a stationary but nonstatic classical spacetime represented by a Lorentzian manifold $(\mathbf{M},\boldsymbol{g}^\mathrm{L})$, the conventional QFT formulation of free-fall field propagation gives rise to *ex nihilo* particle production.[26] This violates local energy-momentum conservation as well as the strong equivalence principle of general relativity since, according to that fundamental principle, all observers in free fall are *inertial*; consequently, the situation should appear to them *locally* the same as in Minkowski space, where such *ex nihilo* particle production is not in evidence.

In the quantum-geometric formulation of QFT, this difficulty can be overcome by applying to all the fields defined on quantum bundles over curved spacetime the *original* form of the path-integration method advanced by Feynman. In that original form[27,28] the path-integral depiction of quantum propagation was formulated in spacetime, rather than in the nowadays more popular momentum representation. Its two central ideas were that such propagation takes place over broken polygonal paths in nonrelativistic or relativistic flat spacetimes, and that all the observed probability transition amplitudes result from the superposition of *spacetime* probability amplitudes over such paths.

The quantum-geometric adaptation[7-10] of these ideas to a curved spacetime represented by a Lorentzian manifold $(\mathbf{M},\boldsymbol{g}^{\mathrm{L}})$ is based on replacing the straight lines of such broken polygonal paths with the arcs of geodesics of the Lorentzian metric $\boldsymbol{g}^{\mathrm{L}}$, and on deriving spacetime propagators by extending the strong equivalence principle from the classical to the quantum regime. The latter feature entails that free-fall quantum-geometric propagation is governed by the Levi-Civita connection determined by $\boldsymbol{g}^{\mathrm{L}}$, so that *ex nihilo* particle production is thereby avoided. Moreover, a central role in such quantum-geometric evolution is played by the *local* proper energy density and 3-momentum of the nongravitational sources associated, as is the case in classical canonical gravity, with a geometrodynamic evolution of spacetime that is mathematically equivalent to a foliation of Lorentzian manifold $(\mathbf{M},\boldsymbol{g}^{\mathrm{L}})$ into a family of reference hypersurfaces Σ_t with unit future-pointing normals $\boldsymbol{n}(x)\in T_x\mathbf{M}$.

The presence of the fundamental length ℓ makes the definition of such geometrically *local* quantities not only mathematically consistent, but also in keeping with the uncertainty principle, since the envisaged spacetime localization is not infinitely precise, but exhibits stochastic fluctuations described by the quantum fluctuations amplitudes in Eq. (4.5). Hence, let us introduce the quantum-geometric field

$$\boldsymbol{\varphi}(x,\boldsymbol{v}):=\varphi(x;(\mathbf{0},\boldsymbol{v}))\ ,\qquad \boldsymbol{v}=v^i\boldsymbol{e}_i(x)\in \boldsymbol{V}_x^+\ , \tag{5.1}$$

whose values are actually independent of the chosen local Poincaré frame $(\boldsymbol{a}(x),\boldsymbol{e}_i(x))\in P\mathbf{M}(\boldsymbol{g}^{\mathrm{L}})$; furthermore, let us consider for any section $\boldsymbol{s}=\{\boldsymbol{a}(x),\boldsymbol{e}_i(x)|\ x\in\mathbf{M}\}$ the local energy-momentum density operators

$$\hat{\boldsymbol{P}}_{j;\boldsymbol{s}}(x)=n^k(x)\int_{v^0>0}:T_{jk}[\boldsymbol{\varphi}(x;\boldsymbol{v})]:\ \delta(v^2-1)\,d^4v\ , \tag{5.2}$$

where $n^k(x)$, $k=0,1,2,3$, are the components of the normal $\boldsymbol{n}(x)$ with respect to the local Lorentz frame $\{\boldsymbol{e}_i(x)\}$, $x\in\Sigma_t$, and the stress-energy operator is the one that appears in Eq. (4.12). If the Poincaré gauge represented by $\boldsymbol{s}$ is adapted to the family of reference hypersurface Σ_t, in the sense that the timelike components of its vierbeins are orthogonal to those hypersurfaces, then the proper energy operator $\boldsymbol{\rho}$ and the 3-momentum operators $\boldsymbol{j}_a$ of the quantum-geometric field $\boldsymbol{\varphi}$ can be defined as in classical canonical gravity:

$$\boldsymbol{\rho}(x)=\hat{\boldsymbol{P}}_{0;\boldsymbol{s}}(x)\ ,\qquad \boldsymbol{j}_a(x)=\hat{\boldsymbol{P}}_{a;\boldsymbol{s}}(x)\ ,\qquad a=1,2,3\ . \tag{5.3}$$

We note that, as opposed to the quantum frame field in Eq. (4.8), the quantum-geometric field in Eq. (5.2) does not depend on the Poincaré gauge variables q^j, $j=0,1,2,3$. The reason is that, although these variables are required for the definition of infinitesimal spacetime translations at each base location $x\in\mathbf{M}$, they cannot play a direct physical role, as it becomes evident as soon as the transition to the special relativistic regime is performed. Indeed, in case that $(\mathbf{M},\boldsymbol{g}^{\mathrm{L}})$ is the Minkowski space $M^4\equiv(\mathbf{R}^4,\boldsymbol{\eta})$, a global Lorentz frame $\mathcal{L}=\{\boldsymbol{e}_i(x_0)|\ i=0,1,2,3\}$ with its origin at $x_0\in M^4$ corresponds to the following cross-section of the Poincaré frame bundle PM^4,

$$\boldsymbol{s}_0(\mathcal{L})=\left\{(\boldsymbol{a}(x),\boldsymbol{e}_i(x))\middle|\ \boldsymbol{a}(x)=-x^i\boldsymbol{e}_i(x)\in T_xM^4\ ,\ x=x^i\boldsymbol{e}_i(x_0)\in M^4\right\}\ , \tag{5.4}$$

due to the fact that each tangent space T_xM^4 can be identified with M^4 itself. Hence, each special-relativistic state vector $\varphi\in\mathbf{F}$ can be then identified with a cross-section $\{\boldsymbol{\Psi}_{1;x}\in\mathcal{F}_x|\ x$

$\in \mathbf{M}\}$ of the quantum bundle $\mathcal{E}(M^4)$ whose local state vectors $\boldsymbol{\Psi}_{1;x}$ have coordinate wave functions that satisfy

$$\Psi_{1;x}(-i\ell v) = \varphi(-a(x) - i\ell v) , \quad -a(x) = (x^0,\ldots,x^3) \in \mathbf{R}^4 , \tag{5.5}$$

in the Poincaré frames of $\mathbf{s}_0(L)$; by analytic continuation the values of $\Psi_{1;x}(q - i\ell v)$ are then uniquely determined at all $q \in \mathbf{R}^4$. In view of the Poincaré gauge invariance of the quantum geometry framework, this identification is invariant under changes of global Lorentz frames L, since all such changes induce corresponding changes in the global Poincaré gauges of the type in Eq. (5.4). Hence, the special relativistic framework of Sec. 3 is recovered as a special case of the present Poincaré gauge invariant framework, and the gauge variables q^j are seen to be physically redundant.

This suggests the construction of the Fock space

$$\hat{\mathcal{F}} = \bigoplus_{n=0}^{\infty} \hat{\mathcal{F}}_n , \quad \hat{\mathcal{F}}_n = \hat{\mathbf{F}} \underset{S}{\otimes} \cdots \underset{S}{\otimes} \hat{\mathbf{F}} , \quad \hat{\mathbf{F}} = \left\{ \hat{f} \,\middle|\, f \in \mathbf{F} \right\} , \tag{5.6}$$

whose state vectors consist of n-exciton field modes $\hat{\Psi}_n \in \hat{\mathcal{F}}_n$ that are related to the corresponding $\Psi_n \in \mathcal{F}_n$ by the equalities

$$\hat{\Psi}_n(v_1,\ldots,v_n) = \Psi_n(-a(x) - i\ell v_1,\ldots,-a(x) - i\ell v_n) , \quad \Psi_n \in \mathcal{F}_n . \tag{5.7}$$

By analytic continuation, these equalities establish isomorphisms between $\hat{\mathcal{F}}_n$ and $\mathcal{F}_n$ for $n = 1,2,\ldots$. In turn, these isomorphisms become isometries if the restrictions of Eq. (3.2) to Lorentz transformations are employed to introduce in $\hat{\mathbf{F}}$ Perelomov-type generalized coherent states as in Eq. (3.3). We then have that

$$\langle \hat{\varphi}_1 | \hat{\varphi}_2 \rangle = Z_{\ell,m} \int_{v^0>0} \hat{\varphi}_1^*(v) \hat{\varphi}_2(v) \delta(v^2 - 1) \, d^4 v = \langle \varphi_1 | \varphi_2 \rangle, \quad \hat{\varphi}_1, \hat{\varphi}_2 \in \hat{\mathbf{F}} , \quad \varphi_1, \varphi_2 \in \mathbf{F} , \tag{5.8}$$

where the value of the renormalization constant $Z_{\ell,m}$ can be determined[10,17] by the method presented in Ref. 15.

We can now define a bundle of localized quantum field excitation modes of $\boldsymbol{\varphi}$,

$$\hat{\mathcal{E}}(\mathbf{M}, \boldsymbol{g}^{\mathrm{L}}) = L\mathbf{M}(\boldsymbol{g}^{\mathrm{L}}) \times_{\mathbf{G}} \hat{\mathcal{F}} , \quad \mathbf{G} = \mathrm{SO}(3,1) , \tag{5.9}$$

which has the Lorentz group as its structure group. On account of (5.7) its Lorentz gauge covariance can be extended into the Poincaré gauge covariance of the Fock-Klein-Gordon bundle in Eq. (4.2), so that the parallel transport in the latter can be transferred to the former. We can use this fact to define the quantum-geometric evolution between two such hypersurfaces $\Sigma_{t'}$ and $\Sigma_{t''}$ of a geometrodynamic spacetime evolution given by a foliation of the Lorentzian manifold $(\mathbf{M}, \boldsymbol{g}^{\mathrm{L}})$ into synchronous reference hypersurfaces Σ_t labeled by a globally defined parameter t.

Let us therefore consider for some integer N the family of reference hypersurfaces Σ_{t_n}, $n = 1,\ldots,N$, which are such that $\Sigma_{t'} = \Sigma_{t_0}$ and $\Sigma_{t''} = \Sigma_{t_N}$, and that $t_n - t_{n-1} = (t'' - t')/N = \varepsilon$. Let us denote by $\mathbf{S}_n$ the segment in the base manifold $(\mathbf{M}, \boldsymbol{g}^{\mathrm{L}})$ between the hypersurfaces $\Sigma_{t_{n-1}}$ and Σ_{t_n}, to which we shall refer as its inflow and outflow hypersurface, respectively. For a choice of cross-section $\boldsymbol{s} = \{\boldsymbol{a}(x), \boldsymbol{e}_i(x) | \, x \in \mathbf{M}\}$ of the Poincaré frame bundle $P\mathbf{M}$ we then introduce above each one of the inflow-outflow hypersurfaces corresponding to $n = 1,\ldots,N-1$ the coherent field-mode sections

$$\hat{\boldsymbol{\Phi}}^{\boldsymbol{s}}_{\varphi_n(x_n)} \in \hat{\mathcal{F}}_{x_n} , \quad \varphi_n(x_n) \in \mathbf{F}_{x_n} , \quad \varphi_n \in \mathbf{F} , \quad x_n \in \Sigma_{t_n} , \tag{5.10}$$

determined[10] from single-boson states φ_n within the typical fibre $\mathbf{F}$ in a manner which naturally extrapolates the construction in Eq. (5.5) from flat to curved spacetimes. Let us then consider all the broken paths between $\Sigma_{t'}$ and $\Sigma_{t''}$ which consist of geodesic arcs $\gamma(x_{n-1},x_n)$ connecting points x_{n-1} on inflow hypersurfaces $\Sigma_{t_{n-1}}$ with points x_n on the respective outflow hypersurfaces Σ_{t_n} of the base-segments $\mathbf{S}_n$, $n = 1,\ldots,N$.

As mentioned earlier, quantum-geometric propagation proceeds by parallel transport that abides by the strong equivalence principle and by the geodesic postulate. Consequently, if $\tau_\gamma(x_n,x_{n-1})$ denotes the operator for parallel transport along $\gamma(x_{n-1},x_n)$ determined by the quantum connection in Eq. (4.16), the quantum-geometric free-fall propagator is given by

$$K(\varphi(x'');\varphi(x')) = \lim_{\varepsilon\to+0} \prod_{n=N}^{1}{}' \int \mathcal{D}\varphi_n \Big\langle e^{i\varepsilon\int d\sigma(x_n)\,\rho(x_n)}\, \hat{\boldsymbol{\Phi}}^{s}_{\varphi_n(x_n)} \Big| \tau_\gamma(x_n,x_{n-1})\, \hat{\boldsymbol{\Phi}}^{s}_{\varphi_{n-1}(x_{n-1})} \Big\rangle, \quad (5.11)$$

where $\rho(x_n)$ is given by Eqs. (5.1)-(5.3), and $d\sigma(x_n)$ is the Riemannian measure element determined by the 3-metric induced by the Lorentzian metric $\boldsymbol{g}^{\mathrm{L}}$ on the spacelike reference hypersurface Σ_{t_n}. The above functional integration is to be carried out over all the coherent field modes above Σ_{t_n} with respect to the functional "measure"

$$\mathcal{D}\varphi_n = \prod\nolimits_{x_n\in\Sigma_{t_n}} \mathcal{D}[\,\varphi_n(x_n)]\ , \quad (5.12)$$

in the sense of Riemannian[10] rather than of Lebesgue integration,[18] so that no transition to imaginary time and the "Euclidean regime" is required – as in the case with the Feynman-Kac formula. Rather, the smoothness of coherent sections secures the existence of such integrals over physical families of stochastic paths in *real* spacetime.

The key formula (4.15), together with other properties of coherent states, enables the recasting of Eq. (5.11) in the form of the path integral

$$K(\varphi(x'');\varphi(x')) = \int \mathrm{D}\varphi \exp[iS(\varphi)]\ , \qquad \mathrm{D}\varphi = \prod\nolimits_{t''>t>t'} \prod\nolimits_{x\in\Sigma_t} \mathcal{D}\varphi(x)\ , \quad (5.13)$$

based on an action that can be expressed terms of a Lagrangian density:

$$S(\varphi) = \int_{t'}^{t''} dt \int_{\Sigma_t} d\sigma(x) \int \delta(v^2-1)\, d^4v\, \mathcal{L}_0(\varphi_t(x,v),\varphi_{t-0}(x,v))\ , \quad (5.14)$$

$$\mathcal{L}_0 = \tfrac{i}{2} Z_{\ell,m}\big[\overline{\varphi}_t(x,v)\,\dot{\varphi}_t(x,v) - \dot{\overline{\varphi}}_t(x,v)\,\varphi_{t-0}(x,v)\big] - T_{00}[\overline{\varphi}_t(x,v)+\varphi_{t-0}(x,v)]\ . \quad (5.15)$$

Thus, in quantum-geometric field theory Lagrangians and action integrals are *derived* from geometric physical principles, rather than being postulated. Nevertheless, it has been shown[10,17] that field interactions can be incorporated without difficulty in the quantum-geometric framework, and that in the special relativistic regime agreement with conventional QFT can be achieved by taking the sharp-point limit $\ell\to 0$ in a *formal* perturbation expansion.

6. The Physical Foundations of Geometric Quantum Gravity

The extrapolation of quantum-geometric propagation to quantum gravity, where matter fields are in *mutual* interaction with a quantum gravitational field $\mathbf{g}^{\mathrm{Q}}$, has to take into account the fact that already in classical gravity such mutual interactions between "matter" fields (including those describing nongravitational radiation) and gravitational fields display diffeomorphism invariance.[29] Thus, if the Lorentzian manifold $(\mathbf{M},\boldsymbol{g}^{\mathrm{L}})$ provides a CGR model for spacetime, and if $\psi:\mathbf{M}\to\mathbf{M}$ is any diffeomorphism of $\mathbf{M}$ onto itself, then the Lorentzian manifold $(\mathbf{M},\boldsymbol{g}'^{\mathrm{L}})$ provides a *physically* equivalent model if we set,

$$\boldsymbol{g}'^{\mathrm{L}} = \psi_* \boldsymbol{g}^{\mathrm{L}} \,, \qquad \boldsymbol{T}'_\alpha = \psi_* \boldsymbol{T}_\alpha \,, \qquad x'_\beta(t) = \psi(x_\beta(t)) \,, \tag{6.1}$$

for all the matter and radiation tensor fields $\boldsymbol{T}_\alpha$, as well as for all causal test-particle world-lines $\{x_\beta(t) \in \mathbf{M}\}$ of that model. Hence, if $Riem^{\mathrm{L}}\mathbf{M}$ is the family of all Lorentzian metrics in $\mathbf{M}$, and if $Diff\mathbf{M}$ is the diffeomorphism group on $\mathbf{M}$, then a CGR model in $\mathbf{M}$ is actually represented by an equivalence class $\mathcal{g}^{\mathrm{L}}$,

$$\mathcal{g}^{\mathrm{L}} = \{ \psi_* \boldsymbol{g}^{\mathrm{L}} \mid \psi \in Diff\mathbf{M} \} \in Riem^{\mathrm{L}}\mathbf{M}/Diff\mathbf{M} \,, \tag{6.2}$$

of Lorentzian metrics in $\mathbf{M}$, which constitutes a gauge orbit $(\mathbf{M}, \mathcal{g}^{\mathrm{L}})$ within the principal bundle $Diff\mathbf{M} \to Riem^{\mathrm{L}}\mathbf{M} \to Riem^{\mathrm{L}}\mathbf{M}/Diff\mathbf{M}$.

The extrapolation of this fact to quantum gravity can be achieved by considering that such a CGR gauge orbit provides a *metrization* in $\mathbf{M}$. This metrization is mathematically represented by the reduction of the general linear frame bundle $GL\mathbf{M}$ to the family of all Lorentz frame bundles $L\mathbf{M}(\boldsymbol{g}^{\mathrm{L}})$ with $\boldsymbol{g}^{\mathrm{L}} \in \mathcal{g}^{\mathrm{L}}$. Physically, such a metrization is tantamount to an operational verification aimed at establishing which *physical* representatives of linear frames $\{\boldsymbol{e}_0(x),\ldots,\boldsymbol{e}_3(x)\}$ in $GL\mathbf{M}$ are actually local Lorentz frames in the *physically existing* metric. Indeed, given the fact that the *existence* of such a metric is a feature of the physical reality around us, rather than a matter of mere choice left to the discretion of the experimenter, or of the observer of a given physical occurrence, from the point of view of CGR such a procedure amounts to verifying which ones of the linear *macro*-frames, consisting of "rigid rods" and "standard clocks," are Lorentzian in the operational sense originally stipulated by Einstein. However, if a relabeling of all frames, test bodies and fields is carried out, this *physical* verification cannot distinguish between all the various choices of *mathematical* labels $\{\boldsymbol{e}_0(x),\ldots\boldsymbol{e}_3(x)\}$, if such relabelings are related by means of diffeomorphisms that give rise to various equivalent Lorentzian metric representatives. Moreover, even in those cases where the mathematical labeling of frames, test bodies and fields is kept fixed, by the strong equivalence principle *physically* equivalent descriptions of natural phenomena are still obtained for physically distinct choices of inertial Lorentz frames, if those frames in free fall are related by Lorentz gauge transformations.

The adaptation of these basic facts to quantum *micro*-frames leads,[9,10] in broad outline, to the following formulation of *quantum* geometrodynamic evolution: the presence of a specific state of the quantum gravitational field $\mathbf{g}^{\mathrm{Q}}$ within a base segment $\mathbf{S}_n$ dictates the quantum-geometric evolution of all the quantum "matter" fields from the inflow hypersurface $\Sigma_{t_{n-1}}$ to the outflow hypersurface Σ_{t_n} of that base-segment; the values of the local quantum energy-momentum density operators of all the quantum "matter" fields in the resulting states of those fields along Σ_{t_n} then *creates* a QGR gauge orbit $(\mathbf{S}_{n+1}, \mathcal{g}^{\mathrm{M}})$ by metrizing the quantum frame bundles over the subsequent base segment $\mathbf{S}_{n+1}$; this determines the formation of local states of the quantum gravitational field $\mathbf{g}^{\mathrm{Q}}$ within that base-segment (whose mean values provide the Lorentzian metrics in $\mathcal{g}^{\mathrm{M}}$), and singles out within the quantum gravitational fibres above its points x the multi-graviton states of quantum gravitational radiation. This process then repeats itself from segment to segment, with the limit to "infinitely-thin" segments being eventually taken.

Of course, the presence of constraints and quantum gravitational self-interaction has to be taken into account before this very sketchy picture of quantum-geometric gravitational evolution can be completed into a legitimate extrapolation to the quantum regime of Einstein's formulation of classical gravity. This involves the extension of the base-segments $\mathbf{S}_n$ into supermanifolds $\boldsymbol{S}_n$ over which a quantum gravitational superbundle can be defined, and on whose sections a quantum gravitational gauge supergroup can act. As we shall see in Sec. 9, this quantum gravitational supergroup incorporates diffeomorphism invariance as well as Poincaré gauge invariance.

The resulting framework shares common features with both covariant and canonical quantum gravity, but it is also fundamentally distinct in many respects. Thus, in geometric quantum gravity equivalence classes g^{M} of mean metrics are generated by the quantum geometrodynamic evolution of gravitational fields in mutual interaction "matter" fields, so that no "background" metric is prescribed – as is the case in covariant quantum gravity. Moreover, the presence of *quantum* (super)-frames also resolves the "issue of time" – which is much-debated in contemporary canonical gravity.[29] This solution emerges from de Broglie's fundamental idea[30] that, on account of its rest mass m, each massive elementary quantum object represents a natural clock with period $T = 2\pi/m$ in Planck natural units. The replacement of global with local frames insures that such a proper time is kept *locally*, in accordance with the fundamental features of Einstein's *general* relativistic conceptualization of an ultimately valid description of nature. On the other hand, the use of bundles of quantum frames or superframes which incorporate the fundamental length ℓ into their very structure eliminates all need for de Broglie's "pilot waves." Thus, the replacement of classical geometries, upon which de Broglie founded his ideas, with quantum geometries, obviates the need for extraneous assumptions about *quantum* reality in the same manner in which the introduction of the geometry of Minkowski space obviated the need for regarding Lorentz contractions of (classically visualized) electrons as real phenomena.

In the subsequent sections we shall outline how these fundamental physical ideas can be described in a mathematically precise manner (cf. Chapter 8 of Ref. 10 for the details).

7. Graviton Fibres and Their Internal Gauges

We initiate the geometro-stochastic quantization of gravity by considering a generic QGR gauge orbit $(\mathbf{S}, g^{\mathrm{M}})$, generated during the course of geometrodynamic evolution. The base-segment $\mathbf{S}$ in such a gauge orbit is a differential manifold, whose boundary $\partial\mathbf{S}$ is composed of two disjoint hypersurfaces Σ' and Σ'' that constitute its inflow and outflow hypersurfaces for quantum data emerging in the course of that evolution, and of an equivalence class g^{M} of mean metrics, which gives rise to the metrized bundle $P^*\mathbf{S}^{\uparrow}(g^{\mathrm{M}})$ of (affine) Poincaré coframes $\boldsymbol{u} = \{\boldsymbol{a}^*(x), \boldsymbol{\theta}^i(x)\}$ above $\mathbf{S}$.

We can associate with each such coframe $\boldsymbol{u} \in P^*\mathbf{S}^{\uparrow}(g^{\mathrm{M}})$ a single-graviton fibre $\mathbf{Z}_{\boldsymbol{u}}$ consisting of the graviton state vectors represented by the wave functions

$$\boldsymbol{f}(\boldsymbol{u};\zeta) = (2\pi)^{-3/2} \int_{V_0^+} \exp(-i\zeta\cdot k)\, \tilde{\boldsymbol{f}}(\boldsymbol{u};k)\, d\Omega_0(k) , \qquad \zeta = q - i\ell v , \tag{7.1}$$

that are generated by all the momentum-space wave functions

$$\tilde{\boldsymbol{f}}(\boldsymbol{u};k) = \tilde{f}_{ij}(k)\, \boldsymbol{\theta}^i(x) \otimes \boldsymbol{\theta}^j(x) , \qquad \tilde{f}_{ij}(k) \equiv \tilde{f}_{ji}(k) , \tag{7.2}$$

in the mass-0 and spin-2 pseudo-unitary representation

$$\tilde{U}(a,\Lambda) : \ \tilde{f}_{ij}(k) \mapsto \tilde{f}'_{ij}(k) = \exp(ia\cdot k)\Lambda_i^{\ i'}\Lambda_j^{\ j'} \tilde{f}_{i'j'}(\Lambda^{-1}k) \tag{7.3}$$

of the orthochronous Poincaré group $\mathrm{ISO}^{\uparrow}(3,1)$. Each such fibre $\mathbf{Z}_{\boldsymbol{u}}$ is a Krein space,[31] i.e., it is a pseudo-Hilbert space, which in the present context carries the indefinite inner product

$$\langle \boldsymbol{f} | \boldsymbol{f}' \rangle = \int_{V_0^+} \eta^{ii'} \eta^{jj'} \tilde{f}^*_{ij}(k)\, \tilde{f}'_{i'j'}(k)\, d\Omega_0(k) , \quad d\Omega_0(k) = \delta(k^2)\, d^4k , \tag{7.4}$$

and the positive-definite $J_{\boldsymbol{u}}$-inner product associated with the coframe $\boldsymbol{u} \in P^*\mathbf{S}^{\uparrow}(\mathbf{g}^{\mathrm{M}})$:

$$(\boldsymbol{f}|\boldsymbol{f}')_{J_u} = \sum_{i,j=0}^{3} \int_{V_0^+} \tilde{f}_{ij}^*(k)\, \tilde{f}'_{ij}(k)\, d\Omega_0(k) \; . \tag{7.5}$$

The former is left invariant by the representation resulting from (7.3),

$$U(a,\Lambda) \; : \; f_{ij}(q-i\ell v) \;\mapsto\; f'_{ij}(q-i\ell v) = \Lambda_i^{\;i'} \Lambda_j^{\;j'} f_{i'j'}(\Lambda^{-1}(q-a)-i\ell\Lambda^{-1}v) \; ; \tag{7.6}$$

whereas, the latter is not invariant under Lorentz boosts, but the majorant topology[31] to which that J_u-inner product gives rise is left invariant by this representation.

Let us now introduce the generalized Lorenz gauge subfibres of $\mathbf{Z}_u$,

$$\mathbf{Z}_u^{(b)} = \left\{ \boldsymbol{f} \in \mathbf{Z}_u \;\middle|\; \partial^i f_{ij} = b_j \;\; , \;\; j = 0,1,2,3 \right\} , \qquad \partial^i = \partial / \partial q^i \; . \tag{7.7}$$

They are obviously left invariant by the internal Lorenz gauge transformations

$$\boldsymbol{f} = f_{ij}\, \boldsymbol{\theta}^i(x) \otimes \boldsymbol{\theta}^j(x) \;\;\mapsto\;\; \boldsymbol{f}' = f'_{ij}\, \boldsymbol{\theta}^i(x) \otimes \boldsymbol{\theta}^j(x) \, , \qquad \boldsymbol{f}, \boldsymbol{f}' \in \mathbf{Z}_u \; , \tag{7.8}$$

$$f'_{ij}(\zeta) = f_{ij}(\zeta) + \partial_i \lambda_j(\zeta) + \partial_j \lambda_i(\zeta) \; , \qquad \partial^i \partial_i \lambda_j(\zeta) = 0 \; , \qquad \partial^i \lambda_i(\zeta) = 0 \; . \tag{7.9}$$

The generic element $\boldsymbol{f}$ of $\mathbf{Z}_u$ can be decomposed in the following unique manner,

$$\boldsymbol{f} = \boldsymbol{f}^{\mathrm{TT}} + \boldsymbol{f}^{\perp} + \boldsymbol{f}^{b} \in \mathbf{Z}_u^{(b)} \, , \qquad \boldsymbol{f}^{\mathrm{TT}} \in \mathbf{Z}_u^{\mathrm{TT}} \, , \qquad \boldsymbol{f}^{\perp} \in \mathbf{N}_u \; , \tag{7.10}$$

where $\boldsymbol{f}^{\mathrm{TT}}$ is the element of the TT-subfibre of $\mathbf{Z}_u$, whose elements satisfy the transverse and traceless conditions that are well-known from linearized gravity, and can be expressed as a superposition of two *physical* (linear or circular) modes of polarization; whereas, the pseudo-orthogonal complement of $\boldsymbol{f}^{\mathrm{TT}}$ belongs to the null-subfibre

$$\mathbf{N}_u = \left\{ \boldsymbol{f} \in \mathbf{Z}_u \,\middle|\, \langle \boldsymbol{f} | \boldsymbol{f} \rangle = 0 \right\} \tag{7.11}$$

of $\mathbf{Z}_u$, which can be shown to be a subspace of the proper Lorenz gauge subfibre corresponding to $b_j(\zeta) \equiv 0$, $j = 0,1,2,3$, in Eq. (7.7).

Let us introduce the following expansion of a superlocal (i.e., local in $P^*\mathbf{S}^{\uparrow}(\mathbf{g}^{\mathrm{M}})$, and not just in $\mathbf{S}$) graviton wave function,

$$\boldsymbol{f}(\boldsymbol{u};\zeta) = (2\pi)^{-3/2} \int_{V_0^+} \exp(-i\,\zeta \cdot k)\, \boldsymbol{\varepsilon}^{[r]}(\boldsymbol{u};k) \otimes \boldsymbol{\varepsilon}^{[s]}(\boldsymbol{u};k)\, \tilde{f}_{[rs]}(k)\, d\Omega_0(k) \; , \tag{7.12}$$

with respect to the graviton polarization null-frame whose elements are given by

$$\boldsymbol{\varepsilon}^{[rs]\zeta}(\boldsymbol{u};\zeta') = \int_{V_0^+} \exp[i(\bar{\zeta} - \zeta') \cdot k]\, \boldsymbol{\varepsilon}^{[r]}(\boldsymbol{u};k) \otimes \boldsymbol{\varepsilon}^{[s]}(\boldsymbol{u};k)\, d\Omega_0(k) \, , \quad r,s = \pm 1, \pm 2. \tag{7.13}$$

This graviton polarization null-frame is obtained from the null polarization cotetrad whose elements within the complexified cotangent space $\overline{T}_x{}^*\mathbf{S}$,

$$\boldsymbol{\varepsilon}^{[\pm 1]}(\boldsymbol{u};k) = \mp \tfrac{1}{\sqrt{2}} \left(\boldsymbol{\varepsilon}^{(1)}(\boldsymbol{u};k) \pm i\, \boldsymbol{\varepsilon}^{(2)}(\boldsymbol{u};k) \right) = \mp \tfrac{1}{\sqrt{2}} \left(\boldsymbol{\theta}^1(x) \pm i\, \boldsymbol{\theta}^2(x) \right) \cdot R(\hat{\mathbf{k}}) \, , \tag{7.14}$$

$$\boldsymbol{\varepsilon}^{[+2]}(\boldsymbol{u};k) = k_i\, \boldsymbol{\theta}^i(x) = \boldsymbol{k} \; , \qquad \boldsymbol{\varepsilon}^{[-2]}(\boldsymbol{u};k) = k_0^{-1} \boldsymbol{\theta}^0(x) - \tfrac{1}{2} k_0^{-2} \boldsymbol{k} \; , \tag{7.15}$$

are the outcome[9,10] of a standard application of Wigner rotations to the linear coframe $\{\boldsymbol{\theta}^i(x)\}$ within the affine coframe $\boldsymbol{u}$. Then the TT-gauge subfibre appearing in Eq. (7.10) can be expressed as follows:[10]

$$\mathbf{Z}_u^{\mathrm{TT}} = \left\{ \sum_{r=-1}^{+1} f_{[rr]\zeta}\, \boldsymbol{\varepsilon}^{[rr]\zeta} \;\middle|\; \boldsymbol{f} = f_{[rs]\zeta}\, \boldsymbol{\varepsilon}^{[rs]\zeta} \in \mathbf{Z}_u^{(0)} \right\} . \tag{7.16}$$

The proper Lorenz gauge subfibre of $\mathbf{Z}_u$ that contains the TT-gauge subfibre in Eq. (7.16) is left invariant by the representation in (7.6). In case that (a,Λ) in Eq. (7.6) is allowed to vary only over the proper Poincaré group $ISO_0(3,1)$, the restriction of the resulting representation of $ISO_0(3,1)$ to the proper Lorenz gauge subfibre of $\mathbf{Z}_{\boldsymbol{u}}$ can be shown[10] to be equivalent the Wigner-type representation of that group given by the following transformations under the action of $(a,\Lambda) \in ISO_0(3,1)$,

$$\alpha_{\pm}(q - i\ell v) \mapsto e^{\pm 2i\theta(\Lambda, i\partial)} \alpha_{\pm}(\Lambda^{-1}(q-a) - i\ell\Lambda^{-1}v) , \quad \partial := (\partial_0, \partial_1, \partial_2, \partial_3) . \tag{7.17}$$

The wave functions in Eq (7.17) are the two graviton circular polarization components

$$\alpha_{\pm}(\zeta) = f_{[\pm 1 \pm 1]\zeta} , \qquad f_{[rs]\zeta} = \eta_{rr'}\eta_{ss'}\left\langle \boldsymbol{\varepsilon}^{[r's']\zeta} \middle| \boldsymbol{f} \right\rangle , \quad r,s = \pm 1, \pm 2 , \tag{7.18}$$

contained in the expansion in Eq. (7.12) of a graviton state vector *above* $\boldsymbol{u} \in P^*\mathbf{S}^{\uparrow}(g^{\mathrm{M}})$.

8. Quantum Gravitational and Berezin-Faddeev-Popov Frames

To carry out the geometro-stochastic second-quantization of gravity, we construct a quantum gravitational fibre $\mathcal{Z}_{\boldsymbol{u}}$ above each Poincaré coframe $\boldsymbol{u} = \{\boldsymbol{a}^*(x), \boldsymbol{\theta}^i(x)\} \in P^*\mathbf{S}^{\uparrow}(\mathbf{g}^{\mathrm{M}})$ by assigning to each $\boldsymbol{u}$ a superlocal vacuum state vector $\boldsymbol{\Psi}_{0;\boldsymbol{u}}$, that spans the one-dimensional superlocal vacuum sector $\mathcal{Z}_{0;\boldsymbol{u}}$. We then use the $\boldsymbol{u}$-dependent $J_{\boldsymbol{u}}$-inner product in Eq. (7.5) to add to $\mathcal{Z}_{0;\boldsymbol{u}}$ the $J_{\boldsymbol{u}}$-direct sums of symmetrized tensor products of the graviton fibre $\mathbf{Z}_{\boldsymbol{u}}$, thus arriving at the quantum gravitational fibre

$$\mathcal{Z}_{\boldsymbol{u}} = \bigoplus_{J\; n=0}^{\infty} \mathcal{Z}_{n;\boldsymbol{u}} , \qquad \mathcal{Z}_{n;\boldsymbol{u}} = \mathbf{Z}_{\boldsymbol{u}} \underset{S}{\otimes} \cdots \underset{S}{\otimes} \mathbf{Z}_{\boldsymbol{u}} . \tag{8.1}$$

Within this quantum gravitational fibre we can define graviton annihilation operators that act on n-graviton state vectors belonging to $\mathcal{Z}_{n;\boldsymbol{u}}$ as follows,

$$\left(g_{ij}^{(-)}(\boldsymbol{u};\zeta)\Psi_{n;\boldsymbol{u}}\right)_{n-1}(\zeta_1,i_1j_1,\ldots,\zeta_{n-1},i_{n-1}j_{n-1}) = \sqrt{n}\; \Psi_{n;\boldsymbol{u}}(\zeta,ij,\zeta_1,i_1j_1,\ldots,\zeta_{n-1},i_{n-1}j_{n-1}), \tag{8.2}$$

and corresponding graviton creation operators which are their $J_{\boldsymbol{u}}$-adjoints.[31] We can then introduce in Lorenz quantum gravitational subfibre of $\mathcal{Z}_{\boldsymbol{u}}$ the elements

$$\boldsymbol{\Phi}_{\boldsymbol{f}} = \exp\left[-\tfrac{1}{2}\langle \boldsymbol{f}|\boldsymbol{f}\rangle + g^{(+)}(\boldsymbol{u};\boldsymbol{f})\right] \boldsymbol{\Psi}_{0;\boldsymbol{u}} , \qquad \boldsymbol{f} \in \mathbf{Z}_{\boldsymbol{u}}^{(0)} , \tag{8.3}$$

of a quantum gravitational frame, defined by the general procedure used in Eq. (4.13). However, due to the presence of an indefinite metric, this quantum gravitational frame provides the continuous resolution of the identity

$$\int_{\mathbf{Z}_{\boldsymbol{u}}^{\mathrm{TT}}} |\boldsymbol{\Phi}_{\boldsymbol{f}}\rangle\, d\boldsymbol{f}\, d\bar{\boldsymbol{f}} \langle \boldsymbol{\Phi}_{\boldsymbol{f}}| = \mathbf{1}_{\boldsymbol{u}}^{\mathrm{TT}} , \tag{8.4}$$

only within the *physical* quantum gravitational fibre above $\boldsymbol{u}$, corresponding to the following $J_{\boldsymbol{u}}$-orthogonal decomposition of the Lorenz quantum gravitational subfibre:

$$\mathcal{Z}_{\boldsymbol{u}}^{(0)} = \mathcal{Z}_{\boldsymbol{u}}^{\mathrm{TT}} \underset{J}{\oplus} \mathcal{N}_{\boldsymbol{u}} , \qquad \mathcal{N}_{\boldsymbol{u}} = \left\{ \boldsymbol{\Psi} \in \mathcal{Z}_{\boldsymbol{u}}^{(0)} \,\middle|\, \langle \boldsymbol{\Psi}|\boldsymbol{\Psi}\rangle = 0 \right\} . \tag{8.5}$$

To treat the remaining degrees of freedom of graviton polarization, representing *un*physical modes, we shall use adaptations[9,10] to the quantum-geometric framework of quantum nonabelian gauge-field theoretical techniques.[32] Thus, we shall introduce above each $\boldsymbol{u} \in P^*\mathbf{S}^{\uparrow}(\mathbf{g}^{\mathrm{M}})$ the graviton ghost and antighost fibres

$$\mathbf{F}_{\boldsymbol{u}} = \left\{ \mathbf{c} = c_i\, \boldsymbol{\theta}^i(x) \mid (c_0,\ldots,c_3) \in \mathbf{K} \right\}, \quad \overline{\mathbf{F}}_{\boldsymbol{u}} = \left\{ \bar{\mathbf{c}} = \bar{c}_i\, \boldsymbol{\theta}^i(x) \,\middle|\, (\bar{c}_0,\ldots,\bar{c}_3) \in \mathbf{K}^{\dagger} \right\} , \tag{8.6}$$

where $\mathbf{K}$ is a Krein space (with $\mathbf{K}^{\dagger}$ denoting its dual) consisting of functions in terms of which we can carry out for the decompositions in Eq. (7.10) the identifications

$$\boldsymbol{f}^{\perp} = f_{ij}^{\perp}\, \boldsymbol{\theta}^{i}(x)\otimes \boldsymbol{\theta}^{j}(x) \leftrightarrow \mathbf{c}^{\perp}\in \mathbf{F}_{\boldsymbol{u}}\ ,\quad f_{ij}^{\perp} = \partial_i c_j^{\perp} + \partial_j c_i^{\perp}\ ,\quad \partial^i c_i^{\perp} = 0\ , \tag{8.7}$$

$$\boldsymbol{f}^{b} = f_{ij}^{b}\, \boldsymbol{\theta}^{i}(x)\otimes \boldsymbol{\theta}^{j}(x) \leftrightarrow \mathbf{c}^{b}\in \mathbf{F}_{\boldsymbol{u}}\ ,\quad f_{ij}^{b} = \partial_i c_j^{b} + \partial_j c_i^{b}\ ,\quad \partial_j \partial^i c_i^{b} = b_j\ . \tag{8.8}$$

We can then construct Faddeev-Popov (FP) ghost and antighost fibres,

$$\mathcal{F}_{\boldsymbol{u}} = \bigoplus_{n=0}^{\infty} (\mathbf{F}_{\boldsymbol{u}} \underset{A}{\otimes} \cdots \underset{A}{\otimes} \mathbf{F}_{\boldsymbol{u}})_n\ ,\qquad \overline{\mathcal{F}}_{\boldsymbol{u}} = \bigoplus_{n=0}^{\infty} (\overline{\mathbf{F}}_{\boldsymbol{u}} \underset{A}{\otimes} \cdots \underset{A}{\otimes} \overline{\mathbf{F}}_{\boldsymbol{u}})_n\ , \tag{8.9}$$

in which we can define in the standard manner the fermionic-type action of the ghost annihilation and creation operators $C^{(-)}(\boldsymbol{u};\zeta)$ and $C^{(+)}(\boldsymbol{u};\zeta)$, respectively, together with that of the conjugate operators $\overline{C}^{(-)}(\boldsymbol{u};\zeta)$ and $\overline{C}^{(+)}(\boldsymbol{u};\zeta)$ acting on antighost states.

By following procedures that emerge from the standard techniques that are employed in Berezin integration,[25] we make the transition to the BFP (i.e., Berezin-Faddeev-Popov) ghost-antighost superfibres $\mathcal{B}_{\boldsymbol{u}} \otimes \overline{\mathcal{B}}_{\boldsymbol{u}}$ obtained by introducing a $J_{\boldsymbol{u}}$-orthonormal basis $\{\mathbf{c}_{\boldsymbol{u};1}, \mathbf{c}_{\boldsymbol{u};2}, \ldots\}$ within each $\mathbf{F}_{\boldsymbol{u}}$, and its conjugate within $\overline{\mathbf{F}}_{\boldsymbol{u}}$, and then setting

$$\mathbf{B}_{\boldsymbol{u}} = \left\{ \boldsymbol{c} = \theta_{\boldsymbol{u}}^{\alpha}\, \mathbf{c}_{\boldsymbol{u};\alpha} \,\middle|\, \theta_{\boldsymbol{u}}^{\alpha} = \left(\mathbf{c}_{\boldsymbol{u};\alpha}|\mathbf{c}\right)_{J_{\boldsymbol{u}}} \vartheta_{\alpha}\ ,\ \ \alpha = 1,2,\ldots\ ,\ \ \mathbf{c}\in \mathbf{F}_{\boldsymbol{u}} \right\}, \tag{8.10}$$

$$\overline{\mathbf{B}}_{\boldsymbol{u}} = \left\{ \overline{\boldsymbol{c}} = \overline{\mathbf{c}}_{\boldsymbol{u};\alpha}\overline{\theta}_{\boldsymbol{u}}^{\alpha} \,\middle|\, \overline{\theta}_{\boldsymbol{u}}^{\alpha} = \left(\overline{\mathbf{c}}_{\boldsymbol{u};\alpha}|\overline{\mathbf{c}}\right)_{J_{\boldsymbol{u}}} \overline{\vartheta}_{\alpha}\ ,\ \ \alpha = 1,2,\ldots\ ,\ \ \overline{\mathbf{c}}\in \overline{\mathbf{F}}_{\boldsymbol{u}} \right\}, \tag{8.11}$$

where ϑ_{α} and $\overline{\vartheta}_{\alpha}$ are the generators of a Grassmann algebra. Within these superfibres we introduce the ghost and antighost BFP superframe elements

$$\boldsymbol{\Phi}_{\boldsymbol{c}} = \exp\!\left(-\tfrac{1}{2}\textstyle\sum_{\alpha=1}^{\infty} \overline{\theta}_{\boldsymbol{u}}^{\alpha}\, \theta_{\boldsymbol{u}}^{\alpha}\right) \exp\!\left(C^{(+)}(\boldsymbol{u};\boldsymbol{c})\right) \boldsymbol{\Psi}_{0;\boldsymbol{u}}\ ,\qquad \boldsymbol{c}\in \mathbf{B}_{\boldsymbol{u}} \tag{8.12}$$

$$\boldsymbol{\Phi}_{\overline{\boldsymbol{c}}} = \exp\!\left(-\tfrac{1}{2}\textstyle\sum_{\alpha=1}^{\infty} \theta_{\boldsymbol{u}}^{\alpha}\, \overline{\theta}_{\boldsymbol{u}}^{\alpha}\right) \exp\!\left(\overline{C}^{(+)}(\boldsymbol{u};\overline{\boldsymbol{c}})\right) \boldsymbol{\Psi}_{0;\boldsymbol{u}}\ ,\qquad \overline{\boldsymbol{c}}\in \overline{\mathbf{B}}_{\boldsymbol{u}}\ . \tag{8.13}$$

The resulting superframes supply continuous resolutions of the identity within the corresponding FP ghost and antighost fibres in Eq. (8.9):

$$\int_{\overline{\mathbf{B}}_{\boldsymbol{u}}\times \mathbf{B}_{\boldsymbol{u}}} \langle \boldsymbol{\Psi}|\boldsymbol{\Phi}_{\boldsymbol{c}}\rangle\, d\overline{\boldsymbol{c}}\, d\boldsymbol{c}\, \langle \boldsymbol{\Phi}_{\boldsymbol{c}}|\boldsymbol{\Psi}'\rangle = \langle \boldsymbol{\Psi}|\boldsymbol{\Psi}'\rangle\ ,\qquad \boldsymbol{\Psi},\boldsymbol{\Psi}'\in \mathcal{F}_{\boldsymbol{u}}\ , \tag{8.14}$$

$$\int_{\mathbf{B}_{\boldsymbol{u}}\times \overline{\mathbf{B}}_{\boldsymbol{u}}} \langle \boldsymbol{\Psi}|\boldsymbol{\Phi}_{\overline{\boldsymbol{c}}}\rangle\, d\boldsymbol{c}\, d\overline{\boldsymbol{c}}\, \langle \boldsymbol{\Phi}_{\overline{\boldsymbol{c}}}|\boldsymbol{\Psi}'\rangle = \langle \boldsymbol{\Psi}|\boldsymbol{\Psi}'\rangle\ ,\qquad \boldsymbol{\Psi},\boldsymbol{\Psi}'\in \overline{\mathcal{F}}_{\boldsymbol{u}}\ . \tag{8.15}$$

The fundamental significance of the quantum gravitational frames and of the BFP superframes introduced in this section to the formulation in Sec. 11 of quantum geometrodynamic evolution is that, in addition to supplying continuous resolutions of the identity, the elements of these frames and superframes are also eigenvectors of corresponding annihilation operators, so that we have:

$$g_{ij}^{(-)}(\boldsymbol{u};\zeta)\,\boldsymbol{\Phi}_{\boldsymbol{f}} = f_{ij}(\zeta)\,\boldsymbol{\Phi}_{\boldsymbol{f}}\ ,\qquad \boldsymbol{f} = f_{ij}\, \boldsymbol{\theta}^{i}(x)\otimes \boldsymbol{\theta}^{j}(x)\in \mathbf{Z}_{\boldsymbol{u}}^{(0)}\ , \tag{8.16}$$

$$C_i^{(-)}(\boldsymbol{u};\zeta)\,\boldsymbol{\Phi}_{\boldsymbol{c}} = c(\zeta,i)\,\boldsymbol{\Phi}_{\boldsymbol{c}}\ ,\qquad \overline{C}_i^{(-)}(\boldsymbol{u};\zeta)\,\boldsymbol{\Phi}_{\overline{\boldsymbol{c}}} = \overline{c}(\zeta,i)\,\boldsymbol{\Phi}_{\overline{\boldsymbol{c}}}\ . \tag{8.17}$$

These features mediate the extension of the techniques of Sec. 5 to quantum gravity.

9. The Quantum Gravitational Supergroup

Each of the TT-gauge subfibres of $\mathbf{Z}_{\boldsymbol{u}}$, defined in Eq. (7.16), is left invariant by the semidirect product of the spacetime translation group T^4 and the rotation group O(3) viewed as a subgroup of the orthochronous Poincaré group $ISO^{\uparrow}(3,1)$, but not by the Lorentz boosts Λ_v within $ISO^{\uparrow}(3,1)$. Hence, we shall carry out the decompositions

$$(a,\Lambda)=(a,R)(0,\Lambda_v)\in ISO^{\uparrow}(3,1)\ ,\ \ (a,R)\in T^4\wedge O(3)\ ,\ \ \Lambda_v\in SO_0(3,1)\ , \tag{9.1}$$

$$(a,\Lambda)^{-1}=(0,\Lambda_v^{-1})(-R^{-1}a,R^{-1})\in ISO^{\uparrow}(3,1)\ ,\ \ (-R^{-1}a,R^{-1})=(a,R)^{-1}\ , \tag{9.2}$$

of a generic orthochronous Poincaré transformation and its inverse. Then, if we introduce the quantum gravitational superfibres

$$\mathcal{Z}_{u}^{\Diamond}=\mathcal{Z}_{u}^{\mathrm{TT}}\otimes\overline{\mathcal{B}}_{u}\otimes\mathcal{B}_{u}\ , \tag{9.3}$$

and consider within them the *quantum gravitational coframes* with elements

$$\boldsymbol{\Phi}^{\dagger}_{f,\bar{\mathbf{c}},\mathbf{b},\mathbf{c}}=\boldsymbol{\Phi}^{\dagger}_{f}\otimes\boldsymbol{\Phi}_{\bar{\mathbf{c}}}\otimes\boldsymbol{\Phi}^{\dagger}_{\mathbf{b},\mathbf{c}}\ ,\quad f\in\mathbf{Z}_{u}^{\mathrm{TT}},\quad \mathbf{b}\in\mathbf{F}_{u}\ ,\quad \bar{\mathbf{c}}\in\overline{\mathbf{F}}_{u}^{\perp}\ ,\quad \mathbf{c}\in\mathbf{F}_{u}^{\perp}\ , \tag{9.4}$$

we can take advantage of Eqs. (9.1) and (9.2) to transfer the action of a Lorentz boost Λ_v upon a (classically conceptualized) coframe $\boldsymbol{u}\in P^*\mathbf{S}^{\uparrow}(\mathbf{g}^{\mathrm{M}})$ into the action upon the corresponding quantum gravitational coframe in Eq. (9.4), carried out via the actions

$$U(0,\Lambda_v):\ \mathbf{c}^{\perp}=\theta_{u}^{\alpha}\mathbf{c}_{u;\alpha}\mapsto\theta'^{\alpha}_{u}\mathbf{c}_{u;\alpha}\ ,\quad \theta'^{\alpha}_{u}=\left(\mathbf{c}_{u;\alpha}\middle|\mathbf{c}^{\perp}\right)_{J_u}U_{\alpha}^{\beta}(0,\Lambda_v)\vartheta_{\beta}\ , \tag{9.5}$$

$$U(0,\Lambda_v):\ \mathbf{c}^{b}=\theta_{u}^{\alpha}\mathbf{c}_{u;\alpha}\mapsto\theta'^{\alpha}_{u}\mathbf{c}_{u;\alpha}\ ,\quad \theta'^{\alpha}_{u}=\left(\mathbf{c}_{u;\alpha}\middle|\mathbf{c}^{b}\right)_{J_u}U_{\alpha}^{\beta}(0,\Lambda_v)\vartheta_{\beta}\ , \tag{9.6}$$

$$U_{\beta}^{\alpha}(0,\Lambda_v)=\left(\mathbf{c}_{u;\beta}\middle|U(0,\Lambda_v)\mathbf{c}_{u;\alpha}\right)_{J_u}\ ,\qquad \alpha,\beta=1,2,\dots\ , \tag{9.7}$$

upon its superlocal ghost states (associated, in accordance with Eqs. (8.7) and (8.8), with the decomposition in Eq. (7.10) in any given generalized internal Lorenz gauge), and to corresponding actions upon its superlocal antighost states. This constitutes the first step in the construction of a quantum gravitational supergroup that represents the counterpart of the CGR gauge group given by the semidirect product

$$\mathcal{G}(\mathbf{M},\mathcal{g}^{\mathrm{L}})=\mathrm{G}_{A}(\mathit{Diff}\,\mathbf{M})\wedge\mathcal{G}(\mathbf{P}(\mathbf{M},ISO^{\uparrow}(3,1)))\ , \tag{9.8}$$

which corresponds to the equivalence class $\mathcal{g}^{\mathrm{L}}$ of physically equivalent Lorentzian metrics of a CGR model over a classical spacetime manifold $\mathbf{M}$ (cf. Sec. 6).

To understand the physical significance of the CGR gauge group in Eq. (9.8), consider the general affine coframe bundle $GA^*\mathbf{M}$ consisting of all the affine coframes $\{\boldsymbol{a}^*(x),\boldsymbol{\theta}^i(x)\}$, $x\in\mathbf{M}$, over the manifold $\mathbf{M}$. Let us then express the notion of diffeomorphism invariance discussed in Sec. 6 in terms of the vertical automorphisms determined by the following action of the diffeomorphisms $\psi\in \mathit{Diff}\,\mathbf{M}$ upon any section $\boldsymbol{s}$ of $GA^*\mathbf{M}$, representing a *moving* coframe in the sense of Cartan:[33]

$$\mathbf{D}(\psi)\ :\ \boldsymbol{s}\ =\{(\boldsymbol{a}^*,\boldsymbol{\theta}^i)\}\ \mapsto\ \boldsymbol{s}'\ =\{(\psi_*(\boldsymbol{a}^*\circ\psi^{-1}),\ \psi_*(\boldsymbol{\theta}^i\circ\psi^{-1})\}\ . \tag{9.9}$$

This action gives rise to a moving coframe $\boldsymbol{s}'$ with the same domain as that of $\boldsymbol{s}$, so that, upon covering $GA^*\mathbf{M}$ with a bundle atlas of local trivializations associated with local sections of $GA^*\mathbf{M}$, we can assign to each $\psi\in \mathit{Diff}\,\mathbf{M}$ a vertical automorphism $\boldsymbol{\psi}\in \mathrm{Aut}_{\mathbf{M}}\mathbf{P}$. In this manner we arrive at an isomorphism between $\mathit{Diff}\,\mathbf{M}$ and a subgroup $\mathrm{G}_A(\mathit{Diff}\,\mathbf{M})$ of the group $\mathrm{Aut}_{\mathbf{M}}\mathbf{P}$ of all vertical automorphisms in the principal bundle $\mathbf{P}=GA^*\mathbf{M}$.

Consider now the metrized affine coframe bundle $P^*\mathbf{M}^{\uparrow}(\mathcal{g}^{\mathrm{L}})$ corresponding to the family of all Lorentz frame bundles $L\mathbf{M}(\mathbf{g}^{\mathrm{L}})$, $\mathbf{g}^{\mathrm{L}}\in\mathcal{g}^{\mathrm{L}}$, described in Sec. 6. Let us call a cross-section of $P^*\mathbf{M}^{\uparrow}(g^{\mathrm{L}})$ a *physical* moving coframe in $\mathbf{M}$ if it coincides with a cross-section $\boldsymbol{s}$ of one of its subbundles $P^*\mathbf{M}^{\uparrow}(\mathbf{g}^{\mathrm{L}})$. We shall denote such a physical moving coframe by $(\boldsymbol{s},\mathbf{g}^{\mathrm{L}})$, so that the corresponding cross-section $\{\boldsymbol{\theta}^{\iota}(x)\,|\,x\in\mathbf{M}\}$ of $L^*\mathbf{M}^{\uparrow}(\mathbf{g}^{\mathrm{L}})$ represents in CGR a vierbein. The following action of the gauge group $\mathrm{Aut}_{\mathbf{M}}P\mathbf{M}^{\uparrow}(\mathbf{g}^{\mathrm{L}})$, which is isomorphic to the group $\mathcal{G}(\mathbf{P}(\mathbf{M},\mathrm{ISO}^{\uparrow}(3,1))$ of Poincaré gauge transformations appearing in Eq. (9.8), and the subgroup $\mathrm{G}_A(\mathit{Diff}\,\mathbf{M})$ of $\mathrm{Aut}_{\mathbf{M}}GA^*\mathbf{M}$ upon any such physical moving coframe $(\boldsymbol{s},\mathbf{g}^{\mathrm{L}})$,

$$(\boldsymbol{\psi},\boldsymbol{\phi}) : (\boldsymbol{s},\mathbf{g}^{\mathrm{L}}) \mapsto (\boldsymbol{s}\circ\boldsymbol{\phi}\circ\boldsymbol{\psi}\,,\psi_*\,\mathbf{g}^{\mathrm{L}}),\quad \boldsymbol{\psi}\in \mathrm{G}_A(\mathit{Diff}\mathbf{M}),\ \boldsymbol{\phi}\in \mathrm{Aut}_{\mathbf{M}}P\mathbf{M}^{\uparrow}(\mathbf{g}^{\mathrm{L}}) \tag{9.10}$$

clearly represents the mathematical embodiment of the kind of generic change of moving frame of reference in CGR under which all *physical* predictions of a CGR model remain invariant. By Eq. (9.10), this combination of Poincaré gauge invariance with the diffeomorphism invariance gives rise to the group multiplication law

$$(\boldsymbol{\psi},\boldsymbol{\phi})(\boldsymbol{\psi}',\boldsymbol{\phi}') = ((\boldsymbol{\psi}\circ\boldsymbol{\phi}')\circ\boldsymbol{\psi}',\boldsymbol{\phi}\boldsymbol{\phi}'),\quad \boldsymbol{\psi},\boldsymbol{\psi}'\in \mathrm{G}_A(\mathit{Diff}\mathbf{M}),\ \boldsymbol{\phi},\boldsymbol{\phi}'\in\mathcal{G}(\mathbf{P}), \tag{9.11}$$

where $\mathbf{P} = \mathbf{P}(\mathbf{M},\mathrm{ISO}^{\uparrow}(3,1))$, thus determining the CGR gauge group in Eq. (9.8).

Due to the fundamentally indeterministic character of quantum theory, which is based on a theory of measurement in which the quantum states of matter fields can undergo reductions that can be only probabilistically predicted, it would be inconsistent to assume that the structure of spacetime can be deterministically predicted at the microscopic level, since any change in those states is bound to affect its quantum geometric structure. Hence, local indeterministic changes in the *mean* metrics $\mathbf{g}^{\mathrm{M}}$ of a quantum gravitational field can occur with the passage of proper time of local observers, so that, as opposed to the situation in CGR, there cannot be a *global* counterpart $(\mathbf{M},g^{\mathrm{M}})$ of the CGR gauge orbit $(\mathbf{M},g^{\mathrm{L}})$, and *global* diffeomorphism invariance loses its *raison d'être* in the quantum gravitational regime. This implies that the only form of causality by which quantum geometrodynamic evolution can abide is the one in which the time-ordering of local events is correlated to the time-ordering of the inflow and outflow hypersurfaces Σ' and Σ'' of base-segments $\mathbf{S}$ *created* during that evolution. In other words, any mean metric $\mathbf{g}^{\mathrm{M}}$ to which the inflow of matter and radiation gives rise has to be such that the futures of all points on Σ' that are classically causal in relation $\mathbf{g}^{\mathrm{M}}$ intersect Σ'', and the pasts of all points on Σ'' that are classically causal in relation to the same $\mathbf{g}^{\mathrm{M}}$ intersect Σ'.

This implies that the quantum metrization of a general affine coframe bundle $GA^*\mathbf{S}$ selects a metrized coframe bundle $P^*\mathbf{S}^{\uparrow}(g^{\mathrm{M}})$ corresponding to a family g^{M} of Lorentzian mean metrics $\mathbf{g}^{\mathrm{M}}$ that are diffeomorphically equivalent under the group $\mathit{Diff}_0\,\mathbf{S}$ of diffeomorphisms of a base-segment $\mathbf{S}$ that leave its inflow and outflow hypersurfaces invariant (rather than under the group $\mathit{Diff}\,\mathbf{S}$ of all diffeomorphisms of $\mathbf{S}$), thus giving rise a QGR gauge orbit $(\mathbf{S},g^{\mathrm{M}})$. Each diffeomorphism $\psi\in\mathit{Diff}_0\mathbf{S}$ gives rise, in the same manner as in the above considered classical context, to a vertical automorphism $\boldsymbol{\psi}\in\mathrm{G}_A(\mathit{Diff}_0\mathbf{S})$. In general, such a $\boldsymbol{\psi}$ acts on physical moving coframes given by cross-sections $\boldsymbol{s}$ of the Poincaré coframe bundle $P^*\mathbf{S}^{\uparrow}(\mathbf{g}^{\mathrm{M}})$, by mapping them, in accordance with Eq. (9.9), into physical moving coframes given by cross-sections $\boldsymbol{s}'$ of some other Poincaré coframe bundle $P^*\mathbf{S}^{\uparrow}(\mathbf{g}'^{\mathrm{M}})$ with $\mathbf{g}'^{\mathrm{M}}\in(\mathbf{S},g^{\mathrm{M}})$. The family $\mathrm{G}_A(\mathit{Diff}_0\mathbf{S})$ of all such vertical automorphisms constitutes a subgroup of the vertical automorphism group $\mathrm{Aut}_{\mathbf{S}}\mathbf{P}$ for $\mathbf{P} = GA^*\mathbf{S}$.

We can proceed as in Eqs. (9.9)-(9.10) to define the combined action of quantum-geometrically causal diffeomorphisms and Poincaré gauge transformations

$$(\boldsymbol{\psi},\boldsymbol{\phi}): (\boldsymbol{s},\mathbf{g}^{\mathrm{M}}) \mapsto (\boldsymbol{s}',\mathbf{g}'^{\mathrm{M}}),\ (\boldsymbol{\psi},\boldsymbol{\phi}) \in \mathcal{G}(\mathbf{S},\mathcal{g}^{\mathrm{M}}) = \mathrm{G}_A(\mathit{Diff}_0\mathbf{S}) \wedge \mathcal{G}(\mathbf{P}(\mathbf{S},\mathrm{ISO}^{\uparrow}(3,1)), \quad (9.12)$$

on any physical moving frame $(\boldsymbol{s},\mathbf{g}^{\mathrm{M}})$, given by a cross-section $\boldsymbol{s}$ of the Poincaré coframe bundle $P^*\mathbf{S}^{\uparrow}(\mathbf{g}^{\mathrm{M}})$ corresponding to some mean metric $\mathbf{g}^{\mathrm{M}} \in (\mathbf{S},\mathcal{g}^{\mathrm{M}})$. The $\boldsymbol{\phi}$-component of this transformation can be related to the supergauge transformation

$$\mathbf{U}_{\boldsymbol{s}}(\boldsymbol{\psi}(x),\boldsymbol{\phi}(x)) = U(a(x),R(x)) \otimes \mathbf{U}(\boldsymbol{s}(x);\overline{\boldsymbol{c}}(x),\boldsymbol{b}(x),\boldsymbol{c}(x)) \ , \quad (9.13)$$

whose first factor is determined on the basis of Eqs. (7.6) and (8.1),

$$U(a,R) = \bigoplus\nolimits_{n=0}^{\infty} U(a,R)^{\otimes n} \ , \qquad (a,R) \in \mathrm{T}^4 \wedge \mathrm{O}(3) \subset \mathrm{ISO}^{\uparrow}(3,1) \ , \quad (9.14)$$

whereas the second factor results from expressing the quantum gravitational coframe in Eq. (9.4) in the form

$$\boldsymbol{\Phi}^{\dagger}_{\bar{\mathbf{c}},\mathbf{b},\mathbf{c}} = \boldsymbol{\Psi}_{0;\boldsymbol{u}} \otimes \boldsymbol{\Psi}^{\dagger}_{0;\boldsymbol{u}} \cdot \mathbf{U}(\boldsymbol{u};\overline{\boldsymbol{c}},\boldsymbol{b},\boldsymbol{c}) \ , \quad (9.15)$$

$$\mathbf{U}(\boldsymbol{u};\overline{\boldsymbol{c}},\boldsymbol{b},\boldsymbol{c}) = \exp\left(\theta\overline{C}(\boldsymbol{u}) + \overline{\theta}C(\boldsymbol{u}) + \theta\overline{\theta}\left[B(\boldsymbol{u}) + \tfrac{1}{2}C(\boldsymbol{u}) \times \overline{C}(\boldsymbol{u})\right]\right) \ , \quad (9.16)$$

$$\overline{\theta}C(\boldsymbol{u}) = \overline{\theta}^{\alpha}C_{\alpha,\mathbf{0}}(\boldsymbol{u}), \quad \theta\overline{C}(\boldsymbol{u}) = \theta^{\alpha}\overline{C}_{\alpha}(\boldsymbol{u}) \ , \quad \theta\overline{\theta}B(\boldsymbol{u}) = \sum\nolimits_{\alpha=1}^{\infty} \theta^{\alpha}\overline{\theta}^{\alpha} B_{\alpha}(\boldsymbol{u}) \ , \quad (9.17)$$

$$\theta\overline{\theta}C(\boldsymbol{u}) \times \overline{C}(\boldsymbol{u}) = \sum\nolimits_{\alpha=1}^{\infty} \theta^{\alpha}\overline{\theta}^{\alpha}\left(4 + \{C_{\alpha,\mathbf{0}}(\boldsymbol{u}),\overline{C}_{\alpha}(\boldsymbol{u})\}\right), \quad B_{\alpha}(\boldsymbol{u}) = B(\boldsymbol{u};\mathrm{b}^{\alpha}\mathbf{c}_{\boldsymbol{u};\alpha}) \ , \quad (9.18)$$

where B denotes a bosonic gauge-fixing field, and C and $\overline{C}$ are the ghost and antighost fields whose creation operators appear in Eqs. (8.12) and (8.13), respectively. Thus, if we start from a TT-gauge fixing with respect to $(\boldsymbol{s},\mathbf{g}^{\mathrm{M}})$, which corresponds to a choice of quantum gravitational coframes whose elements are the duals of the coherent states $\boldsymbol{\Phi}_{\boldsymbol{f}}$ for $\boldsymbol{f} \in \mathbf{F}_{\boldsymbol{s}(x)}$ in the TT-gauge everywhere in $\mathbf{S}$, and take advantage of Eqs. (9.1)-(9.2) to write

$$\boldsymbol{\phi}(\boldsymbol{s}(x)\cdot(a(x),\Lambda(x))) = (0,\Lambda_v^{-1}(x))\,\boldsymbol{\phi}(\boldsymbol{s}(x)\cdot(a(x),R(x)))(0,\Lambda_v(x)) \ , \quad x \in \mathbf{S} \ , \quad (9.19)$$

we can assign to the gauge transformation in Eq. (9.12) the *quantum* gravitational gauge transformation provided by the following change of quantum gravitational coframes:

$$\boldsymbol{\Phi}^{\dagger}_{\boldsymbol{f}} \ \mapsto \ \boldsymbol{\Phi}^{\dagger}_{\boldsymbol{f}} \cdot \mathbf{U}_{\boldsymbol{s}}(\boldsymbol{\psi},\boldsymbol{\phi}) = \boldsymbol{\Phi}^{\dagger}_{\boldsymbol{f}} \cdot U(a,R) \otimes \boldsymbol{\Psi}^{\dagger}_{0;\boldsymbol{u}} \cdot \mathbf{U}(\boldsymbol{u};\overline{\boldsymbol{c}},\boldsymbol{b},\boldsymbol{c}) = \boldsymbol{\Phi}^{\dagger}_{U^{-1}(a,R)\boldsymbol{f},\bar{\mathbf{c}},\mathbf{b},\mathbf{c}} \ . \quad (9.20)$$

By carrying out the resulting the transformations for all physical moving frames $\boldsymbol{s}$ in $P^*\mathbf{S}^{\uparrow}(\mathcal{g}^{\mathrm{M}})$, all diffeomorphisms $\psi \in \mathit{Diff}_0\mathbf{S}$, and all Poincaré gauge transformations $\boldsymbol{\phi} \in \mathcal{G}(\mathbf{P}(\mathbf{S},\mathrm{ISO}^{\uparrow}(3,1))$, we arrive at the action of a quantum gravitational supergroup $Q\mathcal{G}(\mathbf{S},\mathcal{g}^{\mathrm{M}})$ upon all quantum gravitational coframes over $\mathbf{S}$.

10. The Quantum Gravitational Connection

To be able to take into account during the formulation of quantum geometrodynamic evolution the degrees of freedom contained in the Grassmann variables in Eqs. (9.15)-(9.18), we associate with each base-segment $\mathbf{S}$ the supermanifold

$$\boldsymbol{S} = \left\{(x,\theta,\overline{\theta}) \,\middle|\, x \in \mathbf{S}, \ \theta = (\theta^1,\theta^2,\ldots), \ \overline{\theta} = (\overline{\theta}^1,\overline{\theta}^2,\ldots)\right\} \ . \quad (10.1)$$

This base supermanifold contains the *quantum* spacetime base-segment

$$\boldsymbol{S}(\mathcal{g}^{\mathrm{M}}) = \left\{(x,\theta,\overline{\theta}) \,\middle|\, x \in \mathbf{S}, \ \theta = (\theta^1_{\boldsymbol{s}(x)},\ldots), \ \overline{\theta} = (\overline{\theta}^1_{\boldsymbol{s}(x)},\ldots), \ \boldsymbol{s} \in \Sigma(P^*\mathbf{S}^{\uparrow}(\mathcal{g}^{\mathrm{M}}))\right\} \ , \quad (10.2)$$

obtained as $\boldsymbol{s}$ ranges over the family $\Sigma(P^*\mathbf{S}^{\uparrow}(\mathcal{g}^{\mathrm{M}}))$ of all cross-sections of the metrized coframe bundle $P^*\mathbf{S}^{\uparrow}(g^{\mathrm{M}})$. Consequently, in this supermanifold we can introduce the quantum gravitational connection

$$\boldsymbol{\nabla} = \mathbf{d} + \boldsymbol{\Gamma} + \mathbf{C} + \overline{\mathbf{C}} \,, \qquad \mathbf{d} = \boldsymbol{d} + \boldsymbol{\delta} + \overline{\boldsymbol{\delta}} \,, \tag{10.3}$$

which incorporates the BRST and anti-BRST operators $\boldsymbol{\delta}$ and $\overline{\boldsymbol{\delta}}$,

$$\boldsymbol{d} = \boldsymbol{\theta}^j \partial_j = \boldsymbol{dx}^\mu \, \partial/\partial x^\mu \,, \quad \boldsymbol{\delta} = \boldsymbol{d}\overline{\theta}^\alpha \, \partial/\partial\overline{\theta}^\alpha \,, \quad \overline{\boldsymbol{\delta}} = \boldsymbol{d}\theta^\alpha \, \partial/\partial\theta^\alpha \,, \tag{10.4}$$

of the general type that appears in quantum theories for nonabelian fields.[32]

The values of the remaining terms in Eq. (10.3) are obtained by first working in the TT-gauge, and with quantum gravitational frames that do not as yet contain the Grassmann variables in Eq. (10.2), so that by an adaptation[10] of the techniques of Sec. 4 we can set

$$\boldsymbol{\nabla}^{\mathrm{TT}} = \boldsymbol{d} + \boldsymbol{\Gamma}^{\mathrm{TT}} \,, \quad \boldsymbol{\Gamma}^{\mathrm{TT}} = i\tilde{\boldsymbol{\theta}}^j_{\boldsymbol{s}} \boldsymbol{P}_{j;\boldsymbol{s}} + \tfrac{i}{2} \tilde{\boldsymbol{\omega}}^{\boldsymbol{s}}_{ab} \boldsymbol{M}^{ab}_{\boldsymbol{s}} \,, \;\; j = 0,1,2,3 \,, \;\; a,b = 1,2,3 \,, \tag{10.5}$$

where the infinitesimal generators of local spacetime translations and rotations

$$\boldsymbol{P}_{j;\boldsymbol{u}} = \int :T_{jk}[\boldsymbol{g}(\boldsymbol{u};\zeta)]: \, d\hat{\sigma}^k(q) \, d\Omega(v) \,, \qquad j = 0,1,2,3 \,, \tag{10.6}$$

$$\boldsymbol{M}^{ij}_{\boldsymbol{u}} = \int :Q^i_{\boldsymbol{u}} T^{jk}[\boldsymbol{g}] - Q^j_{\boldsymbol{u}} T^{ik}[\boldsymbol{g}] + \boldsymbol{g}^\dagger \, S^{ijk}_{\boldsymbol{u}} \, \boldsymbol{g} : d\hat{\sigma}_k(q) d\Omega(v) \,, \tag{10.7}$$

are derivable from the following quantum gravitational stress-energy tensor:[9,10]

$$T_{ij}[\boldsymbol{g}] = \tfrac{1}{2} \eta_{ij} g_{kl,m} \, g^{kl,m} - g_{km,i} \, g^{km}{}_{,j} \,, \quad g_{ik,j} = \partial g_{ik} / \partial q^j \,, \tag{10.8}$$

$$\boldsymbol{g} = g_{ij} \, \boldsymbol{\theta}^i(x) \otimes \boldsymbol{\theta}^j(x) \,, \quad g_{ij} = g^{(+)}_{ij} + g^{(-)}_{ij} \,, \quad g^{ij} = \eta^{ii'} \eta^{jj'} g_{i'j'} \,. \tag{10.9}$$

The transition to generic gauges can be then achieved by taking advantage of the Maurer-Cartan form $g^{-1}\boldsymbol{d}g$ of the quantum gravitation gauge group, which makes its appearance in the connection form

$${}^{g}\boldsymbol{\omega} = \mathrm{Ad}_{g^{-1}} \boldsymbol{\omega} + g^{-1}\boldsymbol{d}g \,, \qquad g = (\boldsymbol{\psi},\boldsymbol{\phi}) \in \mathcal{G}(\mathbf{S},g^{\mathrm{M}}) \,, \tag{10.10}$$

for the quantum gravitational connection. Hence, in a generic Poincaré gauge $\boldsymbol{s} \in \Sigma(P^*\mathbf{S}^{\uparrow}(\mathbf{g}^{\mathrm{M}}))$ we have (cf. Ref. 10, Sec. 8.7):

$$\boldsymbol{\nabla}_j = \boldsymbol{\partial}_j + \boldsymbol{\Gamma}_j \,, \qquad \boldsymbol{\Gamma}_j = \mathrm{Ad}_{\mathrm{U}^{-1}(g)} \boldsymbol{\Gamma}^{\mathrm{TT};\boldsymbol{s}}_j + \mathrm{U}^{-1}(g) \lambda_j{}^\mu \partial_\mu \mathrm{U}(g) \,, \tag{10.11}$$

$$\boldsymbol{\Gamma}_j(x,\theta,\overline{\theta}) = \boldsymbol{\Gamma}^{\mathrm{TT};\boldsymbol{s}(x)}_j + \theta D_j \overline{C}(x) + \overline{\theta} D_j C(x) + \theta\overline{\theta}\left(D_j B(x) + D_j C(x) \times \overline{C}(x)\right) , \tag{10.12}$$

$$D_j = \boldsymbol{\partial}_j + [\boldsymbol{\Gamma}^{\mathrm{TT};\boldsymbol{s}}_j , \cdot\,] \,, \qquad \boldsymbol{\partial}_j := \partial_{e_j} = \lambda_j{}^\mu \partial_\mu \,. \tag{10.13}$$

The infinitesimal parallel transport in directions related to the Grassmannian degrees of freedom labeled by θ and $\overline{\theta}$ is governed by the following quantum gravitational BRST and anti-BRST connection forms,

$$s = \boldsymbol{\delta} + U^{-1}(g) \, \partial_{\overline{\theta}} U(g) \, \boldsymbol{d}\overline{\theta} \,, \qquad \partial_{\overline{\theta}} = \partial/\partial\overline{\theta} \,, \tag{10.14}$$

$$\overline{s} = \overline{\boldsymbol{\delta}} + U^{-1}(g) \, \partial_\theta U(g) \, \boldsymbol{d}\theta \,, \qquad \partial_\theta = \partial/\partial\theta \,, \tag{10.15}$$

derivable from Eq. (10.10). This yields the gauge potentials

$$C_\alpha(x,\theta,\overline{\theta}) = C_\alpha(x) + \theta^\alpha \overline{B}_\alpha(x) - \tfrac{1}{2}\overline{\theta}^\alpha \{C_\alpha(x),C_\alpha(x)\} + \theta^\alpha\overline{\theta}^\alpha [\overline{B}_\alpha(x),C_\alpha(x)] \,, \tag{10.16}$$

$$\overline{C}_\alpha(x,\theta,\overline{\theta}) = \overline{C}_\alpha(x) + \overline{\theta}^\alpha B_\alpha(x) - \tfrac{1}{2}\theta^\alpha \{\overline{C}_\alpha(x),\overline{C}_\alpha(x)\} - \theta^\alpha\overline{\theta}^\alpha [B_\alpha(x),\overline{C}_\alpha(x)] \,, \tag{10.17}$$

in which we have set, by definition,

$$\overline{B}(x) = -B(x) - C(x) \times \overline{C}(x) \, . \tag{10.18}$$

The quantum gravitational connection described by Eqs. (10.3)-(10.18) governs the parallel transport of quantum gravitational superframes that underlies the quantum geometrodynamic evolution of the gravitational field. For this parallel transport to be *physical* within a base supermanifold $\boldsymbol{S}$ containing Grassmannian degrees of freedom related to unphysical modes of graviton polarization it has to take place in $\boldsymbol{S}$ only along curves whose tangent vectors

$$\boldsymbol{Y} = \dot{x}^{\mu}\partial_{\mu} + \dot{\overline{\theta}}\partial_{\overline{\theta}} + \dot{\theta}\partial_{\theta} \in T_X \boldsymbol{S}(\mathcal{g}^{\mathrm{M}}) \, , \qquad X = (x, \theta, \overline{\theta}) \, , \tag{10.19}$$

satisfy the following subsidiary conditions:

$$s_{\boldsymbol{Y}(t)}\, \boldsymbol{\Phi}^{\boldsymbol{s}}_{f(t)} = \left(\overline{\partial}_{\alpha} + C_{\alpha}(X(t))\right)\dot{\overline{\theta}}^{\alpha}(t)\, \boldsymbol{\Phi}^{\boldsymbol{s}}_{f(t)} = \boldsymbol{0} \, , \tag{10.20}$$

$$\overline{s}_{\boldsymbol{Y}(t)}\, \boldsymbol{\Phi}^{\boldsymbol{s}}_{f(t)} = \left(\partial_{\alpha} + \overline{C}_{\alpha}(X(t))\right)\dot{\theta}^{\alpha}(t)\, \boldsymbol{\Phi}^{\boldsymbol{s}}_{f(t)} = \boldsymbol{0} \, . \tag{10.21}$$

This is a geometric version of the Kugo-Ojima[34] type of subsidiary conditions that is encountered also in the conventional treatment of quantum nonabelian gauge fields.[32] In the present quantum-geometric context it has the following physical meaning: for any fixed choice of gauge for the diffeomorphism group, reflected in a choice of mean metric $\mathbf{g}^{\mathrm{M}}$ from the equivalence class $\mathcal{g}^{\mathrm{M}}$ of diffeomorphically equivalent mean metrics, and for any fixed Poincaré gauge, corresponding to some choice $\mathbf{s}$ of cross-section of the Poincaré coframe bundle $P^{*}\mathbf{S}^{\uparrow}(\mathbf{g}^{\mathrm{M}})$, all *physical* parallel transport underlying quantum-geometric propagation in quantum spacetime has to take place along paths which are horizontal lifts of paths in base-segments $\mathbf{S}$, so that the causality prevailing in each $\mathbf{S}$ is thereby preserved.

11. Quantum Geometrodynamic Evolution

In order to formulate the quantum geometrodynamic evolution of a superlocal quantum gravitational field in mutual interaction with all "matter" quantum fields in existence, we modify the pattern followed in Sec. 5 to suit the present much more general situation. Thus, we first introduce the gravitational exciton-mode fibres

$$\hat{\mathbf{Z}}^{\mathrm{TT}}_{\boldsymbol{u}} = \left\{ \hat{\boldsymbol{f}} \;\middle|\; \boldsymbol{f} \in \mathbf{Z}^{\mathrm{TT}}_{\boldsymbol{u}} \right\} \, , \quad \hat{\boldsymbol{f}}(v) := \boldsymbol{f}(\boldsymbol{u}; -a(x) - i\ell v) \, , \qquad v \in \boldsymbol{V}^{+}, \tag{11.1}$$

whose indefinite inner product is formally given by

$$\left\langle \hat{\boldsymbol{f}} \middle| \hat{\boldsymbol{f}}' \right\rangle = Z_{\ell,0} \int_{\boldsymbol{V}^{+}} \eta^{ii'} \eta^{jj'} \hat{f}^{*}_{ij}(v)\, \hat{f}'_{i'j'}(v)\, d\Omega(v) = \langle \boldsymbol{f} | \boldsymbol{f}' \rangle \, , \tag{11.2}$$

where the renormalization constant $Z_{\ell,0}$ is infinite – so that the above inner product is *de facto* determined by a zero-mass limit.[9,10] The elements of the ensuing quantum gravitational field-mode fibres,

$$\hat{\mathcal{Z}}^{\mathrm{TT}}_{\boldsymbol{u}} = \bigoplus_{n=0}^{\infty} \hat{\mathcal{Z}}^{\mathrm{TT}}_{\boldsymbol{u};n} \quad , \qquad \hat{\mathcal{Z}}^{\mathrm{TT}}_{\boldsymbol{u};n} = \hat{\mathbf{Z}}^{\mathrm{TT}}_{\boldsymbol{u}} \underset{S}{\otimes} \cdots \underset{S}{\otimes} \hat{\mathbf{Z}}^{\mathrm{TT}}_{\boldsymbol{u}} \, , \tag{11.3}$$

can be identified, as in Eq. (5.7), with those of the corresponding TT-subfibres of $\mathcal{Z}_{\boldsymbol{u}}$,:

$$\hat{\Psi}_n(v_1, \ldots, v_n) = \Psi_n(\boldsymbol{u}; -a(x) - i\ell v_1, \ldots, -a(x) - i\ell v_n) \, , \quad \Psi_n \in \hat{\mathcal{Z}}^{\mathrm{TT}}_{\boldsymbol{u};n} \, . \tag{11.4}$$

An analogous procedure can be applied to ghost and antighost field-modes, thus ultimately arriving at the quantum gravitational field-mode superfibres

$$\hat{\mathcal{Z}}^{\lozenge}_{s(x)} = \hat{\mathcal{Z}}^{\mathrm{TT}}_{s(x)} \otimes \hat{\mathcal{B}}_{s(x)} \otimes \hat{\bar{\mathcal{B}}}_{s(x)} . \tag{11.5}$$

Within these superfibres we can implement the continuous expansions

$$\left\langle \hat{\Psi} \middle| \hat{\Psi}' \right\rangle = \int_{\mathbf{Z}^{\mathrm{TT}}_{s(x)}} df\, d\bar{f} \int d\bar{\theta}\, d\theta \left\langle \hat{\Phi}^{s}_{f}(x,\theta,\bar{\theta}) \middle| \hat{\Psi} \otimes \hat{\Psi}' \right\rangle , \quad \hat{\Psi}, \hat{\Psi}' \in \hat{\mathcal{Z}}^{\mathrm{TT}}_{s(x)} \otimes \hat{\mathcal{F}}_{s(x)} , \tag{11.6}$$

by the methods of Berezin integration.

For any given choice of a physical moving frame $\boldsymbol{s} = \{(\boldsymbol{a}^*(x), \boldsymbol{\theta}^i(x)) |\, x \in \mathbf{S}\}$, we define a *quantum gravitational superfield* at each base superlocation X as follows:

$$\mathbf{g}^{\mathrm{Q}}(X,v) = \mathrm{g}^{\mathrm{Q}}_{ij}(X,v)\, \boldsymbol{\theta}^{i}(x) \otimes \boldsymbol{\theta}^{j}(x) , \qquad X = (x,\theta,\bar{\theta}) \in \boldsymbol{S}(g^{\mathrm{M}}) , \tag{11.7}$$

$$\mathrm{g}^{\mathrm{Q}}_{ij}(X,v) = \int_{\mathbf{Z}^{\mathrm{TT}}_{s(x)}} df\, d\bar{f} \int_{\mathbf{Z}^{\mathrm{TT}}_{s(x)}} df' d\bar{f}' \left| \hat{\Phi}^{s}_{f}(X) \right\rangle \mathrm{g}^{\mathrm{Q}}_{ijff'}(x,v) \left\langle \hat{\Phi}^{s}_{f'}(X) \right| , \tag{11.8}$$

$$\mathrm{g}^{\mathrm{Q}}_{ijff'}(x,v) = \left\langle \hat{\Phi}_{f} \middle| \left(\eta_{ij} + g_{ij}(\boldsymbol{s}(x); -i\ell v) \right) \hat{\Phi}_{f'} \right\rangle , \quad \hat{\Phi}_{f}, \hat{\Phi}_{f'} \in \hat{\mathcal{Z}}^{\mathrm{TT}}_{s(x)} . \tag{11.9}$$

We can then easily establish that the quantum gravitational fluctuations around the mean quantum gravitational field values

$$\mathbf{g}^{\mathrm{M}}(x) = \eta_{ij}\, \boldsymbol{\theta}^{i}(x) \otimes \boldsymbol{\theta}^{j}(x) , \qquad x \in \mathbf{S} , \tag{11.10}$$

are provided by the quantum gravitational *radiation* superfield with components

$$\mathrm{g}_{ij}(X,v) = \mathbf{U}^{\dagger}_{s}(X)\, \mathrm{g}_{ij}(x,v)\, \mathbf{U}_{s}(X) , \qquad X = (x,\theta,\bar{\theta}) \in \boldsymbol{S}(g^{\mathrm{M}}) , \tag{11.11}$$

$$\mathrm{g}_{ij}(x,v) = \int_{\mathbf{Z}^{\mathrm{TT}}_{s(x)}} df\, d\bar{f} \int_{\mathbf{Z}^{\mathrm{TT}}_{s(x)}} df' d\bar{f}' \left| \hat{\Phi}_{f} \right\rangle \left(\hat{f}^{*}_{ij}(v) + \hat{f}'_{ij}(v) \right) \left\langle \hat{\Phi}_{f} \middle| \hat{\Phi}_{f'} \right\rangle \left\langle \hat{\Phi}_{f'} \right| . \tag{11.12}$$

To describe the mutual interaction of "matter" fields with the quantum gravitational superfield in Eq. (11.7), these fields have to be defined at all superlocations $X \in \boldsymbol{S}(g^{\mathrm{M}})$. This can be achieved by "lifting," for example, the coherent field-mode sections in Eq. (5.10) from the base-locations x to the base-superlocations X ,

$$\hat{\Phi}^{s}_{\varphi(x)} \mapsto \hat{\Phi}^{s}_{\varphi(X)} \in \hat{\mathcal{F}}^{\mathrm{mat}}_{X} , \quad \varphi(x) \in \mathbf{F}^{\mathrm{mat}}_{x} , \quad \varphi(X) = \mathbf{U}^{-1}_{s}(X)\, \varphi(x) \in \mathbf{F}^{\mathrm{mat}}_{X} , \tag{11.13}$$

when executing a quantum gravitational gauge transition involving Lorentz boosts. More general situations can be handled along similar lines, thus arriving at well-defined proper energy-momentum operators for "matter" fields,

$$\hat{\boldsymbol{P}}^{\mathrm{mat}}_{j;s}(X) = n^{k}(x) \int_{\mathbf{V}^{+}_{x}} :T^{\mathrm{mat}}_{jk}[\boldsymbol{\varphi}^{\mathrm{mat}}(X,\boldsymbol{v})]:\, d\Omega(\boldsymbol{v}) , \qquad j = 0,1,2,3 . \tag{11.14}$$

These operators are obtained from the stress-energy tensors in the semi-classical regime by replacing in the latter Poincaré-covariant derivatives with corresponding quantum gravitationally covariant derivatives, such as

$$\begin{aligned} \boldsymbol{\nabla}^{\mathrm{mat}}_{j} &= \boldsymbol{\partial}_{j} - i\, \tilde{\boldsymbol{\theta}}^{j}_{s} \boldsymbol{P}^{\mathrm{mat}}_{j;s} + \tfrac{i}{2} \tilde{\boldsymbol{\omega}}^{s}_{ab} \boldsymbol{M}^{ab}_{\mathrm{mat};s} \\ &+ \theta D_{j} \bar{C}(x) + \bar{\theta} D_{j} C(x) + \theta\bar{\theta} \left(D_{j} B(x) + D_{j} C(x) \times \bar{C}(x) \right) . \end{aligned} \tag{11.15}$$

The resulting expressions contribute to the real-valued part of the total mean (local) proper energy density and total mean (local) proper 3-momentum density:

$$\bar{\boldsymbol{\rho}}^{\mathrm{tot}}(x) = \left\langle \boldsymbol{\Psi}^{\mathrm{tot}}_{X} \middle| \boldsymbol{\rho}^{\mathrm{tot}}(X)\, \boldsymbol{\Psi}^{\mathrm{tot}}_{X} \right\rangle^{\mathrm{body}} , \qquad \bar{\boldsymbol{j}}^{\mathrm{tot}}_{a}(x) = \left\langle \boldsymbol{\Psi}^{\mathrm{tot}}_{X} \middle| \boldsymbol{j}^{\mathrm{tot}}_{a}(X)\, \boldsymbol{\Psi}^{\mathrm{tot}}_{X} \right\rangle^{\mathrm{body}} . \tag{11.16}$$

In turn, the above values of these densities along each given inflow hypersurface of a base-segment, together with those of the mean Riemannian 3-metric tensor $\boldsymbol{\gamma}^{\mathrm{M}}$ induced by $\mathbf{g}^{\mathrm{M}}$ and of the mean extrinsic curvature tensor $\boldsymbol{K}$, dictate the creation of the infinitesimal QGR gauge orbits $(\mathbf{S}, g^{\mathrm{M}})$. Mathematically, each gauge orbit is obtained by solving[24] the CGR Cauchy problem, with the four Einstein equations

$$R(\boldsymbol{\gamma}^{\mathrm{M}}) - K_{ab}K^{ab} + \left(K^{a}{}_{a}\right)^{2} = 16\pi\bar{\rho}^{\mathrm{tot}}, \quad K^{a}{}_{b} = \gamma^{\mathrm{M}}_{bc}K^{ac}, \tag{11.17}$$

$$\mathrm{D}_{b}K^{b}_{a} - \mathrm{D}_{a}K^{b}{}_{b} = 8\pi\bar{j}^{\mathrm{tot}}_{a}, \quad a = 1,2,3, \tag{11.18}$$

used as constraints on $\boldsymbol{K}$ for given $\boldsymbol{\gamma}^{\mathrm{M}}$, where D_a denotes the $\boldsymbol{e}_a$-component (with respect to the $\boldsymbol{\gamma}^{\mathrm{M}}$-orthonormal spatial frames $\{\boldsymbol{e}_a(x) | \, a = 1,2,3\}$ dual to $\{\boldsymbol{\theta}^{a}(x) | \, a = 1,2,3\}$) of the operator for the covariant differentiation determined by the Levi-Civita connection to which each 3-metric $\boldsymbol{\gamma}^{\mathrm{M}}$ gives rise on the inflow hypersurface of $\mathbf{S}$.

Thus, the framework for geometric quantum gravity postulates no background metrics, and, in fact, it does not assume that a classical spacetime manifold of any kind is *a priori* given. However, it does take for granted that, in accordance with observational data, mean metrics constituting QGR gauge orbits perpetually emerge during the process of quantum-geometric evolution as a result of the metrization of classical frame and coframe bundles.

The self-interactions of the quantum gravitational radiation field in Eq. (11.11) are taken to contribute, following a study by Deser,[35] to the total operator-valued densities in Eq. (11.16) via the self-interaction stress-energy tensor

$$T^{\mathrm{s.int}}_{ij}[\mathbf{g}] = (1/16\pi)\left(\hat{\Gamma}^{l}{}_{ij}\hat{\Gamma}^{k}{}_{kl} - \hat{\Gamma}^{l}{}_{ki}\hat{\Gamma}^{k}{}_{jl}\right), \quad \hat{\Gamma}^{i}{}_{jk} = \tfrac{1}{2}\eta^{il}\left(g_{lj,k} + g_{lk,j} - g_{jk,l}\right). \tag{11.19}$$

The above stress-energy tensor also contributes to the total stress-energy tensor for the energy-momentum of quantum gravitational radiation

$$\hat{\boldsymbol{P}}^{\mathrm{rad}}_{j;\boldsymbol{s}}(X) = n^{k}(x)\int_{\mathbf{V}^{+}_{x}} :T^{\mathrm{rad}}_{jk}[\mathbf{g}(X,\boldsymbol{v})]: \, d\Omega(\boldsymbol{v}), \quad j = 0,1,2,3, \tag{11.20}$$

a term which has to be added to the one in (10.8), so that:

$$T^{\mathrm{rad}}_{ij}[\mathbf{g}] = T_{ij}[\mathbf{g}] + T^{\mathrm{s.int}}_{ij}[\mathbf{g}]. \tag{11.21}$$

The mean values in Eq. (11.16) resulting from Eqs. (11.14) and (11.20) along each inflow hypersurface $\Sigma_{t_{n-1}}$ determine (cf. Ref. 24, Sec. 1.2.5), via the constraint equations (11.17)-(11.18) and the remaining six geometrodynamic Einstein equations for $\mathbf{g}^{\mathrm{M}}$, the QGR gauge orbit $(\mathbf{S}_n, g^{\mathrm{M}})$, including the outflow hypersurface Σ_{t_n} of $\mathbf{S}_n$ at the proper time separation $t_n - t_{n-1} = (t'' - t')/N = \varepsilon$. Hence, the quantum spacetime segment $\boldsymbol{S}_n(g^{\mathrm{M}})$ is thereby determined via Eqs. (10.1) and (10.2).

With all the remaining ingredients of the quantum-geometric framework for gravity now available, the formulation of the quantum-geometric propagator

$$K(f(X''); f(X')) = \lim_{\varepsilon\to+0} \prod_{n=N}^{1}{}' \int \mathcal{D}f_n \left\langle \hat{\boldsymbol{\Phi}}^{\boldsymbol{s}}_{f_n}(X_n) \right| e^{-i\varepsilon\int d\sigma(x_n)\rho^{\mathrm{tot}}(X_n)} \left. \hat{\boldsymbol{\Phi}}^{\boldsymbol{s}}(X_n; f_{n-1}) \right\rangle, \tag{11.22}$$

$$\mathcal{D}f_n = \prod_{x_n\in\Sigma_{t_n}} \mathcal{D}[g_n(x_n)]\,\mathcal{D}[\bar{c}_n(x_n)]\,\mathcal{D}[c_n(x_n)]\mathcal{D}[\varphi^{\mathrm{mat}}_n(X_n)]. \tag{11.23}$$

for gravitational and "matter" superfields, whose exciton modes are collectively denoted by f, can be obtained by suitable adaptation of the type of iterative procedures considered in Sec. 5. The Grassmannian values assumed by the ghost and antighost exciton modes above each base location $x_n \in \Sigma_{t_n}$ are those corresponding to various superlocations $X_n \in \boldsymbol{\Sigma}_{t_n}$.

Thus, the quantum geometrodynamic propagation proceeds, under the subsidiary conditions in Eqs. (10.20) and (10.21), along those paths $\gamma^{\uparrow}(X_{n-1},X_n)$ in the base-supersegments $\mathbf{S}_n(g^{\mathrm{M}})$ which are the lifts of the geodesic arcs $\gamma(x_{n-1},x_n)$ connecting each point x_{n-1} on the inflow hypersurface $\Sigma_{t_{n-1}}$ with the points x_n on the outflow hypersurface Σ_{t_n} of each base-segment $\mathbf{S}_n$.

12. Conclusion

We can now in the position to compare the framework for geometric quantum gravity, outlined in the preceding five sections, with that of the covariant and the canonical frameworks[36] for quantum gravity.

The geometric formulation shares with the covariant one the use of graviton states representing quantum gravitational radiation. On the other hand, the covariant formulation postulates a globally existing background metric – for which, on pragmatic grounds, the usual choice is that of Minkowski metric.[36] In addition to the loss of diffeomorphism as well as Poincaré gauge invariance, such a postulate precludes the *mutual* interaction between matter and geometry, which is physically essential to general relativity.

Consequently, in the geometric framework the *mean* metrics are those *created* by that mutual interaction during the quantum geometrodynamic evolution. This allows the indeterministic quantum phenomena to affect their values locally. Hence, the resulting quantum spacetime has to be viewed as being in a constant process of creation, rather than as a conceptually static object, determined in its minutest details at the instant of its creation.

As in the canonical approach to quantum gravity,[36,37] this *quantum* geometrodynamic evolution requires reference 3-surfaces. Quantum base geometries are then created by the metrization of the general affine frame bundles over those surfaces. However, whereas in the superspace approach to quantum gravity the fluctuations of 3-metrics are envisaged to take place globally over those reference 3-surfaces, in the geometric approach such metric fluctuations take place locally, i.e., within the fibres above their base spacetime locations.

These dissimilarities become more pronounced in relation to recent work on canonical quantum gravity,[29] in which the concept of connection takes precedence over that of metric; whereas in the quantum geometric framework the concept of metric retains a central role, which remains on par with that which it plays in classical general relativity.

In general, during the entire period of development[4,9,10] of the quantum geometry framework, *physical* principles were deemed primary. Hence, suitable mathematical techniques were searched for, and, if not found in existing literature, created in order to fit those physical principles – rather than vice versa. The central physical and mathematical ideas which thereby emerged are those of *quantum (super)frame bundles* and of *quantum geometric evolution*. Quantum frame bundles represent systems of covariance[9,10] that are generalizations of the systems of imprimitivity[4,6] to which complete sets of observables give rise; whereas quantum geometric evolution represents a physically grounded geometric extrapolation of the path integral method to the general relativistic regime. Thus, the entire quantum geometric framework for quantum field theory in curved spacetime and quantum gravity rests on physically well founded extrapolations of orthodox quantum theory.

References

1. E. Prugovečki, *Ann. Phys. (N.Y.)* **110**, 102 (1978).
2. E. Prugovečki, *J. Math. Phys.* **19**, 2260 (1978).
3. E. Prugovečki, *Phys. Rev. D* **18**, 3655 (1978).
4. E. Prugovečki, *Stochastic Quantum Mechanics and Quantum Spacetime* (Reidel,

Dordrecht, 1984; reprinted, with corrections, 1986).
5. M. Born, *Rev. Mod. Phys.* **21**, 463 (1949).
6. E. P. Wigner, in *Quantum Theory and Measurement*, J. A. Wheeler and W. H. Zurek, eds. (Princeton University Press, Princeton, 1983), pp. 260-314.
7. E. Prugovečki, *Class. Quantum Grav.* **4**, 1659 (1987).
8. E. Prugovečki, *Nuovo Cimento A* **97**, 597, 837 (1987).
9. E. Prugovečki, *Quantum Geometry* (Kluwer, Dordrecht, 1992).
10. E. Prugovečki, *Principles of Quantum General Relativity* (World Scientific, Singapore, 1995).
11. S. T. Ali and G. G. Emch, *J. Math. Phys.* **27**, 2936 (1986).
12. B. Kostant, *Quantization and Unitary Representations* (Springer Lecture Notes in Mathematics, vol. 170, New York, 1970).
13. J.-M. Souriau, *Structure des systèmes dynamiques* (Dunod, Paris, 1970).
14. S. T. Ali and E. Prugovečki, *Acta Appl. Math.* **6**, 1, 19, 47 (1986).
15. A. M. Perelomov, *Generalized Coherent States and Their Applications*, (Springer, Berlin, 1986).
16. F. A. Berezin, *Commun. Math. Phys.* **40**, 153 (1975).
17. E. Prugovečki, *Class. Quantum Grav.* **11**, 1981 (1994).
18. E. Prugovečki, *Quantum Mechanics in Hilbert Space*, 2nd edition (Academic Press, New York, 1981).
19. E. Prugovečki, *Nuovo Cimento A* **61**, 85 (1981).
20. B. S. DeWitt, in *Gravitation: An Introduction to Current Research*, L. Witten, ed. (Wiley, New York, 1962).
21. H.-H. von Borzeszkowski and H.-J. Treder, *The Meaning of Quantum Gravity* (Kluwer, Dordrecht, 1988).
22. W. Drechsler, *Fortschr. Phys.* **23**, 449 (1984).
23. N. N. Bogolubov, A. A. Logunov and I. T. Todorov, *Introduction to Axiomatic Quantum Field Theory* (Benjamin, Reading, Mass., 1975).
24. Y. Choquet-Bruhat and J. W. York, in *General Relativity and Gravitation*, vol. 1, A. Held, ed. (Plenum, New York, 1980).
25. F. A Berezin, *The Method of Second Quantization* (Academic Press, New York, 1966).
26. N. D. Birrell and P. C. W. Davies, *Quantum Fields in Curved Space* (Cambridge University Press, Cambridge, 1986).
27. R. P. Feynman, *Rev. Mod. Phys.* **20**, 367 (1948).
28. R. P. Feynman, *Phys. Rev.* **76**, 749, 769 (1949).
29. A. Ashtekar and J. Stachel, eds., *Conceptual Problems in Quantum Gravity* (Birkhäuser, Boston, 1991).
30. L. de Broglie, in *Perspectives in Quantum Theory*, W. Yougrau and A. van der Merwe, eds. (Dover, New York, 1979).
31. J. Bognár, *Indefinite Inner Product Spaces* (Springer, Berlin, 1974).
32. L. Baulieu, *Phys. Reports* **129**, 1 (1985).
33. E. Cartan, *La Méthode du Repère Mobile, la Théorie des Groupes Continus et les Espaces Généralisées* (Act. Sci. et Ind., #194, Hermann, 1935).
34. T. Kugo and I. Ojima, *Prog. Theor. Phys. Suppl.* **66**, 1 (1979).
35. S. Deser, *Gen. Rel. Grav.* **1**, 9 (1970).
36. C. J. Isham, in *Quantum Gravity: An Oxford Symposium*, C.J. Isham, R. Penrose and D.W. Sciama, eds. (Clarendon Press, Oxford, 1975).
37. P. G. Bergmann and A. Komar, in *General Relativity and Gravitation*, vol. 1, A. Held, ed. (Plenum, New York, 1980).
38. E. Prugovečki, "On quantum-geometric connections and propagators in curved spacetime" (University of Toronto preprint – to be published).

Strings and Multi-Strings in Black Hole and Cosmological Spacetimes*

A.L. Larsen and N. Sánchez

Observatoire de Paris, DEMIRM. Laboratoire Associé au CNRS UA 336, Observatoire de Paris et École Normale Supérieure. 61, Avenue de l'Observatoire, 75014 Paris, France.

I wish to dedicate this article to Professor Peter Bergmann, with deep and highest appreciation, and with All Good and Best Greetings to him and to his wife Margot.

Norma Sánchez

Paris/Erice, August 1995

Abstract

Recent results on classical and quantum strings in a variety of black hole and cosmological spacetimes, in various dimensions, are presented. The curved backgrounds under consideration include the 2+1 black hole anti de Sitter spacetime and its dual, the black string, the ordinary $D \geq 4$ black holes with or without a cosmological constant, the de Sitter and anti de Sitter spacetimes and static Robertson-Walker spacetimes. Exact solutions to the string equations of motion and constraints, representing circular strings, stationary open strings and dynamical straight strings, are obtained in these backgrounds and their physical properties (length, energy, pressure) are described. The existence of *multi-string* solutions, describing finitely or infinitely many strings, is shown to be a general feature of spacetimes with a positive or negative cosmological constant. Generic approximative solutions are obtained using the string perturbation series approach, and the question of the stability of the solutions is addressed.

Furthermore, using a canonical quantization procedure, we find the string mass spectrum in de Sitter and anti de Sitter spacetimes. New features as compared to the string spectrum in flat Minkowski spacetime appear, for instance the *fine-structure effect* at all levels beyond the graviton in both de Sitter and anti de Sitter spacetimes, and the *non-existence* of a Hagedorn temperature in anti de Sitter spacetime. We discuss the physical implications of these results. Finally, we consider the effect of spatial curvature on the string dynamics in Robertson-Walker spacetimes.

1 Introduction

The classical and quantum propagation of strings in curved spacetimes has attracted a great deal of interest in recent years. The main complication, as compared to the case of flat Minkowski spacetime, is related to the non-linearity of the equations of motion. It makes it possible to obtain the complete analytic solution only in a very few special cases like conical spacetime [1] and plane-wave/shock-wave backgrounds [2]. There are however also very general results concerning integrability and solvability for maximally symmetric spacetimes [3, 4] and gauged WZW models [5, 6]. These are the exceptional cases; generally the string equations of motion in curved spacetimes are not integrable and even if they are, it is usually an extremely difficult task to actually separate the equations, integrate them and finally write down the complete solution in closed form. Fortunately, there are several different ways to "attack" a system of coupled non-linear partial differential equations.

The systematic study of string dynamics in curved spacetimes and its associated physical phenomena was started in Refs.[7, 8]. Besides numerical methods, which will not be discussed here, approximative [7-10] and exact [11-14] methods for solving the string equations of motion and constraints in curved spacetimes, have been developed. Classical and quantum string dynamics have been investigated in black hole backgrounds [15-18], cosmological spacetimes [7, 11-14, 17-21], cosmic string spacetime [1], gravitational wave backgrounds [2], supergravity backgrounds (which are necessary for fermionic strings) [22], and near spacetime singularities [23]. Physical phenomena like the Hawking-Unruh effect in string theory [8, 24], horizon string stretching [8, 24], particle transmutation [15, 25], string scattering [1, 2, 15], mass spectrum and critical dimension [1, 7, 15, 27], string instability [7, 11-13, 16, 21] and multi-string solutions [11-13, 28] have been found.

In a generic D-dimensional curved spacetime with metric $g_{\mu\nu}$ and coordinates x^μ, $(\mu = 0, 1, ..., D-1)$, the string equations of motion and constraints are:

$$\ddot{x}^\mu - x''^\mu + \Gamma^\mu_{\rho\sigma}(\dot{x}^\rho \dot{x}^\sigma - x'^\rho x'^\sigma) = 0, \tag{1.1}$$

$$g_{\mu\nu}\dot{x}^\mu x'^\nu = g_{\mu\nu}(\dot{x}^\mu \dot{x}^\nu + x'^\mu x'^\nu) = 0, \tag{1.2}$$

where dot and prime stand for derivative with respect to the world-sheet coordinates τ and σ, respectively and $\Gamma^\mu_{\rho\sigma}$ are the Christoffel symbols with respect to the metric $g_{\mu\nu}$. In the following we present recent results [26-29] on solutions to Eqs.(1.1)-(1.2) in a variety of curved spacetimes from cosmology, gravitation and string theory. We obtain explicit (exact and/or approximate) mathematical solutions, discuss the corresponding physical properties, we quantize the solutions in different ways (canonical quantization, semi-classical quantization) and find the physical content: the mass spectrum.

The presentation is organized as follows: In Section 2, we discuss the classical string dynamics in the 2 + 1 black hole anti de Sitter (BH-AdS) spacetime, recently found by Bañados et al [30]. We compare with the string dynamics in ordinary cosmological and black hole spacetimes. This clarifies the geometry (as seen by a string) of the BH-AdS spacetime. In Section 3, generalizing results from Section 2, we derive the quantum string mass spectrum in ordinary D-dimensional anti de Sitter spacetime. We discuss, in particular, the sectors of low and very high mass states. New physical phenomena arise like the fine-structure effect at all levels beyond the graviton and the non-existence of a Hagedorn critical temperature. The results are compared with corresponding results obtained in Minkowski and de Sitter spacetimes. Sections 4 and 5 are devoted to the investigation of the more general underlying structure of solutions to Eqs.(1.1)-(1.2). In Section 4, we find new classes of exact string and multi-string solutions in cosmological and black hole spacetimes, while in Section 5, we consider the effects of a non-zero spatial curvature on the classical and quantum string dynamics in Friedmann-Robertson-Walker (FRW) universes.

2 Classical String Dynamics in 2+1 BH-AdS

Anti de Sitter (AdS) spacetime is often considered to be a spacetime of minor importance in a cosmological context. However, first of all, it *is* a FRW universe and as such should not be neglected, secondly; it serves as a simple and convenient spacetime for comparison and understanding of results obtained in (say) Minkowski or de Sitter spacetimes and third, it has a tendency to show up (in disguise) as a solution in various models of dilaton-gravity and string theory. An example of the latter is represented by the $2+1$ BH-AdS spacetime of Bañados et al [30]. This spacetime background has arised much interest recently [31-35]. It describes a two-parameter family (mass M and angular momentum J) of black holes in 2+1 dimensional general relativity with metric:

$$ds^2 = (M - \frac{r^2}{l^2})dt^2 + (\frac{r^2}{l^2} - M + \frac{J^2}{4r^2})^{-1}dr^2 - Jdtd\phi + r^2d\phi^2. \tag{2.1}$$

It has two horizons $r_\pm = \sqrt{\frac{Ml^2}{2} \pm \frac{l}{2}\sqrt{M^2l^2 - J^2}}$ and a static limit $r_{\text{erg}} = \sqrt{M}l$, defining an ergosphere, as for ordinary Kerr black holes. The spacetime is not asymptotically flat; it approaches anti de Sitter spacetime asymptotically with cosmological constant $\Lambda = -1/l^2$. The curvature is constant $R_{\mu\nu} = -(2/l^2)g_{\mu\nu}$ everywhere, except probably at $r = 0$, where it has at most a delta-function singularity. Notice the weak nature of the singularity at $r = 0$ in 2+1 dimensions as compared with the power law divergence of curvature scalars in $D > 3$ (We will not discuss here the geometry near $r = 0$. For a discusion, see Refs.[33, 35]). The spacetime, Eq.(2.1), is also a solution of the low energy effective action of string theory with zero dilaton field $\Phi = 0$, anti-symmetric tensor field $H_{\mu\nu\rho} = (2/l^2)\epsilon_{\mu\nu\rho}$ (i.e. $B_{\phi t} = r^2/l^2$) and $k = l^2$ [31]. Moreover, it yields an exact solution of string theory in 2+1 dimensions, obtained by gauging the WZWN sigma model of the group

$SL(2,R)\times R$ at level k [31, 32] (for non-compact groups, k does not need to be an integer, so the central charge $c = 3k/(k-2) = 26$ when k=52/23). This solution is the black string background [36]:

$$d\tilde{s}^2 = -(1-\tfrac{\mathcal{M}}{\tilde{r}})d\tilde{t}^2 + (1-\tfrac{Q^2}{\mathcal{M}\tilde{r}})d\tilde{x}^2 + (1-\tfrac{\mathcal{M}}{\tilde{r}})^{-1}(1-\tfrac{Q^2}{\mathcal{M}\tilde{r}})^{-1}\tfrac{l^2 d\tilde{r}^2}{4\tilde{r}^2},$$

$$\tilde{B}_{\tilde{x}\tilde{t}} = \tfrac{Q}{r}, \quad \tilde{\Phi} = -\tfrac{1}{2}\log\tilde{r}l, \tag{2.2}$$

which is related by duality [31, 37] to the 2+1 BH-AdS spacetime, Eq.(2.1). It has two horizons $\tilde{r}_{\pm} = r_{\pm}$, the same as the metric, Eq.(2.1), while the static limit is $\tilde{r}_{\text{erg}} = J/(2\sqrt{M})$.

We first investigate the string propagation in the $2+1$ BH-AdS background by considering the perturbation series around the exact center of mass of the string:

$$x^\mu(\tau,\sigma) = q^\mu(\tau) + \eta^\mu(\tau,\sigma) + \xi^\mu(\tau,\sigma) + ... \tag{2.3}$$

The original method of Refs.[7, 8] can be conveniently formulated in covariant form. It is useful to introduce $D-1$ normal vectors n^μ_R $(R = 1,..,D-1)$, (which can be chosen to be covariantly constant by gauge fixing), and consider comoving perturbations δx_R, i.e. those seen by an observer travelling with the center of mass, thus $\eta^\mu = \delta x^R n^\mu_R$. After Fourier expansion, $\delta x^R(\tau,\sigma) = \sum_n C^R_n(\tau)e^{-in\sigma}$, the first order perturbations satisfy the matrix Schrödinger-type equation in τ:

$$\ddot{C}_{nR} + (n^2\delta_{RS} - R_{\mu\rho\sigma\nu}n^\mu_R n^\nu_S \dot{q}^\rho\dot{q}^\sigma)C^S_n = 0. \tag{2.4}$$

Second order perturbations ξ^μ and constraints are similarly covariantly treated, ξ^μ also satisfying Schrödinger-type equations with source terms, see Ref.[26].

For our purposes here it is enough to consider the non-rotating $(J = 0)$ 2+1 BH-AdS background and a radially infalling string. We have solved completely the c.m. motion $q^\mu(\tau)$ and the first and second order perturbations $\eta^\mu(\tau,\sigma)$ and $\xi^\mu(\tau,\sigma)$ in this background. Eqs.(2.4) become:

$$\ddot{C}_{nR} + (n^2 + \frac{m^2}{l^2})C_{nR} = 0; \quad R = \perp, \parallel \tag{2.5}$$

The first order perturbations are independent of the black hole mass, only the anti de Sitter (AdS) part emerges. All oscillation frequencies $\omega_n = \sqrt{n^2 + m^2/l^2}$ are real; there are no unstable modes in this case. The perturbations:

$$\delta x_R(\tau,\sigma) = \sum_n [A_{nR}e^{-i(n\sigma+\omega_n\tau)} + \tilde{A}_{nR}e^{-i(n\sigma-\omega_n\tau)}] \tag{2.6}$$

are completely finite and regular. This is also true for the second order perturbations which are bounded everywhere, even for $r \to 0$ $(\tau \to 0)$.

We have also computed the conformal generators L_n, (see Ref.[26]), and the string mass:

$$m^2 = 2\sum_n (2n^2 + \frac{m^2}{l^2})[A_{n\|}\tilde{A}_{-n\|} + A_{n\perp}\tilde{A}_{-n\perp}]. \tag{2.7}$$

The mass formula is modified (by the term m^2/l^2) with respect to the usual flat spacetime expression. This is due to the asymptotic (here AdS) behaviour of the spacetime. In ordinary $D \geq 4$ black hole spacetimes (without cosmological constant), which are asymptotically flat, the mass spectrum is the same as in flat Minkowski spacetime [15].

We compare with the string perturbations in the ordinary ($D \geq 4$) black hole anti de Sitter spacetime. In this case Eqs.(2.4) become:

$$\ddot{C}_{nS\perp} + (n^2 + m^2H^2 + \frac{Mm^2}{r^3})C_{nS\perp} = 0, \quad S = 1, 2 \tag{2.8}$$

$$\ddot{C}_{n\|} + (n^2 + m^2H^2 - \frac{2Mm^2}{r^3})C_{n\|} = 0. \tag{2.9}$$

The transverse $\perp$-perturbations are oscillating with real frequencies and are bounded even for $r \to 0$. For longitudinal $\|$-perturbations, however, imaginary frequencies arise and instabilities develop. The ($\mid n \mid = 1$)-instability sets in at:

$$r_{\text{inst.}} = (\frac{2Mm^2}{1 + m^2H^2})^{1/3} \tag{2.10}$$

Lower modes become unstable even outside the horizon, while higher modes develop instabilities at smaller r and eventually only for $r \approx 0$. For $r \to 0$ (which implies $\tau \to \tau_0$) we find $r(\tau) \approx (3m\sqrt{M/2})^{2/3}(\tau_0 - \tau)^{2/3}$ and:

$$\ddot{C}_{nS\perp} + \frac{2}{9(\tau - \tau_0)^2}C_{nS\perp} = 0, \quad S = 1, 2 \tag{2.11}$$

$$\ddot{C}_{n\|} - \frac{4}{9(\tau - \tau_0)^2}C_{n\|} = 0. \tag{2.12}$$

For $\tau \to \tau_0$ the $\|$-perturbations blow up while the string ends trapped into the $r = 0$ singularity. We see the important difference between the string evolution in the 2+1 BH-AdS background and the ordinary 3+1 (or higher dimensional) black hole anti de Sitter spacetime.

We also compare with the string propagation in the 2+1 black string background, Eq.(2.2) (with $J = 0$). In this case, Eqs.(2.4) become:

$$\ddot{C}_{n\perp} + n^2C_{n\perp} = 0, \tag{2.13}$$

$$\ddot{C}_{n\|} + (n^2 - \frac{2m^2M}{lr})C_{n\|} = 0. \tag{2.14}$$

The $\perp$-modes are stable, while $C_{n\|}$ develop instabilities. For $r \to 0$ (which implies $\tau \to \tau_0$) we find $r(\tau) \approx \frac{m^2 M}{l}(\tau_0 - \tau)^2$ and:

$$\ddot{C}_{n\|} - \frac{2}{(\tau_0 - \tau)^2} C_{n\|} = 0, \tag{2.15}$$

with similar conclusions as for the ordinary 3+1 (or higher dimensional) black hole anti de Sitter spacetime.

In order to extract more information about the string evolution in these backgrounds, in particular exact properties, we consider the circular string ansatz:

$$t = t(\tau), \quad r = r(\tau), \quad \phi = \sigma + f(\tau), \tag{2.16}$$

in the equatorial plane ($\theta = \pi/2$) of the stationary axially symmetric backgrounds:

$$ds^2 = g_{tt}(r)dt^2 + g_{rr}(r)dr^2 + 2g_{t\phi}(r)dtd\phi + g_{\phi\phi}(r)d\phi^2. \tag{2.17}$$

This includes all the cases of interest here: The 2+1 BH-AdS spacetime, the black string, as well as the equatorial plane of ordinary Einstein black holes. The string dynamics, determined by Eqs.(1.1)-(1.2), is then reduced to a system of second order ordinary differential equations and constraints, also described as a Hamiltonian system:

$$\dot{r}^2 + V(r) = 0; \quad V(r) = g^{rr}(g_{\phi\phi} + E^2 g^{tt}), \tag{2.18}$$

$$\dot{t} = -Eg^{tt}, \quad \dot{f} = -Eg^{t\phi}; \quad E = -P_t = \text{const.}, \tag{2.19}$$

which in all backgrounds considered here are solved in terms of either elementary or elliptic functions. The dynamics of the circular strings takes place at the r-axis in the $(r, V(r))$-diagram and from the properties of the potential $V(r)$ (minima, zeroes, asymptotic behaviour for large r and the value $V(0)$), general knowledge about the string motion can be obtained. On the other hand, the line element of the circular string turns out to be:

$$ds^2 = g_{\phi\phi}(d\sigma^2 - d\tau^2), \quad \text{i.e. } S(\tau) = \sqrt{g_{\phi\phi}(r(\tau))}, \tag{2.20}$$

$S(\tau)$ being the invariant string size. For all the static black hole AdS spacetimes (2+1 and higher dimensional): $S(\tau) = r(\tau)$.

For the rotating 2+1 BH-AdS spacetime:

$$V(r) = r^2(\frac{r^2}{l^2} - M) + \frac{J^2}{4} - E^2, \tag{2.21}$$

(see Fig.1). $V(r)$ has a global minimum $V_{\text{min}} < 0$ between the two horizons r_+, r_- (for $Ml^2 \geq J^2$, otherwise there are no horizons). The vanishing of $V(r)$ at $r = r_{01,2}$ (see Ref.[26]) determines three different types of solutions: (i) For $J^2 > 4E^2$, there are two positive zeroes $r_{01} < r_{02}$, the string never comes outside the static limit, never falls into $r = 0$ neither (there is a barrier between $r = r_{01}$ and $r = 0$). The mathematical solution oscillates between r_{01} and r_{02} with $0 < r_{01} < r_- < r_+ < r_{02} < r_{\text{erg}}$. It may be interpreted as a string travelling between the different universes described by the maximal analytic extension of the manifold. (ii) For $J^2 < 4E^2$, there is only one positive zero r_0 outside the static limit and there is no barrier preventing the string from collapsing into $r = 0$. The string starts at $\tau = 0$ with maximal size $S^{(ii)}_{\text{max}}$ outside the static limit, it then contracts through the ergosphere and the two horizons and eventually collapses into a point $r = 0$. For $J \neq 0$, it may be still possible to continue this solution into another universe like in the case (i). (iii) $J^2 = 4E^2$ is the limiting case where the maximal string size equals the static limit: $S^{(iii)}_{\text{max}} = l\sqrt{M}$. In this case $V(0) = 0$, thus the string contracts through the two horizons and eventually collapses into a point $r = 0$.

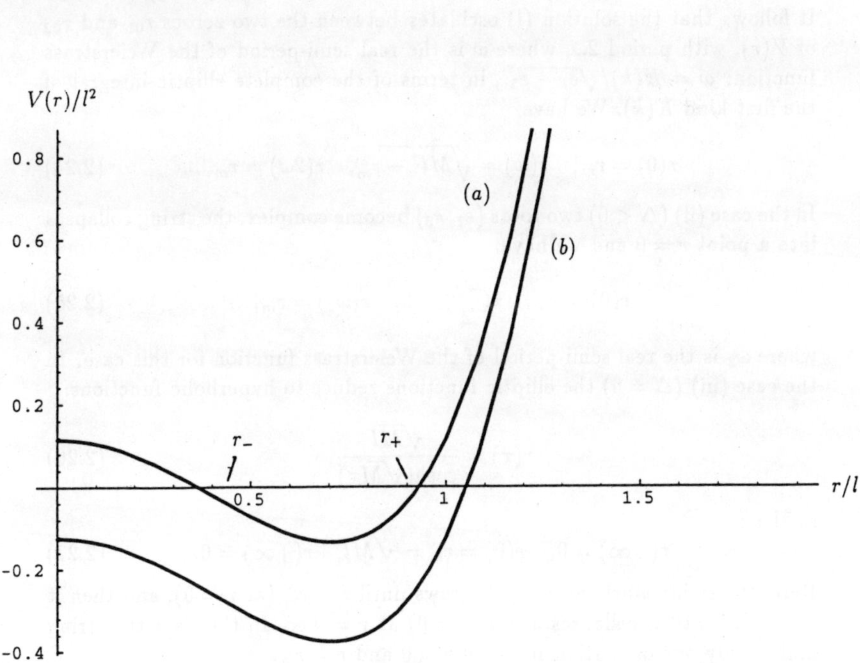

Figure 1. The potential $V(r)$, Eq.(2.21), for a circular string in the $2+1$ BH-AdS spacetime. In (a) we have $J^2 > 4E^2$ and a barrier between the inner horizon and $r = 0$, while (b) represents a case where $J^2 < 4E^2$ and a string will always fall into $r = 0$. The static limit is $r_{\text{erg}} = l$.

The exact general solution in the three cases (i)-(iii) is given by:

$$r(\tau) = \mid r_m - \frac{1}{c_1 \wp(\tau - \tau_0) + c_2} \mid, \tag{2.22}$$

where:

$$r_m = S_{\text{max}} = \sqrt{\frac{Ml^2}{2}} \sqrt{1 + \sqrt{1 - \frac{4V(0)}{Ml^2}}}, \quad V(0) = \frac{J^2}{4} - E^2. \tag{2.23}$$

c_1, c_2 are constants in terms of (l, M, r_m), given in Ref.[26], and $\wp$ is the Weierstrass elliptic $\wp$-function with invariants (g_2, g_3), discriminant Δ and roots (e_1, e_2, e_3), also given in Ref.[26]. The three cases (i)-(iii) correspond to the cases $\Delta > 0$, $\Delta < 0$ and $\Delta = 0$, respectively. Notice that $S^{(ii)}_{\text{max}} > S^{(iii)}_{\text{max}} = l\sqrt{M} > S^{(i)}_{\text{max}}$. In the case (i), $r(\tau)$ can be written in terms of the Jacobian elliptic function $\text{sn}[\tau^*, k]$, $\tau^* = \sqrt{e_1 - e_3}\,\tau$, $k = \sqrt{(e_2 - e_3)/(e_1 - e_3)}$. It follows that the solution (i) oscillates between the two zeroes r_{01} and r_{02} of $V(r)$, with period 2ω, where ω is the real semi-period of the Weierstrass function: $\omega = K(k)/\sqrt{e_1 - e_3}$, in terms of the complete elliptic integral of the first kind $K(k)$. We have:

$$r(0) = r_m, \quad r(\omega) = \sqrt{Ml^2 - r_m^2}, \quad r(2\omega) = r_m, \ldots \tag{2.24}$$

In the case (ii) ($\Delta < 0$) two roots (e_1, e_3) become complex, the string collapses into a point $r = 0$ and we have:

$$r(0) = r_m, \quad r(\frac{\omega_2}{2}) = 0, \quad r(\omega_2) = r_m, \ldots \tag{2.25}$$

where ω_2 is the real semi-period of the Weierstrass function for this case. In the case (iii) ($\Delta = 0$) the elliptic functions reduce to hyperbolic functions:

$$r(\tau) = \frac{\sqrt{M}l}{\cosh(\sqrt{M}\tau)}, \tag{2.26}$$

so that:

$$r(-\infty) = 0, \quad r(0) = r_m = \sqrt{M}l, \quad r(+\infty) = 0. \tag{2.27}$$

Here, the string starts as a point, grows until $r = r_m$ (at $\tau = 0$), and then it contracts until it collapses again ($r = 0$) at $\tau = +\infty$. In this case the string makes only one oscillation between $r = 0$ and $r = r_m$.

Notice that for the static background ($J = 0$), the only allowed motion is (ii), i.e. $r_m > r_{\text{hor}} = \sqrt{M}l$ (there is no ergosphere and only one horizon in this case), with:

$$r_m = \sqrt{\frac{Ml^2}{2}} \sqrt{1 + \sqrt{1 + \frac{4E^2}{Ml^2}}}. \tag{2.28}$$

For $J = 0$, the string collapses into $r = 0$ and stops there. The Penrose diagram of the 2+1 BH-AdS spacetime for $J = 0$ is very similar to the Penrose diagram of the ordinary ($D \geq 4$) Schwarzschild spacetime, so the string motion outwards from $r = 0$ is unphysical because of the causal structure. The coordinate time $t(\tau)$ can be expressed in terms of the incomplete elliptic integral of the third kind II, see Ref.[26]. The string has its maximal size r_m at $\tau = 0$, passes the horizon at $\tau = \tau_{\text{hor}}$ (expressed in terms of an incomplete elliptic integral of the first kind) and falls into $r = 0$ for $\tau = \omega_2/2$, ω_2 being the real semi-period of the Weierstrass function. That is, we have:

$$r(0) = r_m, \quad r(\tau_{\text{hor}}) = \sqrt{Ml}, \quad r(\omega_2/2) = 0,$$
$$t(0) = 0, \quad t(\tau_{\text{hor}}) = \infty, \tag{2.29}$$

and $t(\omega_2/2)$ can be expressed in terms of the Jacobian zeta function Z, see Ref.[26].

We also study the circular strings in the ordinary $D \geq 4$ spacetimes. In the generic 3+1 Kerr anti de Sitter (or Kerr de Sitter) spacetime, the potential $V(r)$ is however quite complicated, see Ref.[26]. The general circular string solution involves higher genus elliptic functions and it is not necessary to go into details here. We will compare with the non-rotating cases, only.

It is instructive to recall [38] the circular string in Minkowski spacetime (Min), for which $V(r) = r^2 - E^2$ (Fig.2a), the string oscillates between its maximal size $r_m = E$, and $r = 0$ with the solution $r(\tau) = r_m \mid \cos\tau \mid$.

In the Schwarzschild black hole (S) $V(r) = r^2 - 2Mr - E^2$ (Fig.2b), the solution is remarkably simple: $r(\tau) = M + \sqrt{M^2 + E^2}\cos\tau$. The mathematical solution oscillates between $r_m = M + \sqrt{M^2 + E^2}$ and $M - \sqrt{M^2 + E^2} < 0$, but because of the causal structure and the curvature singularity the motion can not be continued after the string has collapsed into $r = 0$.

For anti de Sitter spacetime (AdS), we find: $V(r) = r^2(1 + H^2r^2) - E^2$ (Fig.2c). The string oscillates between $r_m = \frac{1}{\sqrt{2}H}\sqrt{-1 + \sqrt{1 + 4H^2E^2}}$ and $r = 0$ with the solution:

$$r(\tau) = r_m \mid \text{cn}[(1 + 4H^2E^2)^{1/4}, k] \mid, \tag{2.30}$$

which is periodic with period 2ω :

$$\omega = \frac{K(k)}{(1 + 4H^2E^2)^{1/4}}, \qquad k = \sqrt{\frac{\sqrt{1 + 4H^2E^2} - 1}{2\sqrt{1 + 4H^2E^2}}}. \tag{2.31}$$

For Schwarzschild anti de Sitter spacetime (S-AdS), we find $V(r) = r^2(1+H^2r^2) - 2Mr - E^2$ (Fig.2d) and:

$$r(\tau) = r_m - \frac{1}{d_1\wp(\tau) + d_2}, \qquad r(0) = r_m \tag{2.32}$$

d_1, d_2 are constants given in terms of (M, H, r_m), see Ref.[26] (r_m is the root of the equation $V(r) = 0$, which has in this case exactly one positive solution). The invariants, the discriminant and the roots are also given explicitly in

Ref.[26]. The string starts with $r = r_m$ at $\tau = 0$, it then contracts and eventually collapses into the $r = 0$ singularity. The existence of elliptic function solutions for the string motion is characteristic of the presence of a cosmological constant. For $\Lambda = 0 = -3H^2$ the circular string motion reduces to simple trigonometric functions. From Fig.2 and our analysis we see that the circular string motion is qualitatively very similar in all these backgrounds (Min, S, AdS, S-AdS): the string has a maximal *bounded* size and then it contracts towards $r = 0$. There are however physical and quantitative differences: in Minkowski and pure anti de Sitter spacetimes, the string truly oscillates between r_m and $r = 0$, while in the black hole cases (S, S-AdS), there is only one half oscillation, the string motion stops at $r = 0$. This also holds for the 2+1 BH-AdS spacetime with $J = 0$ (Fig.1b). Notice also that in all these cases, $V(0) = -E^2 < 0$ and $V(r) \sim r^\alpha$; $(\alpha = 2, 4)$ for $r >> E$.

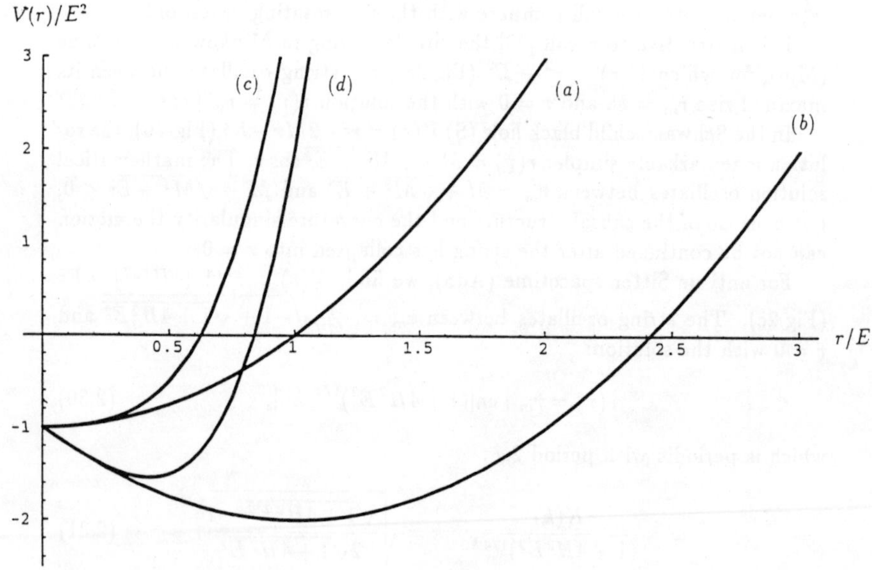

Figure 2. The potential $V(r)$, Eq.(2.18), for a circular string in the equatorial plane of the four $3+1$ dimensional spacetimes: (a) Minkowski (Min), (b) Schwarzschild black hole (S), (c) anti de Sitter (AdS), and (d) Schwarzschild anti de Sitter space (S-AdS).

The similarity can be pushed one step further by considering small perturbations around the circular strings. We find [18]:

$$\ddot{C}_n + (n^2 + \frac{r}{2}\frac{da(r)}{dr} + \frac{r^2}{2}\frac{d^2a(r)}{dr^2} - \frac{2E^2}{r^2})C_n = 0, \tag{2.33}$$

determining the Fourier components of the comoving perturbations. For the spacetimes of interest here, $a(r) = 1 - 2M/r + H^2r^2$ (Min, S, AdS, S-AdS), or $a(r) = r^2/l^2 - 1$ (2+1 BH-AdS), the comoving perturbations are regular except near $r = 0$, where we find (for all cases) $r(\tau) \approx -E(\tau - \tau_0)$ and:

$$\ddot{C}_n - \frac{2}{(\tau - \tau_0)^2}C_n = 0. \tag{2.34}$$

It follows that not only the unperturbed circular strings, but also the comoving perturbations around them behave in a similar way in all these non-rotating backgrounds (2+1 and higher dimensional). This should be contrasted with the string perturbations around the center of mass, which behave differently in these backgrounds. It must be noticed that for rotating ($J \neq 0$) spacetimes, the circular string behaviour is qualitatively different from the non-rotating ($J = 0$) spacetimes. For large J, both in the 2+1 BH-AdS as well as in the 3+1 ordinary Kerr-AdS spacetimes, non-collapsing circular string solutions exist. The potential $V(r) \to +\infty$ for $r \to 0$ and no collapse into $r = 0$ is possible.

The dynamics of circular strings in curved spacetimes is determined by the string tension, which tends to contract the string, and by the local gravity (which may be attractive or repulsive). In all the previous non-rotating backgrounds, the local gravity is attractive (i.e. $da(r)/dr > 0$), and it acts together with the string tension in the sense of contraction. But in spacetimes with regions in which repulsion (i.e. $da(r)/dr < 0$) dominates, the strings can expand with unbounded radius (unstable strings [11, 13]). It may also happen that the string tension and the local gravity be of the same order, i.e. the two opposite effects can balance, and the string is stationary. De Sitter spacetime provides an example in which all such type of solutions exist [11, 13]. In de Sitter spacetime, $V(r)$ is unbounded from below for $r \to \infty$ ($V(r) \sim -r^4$) and unbounded expanding circular strings are present. In addition, an interesting new feature appears in the presence of a positive cosmological constant: the existence of *multi-string* solutions [9-11]. The world-sheet time τ turns out to be a multi (finite or infinite) valued function of the physical time. That is, one single world-sheet where $-\infty \leq \tau \leq +\infty$, can describe many (even infinitely many [13]) different and independent strings (in flat spacetime, one single world-sheet describes only one string). In the S, AdS and S-AdS spacetimes, the multi-string feature for circular strings is absent.

The main conclusions of this section are given in Tables 1, 2. Further details can be found in Ref.[26]

Table 1. String motion described by the string perturbation series approach in the 2+1 BH-AdS, ordinary black hole-AdS, de Sitter (dS) and black string backgrounds.

String Perturbation Series Approach

$$x^\mu(\tau,\sigma) = q^\mu(\tau) + \eta^\mu(\tau,\sigma) + \xi^\mu(\tau,\sigma) + \ldots$$

$$\eta^\mu = n^\mu_R \delta x^R, \quad \delta x^R(\tau,\sigma) = \textstyle\sum_n C^R_n(\tau) e^{-in\sigma}; \;\; R = 1, \ldots, (D-1)$$

$$\ddot{C}_{nR} + (n^2\delta_{RS} - R_{\mu\rho\sigma\nu} n^\mu_R n^\nu_S \dot{q}^\rho \dot{q}^\sigma) C^S_n = 0$$

2+1 Black Hole-ADS	Ordinary ($D \geq 4$) Black Hole-ADS
$\ddot{C}_{n\perp} + (n^2 + \frac{m^2}{l^2}) C_{n\perp} = 0$ $\ddot{C}_{n\parallel} + (n^2 + \frac{m^2}{l^2}) C_{n\parallel} = 0$ No instability, $\omega_n = \sqrt{n^2 + \frac{m^2}{l^2}}$ δx^R, ξ^μ bounded everywhere. $\delta x_R = \sum_n [A_{nR} e^{-i(n\sigma+\omega_n\tau)} + \tilde{A}_{nR} e^{-i(n\sigma-\omega_n\tau)}]$ $m^2 = 2\sum_n (2n^2 + \frac{m^2}{l^2}) \sum_R A_{nR} \tilde{A}_{-nR}$	$\ddot{C}_{n\perp} + (n^2 + m^2H^2 + \frac{Mm^2}{r^3}) C_{n\perp} = 0$ $\ddot{C}_{n\parallel} + (n^2 + m^2H^2 - \frac{2Mm^2}{r^3}) C_{n\parallel} = 0$ Instability, $r_{\text{inst}} = (\frac{2Mm^2}{1+m^2H^2})^{1/3}$ Unbounded perturbations $C_{n\parallel}$ for $r(\tau) \approx (3m\sqrt{M/2})^{2/3} (\tau_0 - \tau)^{2/3} \to 0$: $\ddot{C}_{n\parallel} - \frac{4}{9(\tau_0-\tau)^2} C_{n\parallel} = 0$ $m^2(H=0) = 4\sum_n n^2 \sum_R A_{nR} \tilde{A}_{-nR}$

Ordinary ($D \geq 4$) de Sitter	Black String
$\ddot{C}_{nR} + (n^2 - m^2H^2) C_{nR} = 0$ Unbounded modes for $\mid n \mid < mH$ Instability, $\omega_n = \sqrt{n^2 - m^2H^2}$ $\delta x_R = \sum_n [A_{nR} e^{-i(n\sigma+\omega_n\tau)} + \tilde{A}_{nR} e^{-i(n\sigma-\omega_n\tau)}]$ $m^2 = 2\sum_n (2n^2 - m^2H^2) \sum_R A_{nR} \tilde{A}_{-nR}$	$\ddot{C}_{n\perp} + n^2 C_{n\perp} = 0$ $\ddot{C}_{n\parallel} + (n^2 - \frac{2Mm^2}{lr}) C_{n\parallel} = 0$ Instability, $r_{\text{inst}} = \frac{2Mm^2}{l}$ Unbounded perturbations $C_{n\parallel}$ for $r(\tau) \approx \frac{Mm^2}{l} (\tau_0 - \tau)^2 \to 0$: $\ddot{C}_{n\parallel} - \frac{2}{(\tau_0-\tau)^2} C_{n\parallel} = 0$ $m^2 = 4\sum_n n^2 \sum_R A_{nR} \tilde{A}_{-nR}$

Table 2. Circular exact string solutions in the indicated backgrounds. $S(\tau)$ is the invariant string size.

Exact Circular Strings

$$ds^2 = g_{tt}(r)dt^2 + g_{rr}(r)dr^2 + 2g_{t\phi}(r)dtd\phi + g_{\phi\phi}d\phi^2$$
$$t = t(\tau), \quad r = r(\tau), \quad \phi = \sigma + f(\tau), \quad (\theta = \pi/2)$$
$$\dot{r}^2 + V(r) = 0; \quad V(r) = g^{rr}(E^2 g^{tt} + g_{\phi\phi}), \quad \dot{t} = -Eg^{tt}, \quad \dot{f} = -Eg^{t\phi}$$
$$ds^2 = g_{\phi\phi}(-d\tau^2 + d\sigma^2), \quad \text{i.e.} \quad S(\tau) = \sqrt{g_{\phi\phi}(r(\tau))}$$

2+1 Black Hole ADS	Ordinary ($D \geq 4$) Black Hole ADS
$V(r) = r^2(r^2/l^2 - M) + J^2/4 - E^2$ $V(0) = J^2/4 - E^2, \quad V(r \to \infty) \propto r^4 \to \infty$ $S(\tau) = r(\tau) = r_m - (c_1\wp(\tau - \tau_0) + c_2)^{-1}$ $S_{\max} = r_m = \sqrt{\frac{Ml^2}{2}}\sqrt{1 + \sqrt{1 - \frac{4V(0)}{Ml^2}}}$ For $J^2 > 4E^2$, $S_{\min} > 0$, no collapse. For $J^2 \leq 4E^2$, $S_{\min} = 0$, collapse. No unbounded string size. No multi-string solutions.	$V(r) = r^2(1 + H^2r^2) - 2Mr - E^2$ $V(0) = -E^2, \quad V(r \to \infty) \propto r^4 \to \infty$ $S(\tau) = r(\tau) = r_m - (c_1\wp(\tau - \tau_0) + c_2)^{-1}$ $V(r_m) \equiv 0; \quad S_{\max} = r_m = r_m(M, H, E)$ String contracts from $r(0) = r_m$ until it collapses into the $r = 0$ singularity. No unbounded string size. No multi-string solutions.
Ordinary ($D \geq 4$) Black Hole DS	**Black String**
$V(r) = r^2(1 - H^2r^2) - 2Mr - E^2$ $V(0) = -E^2, \quad V(r \to \infty) \propto -r^4 \to -\infty$ Contracting, expanding or stationary solutions, depending on the balance between the string tension and the local gravity. $M = 0: \ S^2(\tau) = H^{-2}\wp(\tau - \tau_0) + H^{-2}/3$ Contracting, expanding or oscillating. Unbounded string size. Multi-string solutions.	$\tilde{V}(r) = J^2/(4r^2) - M/r^2 + 1/l^2 - E^2$ $\tilde{V}(0) = \infty \ (J \neq 0), \ \tilde{V}(0) = -\infty \ (J = 0)$ $\tilde{S}(\tau) = 1/r(\tau), \quad \tilde{V}(\infty) = 1/l^2 - E^2$ For $J \neq 0$, all solutions are bounded (finite $\tilde{S}$). For $J = 0$ unbounded size solutions exist as well. $J = 0, El = 1: \ \tilde{S}(\tau) = (\mid 2\sqrt{M}\tau \mid)^{-1/2}$ $\tilde{S}(-\infty) = 0, \ \tilde{S}(0) = \infty, \ \tilde{S}(\infty) = 0$ Unbounded string size. Multi-string solutions.

3 Quantum String Spectrum in Ordinary AdS

In the first part of the previous section, the classical string dynamics was studied in the $2+1$ BH-AdS spacetime. In this section we go further in the investigation of the physical properties of strings in AdS spacetime, by performing string quantization. Since the 2+1 BH-AdS spacetime is locally AdS, the results of Section 2 can be extended to classical strings in D-dimensional AdS spacetime as well. In AdS spacetime, the string motion is oscillatory in time and is *stable*; all fluctuations around the string center of mass are well behaved and bounded. Local gravity of AdS spacetime is always negative and string instabilities do not develop. The string perturbation series approach, considering fluctuations around the center of mass, is particularly appropriate in AdS spacetime, the natural dimensionless expansion parameter being $\lambda = \alpha'/l^2 > 0$, where α' is the string tension and the 'Hubble constant' $H = 1/l$. The negative cosmological constant of AdS spacetime is related to the 'Hubble constant' H by $\Lambda = -(D-1)(D-2)H^2/2$.

All the spatial ($\mu = 1, ..., D-1$) modes in D-dimensional AdS, oscillate with frequency $\omega_n = \sqrt{n^2 + m^2\alpha'\lambda}$, which are real for all n (m being the string mass). In this section, we perform a canonical quantization procedure. From the conformal generators L_n, $\tilde{L}_n$ and the constraints $L_0 = \tilde{L}_0 = 0$, imposed at the quantum level, we obtain the mass formula (the quantum version of Eq.(2.7) written in units where α' appears explicitly):

$$m^2\alpha' = (D-1)\sum_{n>0}\Omega_n(\lambda) + \sum_{n>0}\Omega_n(\lambda)\sum_{R=1}^{D-1}[(\alpha_n^R)^\dagger\alpha_n^R + (\tilde{\alpha}_n^R)^\dagger\tilde{\alpha}_n^R], \tag{3.1}$$

where:

$$\Omega_n(\lambda) = \frac{2n^2 + m^2\alpha'\lambda}{\sqrt{n^2 + m^2\alpha'\lambda}}, \tag{3.2}$$

and we have applied symmetric ordering of the operators. The operators α_n^R, $\tilde{\alpha}_n^R$ satisfy:

$$[\alpha_n^R, (\alpha_n^R)^\dagger] = [\tilde{\alpha}_n^R, (\tilde{\alpha}_n^R)^\dagger] = 1, \qquad \text{for all } n > 0. \tag{3.3}$$

For $\lambda << 1$, which is clearly fulfilled in most interesting cases, we have found the lower mass states $m^2\alpha'\lambda << 1$ and the quantum mass spectrum [27]. Physical states are characterized by the eigenvalue of the number operator:

$$N = \frac{1}{2}\sum_{n>0} n \sum_{R=1}^{D-1}[(\alpha_n^R)^\dagger\alpha_n^R + (\tilde{\alpha}_n^R)^\dagger\tilde{\alpha}_n^R], \tag{3.4}$$

and the ground state is defined by:

$$\alpha_n^R \mid 0 >= \tilde{\alpha}_n^R \mid 0 >= 0, \qquad \text{for all } n > 0. \tag{3.5}$$

We find that $m^2\alpha' = 0$ is an *exact* solution of the mass formula in $D = 25$ and that there is a graviton at $D = 25$, which indicates, as in de Sitter space [7], that the critical dimension of AdS is 25 (although it should be stressed that the question whether these spacetimes are solutions to the full β-function equations remains open). As in Minkowski spacetime, the ground state is a tachyon. Remarkably enough, for $N \geq 2$ we find that a generic feature of all excited states beyond the graviton, is the presence of a *fine structure* effect: for a given eigenvalue $N \geq 2$, the corresponding states have different masses. For the lower mass states the expectation value of the mass operator in the corresponding states (generically labelled $\mid j >$) turns out to have the form (see Ref.[27]):

$$< j \mid m^2\alpha' \mid j >_{\mathrm{AdS}} = a_j + b_j\lambda^2 + c_j\lambda^3 + \mathcal{O}(\lambda^4). \tag{3.6}$$

The collective index "j" generically labels the state $\mid j >$ and the coefficients a_j, b_j, c_j are all well computed numbers, different for each state, their precise values are given in Ref.[27]. The corrections to the mass in Minkowski spacetime appear to order λ^2. Therefore, the leading Regge trajectory for the lower mass states is:

$$J = 2 + \frac{1}{2}m^2\alpha' + \mathcal{O}(\lambda^2). \tag{3.7}$$

In Minkowski spacetime the mass and number operator of the string are related by $m^2\alpha' = -4 + 4N$. In AdS (as well as in de Sitter (dS)), there is no such simple relation between the mass and the number operators; the splitting of levels increases considerably for very large N. The *fine structure* effect we find here is also present in de Sitter space. Up to order λ^2, the lower mass states in dS and in AdS are the same, the differences appear to the order λ^3. The lower mass states in de Sitter spacetime are given by Eq.(3.6) but with the λ^3-term getting an opposite sign $(-c_j\lambda^3)$.

For the very high mass spectrum, we find more drastic effects. States with very large eigenvalue N, namely $N >> 1/\lambda$, have masses:

$$< j \mid m^2\alpha' \mid j >_{\mathrm{AdS}} \approx d_j\lambda N^2 \tag{3.8}$$

and angular momentum:

$$J^2 \approx \frac{1}{\lambda}m^2\alpha', \tag{3.9}$$

where d_j are well computed numbers different for each state, their precise values are given in Ref.[27]. Since $\lambda = \alpha'/l^2$, we see from Eq.(3.8) that the masses of the high mass states are *independent* of α'. In Minkowski spacetime, very large N states all have the same mass $m^2\alpha' \approx 4N$, but here in AdS the masses of the high mass states with the same eigenvalue N are *different* by factors d_j. In addition, because of the fine structure effect, states with different N can get mixed up. For high mass states, the level spacing *grows* with N (instead of being constant as in Minkowski spacetime). As a consequence, the density of states $\rho(m)$ as a function of mass grows like $\mathrm{Exp}[(m/\sqrt{\mid \Lambda \mid}\,)^{1/2}]$ (instead of $\mathrm{Exp}[m\sqrt{\alpha'}]$ as in Minkowski spacetime), and *independently* of α'. The partition function for a gas of strings at a temperature β^{-1} in AdS spacetime is well defined for all finite temperatures β^{-1}, discarding the existence of the Hagedorn temperature.

For the low mass states ($m^2\alpha'\lambda << 1$) in anti de Sitter spacetime, our results can be written as [27]:

$$< j \mid m^2(\alpha', l) \mid j >= \frac{4(N-1)}{\alpha'} + \frac{1}{l^2}\sum_{n=0}^{\infty} a_{jn}(\alpha'/l^2)^n, \qquad (3.10)$$

where "j" is a collective index labelling the state $\mid j >$. It is now important to notice that $a_{j0} = 0$ for *all* the low mass states [27], i.e. there is no "constant" term on the right hand side of Eq.(3.10). A non-zero a_{j0}-term would give rise to a α'-independent contribution to the string mass. Its absense, on the other hand, means that the first term on the right hand side of Eq.(3.10) is super-dominant (since, in all cases, $\alpha'/l^2 = \lambda << 1$) and that the string scale is therefore set by $1/\alpha'$.

For the high mass states ($m^2\alpha'\lambda >> 1$) we found instead [27]:

$$< j \mid m^2(\alpha', l) \mid j >\approx \frac{d_j}{l^2}N^2, \qquad \text{for } N >> l^2/\alpha' \qquad (3.11)$$

where the number d_j depends on the state. The right hand side of Eq.(3.11) is exactly like a non-zero dominant a_{j0}-term in Eq.(3.10). For the high mass states the scale is therefore set by $1/l^2$ which is equal to the absolute value of the cosmological constant Λ (up to a geometrical factor) and *independent* of α'. This suggests that for $\lambda << 1$, the masses of *all* string states can be represented by a formula of the form (3.10). For the low mass states $a_{j0} = 0$, while for the high mass states a_{j0} becomes a large positive number.

In the black string background, Eq.(2.2), we have calculated explicitly the first and second order string fluctuations around the center of mass [27]. We then determined the world-sheet energy-momentum tensor and we derived the mass formula in the asymptotic region. The mass spectrum is equal to the mass spectrum in flat Minkowski spacetime. Therefore, for a gas of strings at temperature β^{-1} in the asymptotic region of the black string background, the partition function goes like:

$$Z(\beta) \sim \int^{\infty} dm\ e^{-m(\beta-\sqrt{\alpha'})}, \qquad (3.12)$$

which is only defined for $\beta > \sqrt{\alpha'}$, i.e. there is a Hagedorn temperature:

$$T_{\text{Hg}} = (\alpha')^{-1/2}. \qquad (3.13)$$

In higher dimensional ($D \geq 4$) black hole spacetimes the next step now would be to set up a scattering formalism, where a string from an asymptotic in-state interacts with the gravitational field of the black hole and reappears in an asymptotic out-state [15]. However, this is not possible in the black string background. In the uncharged black string background under consideration here, *all* null and timelike geodesics incoming from spatial infinity pass through the horizon and fall into the physical singularity [36]. No "angular momentum", as in the case of scattering off the ordinary Schwarzschild black hole, can prevent a point particle from falling into the singularity. The string solutions considered here are based on perturbations around the string center of mass which follows, at least approximately, a point particle geodesic. Therefore, a string incoming from spatial infinity inevitably falls into the singularity in the black string background.

Table 3. Characteristic features of the quantum string mass spectrum in anti de Sitter (AdS) and de Sitter (dS) spacetimes.

Quantum Strings	
Anti de Sitter spacetime (AdS)	de Sitter spacetime (dS)
Classical motion is stable and oscillatory in time with real frequencies $\omega_n = \sqrt{n^2 + m^2\alpha'^2 H^2}$	Classical motion is unstable with frequencies $\omega_n = \sqrt{n^2 - m^2\alpha'^2 H^2}$ Unbounded string size and energy for large de Sitter radius, $R \to \infty$.
The mass formula is well defined for all m. There is an infinite number of states with arbitrary high masses. $m^2\alpha' = 0$ is an exact solution at $D = 25$.	Real mass solutions only for $m < 1/(\alpha' H)$. Finite number of states, $N_{\max} \approx 0.15/(\alpha' H^2)$. $m^2\alpha' = 0$ is an exact solution at $D = 25$.
The coupling to the gravitational background produces a Fine structure effect at all levels in the mass spectrum. The number of levels considerably increases with respect to flat space.	Fine structure effect appears in low mass spectrum. Is similar to AdS; the differences appear to order $(\alpha' H^2)^3$
For the high mass states: $< m^2 > \sim \mid \Lambda \mid N^2, \quad J^2 \sim m^2 / \mid \Lambda \mid$ Both are *independent* of α' ! The level spacing grows with N. $\rho(m) \sim \text{Exp}[m/\sqrt{\mid \Lambda \mid})^{1/2}]$, high m. No Hagedorn temperature exists.	The similar region of high mass states does not exist in the de Sitter spacetime.

4 New Classes of Exact String and Multi-String Solutions in Curved Spacetimes

In this section we return to the string equations of motion and constraints, Eqs.(1.1)-(1.2), to look for new classes of exact solutions. In most spacetimes, quite general families of exact solutions can be found by making an appropriate ansatz, which exploits the symmetries of the underlying curved spacetime. In axially symmetric spacetimes, a convenient ansatz corresponds to circular strings, as we saw in Section 2. Such an ansatz effectively decouples the dependence on the spatial world-sheet coordinate σ, and the string equations of motion and constraints reduce to non-linear coupled ordinary differential equations, Eqs.(2.18)-(2.19). In this section we will make instead an ansatz which effectively decouples the dependence on the temporal world-sheet coordinate τ. This ansatz, which we call the "stationary string ansatz" is dual to the "circular string ansatz" in the sense that it corresponds to a formal interchange of the world-sheet coordinates (τ, σ), as well as of the azimuthal angle ϕ and the stationary time t in the target space:

$$\tau \leftrightarrow \sigma, \qquad t \leftrightarrow \phi. \tag{4.1}$$

The stationary string ansatz will describe stationary strings when t (and τ) are timelike, for instance in anti de Sitter spacetime (in static coordinates) and outside the horizon of a Schwarzschild black hole. On the other hand, if t (and τ) are spacelike, for instance inside the horizon of a Schwarzschild black hole or outside the horizon of de Sitter spacetime (in static coordinates), the stationary string ansatz will describe dynamical propagating strings. Considering for simplicity a static line element in the form:

$$ds^2 = -a(r)dt^2 + \frac{dr^2}{a(r)} + r^2(d\theta^2 + \sin^2\theta d\phi^2), \tag{4.2}$$

the stationary string ansatz reads explicitly:

$$t = \tau, \quad r = r(\sigma), \quad \phi = \phi(\sigma), \quad \theta = \pi/2. \tag{4.3}$$

The string equations of motion and constraints, Eqs.(1.1)-(1.2), reduce to two separated first order ordinary differential equations:

$$\phi' = \frac{L}{r^2}, \qquad r'^2 + V(r) = 0; \quad V(r) = -a(r)[a(r) - \frac{L^2}{r^2}], \tag{4.4}$$

where L is an integration constant. The qualitative features of the possible string configurations can be read off directly from the shape of the potential $V(r)$. Thereafter, the detailed analysis of the quantitative features can be performed by explicitly solving the (integrable) system of equations (4.4). The induced line element on the world-sheet is given by:

$$ds^2 = a(r)(-d\tau^2 + d\sigma^2). \tag{4.5}$$

Thus, if $a(r)$ is negative, the world-sheet coordinate τ becomes spacelike while σ becomes timelike and the stationary string ansatz (4.3) describes a dynamical string. If $a(r)$ is positive, the ansatz describes a stationary string.

In this section we solve explicitly Eqs.(4.4) and we analyze in detail the solutions and their physical interpretation in Minkowski, de Sitter, anti de Sitter, Schwarzschild and $2+1$ black hole anti de Sitter spacetimes. In all these cases, the solutions are expressed in terms of elliptic (or elementary) functions. We furthermore analyze the physical properties, energy and pressure, of these solutions. The energy and pressure densities of the strings can be obtained from the $3+1$ dimensional spacetime energy-momentum tensor:

$$\sqrt{-g}T^{\mu\nu} = \frac{1}{2\pi\alpha'}\int d\tau d\sigma(\dot{X}^{\mu}\dot{X}^{\nu} - X'^{\mu}X'^{\nu})\delta^{(4)}(X - X(\tau,\sigma)). \tag{4.6}$$

After integration over a spatial volume that completely encloses the string [39], the energy-momentum tensor takes the form of a fluid:

$$T^{\mu}{}_{\nu} = \text{diag.}(-E, P_1, P_2, P_3). \tag{4.7}$$

In Minkowski spacetime (Min), the potential defined in Eqs.(4.4) is given by:

$$V_{\text{M}}(r) = \frac{L^2}{r^2} - 1, \tag{4.8}$$

and the solution of Eqs.(4.4) describes one infinitely long straight string with "impact-parameter" L. The equation of state, relating energy and pressure densities, takes the well-known form Ref.[28]:

$$dE = -dP_2 = \text{const.}, \quad P_1 = P_3 = 0. \tag{4.9}$$

In anti de Sitter spacetime (AdS), the potential is given by:

$$V_{\text{AdS}}(r) = -(1 + H^2r^2)[1 + H^2r^2 - \frac{L^2}{r^2}]. \tag{4.10}$$

The radial coordinate $r(\sigma)$ is periodic with finite period T_σ, which can be expressed in terms of a complete elliptic integral, see Ref.[28]. For $\sigma \in [0, T_\sigma]$, the solution describes an infinitely long stationary string in the wedge $\phi \in]0, \Delta\phi[$, where:

$$\Delta\phi = 2k\sqrt{\frac{1-2k^2}{1-k^2}}[\Pi(1-k^2,\ k) - K(k)] \ \in\ ]0,\ \pi[\tag{4.11}$$

The elliptic modulus k, which is a function of HL [28], parametrizes the solutions, $k \in]0, 1/\sqrt{2}[$. The azimuthal angle is generally not a periodic function of σ, thus when the spacelike world-sheet coordinate σ runs through the range $]-\infty, +\infty[$, the solution describes an *infinite number* of infinitely long stationary open strings. The general solution is therefore a *multi-string* solution. Until now multi-string solutions were only found in de Sitter spacetime [11-13]. Our results show that multi-string solutions are a general feature of spacetimes with a cosmological constant (positive or negative). The solution in anti de Sitter spacetime describes a *finite* number of strings if the following relation holds:

$$N\Delta\phi = 2\pi M. \tag{4.12}$$

Here N and M are integers, determining the number of strings and the winding in azimuthal angle, respectively, for the multi-string solution, see Fig.3. The equation of state for a full multi-string solution takes the form ($P_3 = 0$):

$$dP_1 = dP_2 = -\frac{1}{2}dE, \quad \text{for } r \to \infty \tag{4.13}$$

corresponding to extremely unstable strings [21].

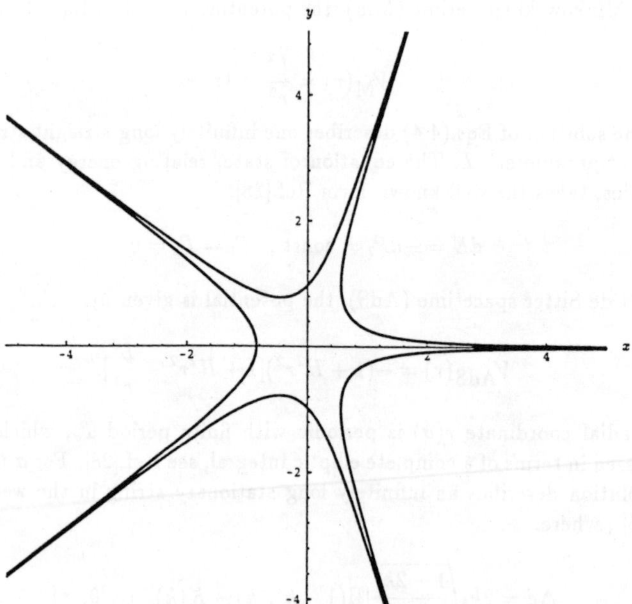

Figure 3. The $(N, M) = (5, 1)$ multi-string solution in anti de Sitter spacetime. The (N, M) multi-string solutions describe N stationary strings with M windings in the Azimuthal angle ϕ, in anti de Sitter spacetime.

In de Sitter spacetime (dS), the potential is given by:

$$V_{\text{dS}}(r) = -(1 - H^2r^2)[1 - H^2r^2 - \frac{L^2}{r^2}]. \tag{4.14}$$

In this case we have to distinguish between solutions inside the horizon (where τ is timelike) and solutions outside the horizon (where τ is spacelike). Inside the horizon, the generic solution describes one infinitely long open stationary string winding around $r = 0$. For special values of the constants of motion, corresponding to a relation, which formally takes the same form as Eq.(4.12), the solution describes a closed string of finite length $l = N\pi/H$. The integer N in this case determines the number of "leaves", see Fig.4. The energy is positive and finite and grows with N. The pressure turns out to vanish identically, thus the equation of state corresponds to cold matter.

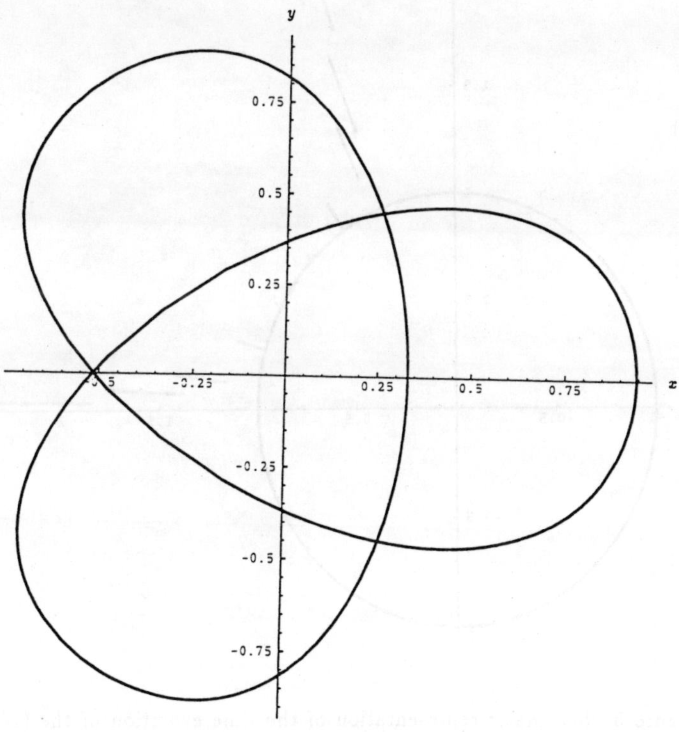

Figure 4. The $(N, M) = (3, 2)$ stationary string solution inside the horizon of de Sitter spacetime. Besides the circular string, this is the simplest stationary closed string configuration in de Sitter spacetime.

Outside the horizon, the world-sheet coordinate τ becomes spacelike while σ becomes timelike, thus we define:

$$\tilde{\tau} \equiv \sigma, \qquad \tilde{\sigma} \equiv \tau, \tag{4.15}$$

and the string solution is conveniently expressed in hyperboloid coordinates. The radial coordinate $r(\tilde{\tau})$ is periodic with a finite period $T_{\tilde{\tau}}$. For $\tilde{\tau} \in [0, T_{\tilde{\tau}}]$, the solution describes a straight string incoming non-radially from spatial infinity, scattering at the horizon and escaping towards infinity again, Fig.5.

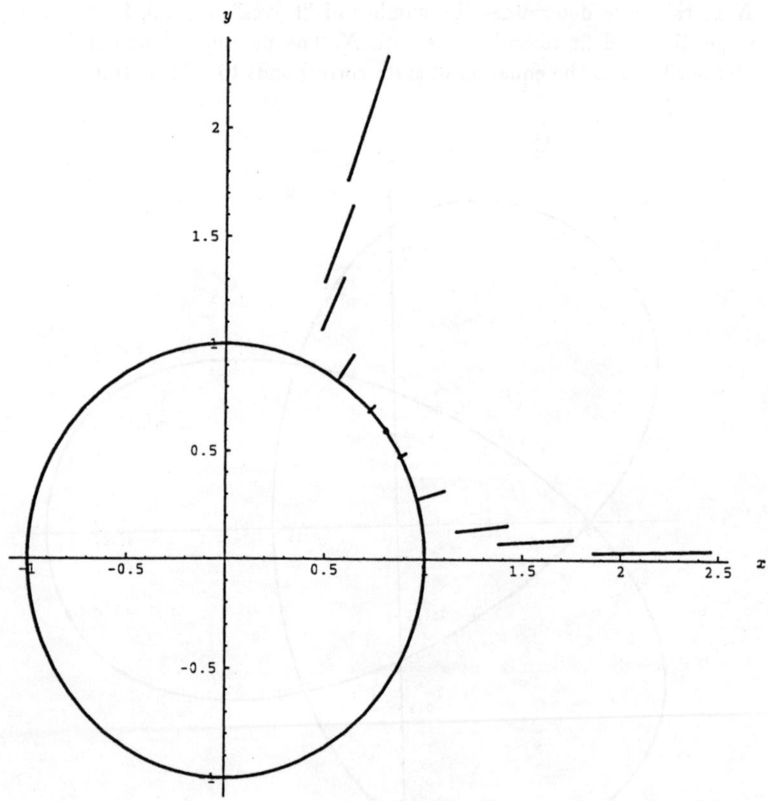

Figure 5. Schematic representation of the time evolution of the $(N, M) = (5, 1)$ dynamical multi-string solution, outside the horizon of de Sitter space-time. Only one of the 5 strings is shown; the others are obtained by rotating the figure by the angles $2\pi/5$, $4\pi/5$, $6\pi/5$ and $8\pi/5$. During the "scattering" at the horizon, the string collapses to a point and re-expands.

The string length is zero at the horizon and grows indefinetely in the asymptotic regions. As in the case of anti de Sitter spacetime, the azimuthal angle is generally not a periodic function, thus when the timelike world-sheet coordinate $\tilde{\tau}$ runs through the range $]-\infty,+\infty[$, the solution describes an *infinite* number of dynamical straight strings scattering at the horizon at different angles. The general solution is therefore a multi-string solution. In particular, a multi-string solution describing a *finite* number of strings is obtained if a relation of the form (4.12) is fulfilled. It turns out that the solution describes *at least* three strings. The energy and pressures of a full multi-string solution have also been computed. In the asymptotic region they fulfill an equation of state corresponding to extremely unstable strings, i.e. like Eq.(4.13).

In the Schwarzschild black hole background (S), the potential is given by:

$$V_{\mathrm{S}}(r) = -(1-2m/r)[1-2m/r-\frac{L^2}{r^2}]. \tag{4.16}$$

No multi-string solutions are found in this case. Outside the horizon the solution of Eqs.(4.4) describes one infinitely long stationary open string. This solution was already derived in Ref.[40], and we shall not discuss it here. Inside the horizon, where τ becomes spacelike while σ becomes timelike, we make the redefinitions (4.15) and the solution is conveniently expressed in terms of Kruskal coordinates, see Ref.[28]. The solution describes one straight string infalling *non-radially* towards the singularity, see Fig.6. At the horizon, the string length is zero and it grows *indefinitely* when the string approaches the spacetime singularity. Thereafter, the solution can not be continued, see Ref.[28].

In the $2+1$ black hole anti de Sitter spacetime (BH-AdS) [30], the potential is given by:

$$V_{\text{BH-AdS}}(r) = -(\frac{r^2}{l^2}-1)[\frac{r^2}{l^2}-1-\frac{L^2}{r^2}]. \tag{4.17}$$

Outside the horizon, the solutions "interpolate" between the solutions found in anti de Sitter spacetime and outside the horizon of the Schwarzschild black hole. The solutions thus describe infinitely long stationary open strings. As in anti de Sitter spacetime, the general solution is a *multi-string* describing *infinitely* many strings. In particular, for certain values of the constants of motion, corresponding to the condition of the form (4.12), the solution describes a finite number of strings. In the simplest version of the $2+1$ BH-AdS background ($M=1,\ J=0$), it turns out that the solution describes *at least* seven strings, see Ref.[28]. Inside the horizon, we make the redefinitions (4.15) and the solution is conveniently expressed in terms of Kruskal-like coordinates, see Ref.[28]. The solution is similar to the solution found inside the horizon of the Schwarzschild black hole, but there is one important difference at $r=0$. As in the Schwarzschild black hole background, the solution describes one straight string infalling non-radially towards $r=0$, and beyond this point the solution can not be continued because of the global structure of the spacetime. At the horizon the string size is zero and during the fall towards $r = 0$, the string size *grows* but stays *finite*. This should be compared with the straight string inside the horizon of the Schwarzschild black hole, where the string size grows indefinetely. The physical reason for this difference is that the point $r = 0$ is not a strong curvature singularity in the $2+1$ BH-AdS spacetime.

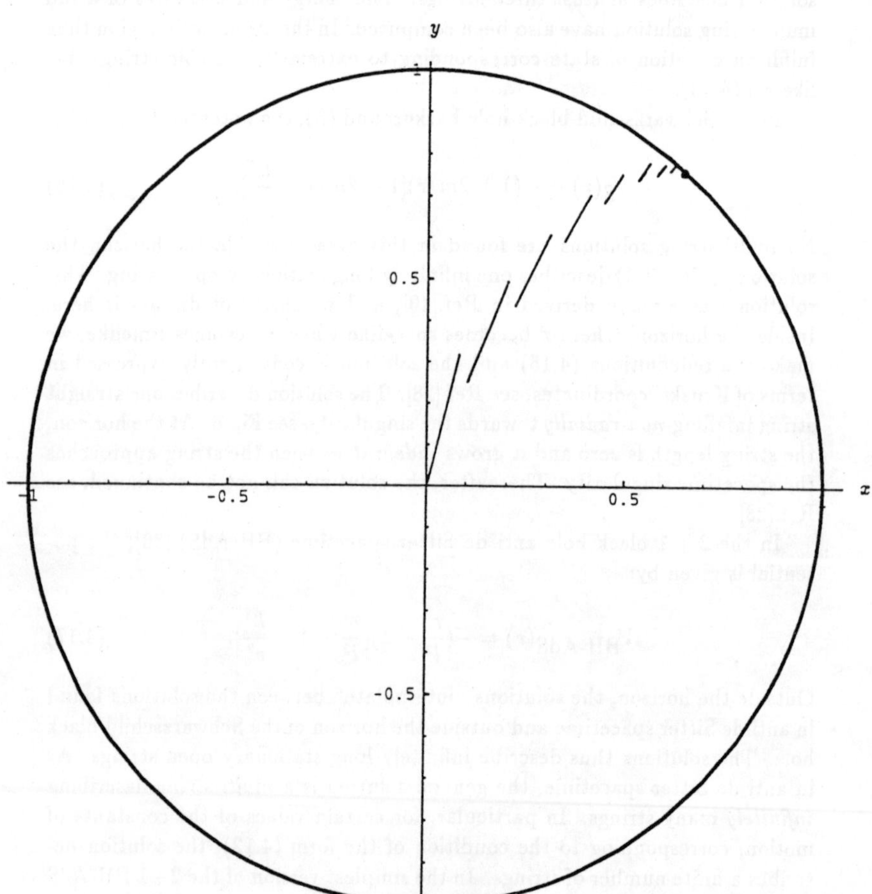

Figure 6. The dynamical straight string inside the horizon of the Schwarzschild black hole. The string infalls non-radially towards the singularity with indefinetely growing length.

We have considered also the Schwarzschild de Sitter and Schwarzschild anti de Sitter spacetimes. These spacetimes contain all the features of the spacetimes already discussed: singularities, horizons, positive or negative cosmological constants. All the various types of string solutions found in the other spacetimes (open, closed, straight, finitely and infinitely long, multi-strings) are therefore present in the different regions of the Schwarzschild-de Sitter and Schwarzschild-anti de Sitter spacetimes. The details are given in Ref.[28].

We close this section with a few remarks on the stability of the solutions. Generally, the question of stability must be addressed by considering small perturbations around the exact solutions. In Ref.[10], a covariant formalism describing physical perturbations propagating along an arbitrary string configuration embedded in an arbitrary curved spacetime, was developed. The resulting equations determining the evolution of the perturbations are however very complicated in the general case, although partial (analytical) results have been obtained in special cases for de Sitter [18, 43] and Schwarzschild black hole [10, 43, 44] spacetimes. The exact solutions found in this section fall essentially into two classes: dynamical and stationary. The dynamical string solutions outside the horizon of de Sitter (or S-dS) and inside the horizon of Schwarzschild (or S-dS, S-AdS) spacetimes, are already unstable at the zeroth order approximation (i.e. without including small perturbations), in the sense that their physical length grows indefinetely. For the stationary string solutions the situation is more delicate. The existence of the stationary configurations is based on an exact balance between the string tension and the local attractive or repulsive gravity. For that reason, it can be expected that the configurations are actually unstable for certain modes of perturbation, especially in strong curvature regions.

The main conclusions of this section are given in Table 4. Further details can be found in Ref.[28].

Table 4. Short summary of the features of the string solutions found in this section. In anti de Sitter spacetime and outside the horizon of de Sitter and $2+1$ BH-AdS spacetimes, the solutions describe a *finite* number of strings provided a condition of the form $N\Delta\phi = 2\pi M$ is fulfilled, where $\Delta\phi$ is the angle betwen the "arms" of the string and (N, M) are integers.

Line element:	$ds^2 = -a(r)dt^2 + \frac{dr^2}{a(r)} + r^2(d\theta^2 + \sin^2\theta d\phi^2)$
Ansatz:	$t = \tau,\quad r = r(\sigma)\quad \phi = \phi(\sigma),\quad \theta = \pi/2$
String solution:	$\phi' = \frac{L}{r^2}\quad r'^2 + V(r) = 0;\quad V(r) = -a(r)[a(r) - \frac{L^2}{r^2}]$
String length element:	$dl = \sqrt{a(r(\sigma))}\, d\sigma$

MINKOWSKI, $a(r) = 1,\quad V(r) = \frac{L^2}{r^2} - 1$

The solution describes one infinitely long stationary straight string.

$\frac{dE}{dl} = -\frac{dP_2}{dl} = \text{const.},\qquad P_1 = P_3 = 0$

ANTI DE SITTER, $a(r) = 1 + H^2r^2,\quad V(r) = -(1 + H^2r^2)[1 + H^2r^2 - \frac{L^2}{r^2}]$

The solution describes a finite or infinite number of infinitely long stationary strings.

$\frac{dE}{dl} = -2\frac{dP_1}{dl} = -2\frac{dP_2}{dl},\quad P_3 = 0,\quad$ (asymptotically for $r \to \infty$)

DE SITTER, $a(r) = 1 - H^2r^2,\quad V(r) = -(1 - H^2r^2)[1 - H^2r^2 - \frac{L^2}{r^2}]$

Inside the horizon, the solution describes one finitely or infinitely long stationary string winding around $r = 0$.

$P_1 = P_2 = P_3 = 0,\quad$ as cold matter, $\quad E$ expressed in terms of elliptic functions.

Outside the horizon, the solution describes a finite or infinite number of dynamical straight strings "scattering at the horizon". The string length vanishes at the horizon, but stretches indefinetely at spatial infinity.

$\frac{dE}{dl} = -2\frac{dP_1}{dl} = -2\frac{dP_2}{dl},\quad P_3 = 0,\quad$ (asymptotically for $r \to \infty$)

SCHWARZSCHILD, $a(r) = 1 - 2m/r,\quad V(r) = -(1 - 2m/r)[1 - 2m/r - \frac{L^2}{r^2}]$

Outside the horizon, the solution describes one infinitely long stationary string.

Inside the horizon, the solution describes one dynamical straight string. The string size is zero at the horizon, and grows *indefinetely* as the string falls towards $r = 0$.

2+1 BLACK HOLE-ADS, $a(r) = \frac{r^2}{l^2} - 1,\quad V(r) = -(\frac{r^2}{l^2} - 1)[\frac{r^2}{l^2} - 1 - \frac{L^2}{r^2}]$

Outside the horizon, the solution describes a finite or infinite number of infinitely long stationary string.

Inside the horizon, the solution describes one dynamical straight string. The string size is zero at the horizon, and grows *finitely* as the string falls towards $r = 0$.

5 Spatial Curvature Effects on the String Dynamics

The propagation of strings in Friedmann-Robertson-Walker (FRW) cosmologies has been investigated using both exact and approximative methods, see for example Refs.[7, 11-14, 17-20] (as well as numerical methods, which shall not be discussed here). Except for anti de Sitter spacetime, which has negative spatial curvature, the cosmologies that have been considered until now, have all been spatially flat. In this section we will consider the physical effects of a non-zero (positive or negative) curvature index on the classical and quantum strings. The non-vanishing components of the Riemann tensor for the generic D-dimensional FRW line element, in comoving coordinates:

$$ds^2 = -dt^2 + a^2(t)\frac{d\vec{x}d\vec{x}}{(1+\frac{K}{4}\vec{x}\vec{x})^2}, \tag{5.1}$$

are given by:

$$R_{itit} = \frac{-aa_{tt}}{(1+\frac{K}{4}\vec{x}\vec{x})^2}, \qquad R_{ijij} = \frac{a^2(K+a_t^2)}{(1+\frac{K}{4}\vec{x}\vec{x})^4}; \quad i \neq j \tag{5.2}$$

where $a = a(t)$ is the scale factor and K is the curvature index. Clearly, a non-zero curvature index introduces a non-zero spacetime curvature; the exceptional case provided by $K = -a_t^2 =$ const., corresponds to the Milne-Universe. From Eqs.(5.2), it is also seen that the curvature index has to compete with the first derivative of the scale factor. The effects of the curvature index are therefore most conveniently discussed in the family of FRW-universes with constant scale factor, the so-called static Robertson-Walker spacetimes. This is the point of view we take in the present section.

We consider both the closed ($K > 0$) and the hyperbolic ($K < 0$) static Robertson-Walker spacetimes, and all our results are compared with the already known results in the flat ($K = 0$) Minkowski spacetime. We determine the evolution of circular strings, derive the corresponding equations of state (using Eq.(4.6)), discuss the question of strings as self-consistent solutions to the Einstein equations [20], and we perform a semi-classical quantization. We also find all the stationary string configurations in these spacetimes.

The radius of a classical circular string in the spacetime (5.1), for $a = 1$, is determined by:

$$\dot{r}^2 + V(r) = 0; \qquad V(r) = (1-Kr^2)(r^2 - b\alpha'^2), \tag{5.3}$$

where b is an integration constant ($b\alpha'^2 = E$, in the notation of Eqs.(2.18)-(2.19)). This equation is solved in terms of elliptic functions and all solutions describe oscillating strings (Fig.7 shows the potential $V(r)$ for $K > 0$, $K = 0$, $K < 0$).

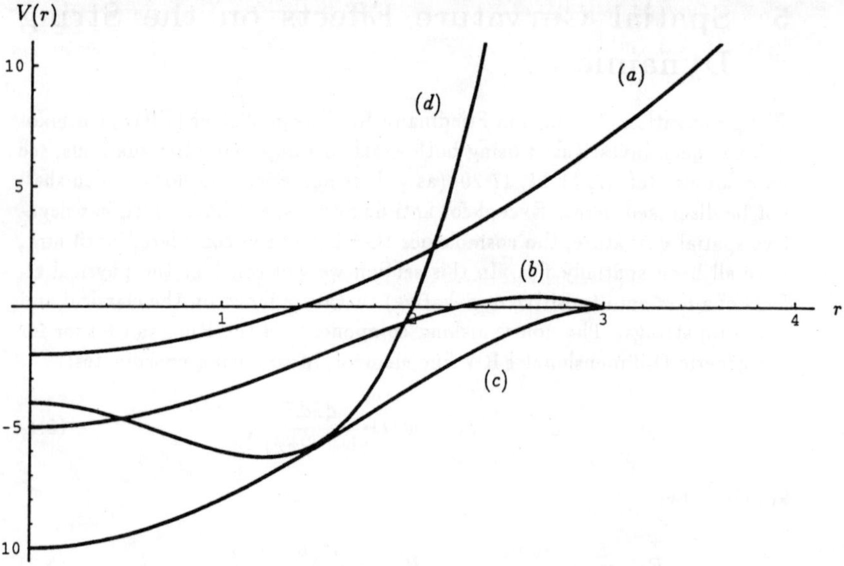

Figure 7. The potential $V(r)$ introduced in Eqs.(5.3) for a circular string in the static Robertson-Walker spacetimes: (a) flat ($K = 0$), (b) closed ($K > 0$ *and* $bK\alpha'^2 \leq 1$), (c) closed ($K > 0$ *and* $bK\alpha'^2 > 1$), (d) hyperbolic ($K < 0$).

For $K > 0$, when the spatial section is a hypersphere, the string either oscillates on one hemisphere or on the full hypersphere. The energy is always positive while the average pressure can be positive, negative or zero, depending on the precise value of an elliptic modulus; the equation of state is given explicitly in Table 5, in the different cases. Interestingly enough, we find that the circular strings provide a self-consistent solution to the Einstein equations with a selected value of the curvature index. Self-consistent solutions to the Einstein equations with string sources have been found previously in the form of power law expanding universes [20].

Table 5. Classical circular strings in the static Robertson-Walker spacetimes. Notice that a self-consistent solution to the Einstein equations, with the string back-reaction included, can be obtained only for $K > 0$.

CLASSICAL CIRCULAR STRING SOLUTIONS

Spacetime: $ds^2 = -dt^2 + \frac{dr^2}{1-Kr^2} + r^2 d\phi^2$

String solution: $\phi = \sigma,\ \ t(\tau) = \sqrt{b}\alpha'\tau,\ \ \dot{r}^2 + V(r) = 0;\ V(r) = (1 - Kr^2)(r^2 - b\alpha'^2)$

String length, energy and pressure: $l(\tau) = 2\pi|r(\tau)|,\ \ T^\mu{}_\nu(\text{string}) = \text{diag.}(-\rho, P, P)$

MINKOWSKI ($K = 0$):

$r(\tau) = \sqrt{b}\alpha' \cos\tau$

$\rho = \sqrt{b},\quad P = -\frac{\sqrt{b}}{2}\cos 2\tau,\quad < P >= 0,$ as cold matter.

Is not a self-consistent solution to the Einstein equations.

CLOSED ($K > 0$):

$$r(\tau) = \begin{cases} \sqrt{b}\alpha' \text{sn}[\tau, k_-] & \text{for } \sqrt{bK}\alpha' \le 1, \quad k_- \equiv \sqrt{bK}\alpha' \in [0,1] \\ \frac{1}{\sqrt{K}}\text{sn}[\tau/k_+, k_+] & \text{for } \sqrt{bK}\alpha' > 1, \quad k_+ \equiv 1/(\sqrt{bK}\alpha') \in]0,1[\end{cases}$$

$$\rho = \sqrt{b},\quad P = \begin{cases} \frac{k_-}{2\sqrt{K}\alpha'}(1 - 2\text{sn}^2[\tau, k_-]) & \text{for } \sqrt{bK}\alpha' \le 1 \\ \frac{1}{2k_+\sqrt{K}\alpha'}(1 - 2k_+^2\text{sn}^2[\tau/k_+, k_+]) & \text{for } \sqrt{bK}\alpha' > 1 \end{cases}$$

$$< P >= \begin{cases} \frac{1}{2\sqrt{K}\alpha'}[\frac{k_-^2 - 2}{k_-} + \frac{2}{k_-}\frac{E(k_-)}{K(k_-)}] & \text{for } \sqrt{bK}\alpha' \le 1, \quad \text{negative.} \\ \frac{1}{2k_+\sqrt{K}\alpha'}[\frac{2E(k_+)}{K(k_+)} - 1] & \text{for } \sqrt{bK}\alpha' > 1, \quad \text{positive or negative.} \end{cases}$$

A self-consistent solution to the Einstein equations is obtained for $k_+ = 0.9089...$

HYPERBOLIC ($K < 0$):

$r(\tau) = \frac{k}{\sqrt{-K}}\text{sd}[\mu\tau, k],\quad \mu \equiv \sqrt{1 - Kb\alpha'^2},\quad k \equiv \sqrt{-Kb\alpha'^2}/\mu \in [0,1[$

$\rho = \sqrt{b},\quad P = \frac{k - 2k(1-k^2)\text{sd}^2[\mu\tau, k]}{2\sqrt{1-k^2}\sqrt{-K}\alpha'}$

$< P >= \frac{1}{2\sqrt{1-k^2}\sqrt{-K}\alpha'}[-k + \frac{2}{k}(1 - \frac{E(k)}{K(k)})],$ positive.

Is not a self-consistent solution to the Einstein equations.

We have semi-classically quantized the circular string solutions of Eq.(5.3). We used an approach developed in field theory by Dashen et. al. [41], based on the stationary phase approximation of the partition function. The result of the stationary phase integration is expressed in terms of the function:

$$W(m) \equiv S_{\rm cl}(T(m)) + m\, T(m), \tag{5.4}$$

where $S_{\rm cl}$ is the action of the classical solution, m is the mass and the period $T(m)$ is implicitly given by:

$$\frac{dS_{\rm cl}}{dT} = -m. \tag{5.5}$$

The bound state quantization condition then becomes [41]:

$$W(m) = 2\pi n, \quad n \in N_0 \tag{5.6}$$

for n 'large'. Parametric plots of $K\alpha' W$ as a function of $K\alpha'^2 m^2$ in the different cases, are shown in Fig.8. The strings oscillating on one hemisphere give rise to a finite number N_- of states with the following mass-formula Ref.[29]:

$$m_-^2 \alpha' \approx \pi\, n, \qquad N_- \approx \frac{4}{\pi K \alpha'}. \tag{5.7}$$

As in flat Minkowski spacetime, the scale of these string states is set by α'. The strings oscillating on the full hypersphere give rise to an infinity of more and more massive states with the asymptotic mass-formula Ref.[29]:

$$m_+^2 \approx K\, n^2. \tag{5.8}$$

The masses of these states are independent of α', the scale is set by the curvature index K. Notice also that the level spacing grows with n. A similar result was found recently for strings in anti de Sitter spacetime [42].

For $K < 0$, when the spatial section is a hyperboloid, both the energy and the average pressure of the oscillating strings are positive. The equation of state is given in Table 5. In this case, the strings can not provide a self-consistent solution to the Einstein equations. After semi-classical quantization, we find an infinity of more and more massive states. The mass-formula is given by (see Ref.[29]):

$$\sqrt{-Km^2\alpha'^2}\, \log \sqrt{-Km^2\alpha'^2} \approx -\frac{\pi}{2} K\alpha'\, n \tag{5.9}$$

Notice that the level spacing grows faster than in Minkowski spacetime but slower than in the closed static Robertson-Walker spacetime.

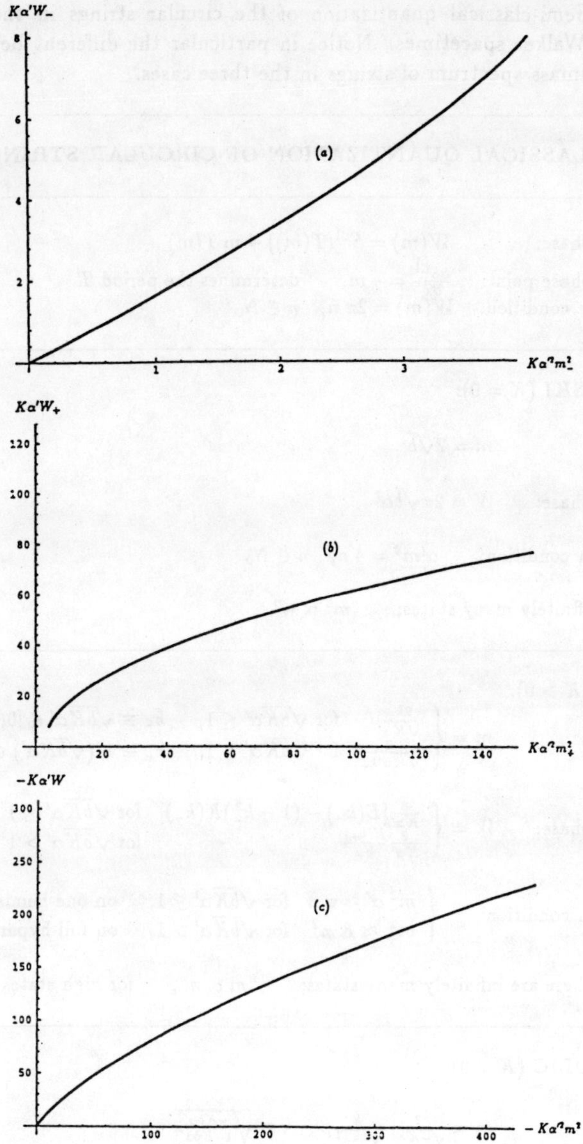

Figure 8. Parametric plot of $K\alpha'W$ as a function of $K\alpha'^2m^2$ in the three cases: (a) $K > 0$, $bK\alpha'^2 \leq 1$ (strings oscillating on one hemisphere, only finitely many states), (b) $K > 0$, $bK\alpha'^2 > 1$ (strings oscillating on the full hypersphere, infinitely many states), (c) $K < 0$ (strings oscillating on the hyperboloid, infinitely many states).

Table 6. Semi-classical quantization of the circular strings in the static Robertson-Walker spacetimes. Notice in particular the different behaviour of the high mass spectrum of strings in the three cases.

SEMICLASSICAL QUANTIZATION OF CIRCULAR STRINGS

Stationary phase: $W(m) = S^{\text{cl}}(T(m)) + m\,T(m)$

Stationary phase point: $\frac{\partial S^{\text{cl}}}{\partial T} = -m,$ determines the period T

Quantization condition: $W(m) = 2\pi\, n,\quad n \in N_0$

MINKOWSKI ($K = 0$):

Mass: $m = 2\sqrt{b}$

Stationary phase: $W = 2\pi\sqrt{b}\alpha'$

Quantization condition: $\alpha' m^2 = 4\, n,\quad n \in N_0$

There are infinitely many states; "$m^2 \propto n$"

CLOSED ($K > 0$):

Mass: $$m = \begin{cases} \frac{2k_-}{\sqrt{K\alpha'}} & \text{for } \sqrt{bK}\alpha' \leq 1, \quad k_- \equiv \sqrt{bK}\alpha' \in [0,1] \\ \frac{2}{\sqrt{K\alpha'}k_+} & \text{for } \sqrt{bK}\alpha' > 1, \quad k_+ \equiv 1/(\sqrt{bK}\alpha') \in]0,1[\end{cases}$$

Stationary phase: $$W = \begin{cases} \frac{8}{K\alpha'}[E(k_-) - (1 - k_-^2)K(k_-)] & \text{for } \sqrt{bK}\alpha' \leq 1 \\ \frac{8}{K\alpha'}\frac{E(k_+)}{k_+} & \text{for } \sqrt{bK}\alpha' > 1 \end{cases}$$

Quantization condition: $$\begin{cases} m_-^2\alpha' \approx \pi\, n & \text{for } \sqrt{bK}\alpha' \leq 1, \quad \text{on one hemisphere.} \\ m_+^2 \approx K\, n^2 & \text{for } \sqrt{bK}\alpha' > 1, \quad \text{on full hypersphere.} \end{cases}$$

Altogether there are infinitely many states; "$m \propto n$", for high states

HYPERBOLIC ($K < 0$):

Mass: $m = \frac{2}{\sqrt{-K\alpha'}}\frac{k}{\sqrt{1-k^2}},\quad k \equiv \sqrt{\frac{-Kb\alpha'^2}{1-Kb\alpha'^2}}$

Stationary phase: $W = \frac{8}{K\alpha'}\frac{E(k)-K(k)}{\sqrt{1-k^2}}$

Quantization condition: $\sqrt{-Km^2\alpha'^2}\log\sqrt{-Km^2\alpha'^2} \approx -\frac{\pi}{2}K\alpha'\, n$

There are infinitely many states; "$m \log m \propto n$", for high states

On the other hand, the stationary strings are determined by Eqs.(4.4):

$$\phi' = \frac{L}{r^2}, \qquad r'^2 + V(r) = 0; \quad V(r) = (1 - Kr^2)(\frac{L^2}{r^2} - 1), \qquad (5.10)$$

where L is an integration constant (Fig.9 shows the potential $V(r)$ for $K > 0$, $K = 0$, $K < 0$).

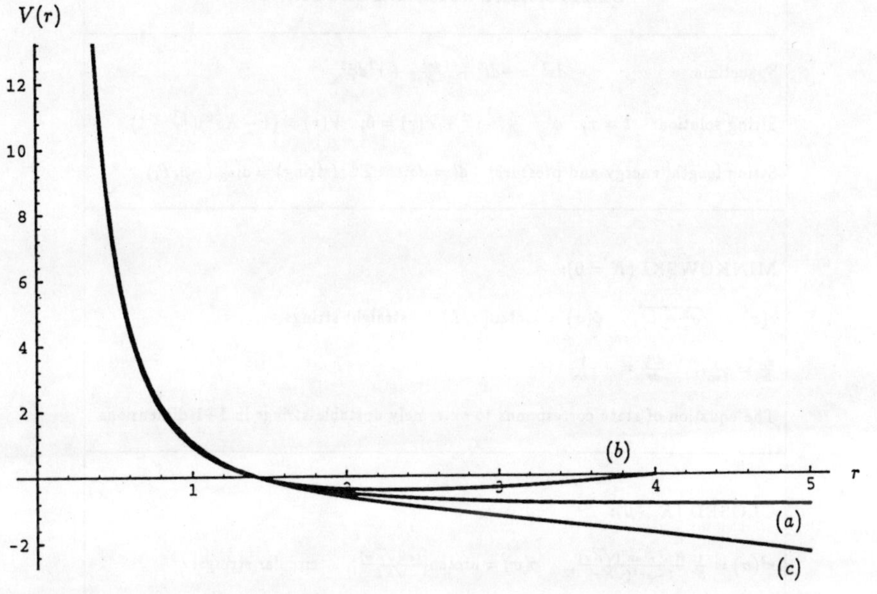

Figure 9. The potential $V(r)$ introduced in Eqs.(5.10) for a stationary string in the static Robertson-Walker spacetimes: (a) flat ($K = 0$), (b) closed ($K > 0$ *and* $KL^2 \leq 1$), (c) hyperbolic ($K < 0$).

For $K > 0$, all the stationary string solutions describe circular strings winding around the hypersphere. The equation of state is of the extremely unstable string type [21]. For $K < 0$, the stationary strings are represented by infinitely long open configurations with an angle between the two "arms" given by:

$$\Delta\phi = \pi - 2\arctan(\sqrt{-KL}). \qquad (5.11)$$

The energy density is positive while the pressure densities are negative. No simple equation of state is found for these solutions.

We also computed the first and second order fluctuations around a static string center of mass, using the string perturbation series approach [7] and its covariant versions [9, 26]. Up to second order, the mass-formula for arbitrary values of the curvature index (positive or negative) is identical to the well-known flat spacetime mass-formula; all dependence on K cancels out.

Table 7. Stationary strings in the static Robertson-Walker spacetimes. Notice that the pressure densities are always negative in all three cases.

STATIONARY STRING SOLUTIONS
Spacetime: $ds^2 = -dt^2 + \frac{dr^2}{1-Kr^2} + r^2 d\phi^2$ String solution: $t = \tau,\quad \phi' = \frac{L}{r^2},\quad r'^2 + V(r) = 0;\quad V(r) = (1 - Kr^2)(\frac{L^2}{r^2} - 1)$ String length, energy and pressure: $dl = d\sigma,\quad T^{\mu}{}_{\nu}(\text{string}) = \text{diag.}(-\rho, P_i)$
MINKOWSKI ($K = 0$): $r(\sigma) = \sqrt{\sigma^2 + L^2},\quad \phi(\sigma) = \arctan(\sigma/L)$, straight strings. $\frac{d\rho}{d\sigma} = \frac{1}{2\pi\alpha'},\quad \frac{dP_y}{d\sigma} = -\frac{1}{2\pi\alpha'}$ The equation of state corresponds to extremely unstable strings in 1+1 dimensions.
CLOSED ($K > 0$): $r^2(\sigma) = \frac{1}{K}\frac{KL^2 + \tan^2(\sqrt{K}\sigma)}{1 + \tan^2(\sqrt{K}\sigma)},\quad \phi(\sigma) = \arctan[\frac{\tan(\sqrt{K}\sigma)}{\sqrt{K}L}]$, circular strings. $\rho = \frac{1}{\sqrt{K}\alpha'},\quad P_x = P_z = -\frac{1}{2\sqrt{K}\alpha'}$ The equation of state corresponds to extremely unstable strings in 2+1 dimensions.
HYPERBOLIC ($K < 0$): $r^2(\sigma) = \frac{1}{K}\frac{KL^2 - \tanh^2(\sqrt{-K}\sigma)}{1 - \tanh^2(\sqrt{-K}\sigma)},\quad \phi(\sigma) = \pm\left\{\frac{\pi}{2} - \arctan[\sqrt{-K}L\tanh^{-1}(\sqrt{-K}\sigma)]\right\}$ Infinitely long open strings, $\Delta\phi = \pi - 2\arctan(\sqrt{-K}L)\ \in\ ]0, \pi]$ $\frac{d\rho}{d\sigma} = \frac{1}{2\pi\alpha'},\quad \frac{dP_x}{d\sigma} = \frac{8KL^2}{\pi\alpha'(1-KL^2)^2}e^{-\sqrt{-K}\lvert\sigma\rvert},\quad \frac{dP_y}{d\sigma} = \frac{-8}{\pi\alpha'(1-KL^2)^2}e^{-\sqrt{-K}\lvert\sigma\rvert}$ The pressure densities are negative but asymptotically go to zero exponentially. No simple equation of state in the general case.

6 Conclusion

We first studied the string propagation in the 2+1 BH-AdS background. We found the first and second order perturbations around the string center of mass as well as the mass formula, and compared with the ordinary black hole AdS spacetime and with the black string background as well. These results were then generalized to ordinary D-dimensional anti de Sitter spacetime. The classical string motion in anti de Sitter spacetime is stable in the sense that it is oscillatory in time with real frequencies and the string size and energy are bounded. Quantum mechanically, this reflects in the mass operator, which is well defined for any value of the wave number n, and arbitrary high mass states (and therefore an infinite number of states) can be constructed. This is to be contrasted with de Sitter spacetime, where string instabilities develop, in the sense that the string size and energy become unbounded for large de Sitter radius. For low mass states (the stable regime), the mass operator in de Sitter spacetime is given by Eq.(3.1) but with

$$\Omega_n(\lambda)_{\text{dS}} = \frac{2n^2 - m^2\alpha'\lambda}{\sqrt{n^2 - m^2\alpha'\lambda}}.$$

Real mass solutions can be defined only up to some *maximal mass* of the order $m^2\alpha' \approx 1/\lambda$ [7]. For $\lambda << 1$, real mass solutions can be defined only for $N \leq N_{\text{max}} \sim 0.15/\lambda$ (where N is the eigenvalue of the number operator) and therefore there exists a *finite* number of states only. These features of strings in de Sitter spacetime have been recently confirmed within a different (semi-classical) quantization approach based on *exact* circular string solutions [42].

The presence of a cosmological constant Λ (positive or negative) increases considerably the number of levels of different eigenvalue of the mass operator (there is a splitting of levels) with respect to flat spacetime. That is, a non-zero cosmological constant *decreases* (although does not remove) the degeneracy of the string mass states, introducing a *fine structure effect*. For the low mass states the level spacing is approximately constant (up to corrections of the order λ^2). For the high mass states, the changes are more drastic and they depend crucially on the sign of Λ. A value $\Lambda < 0$ causes the *growing* of the level spacing linearly with N instead of being constant as in Minkowski space. Consequently, the density of states $\rho(m)$ grows with the exponential of $\sqrt{m}$ (instead of m as in Minkowski space) discarding the existence of a Hagedorn temperature in AdS spacetime, and the possibility of a phase transition. In addition, another important feature of the high mass string spectrum in AdS spacetime is that it becomes independent of α'. The string scale for the high mass states is given by $\mid \Lambda \mid$, instead of $1/\alpha'$ for the low mass states.

We have found new classes of exact string solutions obtained either by the circular string ansatz, Eq.(2.16), or by the stationary string ansatz, Eq.(4.3), in a variety of curved backgrounds including Schwarzschild, de Sitter and anti de Sitter spacetimes. Many different types of solutions have been found: oscillating circular strings, closed stationary strings, infinitely long stationary strings, dynamical straight strings and multi-string solutions describing finitely or infinitely many stationary or dynamical strings. In all the cases we have obtained the exact solutions in terms of either elementary or elliptic functions. Furthermore, we have analyzed the physical properties (length, energy, pressure) of the string solutions.

Finally, we solved the equations of motion and constraints for circular strings in static Robertson-Walker spacetimes. We computed the equations of state and found that there exists a self-consistent solution to the Einstein equations in the case of positive spatial curvature. The solutions were quantized semi-classically using the stationary phase approximation method, and the resulting spectra were analyzed and discussed. We have also found all stationary string configurations in these spacetimes and we computed the corresponding physical quantities, string length, energy and pressure.

Acknowledgements:
A.L. Larsen is supported by the Danish Natural Science Research Council under grant No. 11-1231-1SE

References

[1] H.J. de Vega and N. Sánchez, Phys. Rev. D42 (1990) 3969.
H.J. de Vega, M. Ramón-Medrano and N. Sánchez, Nucl. Phys. B374 (1992) 405.

[2] H.J. de Vega and N. Sánchez, Phys. Lett. B244 (1990) 215, Phys. Rev. Lett. 65C (1990) 1517, IJMP A7 (1992) 3043, Nucl. Phys. B317 (1989) 706 and ibid 731.
D. Amati and C. Klimcik, Phys. Lett. B210 (1988) 92.
M. Costa and H.J. de Vega, Ann. Phys. 211 (1991) 223 and ibid 235.
C. Loustó and N. Sánchez, Phys. Rev. D46 (1992) 4520.

[3] V.E. Zakharov and A.V. Mikhailov, JETP 47 (1979) 1017.
H. Eichenherr, in "Integrable Quantum Field Theories", ed. J. Hietarinta and C. Montonen (Springer, Berlin, 1982).

[4] H.J. de Vega and N. Sánchez, Phys. Rev. D47 (1993) 3394.

[5] I. Bars and K. Sfetsos, Mod. Phys. Lett. A7 (1992) 1091.

[6] H.J. de Vega, J.R. Mittelbrunn, M.R. Medrano and N. Sánchez, Phys. Lett. B232 (1994) 133 and "The Two-Dimensional Stringy Black Hole: a new Approach and a Pathology", PAR-LPTHE 93/14.

[7] H.J. de Vega and N. Sánchez, Phys. Lett. B197 (1987) 320.

[8] H.J. de Vega and N. Sánchez, Nucl. Phys. B299 (1988) 818.

[9] P.F. Mende, in "String Quantum Gravity and Physics at the Planck Energy Scale". Proceedings of the Erice Workshop held in June 1992, ed. N. Sánchez (World Scientific, 1993).

[10] A.L. Larsen and V.P. Frolov, Nucl. Phys. B414 (1994) 129.

[11] H.J. de Vega, A.V. Mikhailov and N. Sánchez, Teor. Mat. Fiz. 94 (1993) 232, Mod. Phys. Lett. A29 (1994) 2745.

[12] F. Combes, H.J. de Vega, A.V. Mikhailov and N. Sánchez, Phys. Rev. D50 (1994) 2754.

[13] H.J. de Vega, A.L. Larsen and N. Sánchez, Nucl. Phys. B427 (1994) 643.

[14] I. Krichever, "Two-Dimensional Algebraic-Geometrical Operators with Self-Consistent Potentials", Landau Institute Preprint, March, 1994, to appear in Funct. Analisis and Appl.

[15] H.J. de Vega and N. Sánchez, Nucl. Phys. B309 (1988) 552 and ibid, 577.

[16] C.O. Loustó and N. Sánchez, Phys. Rev. D47 (1993) 4498.

[17] H.J. de Vega and I.L. Egusquiza, Phys. Rev. D50 (1994) 763.

[18] A.L. Larsen, Phys. Rev. D50 (1994) 2623.

[19] A.L. Larsen and M. Axenides, Phys. Lett. B318 (1993) 47.

[20] H.J. de Vega and N. Sánchez, Phys. Rev. D50 (1994) 7202.

[21] N. Sánchez and G. Veneziano, Nucl. Phys. B333 (1990) 253.
M. Gasperini, N. Sánchez and G. Veneziano, IJMP A6 (1991) 3853, Nucl. Phys. B364 (1991) 365.

[22] H.J. de Vega, M. Ramón-Medrano and N. Sánchez, Nucl. Phys. B374 (1992) 425.

[23] H.J. de Vega and N. Sánchez, Phys. Rev. D45 (1992) 2783.
M. Ramón-Medrano and N. Sánchez, Class. Quantum Grav. 10 (1993) 2007.

[24] N. Sánchez, Phys. Lett. B195 (1987) 160.

[25] H.J. de Vega, M. Ramón-Medrano and N. Sánchez, Nucl. Phys. B351 (1991) 277.

[26] A.L. Larsen and N. Sánchez, Phys. Rev. D50 (1994) 7493.

[27] A.L. Larsen and N. Sánchez, "Mass Spectrum of Strings in Anti de Sitter Spacetime", DEMIRM-Paris-94048, to appear in Phys. Rev. D.

[28] A.L. Larsen and N. Sánchez, "New Classes of Exact Multi-String Solutions in Curved Spacetimes", DEMIRM-Paris-95003, to appear in Phys. Rev. D.

[29] A.L. Larsen and N. Sánchez, "The Effect of Spatial Curvature on the Classical and Quantum Strings", DEMIRM-Paris-95004.

[30] M. Bañados, C. Teitelboim and J. Zanelli, Phys. Rev. Lett. 69 (1992) 1849.

[31] G.T. Horowitz and D.L. Welch, Phys. Rev. Lett. 71 (1993) 328.

[32] N. Kaloper, Phys. Rev. D48 (1993) 2598.

[33] M. Bañados, M. Henneaux, C. Teitelboim and J. Zanelli, Phys. Rev. D48 (1993) 1506.

[34] C. Farina, J. Gamboa and A.J. Seguí-Santonja, Class. Quantum Grav. 10 (1993) L193.
N. Cruz, C. Martínez and L. Peña, Class. Quantum Grav. 11 (1994) 2731.

[35] A. Achúcarro and M.E. Ortiz, Phys. Rev. D48 (1993) 3600.
D. Cangemi, M. Leblanc and R.B. Mann, Phys. Rev. D48 (1993) 3606.

[36] J. Horne and G.T. Horowitz, Nucl. Phys. B368 (1992) 444.

[37] T. Buscher, Phys. Lett. B201 (1988) 466, B194 (1987) 59.

[38] A. Vilenkin, Phys. Rev. D24 (1981) 2082.

[39] H.J. de Vega and N. Sánchez, Int. J. Mod. Phys. A7 (1992) 3043.

[40] V.P. Frolov, V.D. Skarzhinsky, A.I. Zelnikov and O. Heinrich, Phys. Lett. B224 (1989) 255.

[41] R. Dashen, B. Hasslacher and A. Neveu, Phys. Rev. D11 (1975) 3424.

[42] H.J. de Vega, A.L. Larsen and N. Sánchez, "Semi-Classical Quantization of Circular Strings in de Sitter and anti de Sitter Spacetimes", DEMIRM-Paris-94049, submitted to Phys. Rev. D.

[43] A.L. Larsen, "Stable and Unstable Circular Strings in Inflationary Universes", NORDITA-94/14-P, to appear in Phys. Rev. D.

[44] A.L. Larsen, Nucl. Phys. B412 (1994) 372.

WHAT IS SPACE-TIME MADE OF?

E. L. Schucking, *Physics Dept., New York Univ., New York, NY 10003*
Jerome Epstein, *Physics Dept., Lehman College, CUNY, Bronx, NY 10468*
William P. Kowalski,*Miss. Baptist Medical Ctr., Jackson MS, 39202*
Salvatore Lauro, *Physics Dept., St. John's Univ., Jamaica NY, 11439*

ABSTRACT

The Euclidean $\mathbb{R}^3$ as carrier of oriented spheres and oriented planes is a four dimensional manifold invariant under the 10 dimensional Laguerre group, known to physicists as the Poincaré group. Complex two spinors serve as plane coordinates for the S^2's. By generalizing the complex spinors to quaternion and octonion spinors, we obtain Laguerre geometries of S^4's in $\mathbb{R}^5$ and S^8's in $\mathbb{R}^9$ which introduce automorphism groups U_1 , SU_2 , and SU_3 . The symmetry is broken by choosing an imaginary unit as the Higgs field. This splits the higher dimensional space into internal and ordinary space.

I. INTRODUCTION

One might suspect that the failure to quantize gravititation is partially due to the strategy of starting with the pure vacuum field. This attitude may have its roots in the misinterpretation of Einstein's theory of gravitation as "General Relativity". A few words about this historical question: Mach suggested in the 1880's that rotation might not be absolute -- picking up an idea of Bishop Berkeley's. With the advent of Special Relativity in 1905 it becomes essentially impossible to retain this view. If the Earth stood still and the distant stars rotated, they would move with fantastic multiples of the speed of light.

When Einstein created his theory of gravitation, it was perhaps not clear that angular velocity and acceleration of a time-like congruence are invariants. When he introduced the cosmological constant his Einstein cosmos would demonstrate general relativity by showing that mass distribution determines the metric. But that didn't work; there were vacuum solutions. Perhaps the first to see that the gravitational field of matter was only a ripple on the ocean of the vacuum was Hermann Weyl. Without gravitation but with a cosmological constant there was a de Sitter hyperboloid. If one makes the cosmological constant zero, one still has Minkowski space-time.

We believe now that Minkowski space-time has a source. It is a scalar field, a conformal factor, that provides the scale for all matter and thus also for the metric, since space can only be measured by matter. It is thus artificial to split the Higgs field from the gravitational field. They must be quantized together. The Higgs is a scalar under Lorentz transformations, but it carries weak hypercharge and weak isospin. Thus the combined Higgs-gravitational field has internal degrees of freedom that may be essential for quantization. We view space as "space-matter modulo matter". Describing space without potential matter fields is unphysical and leads only to unmeasurable mathematical space. We have to quantize space and matter together.

When Minkowski gave his famous speech before the German Society of Scientists and Physicians in 1908, he advocated the marriage of space and time. Einstein at the time didn't think much of the union but later approved it. We wish to advocate divorce, because it seems that in Quantum Mechanics space and time must be treated differently. But enough complaining; let us make some suggestions for an alternative approach, first at the classical level, hoping that this might provide a fresh start toward quantization.

II. LAGUERRE GEOMETRY

MOTTO: "Le metrè est a longueur dur trajet parcouru dans le vide par la lumière pendant une durée de 1/299,792,458 de seconde."

17th General Conference on Measures and Weights, October 1983.

The Euclidean $\mathbb{R}^3$ is usually considered as a 3-dimensional manifold whose irreducible elements are points. It is given a metric structure through the definition of a distance between points. Prior to Relativity such distances were identified with marks on rigid rods. This procedure is no longer satisfactory due to Lorentz contraction. The new definition of the unit of length reflects Einstein's postulate of the constancy of the light speed and reduces the measurement of length to the measurement of time intervals.

A flash of light issuing from the origin of an inertial frame at t=0 is found at later times t_1 and $t_2 > t_1$ on spheres with radii t_2 and t_1 (c=1). The difference of the radii $t_2 - t_1$ is the geometric picture of the time interval. This suggests a different view of the $\mathbb{R}^3$ as the carrier of the four dimensional manifold of oriented spheres.

A sphere in $\mathbb{R}^3$ is characterized by the 3 coordinates of its center and the fourth coordinate of its radius, where the sign of the radius gives the orientation of the sphere. This point of view is by no means new. It was introduced in 1879 by Laguerre. One naturally includes oriented planes as spheres with infinite radius.

For two oriented spheres with centers $\vec{x}_1$ and $\vec{x}_2$ and radii t_1 and t_2, Laguerre introduced as invariant their tangential distance, s, given by:

$$|s|^2 = |(\vec{x}_1 - \vec{x}_2)\bullet(\vec{x}_1 - \vec{x}_2) - (t_1 - t_2)^2| \qquad (1)$$

The linear transformations in Laguerre's sphere geometry that transform spheres into spheres and planes into planes while keeping $|s|^2$ invariant form his 10 parameter Laguerre group, known to physicists as the Poincaré Group.

By dropping the point as irreducible element of $\mathbb{R}^3$ and replacing it by a sphere, one may gain an advantage for the inevitable quantization of geometry, since spheres are extended objects with built-in Lorentz invariance.

III. LAGUERRE GEOMETRY AND MINKOWSKI SPACE-TIME

Since Minkowski's space-time geometry is better known to physicists than Laguerre's space geometry, it is worthwhile to view Laguerre's geometry from the four dimensional framework. Minkowski's irreducible elements are events, i.e. space-time points, with coordinates $(\vec{x},t)$ Since matter has a persistence tendency and doesn't occur as "instantons", the event is not well approximated as a physical object. Instead the light cone with a given event as vertex provides a better description.

We can describe the light cone with vertex $(\vec{x},t)$ through its intersection with the plane t=0. This gives a sphere with radius |t| (with c=1). We give the sphere an orientation by fitting it with a set of normals pointing either outward or inward, depending on whether t is positive or negative. In this way the $\mathbb{R}^3$ becomes the carrier of oriented spheres. For t=0 we obtain points without orientation. The events of Minkowski space-time are thus mapped one-to-one on the oriented spheres of an $\mathbb{R}^3$.

Further we consider null hyperplanes in Minkowski space-time. They intersect the space section t=0 in a plane. A future directed null vector in the null hyperplane, when parallelly projected along the t-axis onto the section t=0 gives a normal to the plane in $\mathbb{R}^3$ and provides it with an orientation. The $\mathbb{R}^3$ becomes thus the carrier of oriented planes that are in one-to-one correspondence with the null hyperplanes of space-time.

An oriented sphere in $\mathbb{R}^3$ can now be viewed as the envelope of the oriented tangent planes. Two oriented spheres in $\mathbb{R}^3$ that share all their oriented tangent planes with each other will be identical, if they have none in common we call them timelike, and null if only one is shared. If the spheres are not identical, and have more than one tangent plane in common, the spheres lie space-like to each other. In this case, they share an infinite number of oriented tangent planes which form the envelope of a cone. The line segments of the generators of this cone connecting the two spheres all have equal length and measure the tangential distance s, the invariant of the Laguerre group.

IV. LAGUERRE GEOMETRY IN n DIMENSIONS.

The equation of a hyperplane in n dimensional affine space $\mathbb{R}^n$ is given by:

$$u_j x^j + u_{n+1} = 0 \qquad (j = 1, \ldots n) \tag{2}$$

with

$$u_j u_j > 0 \tag{3}$$

(summation convention assumed). For given u_j and u_{n+1} the x^j are the coordinates of a point lying on the hyperplane. On the other hand, for given x^j the set of n+1 numbers $\{u_j, u_{n+1}\}$ are hyperplane coordinates characterizing all hyperplanes that go through the fixed point x^j. The hyperplane coordinates $\{u_j, u_{n+1}\}$ are redundant. That is to say, the set

$$\{u_j/u_0, \; u_{n+1}/u_0\} \tag{4}$$

for arbitrary, positive u_0 gives rise to the same hyperplane. We specialize now to the case,

$$\mathbb{R}^n \text{ oriented} \qquad n \text{ odd} \tag{5}$$

Under affine transformations with positive determinant u_j/u_0 cannot be continuously transformed into $-u_j/u_0$. We distinguish now the hyperplane coordinates in (4) from those with sign reversed as the two different coordinates of two *oriented* hyperplanes which share the same point set of the oriented $\mathbb{R}^n$ but differ in their orientation. We normalize u_0 then by the equation:

$$(u_0)^2 - u_j u_j = 0 \tag{6}$$

To motivate the following let us assume the oriented $\mathbb{R}^n$ has a negative definite Euclidean metric:

$$g_{jk} = -\delta_{jk} \tag{7}$$

Then u_j/u_0 are the covariant components of a unit vector and $-u_j/u_0$ are its contravariant components. Our aim is to describe an oriented sphere with center $x_{(1)}{}^j$ and radius $-x_{(1)}{}^0$ in terms of its oriented tangent hyperplanes. The points of the sphere have coordinates:

$$x^j = x_{(1)}{}^j + (u_j/u_0)x_{(1)}{}^0 \tag{8}$$

where the sign of $x_{(1)}{}^0$ gives the orientation of the sphere, positive for the outer normal and negative for the inward normal. The hyperplanes through these points with covariant normal u_j/u_0 are given by (2). We find after multiplying through by u_0:

$$u_0 x_{(1)}{}^0 + u_j x_{(1)}{}^j + u_{n+1} = 0 \tag{9}$$

or, letting Greek indices run from 0 to n:

$$u_\mu x_{(1)}{}^\mu + u_{n+1} = 0 \qquad \eta^{\mu\nu} u_\mu u_\nu = 0 \tag{10}$$

with positive u_0, characterizing oriented spheres in hyperplane coordinates, with $\eta^{\mu\nu}$ as the Minkowski metric. If a second oriented sphere with n+1 coordinates $x_{(2)}{}^{\mu}$ shares a tangent hyperplane with the sphere $x_{(1)}{}^{\mu}$,we have (because (10) holds for sphere 2 as well):

$$u_{\mu}(x_{(1)}{}^{\mu} - x_{(2)}{}^{\mu}) = 0 \tag{11}$$

This means:

$$(x_{(1)}{}^{\mu} - x_{(2)}{}^{\mu})\eta_{\mu\nu}(x_{(1)}{}^{\nu} - x_{(2)}{}^{\nu}) \leq 0 \tag{12}$$

The two spheres are then null or spacelike with respect to each other.

V. THE PROPER ISOCHRONOUS LAGUERRE GROUP

The coordinates x^{μ} of an oriented sphere appear only linearly in (10). That means the oriented spheres of the $\mathbb{R}^n$ form a linear manifold. Under the proper isochronous Laguerre group we have:

$$u_{\mu} = u'_{\nu} A^{\nu}{}_{\rho}, \quad x'^{\nu} = A^{\nu}{}_{\mu} x^{\mu} + \beta^{\nu}, \quad A^{0}{}_{0} > 0, \quad A^{\mu}{}_{\nu}\eta_{\mu\rho} A^{\rho}{}_{\sigma} = \eta_{\nu\sigma} \tag{13}$$

This gives:

$$u_{\mu}x^{\mu} + u_{n+1} = 0 = u'_{\nu} A^{\nu}{}_{\mu}x^{\mu} + u_{n+1} = u'_{\nu} x'^{\nu} + u_{n+1} - u'_{\nu}\beta^{\nu} \tag{14}$$

Comparing this with

$$u'_{\nu} x'^{\nu} + u'_{n+1} = 0 \tag{15}$$

gives the transformation law of the hyperplane coordinate u_{n+1}:

$$u_{n+1} = u'_{n+1} + u'_{\nu}\beta^{\nu} \tag{16}$$

This shows that u_{n+1} behaves like a scalar under the homogenous transformations and picks up an additive term under translations. It is an easy consequence now that the difference of two spheres $x_{(1)}{}^{\mu}$ and $x_{(2)}{}^{\mu}$, given by:

$$(x_{(1)}{}^{\mu} - x_{(2)}{}^{\mu})\eta_{\mu\nu}(x_{(1)}{}^{\nu} - x_{(2)}{}^{\nu}) = \text{const.} \tag{17}$$

is an invariant of the Laguerre group.

VI. THE THREE DIMENSIONAL SPACE

We introduce generalized Pauli matrices σ^{μ}

$$\sigma^0 = \begin{pmatrix} 1 & 0 \\ 0 & 1 \end{pmatrix} \quad \sigma^1 = \begin{pmatrix} 0 & 1 \\ 1 & 0 \end{pmatrix} \quad \sigma^2 = \begin{pmatrix} 0 & -i \\ i & 0 \end{pmatrix} \quad \sigma^3 = \begin{pmatrix} 1 & 0 \\ 0 & -1 \end{pmatrix}$$

and $\widehat{\sigma}^{\mu}$: (18)

$$\widehat{\sigma}^0 = \begin{pmatrix} 1 & 0 \\ 0 & 1 \end{pmatrix} \quad \widehat{\sigma}^1 = -\begin{pmatrix} 0 & 1 \\ 1 & 0 \end{pmatrix} \quad \widehat{\sigma}^2 = -\begin{pmatrix} 0 & -i \\ i & 0 \end{pmatrix} \quad \widehat{\sigma}^3 = -\begin{pmatrix} 1 & 0 \\ 0 & -1 \end{pmatrix}$$

We have then:

$$\sigma^{\mu}\widehat{\sigma}^{\nu} + \sigma^{\nu}\widehat{\sigma}^{\mu} = \widehat{\sigma}^{\mu}\sigma^{\nu} + \widehat{\sigma}^{\nu}\sigma^{\mu} = \eta^{\mu\nu}\mathbf{1} \tag{19}$$

Let Λ be a 2 by 2 complex matrix with determinant 1:

$$\Lambda = \begin{pmatrix} a & b \\ c & d \end{pmatrix} \qquad a, b, c, d \in \mathbf{C} \qquad ad - bc = 1 \tag{20}$$

and ϕ a complex 2-spinor. Then

$$\phi = \begin{pmatrix} \phi_1 \\ \phi_2 \end{pmatrix} \qquad \phi' = \begin{pmatrix} \phi'_1 \\ \phi'_2 \end{pmatrix} = \Lambda\phi \tag{21}$$

transforms under $SL_2(\mathbf{C})$ which is the covering group of the homogeneous orthochronous, proper Lorentz group.
The adjoint spinor $\phi^{\dagger}$ transforms according to:

$$\phi^{\dagger\prime} = \phi^{\dagger}\Lambda^{\dagger}, \qquad \Lambda^{\dagger} = \begin{pmatrix} \overline{a} & \overline{c} \\ \overline{b} & \overline{d} \end{pmatrix} \qquad \phi^{\dagger} = \begin{pmatrix} \overline{\phi}_1 & \overline{\phi}_2 \end{pmatrix} \tag{22}$$

We turn now to a spinor formulation of Laguerre Geometry. The quadratic equation for the plane coordinates u_{μ} can be solved by putting:

$$\phi\phi^{\dagger} = \sigma^{\mu} u_{\mu} \tag{23}$$

This gives:

$$u_0 = \tfrac{1}{2} \operatorname{tr} \phi\phi^{\dagger} = \tfrac{1}{2}(|\phi_1|^2 + |\phi_2|^2) \tag{24}$$

and guarantees $u_0 > 0$ for a non-vanishing spinor ϕ. Because

$$\phi'\phi'^{\dagger} = \Lambda\sigma^{\mu}\Lambda^{\dagger} u_{\mu} = u_{\mu}(A^{-1\ \dagger})^{\mu}{}_{\nu}\sigma^{\nu} = u'_{\nu}\sigma^{\nu} \tag{25}$$

with (13) we have for u_{μ} the transformation law of a covector which preserves the equation:

$$\det |\phi\phi^{\dagger}| = (u_0)^2 - (u_1)^2 - (u_2)^2 - (u_3)^2 = 0 \tag{26}$$

In order to achieve also invariance under space reflections we introduce a second kind of 2-component spinor

$$\hat{\phi} = \begin{pmatrix} \hat{\phi}_1 \\ \hat{\phi}_2 \end{pmatrix} \qquad \hat{\phi}' = \begin{pmatrix} \hat{\phi}'_1 \\ \hat{\phi}'_2 \end{pmatrix} = \Lambda^{-1\dagger}\ \hat{\phi} \tag{27}$$

which transforms also under SL2($\mathbf{C}$). The two representations, Λ and $\Lambda^{-1\dagger}$, are inequivalent. The adjoint spinor $\hat{\phi}^{\dagger}$ transforms according to:

$$\hat{\phi}'^{\dagger} = \hat{\phi}^{\dagger}\Lambda^{-1}, \qquad \hat{\phi}^{\dagger} = \begin{pmatrix} \bar{\hat{\phi}}_1 & \bar{\hat{\phi}}_2 \end{pmatrix} \tag{28}$$

We represent now the oriented sphere x^{μ} by the Hermitian 2 x 2 matrix $\hat{X}$ as:

$$\hat{X} = \hat{\sigma}_{\mu} x^{\mu} \tag{29}$$

Under homogeneous proper orthochronous Lorentz transformations:

$$\hat{X}' = \Lambda^{-1\dagger}\,\hat{X}\Lambda^{-1} = \Lambda^{-1\dagger}\,\hat{\sigma}_{\mu}\,\Lambda^{-1} x^{\mu} = \hat{\sigma}_{\nu} A^{\nu}{}_{\mu} x^{\mu} = \hat{\sigma}_{\nu} x'^{\nu} \tag{30}$$

This gives the transformation law:

$$x'^{\nu} = A^{\nu}{}_{\mu} x^{\mu} \tag{31}$$

One confirms easily from (19) that:

$$\mathrm{tr}\ \sigma^\mu \hat{\sigma}^\nu = 2\eta^{\mu\nu} \tag{32}$$

and therefore:

$$\tfrac{1}{2}\mathrm{tr}(\ \phi\phi^\dagger \hat{X}) = \mathrm{tr}\ \tfrac{1}{2}\sigma^\mu u_\nu \hat{\sigma}_\nu x^\nu = u_\nu x^\nu \tag{33}$$

The equation for a sphere in plane coordinates becomes thus:

$$\tfrac{1}{2}\mathrm{tr}(\ \phi\phi^\dagger \hat{X}) + u_4 = 0 \tag{34}$$

VII. SPACE INVERSION

Instead of (33), we could have written:

$$\tfrac{1}{2}\mathrm{tr}\ (\hat{\phi}\hat{\phi}^\dagger X) + u_4 = 0 \tag{35}$$

where, analogous to (29):

$$X = \sigma_\mu x^\mu \tag{36}$$

and transforms as

$$X' = \Lambda X \Lambda^\dagger = \sigma_\nu x'^\nu \tag{37}$$

Combining (34) and (35) we get:

$$\tfrac{1}{4}\mathrm{tr}(\phi\phi^\dagger \hat{X} + \hat{\phi}\hat{\phi}^\dagger X) + u_4 = 0 \tag{38}$$

which is invariant under total space reflection.
We introduce 4-component spinors Ψ by combining ϕ and $\hat{\phi}$:

$$\Psi = \begin{pmatrix} \hat{\phi} \\ \phi \end{pmatrix} \tag{39}$$

The space inversion Π is then given by interchange of ϕ and $\hat{\phi}$:

$$\Psi' = \Pi\Psi = \begin{pmatrix} 0 & 1 \\ 1 & 0 \end{pmatrix}\begin{pmatrix} \hat{\phi} \\ \phi \end{pmatrix} = \begin{pmatrix} \phi \\ \hat{\phi} \end{pmatrix} \tag{40}$$

Correspondingly we get for the sphere coordinates:

$$\begin{pmatrix} \hat{X}' & \mathbf{0} \\ \mathbf{0} & X' \end{pmatrix} = \begin{pmatrix} \mathbf{0} & \mathbf{1} \\ \mathbf{1} & \mathbf{0} \end{pmatrix} \begin{pmatrix} \hat{X} & \mathbf{0} \\ \mathbf{0} & X \end{pmatrix} \begin{pmatrix} \mathbf{0} & \mathbf{1} \\ \mathbf{1} & \mathbf{0} \end{pmatrix} = \begin{pmatrix} X & \mathbf{0} \\ \mathbf{0} & \hat{X} \end{pmatrix} \tag{41}$$

The bold face **0** and **1** indicate 2 x 2 submatrices.

VIII. THE LAGUERRE GROUP

The space-time translations are given by:

$$x'^{\mu} = x^{\mu} + \beta^{\mu} \qquad \beta^{\mu} \in \mathbb{R} \tag{42}$$

We can form the Hermitian matrices:

$$B = \sigma_{\mu}\beta^{\mu} \qquad \hat{B} = \hat{\sigma}_{\mu}\beta^{\mu} \tag{43}$$

This enables us to describe the space-time translations by:

$$X' = X + B \qquad \hat{X}' = \hat{X} + \hat{B} \tag{44}$$

The covering group of the Laguerre group becomes then:

$$X' = \Lambda X \Lambda^{\dagger} + B \qquad \hat{X}' = \Lambda^{-1\dagger} \hat{X} \Lambda^{-1} + \hat{B} \tag{45}$$

We show a matrix representation of the group in terms of 4 x 4 matrices:

$$\begin{pmatrix} \mathbf{1} & X' \\ \mathbf{0} & \mathbf{1} \end{pmatrix} = \begin{pmatrix} \Lambda & \frac{1}{2}B\Lambda^{-1\dagger} \\ \mathbf{0} & \Lambda^{-1\dagger} \end{pmatrix} \begin{pmatrix} \mathbf{1} & X \\ \mathbf{0} & \mathbf{1} \end{pmatrix} \begin{pmatrix} \Lambda^{-1} & \frac{1}{2}\Lambda^{-1}B \\ \mathbf{0} & \Lambda^{\dagger} \end{pmatrix} \tag{46}$$

The last matrix is, in fact:

$$\begin{pmatrix} \Lambda^{-1} & \frac{1}{2}\Lambda^{-1}B \\ \mathbf{0} & \Lambda^{\dagger} \end{pmatrix} = \begin{pmatrix} \mathbf{1} & \mathbf{0} \\ \mathbf{0} & -\mathbf{1} \end{pmatrix} \begin{pmatrix} \Lambda & \frac{1}{2}B\Lambda^{-1\dagger} \\ \mathbf{0} & \Lambda^{-1\dagger} \end{pmatrix}^{-1} \begin{pmatrix} \mathbf{1} & \mathbf{0} \\ \mathbf{0} & -\mathbf{1} \end{pmatrix} \tag{47}$$

We have also correspondingly:

$$\begin{pmatrix} 1 & \hat{X}' \\ 0 & 1 \end{pmatrix} = \begin{pmatrix} \Lambda^{-1\dagger} & \frac{1}{2}\hat{B}\Lambda \\ 0 & \Lambda \end{pmatrix} \begin{pmatrix} 1 & \hat{X} \\ 0 & 1 \end{pmatrix} \begin{pmatrix} \Lambda^{\dagger} & \frac{1}{2}\Lambda^{\dagger}\hat{B} \\ 0 & \Lambda^{-1} \end{pmatrix} \tag{48}$$

A covering group for the orthochronous Laguerre group is then obtained by the sum of the representations (46) and (48) to which we add the 8 x 8 matrix:

$$\begin{pmatrix} 1 & X & 0 & 0 \\ 0 & 1 & 0 & 0 \\ 0 & 0 & 1 & \hat{X} \\ 0 & 0 & 0 & 1 \end{pmatrix} = \begin{pmatrix} 0 & 0 & 1 & 0 \\ 0 & 0 & 0 & 1 \\ 1 & 0 & 0 & 0 \\ 0 & 1 & 0 & 0 \end{pmatrix} \begin{pmatrix} 1 & \hat{X} & 0 & 0 \\ 0 & 1 & 0 & 0 \\ 0 & 0 & 1 & X \\ 0 & 0 & 0 & 1 \end{pmatrix} \begin{pmatrix} 0 & 0 & 1 & 0 \\ 0 & 0 & 0 & 1 \\ 1 & 0 & 0 & 0 \\ 0 & 1 & 0 & 0 \end{pmatrix} \tag{49}$$

which interchanges the two representations. The representation so generated is irreducible. The equation for a sphere in plane coordinates becomes now with (34):

$$\begin{aligned} &\tfrac{1}{2}\mathrm{tr}(\phi'\phi'^{\dagger}\ \hat{X}') + u'_4 = \tfrac{1}{2}\mathrm{tr}[\Lambda\phi\phi^{\dagger}\ \Lambda^{\dagger}\ (\Lambda^{-1\dagger} X \Lambda^{-1} + \hat{B})] + u'_4 \\ &\quad = \tfrac{1}{2}\mathrm{tr}(\phi\phi^{\dagger}\hat{X}) + \tfrac{1}{2}\mathrm{tr}(\phi'\phi'^{\dagger}\hat{B}) + u'_4 = \tfrac{1}{2}\mathrm{tr}(\phi\phi^{\dagger}\hat{X}) + u_4 = 0 \end{aligned} \tag{50}$$

We obtain thus a transformation law for the coordinate u_4:

$$u_4 = u'_4 + \tfrac{1}{2}\mathrm{tr}(\phi'\phi'^{\dagger}\ \hat{B}) \tag{51}$$

In the case of complex spinors the spinor $\hat{\phi}$ is obtained from ϕ through the transformation:

$$\hat{\phi}_1 = -\bar{\phi}_2 \qquad \hat{\phi}_2 = \bar{\phi}_1 \tag{52}$$

This gives:

$$\hat{\phi}\hat{\phi}^{\dagger} = \hat{\sigma}^{\mu} u_{\mu} \tag{53}$$

We obtain then, instead of (34), for the equation of a sphere:

$$\tfrac{1}{2}\mathrm{tr}(\hat{\phi}\hat{\phi}^{\dagger}\ X) + u_4 = 0 \tag{54}$$

Combining (54) with (34) gives:

$$\tfrac{1}{4}\operatorname{tr}(\hat{\phi}\hat{\phi}^{\dagger}X+\phi\phi^{\dagger}\hat{X})+u_{4}=0 \qquad (55)$$

which is invariant under space inversion.

IX. DISCUSSION

The preceding sketch of Laguerre geometry has made it clear that it is equivalent to Minkowski geometry. However, when one considers extensions of space-time that include internal symmetry spaces and their quantization, it appears that Laguerre's geometry lends itself more naturally and more definitively to generalization from a physicist's point of view. The physicist, as opposed to the mathematician, who simply assumes that space carries a field of points labeled by coordinates, measures space in terms of matter fields defining length. Space for the physicist is the quotient of space-matter modulo matter. Matter, however, is described by 2-component complex spinor fields, together with the Higgs field that determines (if one believes the standard model) the Compton wavelength (i.e. the mass) of all particles.

The 2-component spinors of Laguerre geometry are chiral and so are the spheres and hyperplanes, while the points of Minkowski space-time are not. The ingredients of Laguerre geometry in the Minkowski picture are the light cones and null hyperplanes of space-time. For the physicist who puts c=1, these are natural objects when described by 2-component spinors as the sequence of wavefronts of spin 1/2 particles for vanishing rest mass.

The need for generalization of the Laguerre structure for $\mathbb{R}^3$ becomes clear when we perceive the Higgs scalar as a conformal factor of the metric. The minimal Higgs carries weak isospin and weak hypercharge of the group SU_2 x U_1. In a geometrical model the product of this group with the Poincaré group has to appear. The SU_2 x U_1. has to be broken down subsequently to a U_1 symmetry. It is here suggested that this aim can be achieved easily from an algebraic point of view. The field of complex numbers is a subfield of the skew field of quaternions. Replacing complex spinors by quaternion spinors will achieve just this goal.

We have for the homogeneous transformations:

$$\phi' = \Lambda\phi\,\bar{p} \qquad\qquad \hat{\phi}' = \Lambda^{-1\dagger}\hat{\phi}\,\bar{p} \tag{56}$$

where: $\Lambda = \begin{pmatrix} a & b \\ c & d \end{pmatrix}$ det Λ=1, a b, c, d, p $\in$ $\mathbf{H}$ $p\bar{p} = 1$ (57)

($\mathbf{H}$ = set of quaternions)

The "phase factor" here is the unit quaternion $\bar{p}$ and is non-trivial due to the non-commutativity of the quaternions. As usual, the vector $X = \phi\phi^\dagger$ is unaffected by the phase. We will show later that the group in question is the direct product of $Spin_+(1,5)$ with SU_2 (the group of the unit quaternions). The quaternion spinors now describe two fermions and their antiparticles (both left and right handed), and we have three imaginary units (conventionally "i", "j", "k",) all on the same level. The free choice of direction in this abstract 3-space is interpreted as the origin of the SU_2 internal symmetry. The U_1 symmetry arises from a phase factor in Λ that does not affect the quaternionic determinant (which is always a non-negative real number). The symmetry is then broken by choosing an imaginary unit (a direction), say "i", and keeping those transformations that commute with this choice.

This leaves the covering group, of the orthochronous Lorentz group times a U_1 group. The gauging of the SU_2 group is interpreted as the phenomenon that the choice of an imaginary unit direction becomes space-time dependent. The remaining U_1 group rotates in the plane of the quaternionic units j and k. The Higgs field being quaternionic will be a real multiple of the imaginary unit. We shall say more about the quaternionic case later.

X. A REMARK ABOUT QUANTIZATION

In Minkowski's 4-dimensional geometry the time and space coordinates play similar roles. Not so in Quantum Mechanics. Time remains a c-number and the uncertainty relation:

$$\Delta E \cdot \Delta t \geq \hbar \tag{58}$$

is not on the same level as the other uncertainty relations. In Laguerre geometry, a time coordinate is introduced through the spatial position of a plane wave front. We have for the equation of a plane in $\mathbb{R}^3$

$$u_j x^j + u_0 x^0 + u_{n+1} = 0 \qquad u_j u_j - u_0^2 = 0 \tag{59}$$

Keeping u_j and u_0 fixed, (they determine, e.g., the null four vector of a plane wave) we consider the one-dimensional set of parallel planes filling $\mathbb{R}^3$ by taking u_{n+1} variable:

$$-\infty < u_{n+1} < \infty \tag{60}$$

This gives the time coordinate x^0 as:

$$x^0 = -\left(\frac{u_{n+1}}{u_0} + \frac{u_j}{u_0} x^j\right) \qquad u_0 > 0 \tag{61}$$

The occurrence of the frequency u_0 in the denominator accounts for the time dilation in the inertial frame.

The differential form:

$$\omega = u_\mu dx^\mu \tag{62}$$

is invariant under Poincaré transformations and gives rise to:

$$d\omega = du_\mu \wedge dx^\mu . \tag{63}$$

This form is closed and non-singular and defines thus a symplectic structure. This turns u_α and x^α into canonically conjugate variables and forms the starting point for a canonical quantization. The u_α have to be expressed quadratically through Fourier transformed massless spin 1/2 quantum fields with anti-commutation relations. This assures the positivity of u_0. Space-time will be defined as the quotient of fields with respect to their internal symmetry quantum numbers. These are certainly some points in a quantization program yet to be clarified. We return now to classical theory.

XI. QUATERNION TWO-SPINORS

The condition:

$$\det \Lambda = \det \begin{pmatrix} a & b \\ c & d \end{pmatrix} = 1 \qquad a, b, c, d \in \mathbf{H} \tag{64}$$

is defined by the formula:

$$\det \Lambda = \left|ad - aca^{-1}b\right|^2 = 1. \qquad (65)$$

If the elements are commuting (i.e. real or complex), this reduces to the *square* of the usual expression. For vanishing a, one gets simply (by going to a=0 through the reals):

$$\det \Lambda = |bc|^2 \qquad (66)$$

This is demonstrated as follows: The quaternion algebra has a representation in 2 x 2 matrices of complex numbers. Determinants for n x n quaternionic matrices can be found by representing the matrix as a 2n x 2n complex matrix and taking the determinant in the usual way. This gives for n=1:

$$\det a = \det(a_0 + a_1 i + a_2 j + a_3 k) = \det \begin{pmatrix} a_0 - ia_3 & -a_2 - ia_1 \\ a_2 - ia_1 & a_0 + ia_3 \end{pmatrix} \qquad (67)$$

This gives: $\det a = |a|^2.$ (68)

We have thus, for $a \neq 0$, using the multiplicative law for determinants:

$$\det \begin{pmatrix} a & b \\ c & d \end{pmatrix} = \det \begin{pmatrix} 1 & 0 \\ ca^{-1} & 1 \end{pmatrix} \begin{pmatrix} a & b \\ 0 & d - ca^{-1}b \end{pmatrix}$$

$$= (\det a)(\det [d - ca^{-1}b]) \qquad (69)$$

This leads directly to (63).This definition of a determinant for quaternionic matrices makes it clear that the group SL_2 (**H**) depends on 15 parameters as the determinant is just one real condition.

To identify SL_2 (**H**) with one of the real classical groups we notice that the Hermitian quaternionic matrices X form real 6-vectors:

$$X = \sigma^{\mu} x_{\mu} \qquad \mu = 0\,..\,5. \qquad (70)$$

We generalize of the Pauli matrices, adding:

$$\sigma^4 = -\hat{\sigma}^4 = \begin{pmatrix} 0 & -j \\ j & 0 \end{pmatrix} \qquad \text{and} \qquad \sigma^5 = -\hat{\sigma}^5 = \begin{pmatrix} 0 & -k \\ k & 0 \end{pmatrix} \tag{71}$$

The relation (19) holds now for indices running from 0 to 5.
We note further that for matrices: $A = (a_{jk})$ and $B = (b_{rl})$, we have

$$(AB)^\dagger = (a_{jk} b_{kl})^\dagger = \overline{a_{lk} b_{kj}} = \bar{b}_{kj} \bar{a}_{lk} = (b_{jk})^\dagger (a_{kl})^\dagger = B^\dagger A^\dagger \tag{72}$$

If we now transform a Hermitian quaternionic matrix X by:

$$X' = \Lambda X \Lambda^\dagger \quad \text{with} \quad \Lambda \in SL_2(\mathbb{H}) \qquad \text{and} \quad X = X^\dagger \tag{73}$$

we find that X' is again Hermitian:

$$X'^\dagger = (\Lambda X \Lambda^\dagger)^\dagger = \Lambda X^\dagger \Lambda^\dagger = \Lambda X \Lambda^\dagger = X' \tag{74}$$

It follows from (65) that the determinants of $\Lambda^\dagger$ and Λ agree and we have, with $\det \Lambda = 1$ that

$$\det X' = \det X \tag{75}$$

Since according to (70):

$$X = \begin{pmatrix} x_0 + x_3 & x_1 - ix_2 - jx_4 - kx_5 \\ x_1 + ix_2 + jx_4 + kx_5 & x_0 - x_3 \end{pmatrix} \tag{76}$$

We have from (65) and (76):

$$\det X = \left(x_0^{\,2} - x_1^{\,2} - x_2^{\,2} - x_3^{\,2} - x_4^{\,2} - x_5^{\,2}\right)^2 = \left(\eta^{\mu\nu} x_{\mu\nu}\right)^2 \tag{77}$$

Writing:
$$X' = \sigma^\nu x'_\nu \tag{78}$$

We have, with
$$\Lambda \sigma^\mu \Lambda^\dagger = \sigma^\nu A^{-1\mu}{}_\nu \qquad A^\mu{}_\nu \in \mathbb{R} \tag{79}$$

that:
$$X' = \Lambda X \Lambda^+ = \Lambda \sigma^\mu \Lambda^\dagger x_\mu = \sigma^\nu x_\mu A^{-1\mu}{}_\nu = \sigma^\nu x'_\nu \tag{80}$$

This gives for the 6-vector x_μ the expected transformation law:

$$x'_\nu = x_\mu A^{-1\mu}{}_\nu \tag{81}$$

with:
$$\eta^{\lambda\nu} x'_\lambda x'_\nu = \eta^{\rho\mu} x_\rho x_\mu \tag{82}$$

The transformations $A^\mu{}_\nu$ belong to the group **O**(1,5,**R**). The invariance of (82) follows from (77) and the law of inertia of quadratic forms under real transformations (81). By taking the trace of (79) for μ=0, we obtain:

$$\mathrm{tr}\ \Lambda\Lambda^+ = 2(A^{-1})^0{}_0 \tag{83}$$

With (57) this gives:

$$\mathrm{tr}\ \Lambda\Lambda^\dagger = \mathrm{tr}\begin{pmatrix} a & b \\ c & d \end{pmatrix}\begin{pmatrix} \bar{a} & \bar{c} \\ \bar{b} & \bar{d} \end{pmatrix} = |a|^2+|b|^2+|c|^2+|d|^2 = 2(A^{-1})^0{}_0 \tag{84}$$

This shows that the transformations must be orthochronous. The $A^\mu{}_\nu$ belong to the group **O**↑(1,5,**R**).We have thus found that the 6 x 6 matrices $A^\mu{}_\nu$ of the group **O**↑(1,5,**R**) provide a representation of the quaternionic group SL_2 (**H**).

However, it is easy to prove that the simultaneous reflection of all five space-like coordinates cannot be obtained in this way. For a total space reflection we would need a matrix Λ such that:

$$\Lambda\sigma^\mu\Lambda^\dagger = \hat{\sigma}^\mu \quad \text{with} \quad \sigma^0 = \hat{\sigma}^0 \qquad \sigma^j = -\hat{\sigma}^j \qquad (j = 1\,..\,5) \tag{85}$$

The case μ=0 would require that Λ be unitary, while the cases μ=j would force Λ=0.

Since the matrices Λ and -Λ of SL_2 (**H**) give rise to the same representation of **SO**↑(1,5;**R**), we find that SL_2(**H**) is the covering group of **SO**↑(1,5;**R**) also known as $\mathbf{Spin}_+(1,5)$ or $\mathbf{Spin}^0(1,5)$.

XII. THE LORENTZ "PINUP" GROUP

It is well known that the group SL_2(**C**) is the covering group for the group **SO**↑(1,3;**R**). Complex 2-spinors transform with SL_2(**C**). This group has two inequivalent 2 dimensional representations, and thus two kinds of 2 component spinors (often called "dotted" and "undotted"). If the undotted spinor ϕ transforms with the matrix Λ ($\Lambda \in SL_2(\mathbf{C})$), then the dotted spinor $\hat{\phi}$ transforms with $\left(\Lambda^{-1}\right)^{\dagger}$.

The group $\mathbf{O}\uparrow(1,3;\mathbb{R})$ of orthochronous transformations contains the total space reflection. It has a covering group $\mathbf{Pin}\uparrow(1,3;\mathbb{R})$ that we call the "pinup" group. To obtain a spinor representation of the pinup group, we have to combine the "dotted" and "undotted" Weyl spinors into 4-component Dirac spinors. The totally space reflected spinor $\hat{\phi}$ is then defined as

$$\hat{\phi} \equiv -i\sigma^2\overline{\phi} \qquad -i\sigma^2 = \begin{pmatrix} 0 & -1 \\ 1 & 0 \end{pmatrix}. \tag{86}$$

The total space reflection is then represented through the interchange of ϕ and $\hat{\phi}$.

The 4 x 4 matrix $\begin{pmatrix} \mathbf{0} & \mathbf{1} \\ \mathbf{1} & \mathbf{0} \end{pmatrix}$ is identified as the operation of total space reflection. Together with the matrices: $\begin{pmatrix} \Lambda^{-1\dagger} & \mathbf{0} \\ \mathbf{0} & \Lambda \end{pmatrix}$ this generates an irreducible representation of the pinup group.

The reason for the choice (83) is that $\hat{\phi}$ will transform with a representation equivalent to $\Lambda^{-1\dagger}$ if ϕ transforms with Λ, also that the null vector given by the Hermitian matrix $\phi\phi^{\dagger}$ goes into the totally space-reflected null vector $\hat{\phi}\hat{\phi}^{\dagger}$.

XIII. THE QUATERNION PINUP GROUP

The group SL_2 ($\mathbf{H}$) has two inequivalent representations in quaternionic 2 x 2 matrices of determinant 1. A quaternion 2-spinor ϕ (undotted) transforms as:

$$\phi' = \Lambda\phi \tag{87}$$

while a dotted quaternion 2-spinor $\hat{\phi}$ transforms as:

$$\hat{\phi}' = \Lambda^{-1\dagger}\hat{\phi} \tag{88}$$

The total space reflection now involves the reversal of all 5 space-like directions.

The map connecting ϕ and $\hat{\phi}$ is now not so simple as (52) due to the lack of commutativity. We will find a unit vector quaternion, n, such that:

$$\hat{\phi}_1 = -n\overline{\phi}_2\overline{n} \qquad \hat{\phi}_2 = n\overline{\phi}_1\overline{n} \tag{89}$$

The map for complex spinors involves the transition to the complex conjugate which is an involutary antiautomorphism of the complex numbers. For the quaternion skew field the general antiautomorphism combines conjugation with the similarity transformation of the group $\mathbf{SO_3}$. The map from ϕ to $\hat{\phi}$ is chosen as follows: the spinor ϕ determines the future directed null six-vector:

$$\phi\phi^{\dagger} = \begin{pmatrix} \phi_1\overline{\phi}_1 & \phi_1\overline{\phi}_2 \\ \phi_2\overline{\phi}_1 & \phi_2\overline{\phi}_2 \end{pmatrix} \tag{90}$$

For the total space reflection we must have:

$$\hat{\phi}\hat{\phi}^{\dagger} = \begin{pmatrix} \hat{\phi}_1\overline{\hat{\phi}}_1 & \hat{\phi}_1\overline{\hat{\phi}}_2 \\ \hat{\phi}_2\overline{\hat{\phi}}_1 & \hat{\phi}_2\overline{\hat{\phi}}_2 \end{pmatrix} = \begin{pmatrix} \phi_2\overline{\phi}_2 & -\phi_1\overline{\phi}_2 \\ -\phi_2\overline{\phi}_1 & \phi_1\overline{\phi}_1 \end{pmatrix} \tag{91}$$

which gives:
$$n\overline{\phi}_1\phi_2\overline{n} = \phi_2\overline{\phi}_1 \tag{92}$$

When ϕ_1 and ϕ_2 do not commute, one can choose:

$$n = a/|a| \qquad \text{with} \qquad a = \phi_1\overline{\phi}_2 + \overline{\phi}_2\phi_1 - \phi_2\overline{\phi}_1 - \overline{\phi}_1\phi_2 \tag{93}$$

Then, analogous to the complex case, the 2-spinors are combined into a 4 component quaternion spinor, and the matrices:

$$\begin{pmatrix} 0 & 1 \\ 1 & 0 \end{pmatrix} \quad \text{and} \quad \begin{pmatrix} \Lambda^{-1\dagger} & 0 \\ 0 & \Lambda \end{pmatrix}$$

generate the quaternion pinup group. The equations (29), (30), (70), (73) continue to hold, the only differences being that the elements are (non-commuting) quaternions and the indices run from 0 to 5.One can show that the $\hat{x}_\mu$ transform just as the x_μ of (81). under $SO\uparrow(1,5;\mathbb{R})$. For the proof one needs a theorem that the scalar part of the trace of a product of two quaternion matrices A and B is:

$$\mathrm{sc}\ \mathrm{tr}\, AB = \mathrm{sc}\ a_{jk} b_{kj} = \mathrm{sc}\ b_{kj} a_{jk} = \mathrm{sc}\ \mathrm{tr}\ BA \tag{94}$$

The map

$$X \rightarrow \hat{X} \tag{95}$$

gives rise to total space reflection of a six vector. Through polar decomposition of the matrix Λ one can give an explicit expression in terms of 15 real parameters:

$$x_j\ , \qquad F_{jk} = - F_{kj}, \qquad j,k = 1\ \ .\ \ .\ \ 5 \tag{96}$$

where the x_j are the five boosts and the F_{jk} generate the five dimensional rotations. We have:

$$\Lambda = (\exp \tfrac{1}{2} x_j \sigma^j)(\exp \tfrac{1}{4} F_{jk} \sigma^j \sigma^k) \tag{97}$$

This gives:

$$\Lambda^{-1\dagger} = (\exp -\tfrac{1}{2} x_j \sigma^j)(\exp \tfrac{1}{4} F_{jk} \sigma^j \sigma^k) \tag{98}$$

as follows immediately from the properties of the generalized Pauli matrices. The sign reversal of the rapidity shows the behavior of velocity under total space reflection.

XIV. THE ENHANCED GROUP

If one sees quaternion spinors as a natural generalization of complex spinors, one doubles the number of fundamental fermions, and it is suggestive to think of them as electrons and neutrinos. This leads naturally to left-handed and right-handed neutrinos. This suggestion gains in persuasive power when one considers that quaternions immediately allow one to bring in the group SU_2 as a “phase” of unit quaternions.

We consider here spinors as 2-vectors of a right vector space over the skew field of the quaternions. Multiplying from left or right with a quaternion gives different results. The natural transformation law is:

$$\phi' = \Lambda\phi\,\bar{p} \qquad p \in \mathbf{H} \qquad p\bar{p} = 1. \tag{99}$$

This operation of quaternion multiplication from the right commutes with the left multilplication by Λ. We are thus led naturally to the enhanced group $(SL_2(\mathbf{H}) \times SU_2)/Z_2$, as the unit quaternions have SU_2 as a symmetry group. We take the quotient with the two element group Z_2 because multiplication with -1 from the right or left gives the same result. It is also clear that the SU_2 group can be gauged by making p space-time dependent.

The group $SL_2(\mathbf{H})$ contains the Lorentz transformations as a subgroup.The elements of $SL_2(\mathbf{H})$ that transform space-time into itself are the group:

$$\mathbf{Spin}_+(1,3) \times U_1 \subset SL_2(\mathbf{H}) \tag{100}$$

The U_1 group is the rotation in the 4,5 plane (the j,k plane). Thus the U_1 group appears automatically as an internal symmetry that commutes with Lorentz transformations. We shall identify the U_1 group with hypercharge rotations and the SU_2 group as the group of weak isospin rotations in the electroweak theory. But first we discuss symmetry breaking.

XV. SYMMETRY BREAKING

To go from quaternion spinors to complex spinors it is necessary to pick an imaginary unit -- a direction in the 3-space of the unit vector quaternions. Gauge theory suggests that this choice could be space-time dependent. We postulate therefore a field, F, such that:

$$F^2 = -f^2\,\mathbf{1} \qquad f \text{ real} > 0 \tag{101}$$

and we put

$$\frac{1}{f}F = D \qquad D^2 = -1 \tag{102}$$

We consider here only the special case where F and D are constant, leaving gauging for future development.

It is clear that F has to commute with orthochronous Lorentz transformations and thus must be a scalar field. We try to identify F with the Higgs field which is a Lorentz scalar and transforms internally as $SU_2 \times U_1$. The U_1 dependence precludes that it is simply a quaternion, i.e. a scalar under $SL_2(\mathbf{H})$. It cannot be a vector either because the squaring operation is not invariant for vectors and does not lead to something negative definite in (101). The next simplest bosonic choice is a skew tensor -- a 2 x 2 quaternionic matrix with vanishing scalar part of the trace. It transforms as:

$$F' = \Lambda F \Lambda^{-1} \qquad \text{sc tr } F = 0 \tag{103}$$

It is clear that the trace condition is invariant as well as f^2. It turns out that (101) is equivalent to requiring that F represents a non-vanishing space-like bivector in the six dimensional space-time. There is also a space reflected $\hat{F}$ that transfolrms as

$$\hat{F}' = \Lambda^{-1\dagger} \hat{F} \Lambda^{\dagger}, \qquad \hat{F} = -F^{\dagger} \tag{104}$$

By using the chiral representation:

$$\gamma^{\mu} = \begin{pmatrix} 0 & \hat{\sigma}^{\mu} \\ \sigma^{\mu} & 0 \end{pmatrix}, \qquad \mu = 0 \ldots 5 \tag{105}$$

with
$$\gamma^{\mu}\gamma^{\nu} + \gamma^{\nu}\gamma^{\mu} = 2\eta^{\mu\nu} \tag{106}$$

We have
$$F \equiv \begin{pmatrix} \hat{F} & 0 \\ 0 & F \end{pmatrix} = F_{\mu\nu}\gamma^{\mu}\gamma^{\nu}, \qquad F_{\mu\nu} = -F_{\nu\mu} \tag{107}$$

This gives the relation between the quaternionic matrices and the 15 real components of $F_{\mu\nu}$.

The symmetry of $SL_2(\mathbf{H})$ is now broken by restricting the transformations to those that leave F invariant.

$$F' = F \tag{108}$$

By a suitable transformation of $SL_2(\mathbf{H})$ $\mathbb{F}$ can always be specialized to:

$$\mathbb{F} = \gamma^4\gamma^5 = \mathrm{f}\,\mathrm{i}\,\mathbf{1} \tag{109}$$

The quaternionic matrices Λ that commute with $\mathbb{F}$ consist of those matrix elements that do not contain the quaternionic units j and k. This leaves 2 x 2 complex matrices with absolute value of the determinant equal to 1, giving $SL_2(\mathbf{C}) \times U_1$. The six dimensional space-time is thus split into a four dimensional space-time and a two dimensional space-like internal symmetry plane, the x^4-x^5 plane. The choice of an imaginary unit, which has been called "I", breaks the symmetry of the enhanced group $[SL_2(\mathbf{H}) \times SU_2]/Z_2$ down to:

$$[SL_2(\mathbf{H}) \times SU_2]/Z_2 \rightarrow [\mathbf{Spin}_+(1,3) \times U_1 \times SU_2]/Z_2 \tag{110}$$

To have a unique imaginary unit "i", we must have:

$$\mathrm{i}(\exp \tfrac{1}{2}\mathrm{i}\lambda)(\exp -\tfrac{1}{2}[\mathrm{i}\mu + \mathrm{j}\nu + \mathrm{k}\rho]) = \mathrm{i}\,; \qquad \lambda, \mu, \nu, \rho \in \mathbb{R}. \tag{111}$$

This breaks the $U_1 \times SU_2$ symmetry to the U_1 group:

$$e^{\mathrm{i}\tau}, \qquad \tau = (\lambda-\mu)/2 \tag{112}$$

This U_1 group will be identified with the electromagnetic gauge group.

XVI. OUTLOOK

Further developments will be given elsewhere. Here we have only begun to sketch a theory of space-time leptons as oriented S^4's and hyperplanes in $\mathbb{R}^5$. Introducing quarks will require the step to octonion spinors and S^8's in $\mathbb{R}^9$, giving us 10 space-time dimensions. Octonions bring in the group G_2 as automorphisms and 7 imaginary units. Picking one of them reduces the group to SU_3. All this deals with one generation only. To bring in all three generations one is led to the Moufang plane and the Albert algebra. Then comes gauging, gravitation, and quantization. There are still many problems along this road, and if they could be solved, the answer to the question in the title of this paper would be "spheres not strings".

XVII. LITERATURE

Laguerre geometry is treated in: W. Blaschke & G. Thomsen: *Vorlesungen Ueber Differentialgeometrie III.* Springer (Berlin) 1929

Quaternion spinors applied to Lorentz transformations are discussed in: P.A.M. Dirac: **Proc. Roy. Irish. Acad.** Vol. L, Sect. A, p. 261, (1945).

Applications to the deSitter group and SO(1,5) are given in W.P. Kowalsky: *Quaternion Representations of the DeSitter and the Lorentz Groups*, N.Y.U. Ph.D. Thesis (1993), unpublished.

Quaternion field theory is covered by: S.L. Adler: *Quaternionic Quantum Mechanics and Quantum Fields*, Oxford Univ. Pr. (1994).

For Octonions, compare M. Postnikov: *Lectures in Geometry, Semester V*, Mir Publishers (1986).

ACKNOWLEDGEMENT

We gratefully acknowledge much valuable help with the computer typesetting from Prof. Donald Wade of Nassau Community College.

A POINCARE GAUGE THEORY OF GRAVITATION IN MINKOWSKI SPACETIME

C. WIESENDANGER
School of Theoretical Physics, Dublin Institute for Advanced Studies
10 Burlington Road, Dublin 4, Ireland
E-mail: wie@stp.dias.ie

ABSTRACT

The conventional role of spacetime geometry in the description of gravity is pointed out. Global Poincaré symmetry as an inner symmetry of field theories defined on a fixed Minkowski spacetime is discussed. Its extension to local **P** gauge symmetry and the corresponding **P** gauge fields are introduced. Their minimal coupling to matter is obtained. The scaling behaviour of the partition function of a spinor in **P** gauge field backgrounds is computed. The corresponding renormalization constraint is used to determine a minimal gauge field dynamics.

1. Introduction

The great success of modern particle physics in the description of the microscopic interactions of elementary particles relies much on the concept of gauge field theories. Only within this framework it was possible to formulate consistent renormalizable quantum field theories for interacting fields in four dimensions [1]. This provided a strong motivation to review the gravitational interaction and its beautiful description in the general theory of relativity both at the classical and the quantum level from the point of view of gauge field theories [2].

One central task thereby is to identify the correct gauge group for gravity. To settle this question rather different propositions were made in the literature on gauge approaches to gravity and yielded much new insight in the structure of gravitational interactions [2]. The earliest attempt was centered about the Lorentz group **L** [3] and was enlarged to an analysis of the full Poincaré group **P** only a few years later [4]. Since then there were many other contributions based on **P** [5]-[18], the translation group **T** [19]-[22] or even larger groups, e.g. [23]-[36].

Another important task is to analyze the status of the spacetime manifold in the description of gravity [37], [38]. On one hand gauge theories in elementary particle physics may be quantized without spoiling renormalizability only on Minkowski spacetime. As soon as a non-trivial background geometry is introduced the usual perturbative approaches face serious difficulties [39]. On the other hand the description of gravity in the general theory of relativity is just given in geometrical terms affecting the structure of spacetime itself. Accordingly one may ask whether some new light is shed on the problems in quantizing gravity using a complementary description which disentangles the structure of spacetime from gravitational physics.

To obtain such a description it is important to notice the purely conventional role of spacetime geometry in the description of the behaviour of matter pointed out by Poincaré already [40]. In fact two equivalent points of view are possible [41].

Either, one defines the line element ds^2 to be of Minkowskian form. Accordingly, in a gravitational field material rods will shrink and clocks slow down w.r.t. this metric. Hence, one defines the geometry of spacetime to be Minkowskian whereas the behaviour of physical rods and clocks has to be determined by experiments.

Or, one defines rods or clocks to have one and the same length or period at any point of spacetime. Accordingly, a measurement of the line element ds^2 using these rods and clocks will yield that spacetime is curved in general. This is the convention Einstein introduced to describe gravitation.

The general theory of relativity and its extensions are based on the second point of view [42] which is very natural as long as one is interested in the macroscopic aspects of gravitation [43]. Its limitation shows up at the quantum level. To extend a picture so intimately related to classical concepts such as rods and clocks to a simple microscopic understanding of gravitation is very difficult. In microphysics spacetime geometry enters only as a background concept necessary in defining a field theory. It cannot be subject to direct measurements in this context.

Hence, at the quantum level one is naturally led to the first point of view avoiding the problematic interrelation of spacetime structure and gravitational phenomena. Here free matter is described by local, causal fields defined on Minkowski spacetime and its interactions are introduced using the gauge principle which allows a far-reaching generalization of the connection between conservation laws and global symmetry requirements [1].

To obtain a gauge theory of gravitation we ensure the conservation of energy-momentum and angular momentum by the requirement of global covariance of the free matter field theory under the Poincaré group. We give a complementary formulation of **P** symmetry and its consequences in the form of an inner symmetry (section 2) suggesting an analogy to the description of the action of inner symmetry groups as groups of generalized 'rotations' in field space [1]. In particular the coordinate system used to specify the spacetime events is not affected anymore by **P** transformations.

We next introduce local **P** gauge transformations and demand the invariance of physical processes under those (section 3). This necessarily leads to the existence of gauge fields with definite behaviour under local **P** gauge transformations. Their coupling to any other field is essentially fixed as in the case of other gauge field theories (section 4).

To restrict the classical gauge field dynamics we demand consistency with renormalization properties of matter fields. In a renormalizable theory the anomalous contribution to a change of a matter partition function under rescaling may be absorbed in the classical actions for the different fields (e.g. [44]). As an example we determine the change of the one-loop partition function for a Dirac spinor under rescaling (sections 5 and 6) and may accordingly fix a minimal gauge field action (section 6).

We work on Minkowski spacetime ($\mathbf{R}^4$,η) with Cartesian coordinates throughout, such that $\eta = \text{diag}(1,-1,-1,-1)$. Indices $\alpha, \beta, \gamma, ...$ from the first half of the Greek alphabet denote quantities defined on ($\mathbf{R}^4$,η) which transform covariantly w.r.t. the Lorentz group. They are correspondingly raised and lowered with η.

2. Global Poincaré symmetry as an inner symmetry

Let us consider a set of fields $\varphi_j(x)$ with $j = 1,..,n$ which are defined on Minkowski spacetime ($\mathbf{R}^4$,η) and belong to some representations of the Poincaré group. Their dynamics shall be specified by the action

$$S_M = \int d^4x \; \mathcal{L}_M(\varphi_j, \partial_\alpha \varphi_j) \tag{1}$$

where the Lagrangian $\mathcal{L}_M$ is assumed to be real.

The usual conception of global Poincaré transformations, acting partly on spacetime and partly in inner field space, is expressed in the transformation formulae

$$\begin{aligned} x^\alpha \longrightarrow x'^\alpha &= x^\alpha + \varepsilon^\alpha + \omega^\alpha{}_\beta x^\beta, \\ \varphi_j(x) \longrightarrow \varphi'_j(x') &= \varphi_j(x) - \frac{i}{4}\omega^{\alpha\beta}\Sigma_{\alpha\beta}\, \varphi_j(x). \end{aligned} \tag{2}$$

$\delta x^\alpha = \varepsilon^\alpha + \omega^\alpha{}_\beta x^\beta$ is the change of x under the combination of a global infinitesimal spacetime translation and an infinitesimal Lorentz rotation, $\delta\varphi_j = -\frac{i}{4}\omega^{\alpha\beta}\Sigma_{\alpha\beta}\,\varphi_j(x)$ the corresponding change of φ_j in field space. $\Sigma_{\gamma\delta}$ are the representations of the generators of the Lie algebra $\mathbf{so(1,3)}$ in inner field space normalized to fulfil the commutation relations

$$[\Sigma_{\gamma\delta}, \Sigma_{\epsilon\zeta}] = 2i\{\eta_{\delta\epsilon}\Sigma_{\gamma\zeta} - \eta_{\delta\zeta}\Sigma_{\gamma\epsilon} + \eta_{\gamma\epsilon}\Sigma_{\zeta\delta} - \eta_{\gamma\zeta}\Sigma_{\epsilon\delta}\}. \tag{3}$$

If the action Eq. (1) does not change under the continuous transformations Eqs. (2) the field theory is globally Poincaré invariant. Accordingly, Noether's theorem yields a conserved vector quantity

$$J^\gamma = \Theta^\gamma{}_\alpha \cdot \varepsilon^\alpha + \frac{1}{2}\mathcal{M}^\gamma{}_{\alpha\beta} \cdot \omega^{\alpha\beta}. \tag{4}$$

As ε^α and $\omega^{\alpha\beta}$ vary independently the canonical energy momentum tensor

$$\Theta^\gamma{}_\alpha = \frac{\partial \mathcal{L}_M}{\partial(\partial_\gamma \varphi_j)} \cdot \partial_\alpha \varphi_j - \eta^\gamma{}_\alpha \cdot \mathcal{L}_M \tag{5}$$

and the canonical angular momentum tensor

$$\mathcal{M}^{\gamma}{}_{\alpha\beta} = \Theta^{\gamma}{}_{\alpha} x_{\beta} - \Theta^{\gamma}{}_{\beta} x_{\alpha} + \frac{i}{2}\frac{\partial \mathcal{L}_M}{\partial(\partial_\gamma \varphi_j)} \Sigma_{\alpha\beta}\varphi_j \tag{6}$$

are individually conserved.

As a complementary conception we now introduce global infinitesimal **P** gauge transformations

$$\begin{aligned} x^{\alpha} \longrightarrow x'^{\alpha} &= x^{\alpha} \\ \varphi_j(x) \longrightarrow \varphi'_j(x) &= \big((\mathbf{1}+\Theta)\varphi_j\big)(x) \end{aligned} \tag{7}$$

where the infinitesimal hermitean gauge operators are given by

$$\Theta = -\{\varepsilon^{\gamma} + \omega^{\gamma\delta} x_{\delta}\}\partial_{\gamma} - \frac{i}{4}\omega^{\gamma\delta}\Sigma_{\gamma\delta}. \tag{8}$$

Much in analogy to non-abelian gauge field theory **P** acts as a Lie group of generalized 'phase rotations' $(\mathbf{1}+\Theta)$ in field space only and leaves the spacetime coordinates x unchanged. Note that one can also decompose Θ w.r.t. the the **p** algebra generators $p_\gamma = i\,\partial_\gamma$ and $m_{\gamma\delta} = i(x_\gamma\partial_\delta - x_\delta\partial_\gamma) + \frac{1}{2}\Sigma_{\gamma\delta}$ which emphasizes the aforementioned useful analogy even more [45].

Eqs. (7) define again a symmetry transformation of globally Poincaré invariant actions as the corresponding Lagrangian just picks up a total divergence under a **P** gauge transformation

$$\begin{aligned} \mathcal{L}_M(\varphi'_j(x), \partial_\alpha\varphi'_j(x)) &= \mathcal{L}_M(\varphi_j(x), \partial_\alpha\varphi_j(x)) \\ &\quad -\{\varepsilon^{\gamma} + \omega^{\gamma}{}_{\beta} x^{\beta}\} \cdot \partial_\gamma \mathcal{L}_M(\varphi_j(x), \partial_\alpha\varphi_j(x)) \end{aligned} \tag{9}$$

which does not contribute to the action integral.

Hence, we are led to a complementary conception of Poincaré symmetry as a purely inner symmetry. The corresponding Noether symmetry current is found to be the same J^γ as in Eq. (4). This shows that the two global conceptions are equivalent w.r.t. their physical consequences. On the other hand it is conceptually easy to generalize global **P** gauge transformations to local ones and to build up the corresponding gauge theory in analogy to the non-abelian case because the structure of spacetime and the action of the gauge group on the fields remain strictly separated.

3. Local P gauge invariance and the covariant derivative $\tilde{\nabla}_\alpha$

Let us extend **P** to a Lie group of local infinitesimal gauge transformations by allowing $\varepsilon(x)$ and $\omega(x)$ to vary with x. We thus consider from now on

$$\Theta(x) = -\{\varepsilon^{\gamma}(x) + \omega^{\gamma\delta}(x) x_{\delta}\}\partial_{\gamma} - \frac{i}{4}\omega^{\gamma\delta}(x)\Sigma_{\gamma\delta}. \tag{10}$$

Note that the algebra of the $\Theta(x)$'s does close again. There is a new element of non-commutativity in their algebra as, contrary to the usual case, the local parameters $\varepsilon(x)$ and $\omega(x)$ do not commute with the generators ∂_γ of the algebra. The emerging ordering problem is overcome by the convention that $\Theta(x)$ in its above form only acts to the right. This convention is motivated by demanding equivalence of the algebra of the $\Theta(x)$'s to the diffeomorphism times **so(1,3)** algebra. The formulae (7) still define the representation of **P** in the space of fields.

Next, to recast a given matter theory in a locally **P** gauge invariant form we must introduce a covariant derivative $\tilde{\nabla}_\alpha$ which is defined by the requirement

$$\tilde{\nabla}'_\alpha\,(\mathbf{1}+\Theta(x)) = (\mathbf{1}+\Theta(x))\,\tilde{\nabla}_\alpha. \tag{11}$$

Here $\tilde{\nabla}'_\alpha$ denotes the gauge transformed covariant derivative. Because $\tilde{\nabla}_\alpha$ transforms as a Lorentz vector we have to supplement the generators $\Sigma_{\gamma\delta}$ of **so(1,3)** in matter field space occurring in the decomposition of $\Theta(x)$ with the corresponding generators $\Sigma_{\gamma\delta}$ acting on vectors to obtain the appropriate product representation as we will automatically do wherever necessary from now on.

We find that the Lagrangian with covariant derivatives $\tilde{\nabla}_\alpha$ replacing the usual ones behaves under local infinitesimal **P** transformations as

$$\begin{aligned}\mathcal{L}_M(\varphi'_j(x),\tilde{\nabla}'_\alpha\varphi'_j(x)) &= \mathcal{L}_M(\varphi_j(x),\tilde{\nabla}_\alpha\varphi_j(x))\\ &-\{\varepsilon^\gamma(x)+\omega^{\gamma\delta}(x)x_\delta\}\cdot\partial_\gamma\mathcal{L}_M(\varphi_j(x),\tilde{\nabla}_\alpha\varphi_j(x)).\end{aligned} \tag{12}$$

Note that this does not yet ensure the local **P** invariance of the original action $S_M = \int\mathcal{L}_M$ because the second term in Eq. (12) is no longer a pure divergence as it was in the case of global infinitesimal transformations.

Now, to fulfil Eq. (11) we use the ansatz

$$\partial_\alpha \longrightarrow \tilde{\nabla}_\alpha = e_\alpha{}^\gamma\partial_\gamma + \frac{i}{4}B_\alpha{}^{\gamma\delta}\Sigma_{\gamma\delta} \tag{13}$$

decomposing $\tilde{\nabla}_\alpha$ w.r.t. ∂_γ and $\Sigma_{\gamma\delta}$ in the same way as the local gauge operator $\Theta(x)$ in Eq. (10). This ansatz is compatible with the required behaviour Eq. (11) of the covariant derivative under local **P** gauge transformations provided the 16 compensating translation gauge fields $e_\alpha{}^\gamma$ transform as

$$\begin{aligned}\delta e_\alpha{}^\gamma &\equiv e'_\alpha{}^\gamma - e_\alpha{}^\gamma = e_\alpha{}^\zeta\cdot\partial_\zeta\{\varepsilon^\gamma+\omega^{\gamma\delta}x_\delta\}\\ &- \{\varepsilon^\zeta+\omega^{\zeta\eta}x_\eta\}\cdot\partial_\zeta e_\alpha{}^\gamma + \omega_\alpha{}^\zeta e_\zeta{}^\gamma\end{aligned} \tag{14}$$

and the 24 Lorentz gauge fields $B_\alpha{}^{\gamma\delta}$ as

$$\begin{aligned}\delta B_\alpha{}^{\gamma\delta} &\equiv B'_\alpha{}^{\gamma\delta} - B_\alpha{}^{\gamma\delta} = e_\alpha{}^\zeta\cdot\partial_\zeta\omega^{\gamma\delta} - \{\varepsilon^\zeta+\omega^{\zeta\eta}x_\eta\}\cdot\partial_\zeta B_\alpha{}^{\gamma\delta}\\ &+ \omega_\alpha{}^\zeta B_\zeta{}^{\gamma\delta} + \omega^\gamma{}_\zeta B_\alpha{}^{\zeta\delta} + \omega^\delta{}_\zeta B_\alpha{}^{\gamma\zeta}.\end{aligned} \tag{15}$$

As in our conception coordinate and **P** gauge transformations are strictly separated we emphasize that the introduction of $e_\alpha{}^\gamma$ and $B_\alpha{}^{\gamma\delta}$ has neither implications on the structure of the underlying spacetime which we assumed to be $(\mathbf{R}^4,\eta)$ endowed with the Minkowski metric η. Nor has it implications on the maximal symmetry group of $(\mathbf{R}^4,\eta)$, which is the Poincaré group if we still restrict ourselves to the use of Cartesian coordinates only.

With the abbreviations

$$d_\alpha \equiv e_\alpha{}^\gamma \partial_\gamma, \qquad B_\alpha \equiv \frac{i}{4} B_\alpha{}^{\gamma\delta}\Sigma_{\gamma\delta}, \tag{16}$$

where $\Sigma_{\gamma\delta}$ must be properly adjusted to the Lorentz group representation it acts upon, we write

$$\tilde{\nabla}_\alpha = d_\alpha + B_\alpha \tag{17}$$

from now on. d_α is just the translation covariant derivative introduced in [20]. We finally remark that $\tilde{\nabla}_\alpha$ may alternatively be decomposed w.r.t. the **p** algebra generators p_γ and $m_{\gamma\delta}$ yielding the most convenient starting point for perturbation expansions [45].

4. The field strength operator. Minimal coupling to matter fields

Before turning to the field strength operator itself we introduce the non **P** covariant translation field strength

$$[d_\alpha, d_\beta] \equiv H_{\alpha\beta}{}^\gamma d_\gamma \tag{18}$$

as in [20]. $H_{\alpha\beta}{}^\gamma$ is expressed in terms of $e_\alpha{}^\gamma$ as [20]

$$H_{\alpha\beta}{}^\gamma = e^{-1\gamma}{}_\epsilon (e_\alpha{}^\zeta \cdot \partial_\zeta e_\beta{}^\epsilon - e_\beta{}^\zeta \cdot \partial_\zeta e_\alpha{}^\epsilon) \tag{19}$$

where $e^{-1\gamma}{}_\epsilon$ is the matrix inverse to $e_\alpha{}^\epsilon$, i.e. $e_\alpha{}^\epsilon \cdot e^{-1\gamma}{}_\epsilon = \delta_\alpha{}^\gamma$.

This allows us now to obtain the field strength operator and its decomposition in a simple way. Taking into account the vector character of $\tilde{\nabla}_\alpha$ a little algebra yields

$$\begin{aligned} S_{\alpha\beta} \equiv [\tilde{\nabla}_\alpha, \tilde{\nabla}_\beta] &= H_{\alpha\beta}{}^\gamma d_\gamma - (B_{\alpha\beta}{}^\gamma - B_{\beta\alpha}{}^\gamma) d_\gamma \\ &+ d_\alpha B_\beta - d_\beta B_\alpha + [B_\alpha, B_\beta]. \end{aligned} \tag{20}$$

Introducing the tensor coefficients of d_γ

$$T_{\alpha\beta}{}^\gamma \equiv B_{\alpha\beta}{}^\gamma - B_{\beta\alpha}{}^\gamma - H_{\alpha\beta}{}^\gamma \tag{21}$$

we may rewrite $S_{\alpha\beta}$ as

$$[\tilde{\nabla}_\alpha, \tilde{\nabla}_\beta] = -T_{\alpha\beta}{}^\gamma \tilde{\nabla}_\gamma + \frac{i}{4}\tilde{R}^{\gamma\delta}{}_{\alpha\beta}\Sigma_{\gamma\delta}, \tag{22}$$

where $\tilde{R}^{\gamma\delta}{}_{\alpha\beta}$ is found to be

$$\begin{aligned}\tilde{R}^{\gamma\delta}{}_{\alpha\beta} &\equiv d_\alpha B_\beta{}^{\gamma\delta} - d_\beta B_\alpha{}^{\gamma\delta} + B_\alpha{}^{\delta\epsilon} B_{\beta\epsilon}{}^{\gamma} \\ &- B_\beta{}^{\delta\epsilon} B_{\alpha\epsilon}{}^{\gamma} - H_{\alpha\beta}{}^{\epsilon} B_\epsilon{}^{\gamma\delta}. \end{aligned} \tag{23}$$

As $S_{\alpha\beta}$ has a decomposition w.r.t. $\tilde{\nabla}_\delta$ and $\Sigma_{\gamma\delta}$ it acts in general not only as a matrix but also as a first order differential operator in field space.

From Eq. (21) we see that only if $B_\alpha{}^{\gamma\delta}$ is related to $H_{\alpha\beta}{}^{\gamma}$ the tensor $T_{\alpha\beta}{}^{\gamma}$ may vanish. Denoting this particular $B_\alpha{}^{\gamma\delta}$ with $C_\alpha{}^{\gamma\delta}$ the required relation becomes

$$C_{\alpha\beta}{}^{\gamma} - C_{\beta\alpha}{}^{\gamma} = H_{\alpha\beta}{}^{\gamma}. \tag{24}$$

We may now solve for $C_\alpha{}^{\gamma\delta}$ in terms of $H_{\alpha\beta}{}^{\gamma}$ with the result

$$C_\alpha{}^{\gamma\delta} = \frac{1}{2}\left(H_\alpha{}^{\gamma\delta} - H_\alpha{}^{\delta\gamma} - H^{\gamma\delta}{}_{\alpha}\right). \tag{25}$$

Whenever $B_\alpha{}^{\gamma\delta} = C_\alpha{}^{\gamma\delta}$, i.e. $T_{\alpha\beta}{}^{\gamma} = 0$ we omit the tilde, hence writing

$$\nabla_\alpha \equiv d_\alpha + C_\alpha. \tag{26}$$

Obviously we obtain now for $S_{\alpha\beta}$ a matrix only

$$[\nabla_\alpha, \nabla_\beta] = \frac{i}{4} R^{\gamma\delta}{}_{\alpha\beta} \Sigma_{\gamma\delta}. \tag{27}$$

By construction $S_{\alpha\beta}$ transforms homogeneously under infinitesimal local **P** gauge transformations

$$S'_{\alpha\beta}(1 + \Theta(x)) = (1 + \Theta(x))\, S_{\alpha\beta} \tag{28}$$

leading to

$$\delta T_{\alpha\beta}{}^{\gamma} = \Theta(x)\, T_{\alpha\beta}{}^{\gamma}, \qquad \delta \tilde{R}^{\gamma\delta}{}_{\alpha\beta} = \Theta(x)\, \tilde{R}^{\gamma\delta}{}_{\alpha\beta}. \tag{29}$$

$T_{\alpha\beta}{}^{\gamma}$ and $\tilde{R}^{\gamma\delta}{}_{\alpha\beta}$ transform homogeneously under infinitesimal local **P** gauge transformations. We emphasize that $T_{\alpha\beta}{}^{\gamma} = 0$ is indeed a gauge covariant statement as we implicitly assumed above introducing $C_\alpha{}^{\gamma\delta}$.

It is very convenient to introduce the homogeneously transforming difference of the two gauge fields

$$K_\alpha{}^{\gamma\delta} \equiv B_\alpha{}^{\gamma\delta} - C_\alpha{}^{\gamma\delta} \tag{30}$$

which is related to $T_{\alpha\beta}{}^{\gamma}$ as

$$K_{\alpha\beta}{}^{\gamma} - K_{\beta\alpha}{}^{\gamma} = T_{\alpha\beta}{}^{\gamma} \tag{31}$$

with the obvious inversion

$$K_\alpha{}^{\gamma\delta} = \frac{1}{2}\left(T_\alpha{}^{\gamma\delta} - T_\alpha{}^{\delta\gamma} - T^{\gamma\delta}{}_{\alpha}\right). \tag{32}$$

We can express now $\tilde{R}^{\gamma\delta}{}_{\alpha\beta}$ in terms of $R^{\gamma\delta}{}_{\alpha\beta}$, which we take as our fundamental field strength variable, and $K_\alpha{}^{\gamma\delta}$ as

$$\begin{aligned}\tilde{R}^{\gamma\delta}{}_{\alpha\beta} &= R^{\gamma\delta}{}_{\alpha\beta} + \nabla_\alpha K_\beta{}^{\gamma\delta} - \nabla_\beta K_\alpha{}^{\gamma\delta} \\ &+ K_\alpha{}^{\delta\epsilon} K_{\beta\epsilon}{}^{\gamma} - K_\beta{}^{\delta\epsilon} K_{\alpha\epsilon}{}^{\gamma}. \end{aligned} \tag{33}$$

Next, let us turn to the extension of globally **P** invariant matter actions to locally gauge invariant ones. In the previous section we have obtained the covariant derivative $\tilde{\nabla}_\alpha$ yielding the behaviour Eq. (12) of a matter field Lagrangian under local **P** gauge transformations which is not yet sufficient for the original action $S_M = \int \mathcal{L}_M$ to be locally **P** gauge invariant. Completing the Lagrangian with $\det e^{-1}$

$$\det e^{-1} \cdot \mathcal{L}_M(\varphi_j, \tilde{\nabla}_\alpha \varphi_j) \tag{34}$$

we find that the combination Eq. (34) changes under a local **P** gauge transformation by a pure divergence only

$$\begin{aligned}\det e'^{-1} \cdot \mathcal{L}_M(\varphi'_j, \tilde{\nabla}'_\alpha \varphi'_j) &= \det e^{-1} \cdot \mathcal{L}_M(\varphi_j, \tilde{\nabla}_\alpha \varphi_j) \\ &- \partial_\gamma \left(\{\varepsilon^\gamma(x) + \omega^{\gamma\delta}(x) x_\delta\} \det e^{-1} \cdot \mathcal{L}_M(\varphi_j, \tilde{\nabla}_\alpha \varphi_j) \right). \end{aligned} \tag{35}$$

Therefore the minimally extended locally **P** gauge invariant matter action finally becomes

$$S_M = \int d^4x \, \det e^{-1}(x) \cdot \mathcal{L}_M(\varphi_j(x), \tilde{\nabla}_\alpha \varphi_j(x)). \tag{36}$$

Of course, S_M remains invariant if we change from one to another inertial system by global coordinate translations or Lorentz rotations.

It is the conception of **P** symmetry as an inner symmetry together with the gauge principle which has led us to this minimal coupling prescription. In this conception the gauge fields and their transformation behaviour do not interfere with the spacetime structure ($\mathbf{R}^4$,η) fixed by an a priori convention and the underlying geometry remains separated from the physics described by the **P** gauge fields in the same manner it remains separated from the physics described by any usual matrix gauge field.

We remark that a geometric re-interpretation of the gauge fields and their corresponding field strengths introduced above may be given in the framework of Riemannian geometry [45]. But then the gauge group **P** and the requirement of local **P** gauge invariance are replaced by the groups of general (infinitesimal) coordinate transformations and local $SO(1,3)$ frame rotations and the requirement of invariance under these groups [42], [43] and the geometry of spacetime is necessarily linked with these complementary symmetry requirements.

5. Dirac partition function in gauge field backgrounds

In Yang-Mills gauge field theory one may fix the gauge field dynamics quite uniquely by demanding gauge invariance of the action and using dimensional arguments related to the renormalization properties of the theory. In a similar fashion we attempt here to obtain information on the classical **P** gauge field dynamics studying quantized matter fields and their renormalization properties in gauge field backgrounds. In the Yang-Mills case our arguments lead straightforward to the usual Yang-Mills action.

The assumption that the interactions of the **P** gauge fields with the different matter fields are renormalizable imposes strong conditions on the classical gauge field dynamics. For let us suppose that a given theory for a matter field and the gauge fields $e_\alpha{}^\gamma$ and $B_\alpha{}^{\gamma\delta}$ is perturbatively renormalizable. Then we know that the change of the partition function of the whole system under rescaling can be absorbed in its classical action yielding at most a nontrivial scale dependence of the different couplings, masses and wavefunction normalizations [44]. Hence, the explicit computation of the change of the one-loop matter partition functions under rescaling will allow us to constrain the classical gauge field dynamics [20].

For brevity we restrict ourselves to the non-trivial case of a Dirac spinor field, the analogous discussions of a scalar and a vector field theory are found in [45]. The globally **P** invariant action for a Dirac spinor with real Lagrangian is given by

$$S_M = \int d^4x \left\{ \frac{i}{2}\overline{\psi}\gamma^\alpha(\partial_\alpha\psi) - \frac{i}{2}\overline{(\partial_\alpha\psi)}\gamma^\alpha\psi - m\overline{\psi}\psi \right\}. \tag{37}$$

The Dirac matrices fulfil the usual Clifford algebra $\{\gamma_\alpha, \gamma_\beta\} = 2\eta_{\alpha\beta}$ and the **so(1,3)** generators become $\Sigma_{\alpha\beta} = \frac{i}{2}[\gamma_\alpha, \gamma_\beta]$. The minimal extension prescription yields the locally **P** gauge invariant action

$$S_M = \int d^4x \; \det e^{-1} \left\{ \frac{i}{2}\overline{\psi}\gamma^\alpha(\tilde{\nabla}_\alpha\psi) - \frac{i}{2}\overline{(\tilde{\nabla}_\alpha\psi)}\gamma^\alpha\psi - m\overline{\psi}\psi \right\}. \tag{38}$$

Due to spin $B_\alpha{}^{\gamma\delta}$ enters the action. Partially integrating $\tilde{\nabla}_\alpha$ in the second term above leads to the usual form of the Dirac action

$$S_M = \int d^4x \; \det e^{-1}\overline{\psi} \left\{ i\gamma^\alpha(\tilde{\nabla}_\alpha - \frac{1}{2}T_{\gamma\alpha}{}^\gamma) - m \right\} \psi. \tag{39}$$

Note the occurrence of the tensor T ensuring the hermiticity of the **P** covariant Dirac operator w.r.t. $(\chi, \psi)_e = \int d^4x \; \det e^{-1}\overline{\chi}\cdot\psi$.

The spinor partition function in the given gauge field background is given by the Grassmann functional integral

$$\mathcal{Z}_\psi[e, B] = \int \mathcal{D}\overline{\psi}\mathcal{D}\psi \; e^{iS_M(\overline{\psi},\psi;e,B)}. \tag{40}$$

Note that we omit possible normalizations in order to obtain the most general renormalization structure later. As S_M is already of the usual quadratic form we may perform the Grassmann integral and formally obtain

$$\mathcal{Z}_\psi[e, B] = e^{\frac{1}{2}\log\det M_\psi(e,B)}. \tag{41}$$

The hyperbolic fluctuation operator $M_\psi(e,B)$ is obtained as usual by squaring the Dirac operator introduced in Eq. (39)

$$M_\psi(e,B) \equiv -\gamma^\alpha(\tilde{\nabla}_\alpha - \frac{1}{2}T_{\gamma\alpha}{}^\gamma)\cdot\gamma^\beta(\tilde{\nabla}_\beta - \frac{1}{2}T_{\delta\beta}{}^\delta) - m^2 \tag{42}$$

and is hermitean w.r.t. $(\ ,\ )_e$ due to the occurrence of T. To make contact to the heat kernel techniques fully described in [45] we have to recast M_ψ. Using $[\tilde{\nabla}_\alpha, \gamma^\beta] = 0$ and $\gamma^\alpha\gamma^\beta = \eta^{\alpha\beta} - i\Sigma^{\alpha\beta}$ we obtain

$$\begin{aligned} M_\psi(e,B) &= -(\tilde{\nabla}_\alpha - \frac{1}{2}T_{\gamma\alpha}{}^\gamma)(\tilde{\nabla}^\alpha - \frac{1}{2}T_\delta{}^{\alpha\delta}) \\ &+ \frac{i}{2}\Sigma^{\alpha\beta}(-T_{\alpha\beta}{}^\delta\tilde{\nabla}_\delta + \frac{i}{4}\tilde{R}^{\gamma\delta}{}_{\alpha\beta}\Sigma_{\gamma\delta} - \tilde{\nabla}_\alpha K_{\delta\beta}{}^\delta) - m^2. \end{aligned} \tag{43}$$

Next we write $T_\alpha \equiv \frac{i}{4}T^{\gamma\delta}{}_\alpha\Sigma_{\gamma\delta}$ and absorb the first order derivative term $-2T_\alpha\tilde{\nabla}^\alpha$ in the second order one. Together with the use of the Jacobi identities for the covariant derivative $\tilde{\nabla}_\alpha$ we then find the manifestly hermitean result

$$\begin{aligned} M_\psi(e,B) &= -(\tilde{\nabla}_\alpha + T_\alpha - \frac{1}{2}T_{\gamma\alpha}{}^\gamma)(\tilde{\nabla}^\alpha + T^\alpha - \frac{1}{2}T_\delta{}^{\alpha\delta}) \\ &+ T_\alpha T^\alpha + \frac{i}{2}\Sigma^{\alpha\beta}(\frac{i}{4}\tilde{R}^{\gamma\delta}{}_{\alpha\beta}\Sigma_{\gamma\delta} + \tilde{R}^\delta{}_{\alpha\delta\beta}) - m^2. \end{aligned} \tag{44}$$

This is the form relevant for further computation.

As we are interested in the behaviour of $\mathcal{Z}_\psi[e,B]$ under rescaling the most suited renormalization of the ultraviolet divergent determinant in Eq. (41) is based on the ζ-function as it is a manifestly gauge invariant technique.

One can define the ultraviolet regularized functional determinant of an operator M satisfying certain conditions to be [46], [47]

$$\log\det M \equiv -\lim_{u\to 0}\frac{d}{du}\zeta(u;\mu;M) \tag{45}$$

where the generalized ζ-function belonging to M is given by the Mellin transformed of the heat kernel

$$\zeta(u;\mu;M) = \frac{i\mu^{2u}}{\Gamma(u)}\int_0^\infty ds\,(is)^{u-1}\,\mathrm{Tr}\,e^{-isM}. \tag{46}$$

The scale μ at which parameters such as couplings, masses and wavefunction normalizations have to be adjusted is introduced in order to keep the determinant dimensionless.

Hence, with the use of Eq. (45) the spinor partition function normalized at scale μ becomes

$$\mathcal{Z}_\psi[\mu;e,B] = e^{-\frac{1}{2}\zeta'(0;\mu;M_\psi(e,B))}. \tag{47}$$

According to the formula

$$\zeta'(0;\tilde{\mu};M) = \zeta'(0;\mu;M) + 2\log\lambda\cdot\zeta(0;\mu;M). \tag{48}$$

we finally obtain the change of $\mathcal{Z}_\psi$ corresponding to a change of scale $\tilde{\mu} = \lambda\mu$

$$\mathcal{Z}_\psi[\tilde{\mu};e,B] = \mathcal{Z}_\psi[\mu;e,B]\cdot e^{-\log\lambda\cdot\zeta(0;\mu;M_\psi(e,B))}. \tag{49}$$

6. Scaling of the Dirac partition function and the dynamics of the gauge fields

In the previous section we expressed the change of the one-loop spinor partition function under rescaling in terms of the ζ-function belonging to the corresponding fluctuation operator. Renormalizability of any theory including dynamical gauge fields requires now at least that this anomalous contribution, which is a local polynomial in $e_\alpha{}^\gamma$ and $B_\alpha{}^{\gamma\delta}$ and their derivatives, may be absorbed in the classical action for the gauge fields $e_\alpha{}^\gamma$ and $B_\alpha{}^{\gamma\delta}$ [44]. Hence, to determine a minimal gauge field dynamics consistent with renormalizability we finally have to obtain explicitly the value of the ζ-function at zero.

In the representation Eq. (46) for $\zeta(u;\mu;M)$ it is the singular part of the s-integration which yields a nonvanishing value for $\zeta(0;\mu;M)$. As this singular part comes from the small s-region we may use the corresponding expansion for the trace of the heat kernel [44]

$$\operatorname{Tr} e^{-isM} \overset{s\to 0}{\approx} \frac{i}{(4\pi i s)^{\frac{d}{2}}}\sum_{k=0}^{\infty}(is)^k\int d^dx\,\det e^{-1}\operatorname{tr} c_k(x) \tag{50}$$

given in terms of the well-known Seeley–DeWitt coefficient functions c_k. Performing the s-integration in (46) singles out the contribution for $k=\frac{d}{2}$ from the infinite sum and one obtains [44]

$$\zeta(0;\mu;M) = \frac{i}{(4\pi)^{\frac{d}{2}}}\int d^dx\,\det e^{-1}\operatorname{tr} c_{\frac{d}{2}}(x). \tag{51}$$

The computation of the $c_{\frac{d}{2}}$ has been done in various ways in the literature [49] - [53] and we restrict ourselves to give the result relevant to our case, where $d=4$ and tr_D denotes the Dirac trace [45]

$$\begin{aligned}\operatorname{tr}_D c_2 \;=\;& 4U_m + (\frac{1}{6}R^{\alpha\beta}{}_{\alpha\beta} - m^2)\cdot(4V_3 - V_1^{\gamma\delta}{}_{\gamma\delta})\\ & -\frac{1}{6}\nabla_\alpha\nabla^\alpha(4V_3 - V_1^{\gamma\delta}{}_{\gamma\delta}) - \frac{1}{24}F_{\alpha\beta\gamma\delta}\cdot F^{\alpha\beta\gamma\delta}\\ & - V_3\cdot V_1^{\gamma\delta}{}_{\gamma\delta} - \frac{1}{8}V_{2\alpha\beta}\cdot V_2^{\alpha\beta} + \frac{1}{8}V_1^{\alpha\beta}{}_{\alpha\beta}\cdot V_1^{\gamma\delta}{}_{\gamma\delta}\\ & + 2V_3^2 + \frac{1}{8}V_{1\alpha\beta\gamma\delta}\cdot(V_1^{\alpha\beta\gamma\delta} + V_1^{\gamma\delta\alpha\beta} + V_1^{\gamma\beta\delta\alpha}).\end{aligned} \tag{52}$$

The term without T-dependence

$$\begin{aligned} U_m &= -\frac{1}{30}\nabla_\gamma \nabla^\gamma R^{\alpha\beta}{}_{\alpha\beta} + \frac{1}{72} R_{\alpha\beta}{}^{\alpha\beta} \cdot R_{\gamma\delta}{}^{\gamma\delta} \\ &+ \frac{1}{180} R_{\alpha\beta\gamma\delta} \cdot R^{\alpha\beta\gamma\delta} - \frac{1}{180} R_{\alpha\gamma}{}^{\alpha}{}_{\delta} \cdot R_{\beta}{}^{\gamma\beta\delta} \\ &- \frac{1}{6} m^2 \cdot R^{\alpha\beta}{}_{\alpha\beta} + \frac{1}{2} m^4 \end{aligned} \tag{53}$$

already occurs in the case of a scalar field, whereas the terms containing T

$$\begin{aligned} V_{1\gamma\delta\alpha\beta} &= \tilde{R}_{\gamma\delta\alpha\beta} + \frac{1}{2} T_{\gamma\delta\eta} T_{\alpha\beta}{}^{\eta}, \\ V_{2\alpha\beta} &= \frac{1}{2}(V_1^{\eta}{}_{\alpha\eta\beta} - V_1^{\eta}{}_{\beta\eta\alpha}), \\ V_3 &= \frac{1}{2}\tilde{\nabla}_\alpha T_\gamma{}^{\alpha\gamma} - \frac{1}{4} T_{\gamma\alpha}{}^{\gamma} T_\delta{}^{\alpha\delta} \end{aligned} \tag{54}$$

and

$$\begin{aligned} F_{\gamma\delta\alpha\beta} &= \tilde{R}_{\gamma\delta\alpha\beta} + \tilde{\nabla}_\alpha T_{\gamma\delta\beta} - \tilde{\nabla}_\beta T_{\gamma\delta\alpha} \\ &+ T_{\alpha\beta}{}^{\eta} T_{\gamma\delta\eta} + T^{\eta}{}_{\gamma\alpha} T_{\eta\delta\beta} - T^{\eta}{}_{\gamma\beta} T_{\eta\delta\alpha} \end{aligned} \tag{55}$$

are due to the spin of the Dirac field. As the result Eq. (52) expressed in terms of the natural variables R, T becomes algebraically tedious we give it in these variables only for the case $T = 0$ [48]

$$\begin{aligned} \mathrm{tr}\,{}_D c_2 &= \frac{1}{30}\nabla_\gamma \nabla^\gamma R^{\alpha\beta}{}_{\alpha\beta} + \frac{1}{72} R_{\alpha\beta}{}^{\alpha\beta} \cdot R_{\gamma\delta}{}^{\gamma\delta} \\ &- \frac{7}{360} R_{\alpha\beta\gamma\delta} \cdot R^{\alpha\beta\gamma\delta} - \frac{1}{45} R_{\alpha\gamma}{}^{\alpha}{}_{\delta} \cdot R_{\beta}{}^{\gamma\beta\delta} \\ &+ \frac{1}{3} m^2 \cdot R^{\alpha\beta}{}_{\alpha\beta} + 2m^4. \end{aligned} \tag{56}$$

Insertion of the results Eq. (52) or Eq. (56) into Eq. (51) with $d = 4$ finally yields $\zeta(0; \mu; M_\psi(e, B))$.

We have obtained now the anomalous contribution to the rescaled spinor partition function as a local **P** gauge invariant polynomial in the fields $e_\alpha{}^\gamma$ and $B_\alpha{}^{\gamma\delta}$, which also must be present in any classical gauge field dynamics consistent with renormalizability of the spinor partition function. The analysis of the corresponding scalar and vector field cases leads to results of the same general structure [45]. Hence, we are finally led to construct a minimal action for the gauge fields just in terms of these **P** gauge invariant expressions.

For $T \neq 0$ we restrict ourselves to the contributions of $O(\partial^0, \partial^2)$ in the derivatives and obtain as minimal classical action to this order

$$\begin{aligned} S_G(e, B) &= \int \det e^{-1} \{ \Lambda - \frac{1}{\kappa^2} \cdot R_{\alpha\beta}{}^{\alpha\beta} + \beta_1 \cdot T_{\gamma\alpha}{}^{\gamma} T_\delta{}^{\alpha\delta} \\ &+ \beta_2 \cdot T_{\alpha\beta\gamma} T^{\alpha\beta\gamma} + \beta_3 \cdot T_{\alpha\beta\gamma} T^{\gamma\alpha\beta} + O(\partial^4) \}, \end{aligned} \tag{57}$$

skipping possible total divergence terms. Here we have to introduce different couplings $\kappa, \beta_1, \beta_2, \beta_3$ and the constant Λ which are independently renormalized by the one-loop contribution we determined above. Note that our reasoning automatically enforces a cosmological constant as to be expected from general renormalization considerations. The action Eq. (57) describes the classical gauge field dynamics correctly at sufficiently low momentum scales and small values of the couplings. Nevertheless, only a dynamics containing the huge number of different $O(\partial^4)$ terms as well, coming along with the same number of independent couplings, will be consistent with renormalizability (see also [39]). Note that no terms of $O(\partial^6)$ or higher are demanded by our reasoning.

If we set $T = 0$ the minimal classical action must contain the terms

$$\begin{aligned} S_G(e) &= \int \det e^{-1}\{\Lambda - \frac{1}{\kappa^2} \cdot R_{\alpha\beta}{}^{\alpha\beta} + \alpha_1 \cdot R_{\alpha\beta}{}^{\alpha\beta} \cdot R_{\gamma\delta}{}^{\gamma\delta} \\ &+ \alpha_2 \cdot R_{\alpha\gamma}{}^{\alpha}{}_{\delta} \cdot R_{\beta}{}^{\gamma\beta\delta} + \alpha_3 \cdot R_{\alpha\beta\gamma\delta} \cdot R^{\alpha\beta\gamma\delta}\}, \end{aligned} \tag{58}$$

if discarding total divergences. The couplings $\kappa, \alpha_1, \alpha_2, \alpha_3$ and the constant Λ obtain again contributions from the one-loop scale anomaly which has been determined above. We emphasize that S_G is an action for gauge fields defined on Minkowski spacetime $(\mathbf{R}^4, \eta)$ and is invariant on one hand under local $\mathbf{P}$ gauge transformations, on the other hand under global Poincaré transformations reflecting the symmetries of the underlying spacetime.

Important aspects of the geometric version of the quantized theory (58) such as one-loop divergences and β-functions and its unitarity problems are discussed in [39] and references given there.

7. Conclusions

To disentangle the structure of spacetime from the description of gravity we have given a complementary conception of Poincaré symmetry as a purely inner symmetry. Its extension to local $\mathbf{P}$ gauge symmetry has led us to introduce gauge fields defined on a fixed Minkowski spacetime. Their coupling to any other field has come out to be essentially fixed. We then constrained their dynamics imposing consistency with renormalization properties of a Dirac field in gauge field backgrounds. In an appropriate low energy limit the resulting gauge field action has been shown to reduce to a form yielding the same observational predictions as made in general relativity which confirms us to have obtained a sensible description of gravity within the present framework.

In our conception there is no direct interrelation between gravity and the structure of spacetime. Although it may be convenient to introduce a second, 'effective' metric on Minkowski spacetime to answer questions about the behaviour of rods and clocks in a classical context [20] we essentially deal here with a field theoretical description of gravitation free of any non-trivial geometrical aspects as proposed to investigate in the introduction. Unfortunately, the resulting theory is in many aspects still too close to the geometric approach and is far from leading to a convincing quantization. This shows up most clearly in the necessity of including terms quadratic in the field strength in the classical gauge field action. Although the corresponding quantum theory is known to be renormalizable, the occurrence of negative energy or negative norm ghost states has destroyed up to now any attempt of establishing unitarity and hence a physical interpretation of the theory (see also [39]).

8. Acknowledgments

This work has partially been supported by Schweizerischer Nationalfonds. It is a pleasure to thank the organizers of the Erice school 1995 on Quantum Gravity for the warm and fruitful hospitality during the course. The discussions with F. W. Hehl in Erice and his friendly interest in this work were an important encouragement. I am indebted to C. Ford and F. Krahe for intensive discussions, to L. O'Raifeartaigh for clarifying some group theoretical aspects.

9. References

1. L. O'Raifeartaigh, *Group Structure of Gauge Theories* (Cambridge University Press, Cambridge, 1986).
2. D. Ivanenko and G. Sardanashvily, *Phys. Rep.* **94** (1983) 3.
3. R. Utiyama, *Phys. Rev.* **101** (1956) 1597.
4. T. W. Kibble, *J. Math. Phys.* **2** (1961) 212.
5. K. Hayashi, *Prog. Theor. Phys.* **39** (1968) 494.
6. K. Hayashi and A. Bregman, *Ann. Phys.* **75** (1973) 562.
7. K. Hayashi and T. Shirafuji, *Prog. Theor. Phys.* **64** (1980) 866.
8. F. W. Hehl, P. von der Heyde, G. Kerlick and J. Nester *Rev. Mod. Phys.* **48** (1976) 393.
9. F. W. Hehl, in *Cosmolgy and Gravitation*, ed. P. Bergmann and V. De Sabbata (Plenum, New York, 1980).
10. F. W. Hehl, J. Nitsch and P. von der Heyde, in *General Relativity and Gravitation*, ed. A. Held (Plenum, New York, 1980).
11. Y. Ne'eman, in *Lect. Notes In Math.* **676**, ed. K. Bleuler et. al. (Springer, Berlin, 1979).
12. J. Hennig and J. Nitsch, *Gen. Rel. Grav.* **13** (1981) 947.
13. Y. M. Cho, *Phys. Rev.* **D14** (1976) 3335.
14. K. A. Pilch, *Lett. Math. Phys.* **4** (1980) 49.
15. A. Trautman, in *General Relativity and Gravitation*, ed. A. Held (Plenum, New York, 1980).
16. T. Kawai, *Gen. Rel. Grav.* **18** (1986) 995; *Gen. Rel. Grav.* **19** (1987) 1285 E.

17. T. Kawai, in *Proceedings of the Sixth Marcel Grossmann Meeting on General Relativity* (World Scientific, Singapore, 1992).
18. H. Kikukawa, J. Kikukawa and A. Nakamura, *Prog. Theor. Phys.* **89** (1993) 1063.
19. K. Hayashi and T. Nakano, *Prog. Theor. Phys.* **38** (1967) 491.
20. C. Wiesendanger, *Class. Quantum Grav.* **12** (1995) 585.
21. Y. M. Cho, *Phys. Rev.* **D14** (1976) 2521.
22. H. Kikukawa, J. Kikukawa and A. Nakamura, *Prog. Theor. Phys.* **89** (1993) 523.
23. C. N. Yang, *Phys. Rev. Lett.* **33** (1974) 445.
24. F. W. Hehl, G. D. Kerlick and P. von der Heyde, *Phys. Lett.* **B63** (1976) 446.
25. F. W. Hehl, E. A. Lord and Y. Ne'eman, *Phys. Rev.* **D17** (1978) 428.
26. R. D. Hecht and F. W. Hehl, in *General Relativity and Gravitational Physics*, ed. R. Cianci et. al. (World Sientific, Singapore, 1991).
27. F. W. Hehl, J. D. McCrea, E. W. Mielke and Y. Ne'eman, preprint TAUP 192-94 University of Cologne (1994).
28. Y. Ne'eman and Dj. Sijacki, *Ann. Phys.* **120** (1979) 292.
29. A. A. Tseytlin, *Phys. Rev.* **D26** (1982) 3327.
30. T. Shirafuji and M. Suzuki, *Prog. Theor. Phys.* **80** (1988) 711.
31. F. Mansouri and L. Chang, *Phys. Rev.* **D13** (1976) 3192.
32. J. P. Harnard and R. B. Pettitt, *J. Math. Phys.* **17** (1976) 1827.
33. W. Drechsler, *J. Math. Phys.* **18** (1977) 1358.
34. W. Drechsler, *Ann. Inst. H. Poincaré* **37** (1982) 155.
35. L. K. Norris, R. O. Fulp and W. R. Davies, *Phys. Lett.* **A79** (1980) 278.
36. E. A. Lord and P. Goswami, *J. Math. Phys.* **27** (1986) 2415.
37. L. P. Grishchuk, A. N. Petrov and A. D. Popova, *Comm. Math. Phys.* **94** (1984) 379.
38. A. N. Petrov, *Class. Quantum Grav.* **10** (1993) 2663.
39. I. L. Buchbinder, S. D. Odintsov and I. L. Shapiro , *Effective Action in Quantum Gravity* (IOP, Bristol, 1992).
40. H. Poincaré, *La sience et l' hypothèse* (Flammarion, Paris, 1968).
41. P. Mittelstaedt, *Philosophische Probleme der modernen Physik* (BI, Mannheim, 1976).
42. R. U. Sexl and H. K. Urbantke, *Gravitation und Kosmologie* (BI, Mannheim, 1987).
43. N. Straumann, *General Relativity and Relativistic Astrophysics* (Springer, Berlin, 1984).
44. C. Wiesendanger and A. Wipf, *Ann. Phys.* **233** (1994) 125.
45. C. Wiesendanger, preprint DIAS-STP-95-18 Dublin Institute for Advanced Studies (1995).

46. J. S. Dowker and R. Critchley, *Phys. Rev.* **D13** (1976) 3224.
47. S. W. Hawking, *Commun. Math. Phys.* **55** (1977) 133.
48. S. M. Christensen and M. J. Duff, *Nucl. Phys.* **B154** (1979) 301.
49. B. S. DeWitt, *Dynamical Theory of Groups and Fields* (Gordon and Breach, New York, 1965).
50. R. T. Seeley, *Proc. Symp. Pure Math., Am. Math. Soc.* **10** (1967) 288.
51. P. B. Gilkey, *Invariance theory, the heat equation and the Atiyah-Singer index theorem* (Publish or Perish, Boston, 1984).
52. A. O. Braginsky and G. A. Vilkovisky, *Phys. Rep.* **119** (1985) 1.
53. V. P. Gusynin and V. V. Kornyak, *J. Symbolic Computation* **17** (1994) 283.

GENERAL RELATIVITY ON SPINOR-TENSOR MANIFOLD

Yu Xin (Alfred Yu)
Department of Applied Mathematics,
The Hong Kong Polytechnic University
Hunghom, Kowloon
Hong Kong

ABSTRACT

This is a nondualistic General Theory of Relativity formulated on a space-time manifold whose geometry is determined by the real representation of $U(2)$, the spinorial group that is fundamental in Quantum Theory. This is thus a realization of the concept first proposed by Einstein and Meyer and followed by the Twistor School that spinors, responsible for half integral spin in Quantum Theory should, in some sense, be the foundation of the space-time manifold.

The real representation of $U(2)$ (The structure group of the space-time fibre bundle) being the subgroup of the intersection of $O(4)$ and $SP(4)$ (a dual symmetry suggestive of boson-fermion unification) compels the formulation of a Theory of General Relativity (if Einstein's logic is followed) that contains, as a natural consequence of the space-time geometry, the electromagnetic field, spin, charge and mass as well as gravitation. A generalized form of Dirac's equation on curved space-time is shown to be derivable from the nondualistic Einstein's equation by factorization. It contains moreover, a Gravitodynamics analogous to the Lorentz-Maxwell Electrodyamics as a natural consequence of the basic geometric structure. This gravitodynamics embodies a gravitional radiation theory that predicts the result of the orbital decay of the PSR1913+16 as accurately as that based on Einstein's Theory.

1. Introduction

The construction of a physical theory, if it is not to run into a tautology or an infinite regression of definitions begins with a set of physical quantities called irreducible primitives (which we shall here call the Primitive Category). The Primitive Category (π) has either been operationally defined or regarded as physically obvious. The structure of the theory is then designed via a covariant functor from π to μ which is a Model Category, supposed to be the model of the system under consideration. The theory is then completed according to certain assumed set of physical principles and its validity is to be verified or else falsified by experiments.

The big bang cosmology forms much of the physical and conceptual basis of Superstring and other Supersymmetric Theories and that makes GR (or its logical extensions) still a viable foundation for a unification scheme despite its devaluation since the days of Niels Bohr and Heisenberg. One of the principal objections to such an approach pioneered by Weyl[1], Eddington[2] and Einstein[3,4,5,6] appears to be (according to Pauli[7]) 'the fundamental character of the duality between the measured field and the test body used as measuring instrument'. Now, the test body is presumably a point particle defined by its position and momentum; but according to Heinsenberg's Uncertain Principle, no such test body can be precisely defined. In order to avoid such self-contradictory duality it is much simpler to regard the (differentiable) space-time manifold as the Primitive Category whose manifestations via the functor and laws still to be constructed constitute the Model Category of the universe. The 'particle' is then according to Eddington[8] a 'carrier of variates' which, like the physical fields are to be derived from the geometry of space-time. The 'Fundamental Theory' by A.S. Eddington[8] based on the continuous field approach aims to explain, apart from the gravitational and electromagnetic fields, also the atomicity of matter indicated by such fundamental parameters as the fine structure constant. Though largely ignored nowadays the approach by Eddington remains compelling. On the Quantum Theory (QT), on the other hand, Dirac[9] has this to say, 'We have a theory in which infinite factors appear when we try to solve the equations. These infinite factors are swept into a renormalization procedure. The result is a theory which is not based on strict mathematics, but is rather a set of working rules, ...'. Indeed, views have been expressed (Penrose[10,11,12]) that when unifying GR with QT, it is the structure of QT that has to be modified in contradistinction to most workers in quantum gravity (QG). It is most likely, nonetheless, that unless the different fields to be unified can be simply added, both GR and QT would assume generalized forms in the unified scheme. All that having been said, however, according to J.A. Stratton[13], 'A theory is not an absolute truth but a self-consistent analytical formulation of the relations governing a group of natural phenomena.' The value of a theory lies, among other things, with its internal logic, simplicity and experimental predictions or else falsifiability.

We take that attitude in formulating the present theory. This is an attempt to explore to the fullest extent Einstein's ideas which are now based, instead of a Pseudo- Riemannian space-time, on a space-time structure in which the usual space-time symmetry and the 'internal' symmetries of a spinor space are just the different manifestations of the one and the same spinor group - the real representation of $U(2)$. The purpose of this approach is to deduce, from the basic symmetries and a minimum number of physical principles as many physical consequences as possible and to see how much of the physcial world can be realized by the present theory.

The unification between GR and QT proposed here differs from most, if not all, previous attempts not so much in its technique or approach as in its very concept of unification. In the usual approaches there are two well defined and distinct structures to be united in ways that differ from school to school. In the present theory, there is only one structure, the spinor structure (which is both the primitive category and the model category) and whether it represents the 'internal' abstract space or 'external' space-time depends on its representation. In its two dimensional complex representation it is the 'internal' space and in its representation in four dimensional real vector space it is the tangent 'space-time'. Whether nature follows this way depends on the physical consequences that follow from it. Thus we have not called the present scheme a 'Unified Theory' because there are no two separate entities to be united. There is only one entity but it has two differnet manifestations from which, according to the present scheme, the phenomenal world springs.

2. Fundamental Concepts of Spinor-Tensor Manifold

2.1. Motivation

It is now almost axiomatic that spinors should play a fundamental role in any theory that attempts to explain basic features of elementary particles in space-time. The way spinors are embedded in a theory, however, differs from school to school. But one basic feature that appears to be common amongst them (with the exceptions of Eddington[8] and the Twistor Theory) is the spinor structure consisting of a 2 to 1 (Homomorphism between a spinor group (say, $U(2)$) and the Lorentz group of space-time. Whilst $U(2)$ yields all representations, including those with half- integral spins the Lorentz groups yield only the representations with integral spins and hence tangent vectors of space-time do not correspond to the spinors (vectors in the two dimensional complex space) representing half-integral spin. That may well be the fact that makes unification between QT and GR difficult.

There is, however, a radically different approach to the formulation of space-time that reflects spinor symmetries. In his Chronometric Theory of the Cosmos I.E. Segal (1976)[14] has shown that whilst reflecting fundamental particle symmetry, $U(2)$ (and its subgroup $SU(2)$) should play an essential role in the definition of space-time if local causality is to be preserved. In particular, the Lie algebra $\mathbf{U}(2)$ of the spinor group $U(2)$ can be identified with $\mathbf{H}(2)$, the Linear Space of 2 x 2 Complex Hermitian Matrices and $\mathbf{H}(2)$ in its turn is isomorphic to the Minkowski Space. The identification of $\mathbf{U}(2)$ with $\mathbf{H}(2)$ then uniquely defines, via a convex cone $C \epsilon \mathbf{U}(2)$, an invariant infinitesimal causal orientation. Thus the isomorphism between the Minkowski tangent space and $\mathbf{H}(2)$ which can be identified with $\mathbf{U}(2)$ suggests that in order to allow spinors to form the Bedrock of space-time, $\{U(2), \xi\}$ (where ξ is a two dimensional complex vector space) should be the primitive category; indeed, in its real representation, it is the space-time structure. There are no two distinct categories, the spinorial and the space-time, to be related: there is only one category. But in its two dimensional complex representation it is called the (internal) spinor space and in its four dimensional real representation it is called the (tangent) space-time. The real representation is in fact a faithful representation. The concept that 'internal' and 'external' spaces are just two different representations of one single primitive structure also receives support from Mach's Principle which states in effect that space-time has no meaning in the absence of matter. On the other hand our daily conception has it that matter 'resides', so to speak, in space-time. The duality between matter and space-time is quite unnecessary from the Machian point of view. It is conceptually much simpler to regard space-time as the manifestation of matter and conversely. Thus if the spinor is taken to be the basic building block of matter then the primitive category $\{U(2), \xi\}$ also represents the space-time structure.

2.2. Technical Details

(1) The usual complex structure on space-time

We shall denote the category $\{U(2), \xi\}$ the spinor space where ξ is the 2- dimensional complex vector space and $U(2)$, its automorphism group and the category $\{L(4), x\}$ Minkowski space-time where $\mathbf{x}$ is the 4-dimensional real vector (tangent) space and $L(4)$ the Lorentz group. The 'usual spinor structure' on space-time is defined by

$$\xi = \xi(\mathbf{x}) \tag{2.1}$$

such that there is a 2-1 homomorphism between $U(2)$ and $L(4)$:

$$U\sigma_\alpha U^* = a^\mu_\alpha \sigma_\mu \tag{2.2}$$

where

$$U, U^* \epsilon U(2) \tag{2.3}$$

$$(a^\mu_\alpha) \epsilon L(4) \tag{2.4}$$

and

$$\sigma_\alpha \quad \text{are the Pauli (2x2) spin matrices} \tag{2.5}$$

Hence, there are two distinct spaces: $\{U(2), \xi\}$ and $\{L(4), \mathbf{x}\}$ and they are related via (2.1) and (2.2). On insisting on the complex representation of $U(2)$, the homomorphism between the two spaces is 2 to 1 and this lack of isomorphism may be the basic obstacle in the path of unification between QT and GR. A modern attempt at the inclusion of spinors and hence QT has been made by Sachs[15] with the above structure but the theory is basically dualistic.

(2) The Spinorial Space-time Structure

Given the primitive Category $\{U(2), \xi\}$ we can construct a unique category $\{\mu(4), V\}$ (called the Spinorial Space-time Structure) as follows.

The complex unitary group $U(2)$ has a real representation $\mu(4)$ in a real vector space V_0 of dimension 4 (Kobayashi and Nomizu[16]; Goldberg[17]) so that $\mu(4)$ leaves concurrently invariant the quadratic form

$$S^2 = (x^0)^2 + (x^1)^2 + (x^2)^2 + (x^3)^2 \tag{2.6}$$

and the 2-form

$$\sigma = x^0 \wedge x^2 + x^1 \wedge x^3 \tag{2.7}$$

where

$$\{x^0, x^1, x^2, x^3\} \epsilon V_0 \tag{2.8}$$

has been constructed from the 2-dimensional complex space $\{\xi\}$:

$$\xi^1 = x^0 + ix^2; \quad i = \sqrt{-1} \tag{2.9}$$

$$\xi^2 = x^1 + ix^3 \tag{2.10}$$

is an orthonormal basis. That is,

$$\mu(4) \subset O(4) \cap Sp(4) \tag{2.11}$$

where $Sp(4)$ denotes the automorphism group of a 4-dimensional real sympletic vector space. Indeed it has been further shown that

$$\mu(4) = O(4) \cap SP(4) \cap G(4) \tag{2.12}$$

where $G(4)$ is the real representation of the complex general linear group $G(2, C)$ and $G(4)$ leaves invariant the bilinear form

$$b = (x^2 y_0 - x^0 y_2) + (x^3 y_1 - x^1 y_3) \tag{2.13}$$

where

$$Y_0, Y_1, Y_2, Y_3 \quad \epsilon \quad V_0^* \tag{2.14}$$

and V_0^* is the vector space dual to V_0. In arbitrary coordinates $\mu(4)$ leaves concurrently invariant the following forms:

$$s^2 = g_{\mu\nu} x^\mu x^\nu \tag{2.15}$$

$$\sigma = \frac{1}{2!} \sigma_{\mu\nu} x^\mu \wedge x^\nu \tag{2.16}$$

where

$$\sigma_{\mu\nu} = \sigma_{[\mu\nu]} = -\sigma_{\nu\mu} \tag{2.17}$$

and

$$b = b^\mu_\nu x^\nu y_\mu \tag{2.18}$$

where we have the relations

$$\sigma_{\mu\nu} = b^\alpha_\mu g_{\mu\nu}, \quad g_{\mu\nu} = \sigma_{\alpha\mu} b^\alpha_\nu, \quad b^\mu_\nu = \sigma_{\alpha\nu} g^{\alpha\mu} \tag{2.19}$$

where

$$[g^{\mu\nu}] = [g_{\mu\nu}]^{-1} \tag{2.20}$$

The quadratic form (2.6) can be converted into a Lorentz quadratic form by putting

$$\omega^0 = ix^0, \quad \omega^1 = x^1, \quad ,\omega^2 = x^2, \quad \omega^3 = x^3 \tag{2.21}$$

and so the automorphism group $L(4)$ of the vector space V is the Lorentz group. The vector space V is related to V_0 via (2.21). By this construction (2.12) can now be written as

$$\mu(4) = L(4) \cap Sp(4) \cap G(4) \tag{2.22}$$

which leaves concurrently invariant the forms

$$s^2 = g_{\mu\nu}\omega^\mu\omega^\nu \tag{2.23}$$

$$\sigma = \frac{1}{2!}\sigma_{\mu\nu}\omega^\mu \wedge \omega^\nu \tag{2.24}$$

and

$$b = b^\mu_\nu \alpha_\mu \omega^\nu \tag{2.25}$$

where α_μ form the orthonormal coordinates of the dual vector.

(3) The Spinorial Space-time Manifold

The Spinorial Space-time Manifold M is defined to be the smooth differentiable manifold of dimension 4 so that its tangent space M_p at any point $p \epsilon U_p \subset M$ is identified with $\{\mu(4), V\}$ as defined by (2.22) where $\{\mu(4), V\}$ is isomorphic to $\{U(2), \xi\}$ in a functorial sense. A fibre-bundle structure can then be defined on M in the Usual way with $\mu(4)$ as its structural group. And then an affine connection can be defined on the $\mu(4)$ - bundle in the usual way. The underlying manifold is called the Pseudo-Riemannian sympletic manifold with a Pseudo-Riemannian Sympletic Connection. As spinors and tensors (of rank 1) are just two different representations of the same tangent space, M is also called the Spinor-Tensor Manifold.

(4) Induced Clifford Tangent Algebra and Spinor Structure Invariance

As a consequence of (2.23) and (2.24) we have now the invariant differential quadratic and 2-forms defined on M_p:

$$ds^2 = g_{\mu\nu}dx^\mu dx^\nu = \delta_{\mu\nu}\omega^\mu\omega^\nu \tag{2.26}$$

$$\sigma = \frac{1}{2!}\sigma_{\mu\nu}dx^\mu \wedge dx^\nu = \omega^0 \wedge \omega^2 + \omega^1 \wedge \omega^3 \tag{2.27}$$

where ω^μ are the coordinates is an orthonormal basis of the infinitesimal tangent vector $d\mathbf{r}$ on M_p and are differential one forms. Thus,

$$d\mathbf{r} = \gamma_\mu \omega^\mu \tag{2.28}$$

where γ_μ form an orthonormal basis on M_p. In order that (2.26) may take on a Lorentz form the basis vector γ_μ must obey

$$\gamma_0^2 = 1, \quad \gamma_\mu^2 = -1, \quad \gamma_\mu \cdot \gamma_\nu = 0 (\mu \neq \nu) \tag{2.29a}$$

That is,

$$\gamma_\mu \cdot \gamma_\nu = \frac{1}{2}(\gamma_\mu \gamma_\nu + \gamma_\nu \gamma_\mu) = g_{\mu\nu} \tag{2.29b}$$

Expression (2.27) then induces an exterior product given by

$$\gamma_\mu \wedge \gamma_\nu = \frac{1}{2}(\gamma_\mu \gamma_\nu - \gamma_\nu \gamma_\mu) = \sigma_{\mu\nu} \tag{2.30}$$

In so far as $\{\gamma_\mu\}$ obeys the above relations it can be considered as the set of Dirac matrices as already demonstrated by Hestenes[18,19]. Hence, the invariant forms (2.26) and (2.27) induces in a natural way a Dirac algebra (or real Clifford algebra) on the tangent space M_p. In particular, for any two vectors $\mathbf{u}, \mathbf{v} \epsilon M_p$ we have the 'geometric product' (Hestenes[19]) given by

$$\mathbf{uv} = \mathbf{u} \cdot \mathbf{v} + \mathbf{u} \wedge \mathbf{v} \tag{2.31}$$

where

$$\mathbf{u} \cdot \mathbf{v} \equiv \frac{1}{2}(\mathbf{uv} + \mathbf{vu}) \tag{2.32}$$

and

$$\mathbf{u} \wedge \mathbf{v} = \frac{1}{2}(\mathbf{uv} - \mathbf{vu}) \tag{2.33}$$

If $\mathbf{u}$ and $\mathbf{v}$ are unit vectors then the object

$$\mathbf{s} = \mathbf{uv} \tag{2.34}$$

as defined by (2.31) is called a 'simple rotor' and the real Dirac Spinor can be built from it as already shown by Hestenes[20,21,21]. Indeed the geometric product contains the dual symmetries of Bosons ($g_{\mu\nu}$) and Fermions ($\sigma_{\mu\nu}$) and hence the present geometric structure could also be said to contain 'Supersymmetry', an idea already contained in the work by de Sabbata[23].

Hence, the simultaneous invariance of the forms (2.26) and (2.27) implies the preservation of the (real) spinor structure on M_p as it 'slides' on M as the 'motion' of M_p on M is determined by the Pseudo-Riemannian Sympletic Connection which preserves the spinor structure. The invariance of (2.26) and (2.27) together with the convariant derivative determined by this connection gives, by means of a straightforward computation,

$$\nabla_\lambda g_{\mu\nu} = 0 \quad \text{and} \quad \nabla_\lambda \sigma_{\mu\nu} = 0 \tag{2.35}$$

which is an alternative statement of (real) spinor structure preservation. The formulation of any dynamical field theory has to be compatible with it as is to be the case in the present theory.

3. General Relativity on the Spinorial Space-time Manifold

3.1. Motivation

It is well known that Einstein considered the LHS of his equation to be built of marble and the RHS, of sand. For, according to M. Bunge[24], '... in GR one does not know one of the two sides of the equation, namely $T_{\mu\nu}$, even if told by some other theory wherefore Einstein called $T_{\mu\nu}$ an asylum ignorantiae. One can therefore easily understand Einstein's tireless efforts to build a unified theory ..... it is not just the dream of the all encompassing formula but an attempt to fill a genuine gap, namely the RHS of the field equations of GR ...'. Indeed, in the Einstein frame work, the particle is to be integrated into the very structure of the field; for, in the words of Tonnelat[25], 'Einstein tries to explain the contribution of matter ($T_{\mu\nu}$) in terms of the purely geometrical contribution ($S_{\mu\nu}$ of the LHS) of the generalized field. For this to be possible, it must be possible, if need be, for $S_{\mu\nu}$ to express the intervention of an electromagnetic field. This, in the view expressed by GR, a truly non-dualistic theory (the elimination of the sources of the field).' Hence, the starting point for the present theory is the Einstein equation for 'empty' space (as Einstein[26] has done for his 'Relativities Theory of the Non-symmetric Field'. It is not the purpose here to propose, however, an aimless generalization of GR. We simply reformulate a nondualistic theory of GR, in accordance with Einstein-Hilbert procedure, on the spinor-tensor Manifold defined above provided that the spinor preservation condition (2.35) is not violated.

3.2. Preliminaries

General Relativity (in 'empty' space) consists of a Pseudo-Riemannian manifold $\{L(4), (M, g)\}$ with the metric $g_{\mu\nu}$ obeying

$$\overset{\circ}{\nabla}_{\lambda} g_{\mu\nu} = 0 \tag{3.1}$$

and the Einstein equation (with Cosmological constant)

$$\overset{\circ}{R}_{\mu\nu} - \frac{1}{2} g_{\mu\nu} \overset{\circ}{R} = \lambda g_{\mu\nu} \tag{3.2}$$

In the above, objects capped with a small '∘' refer to the Levi-Civita connection. This is to distinguish the corresponding objects (without the capped zero) used later which refer to the Pseudo-Riemannian Sympletic Connection. The above scheme may be obtained in two different ways as follows.

(1) The first way is to impose (3.1) as the Pseudo-Riemann condition and then apply 'Einstein's Variational Principle', according to Schrodinger[26], to the action

$$A = \int \sqrt{-g}(\overset{\circ}{R} - 2\lambda) d\tau \tag{3.3}$$

where

$$g = \det(g_{\mu\nu}) \tag{3.4}$$

and

$$\overset{\circ}{R} = g^{\mu\nu}\overset{\circ}{R}_{\mu\nu} \tag{3.5}$$

is the Ricci scalar.
(2) The second way is not to assume (3.1) in the first instance but to apply 'Palatini's Variational Principle' to (3.3) hence obtaining both (3.1) and (3.2) at the same time. The advantage of Palatini's procedure is not just its simplicity but that it determines the type of connection as well, given by (3.1) in the present case. In other words, it also limits the choice of the action A, if, from physical considerations certain geometry is desirable. Thus, if (3.1) is regarded as a criterion for a real physical space-time, then the action (3.3) is a correct choice (and may be the simplest one) if Palatini's procedure is followed.

3.3. Theoretical Foundation

According to the foregoing, we seek an action A such that (1) when Palatini's procedure is followed, it yields (2.35) and (2) it is a natural extension of (3.2) in the presence of a spinor structure.

To do this we denote the dual of (2.27)

$$F = \frac{1}{2!}F_{\mu\nu}dx^{\mu} \wedge dx^{\nu} =^{*} \sigma \tag{3.6}$$

i.e.,

$$F_{\mu\nu} = \frac{1}{2}\epsilon_{\mu\nu\alpha\beta}\sigma^{\alpha\beta} =^{*} \sigma_{\mu\nu} \tag{3.7}$$

where

$$(\sigma^{\mu\nu}) = (\sigma_{\mu\nu})^{-1}, \tag{3.8}$$

and propose the action density by

$$A = \int \{\sqrt{-g}g^{\mu\nu}R_{\mu\nu} + \lambda\sqrt{\det(g_{\mu\nu} + F_{\mu\nu})} + \mu\sqrt{f}f^{\mu\nu}R_{\mu\nu}\}d\tau \tag{3.9}$$

where $R_{\mu\nu} = R^{\alpha}_{\mu\nu\alpha}$ is the Ricci Tensor derived from the curvature 2 form of the $\mu(4)$ - manifold and $f^{\mu\nu}$ is the matrix inverse of $F_{\mu\nu}$ i.e.,

$$f^{\mu\alpha}F_{\nu\alpha} = \delta^{\mu}_{\nu} \tag{3.10}$$

and

$$f = \det(F_{\mu\nu}) \tag{3.11}$$

The geometry of the space-time manifold is to be determined by the Noetherian symmetries

$$\frac{\delta A}{\delta g^{\mu\nu}} = 0 \tag{3.12}$$

$$\frac{\delta A}{\delta f^{\mu\nu}} = 0 \tag{3.13}$$

where Hamiltonian derivatives in the sense of Palatini is understood. The above symmetries imply also conservation laws in some sence.

The choice for A in (3.9), as is evident, satisfies the following principles:
(1) The principle of natural extension of (3.3) to include $F_{\mu\nu}$.
(2) The principle of symmetry between $g_{\mu\nu}$ and $F_{\mu\nu}$ i.e., that $g_{\mu\nu}$ and $F_{\mu\nu}$ should enter into A symmetrically and
(3) The principle that the spinorial symmetry (2.35) should be derivable from (3.12) and (3.13).

The last principle is of such paramount importance that a special name is given to it: The Principle of Kinematic-Dynamic Symmetry. It states that the spinorial invariance given by (2.35) describing the fundamental Kinematic Symmetry of the world picture should be derivable from the dynamical symmetries given by (3.12) and (3.13). That (3.9) satisfies the Principle of Kinematic-Dynamic Symmetry will be demonstrated below. It would be difficult to find a simpler expression for A that satisfies all the three principles. The action density used by Einstein[6] in his theory of the Non-symmetric field, for example, satisfies (1) and (2) in a sense but not (3). Hence there is nothing in Einstein's Theory that would yield a description of the spinorial structure.

Below is a demonstration as to what kind of a world picture the scheme (3.9), (3.11), (3.13) contains.

3.4. Principle of Equivalence (PE)

In order that experimental evidence forms the foundation of the theory, the Principle of Equivalence has to be imposed as well as the foregoing symmetry principles. This is specifically required as the $\mu(4)$ - connection contains a torsion tensor whose presence has often been suspected of violating the Equivalence Principle. The implication of PE for the space-time structure is shown as follows.

The PE implies that the tangent space M_p is to be a Minkowski space and hence at M_p

$$(g_{\mu\nu})_p = \eta_{\mu\nu}, \quad (\partial_\lambda g_{\mu\nu})_p = 0 \tag{3.14}$$

where $\eta_{\mu\nu}$ is the Minkowski metric. Now the coefficients of the affine connection is given by (Schouten, p. 141)[27]

$$\Gamma^{\lambda}_{\mu\nu} = \{^{\nu}_{\mu\lambda}\} + T_{\mu\lambda}^{\cdot\cdot\nu} \tag{3.15}$$

where the first term on the RHS is the coefficients ofthe Levi-Civita connection and the second term, called the 'contorsion' is given by

$$T_{\mu\lambda}^{\cdot\cdot\nu} = g^{\nu\rho}\{S_{\mu\rho\lambda} - S_{\rho\lambda\mu} + S_{\lambda\mu\rho}\} \tag{3.16}$$

where $S_{\mu\nu}^{\cdot\cdot\lambda}$ is the torsion tensor.

The first member of (2.35) then gives

$$\nabla_\lambda g_{\alpha\beta} = \overset{\circ}{\nabla}_\lambda g_{\alpha\beta} - T_{\gamma\alpha}^{\cdot\cdot\lambda} g_{\lambda\beta} - T^{\cdot\cdot\lambda}{}_{\gamma\beta} g_{\alpha\beta} = 0 \tag{3.17}$$

which is valid at p also. Thus substitution of (3.14) into (3.17) gives

$$\{T_{\gamma\alpha\beta} + T_{\gamma\beta\alpha}\}_p = 0 \tag{3.18}$$

since (3.14) gives

$$\{\overset{\circ}{\nabla}_\lambda g_{\alpha\beta}\}_p = 0 \tag{3.19}$$

As the above are tensor relations, they must be valid in all coordinate systems and at all points and hence we have

$$T_{\gamma\alpha\beta} = T_{\gamma[\alpha\beta]} \tag{3.20}$$

and

$$\overset{\circ}{\nabla}_\lambda g_{\alpha\beta} = 0 \tag{3.21}$$

That is, the imposition of PE implies the following equivalence:

$$\{\nabla_\lambda g_{\alpha\beta} = 0, PE\} \Leftrightarrow (3.20), (3.21) \tag{3.22}$$

Now, (3.16),(3.20)$\Rightarrow$

$$S_{\mu\lambda\nu} = S_{[\mu\lambda\nu]} \tag{3.23}$$

The complete anti-symmetry for the torsion tensor in (3.25) has been obtained as the result of imposition of PE on the space-time structure and not as the result of a priori assumptions concerning the possible physical meaning of the torsion tensor. This all important result (first stated by Yu[28]) helps to simplify many formulae involving curvature and torsion. In particular, we have (Schouten, p.141)[27]

$$R_{(\mu\lambda)} = \overset{\circ}{R}_{\mu\lambda} + \overset{\circ}{\nabla}_\alpha S_{\mu\beta}^{\cdot\cdot\alpha} S_{\alpha\lambda}^{\cdot\cdot\beta} \tag{3.24}$$

$$R_{[\mu\lambda]} = -\overset{\circ}{\nabla}_\alpha S_{\mu\lambda}^{\cdot\cdot\alpha} \tag{3.25}$$

In view of (3.23) we can always put

$$S_{\lambda\mu\nu} = \epsilon_{\lambda\mu\nu\alpha} h^\alpha \tag{3.26}$$

where $\epsilon_{\lambda\mu\nu\alpha}$ is a unit alternating tensor and $h^\alpha \epsilon M_p$ is a 4-vector. Thus,

$$S_{\mu\alpha\beta}S^{\lambda\alpha\beta} = \epsilon_{\mu\alpha\beta\rho}\epsilon^{\lambda\alpha\beta\sigma}h^\rho h_\sigma = -2(\delta_\mu^\lambda h_\alpha h^\alpha - h^\lambda h_\mu) \tag{3.27}$$

Therefore

$$R = g^{\mu\lambda}R_{(\mu\lambda)} = \overset{\circ}{R} + 6h_\alpha h^\alpha \tag{3.28}$$

Thus, with the equivalence relation (3.22) and (3.28), the action density in (3.9) becomes

$$A = \int\{\sqrt{-g}(\overset{\circ}{R} + 6h_\alpha h^\alpha) + \lambda[\det(g_{\mu\nu} + F_{\mu\nu})] + \mu\sqrt{f}\tilde{R}\}d\tau \tag{3.29}$$

where

$$\tilde{R} \equiv f^{\mu\nu}R_{\mu\nu} = f^{\mu\nu}R_{[\mu\nu]} \tag{3.30}$$

3.5. Principle of Kinematic Dynamic Symmetry

Application of the dynamic symmetries (3.12) and (3.13) to (3.29) gives (after Schrodinger[26])

$$\overset{\circ}{\nabla}_\lambda g_{\mu\nu} = 0 \tag{3.31}$$

$$\overset{\circ}{R}_{\mu\nu} - \frac{1}{2}g_{\mu\nu}\overset{\circ}{R} = -6(h_\mu h_\nu - \frac{1}{2}g_{\mu\nu}h_\alpha h^\alpha) + \frac{\alpha}{L}[\frac{1}{2}g_{\mu\nu}(1-G^2) + (F_{\mu\alpha}F_\nu^\alpha - \frac{1}{4}g_{\mu\nu}F_{\alpha\beta}F^{\alpha\beta})] \tag{3.32}$$

$$\nabla_\lambda F_{\mu\nu} = S_{\lambda\mu}^{\cdot\cdot\alpha}F_{\nu\alpha} \tag{3.33}$$

and

$$R_{[\mu\nu]} - \frac{1}{2}F_{\mu\nu}\tilde{R} = \frac{\beta}{L}(F^{\rho\sigma} - G * F^{\rho\sigma})F_{\mu\rho}F_{\nu\sigma} \tag{3.34}$$

where the constants α and β are numerical multiples of λ amd μ respectively and

$$L = \sqrt{(1 - G^2 + \frac{1}{2}g^{\mu\rho}g^{\nu\sigma}F_{\mu\nu}F_{\rho\sigma})} \tag{3.35}$$

$$G = \sqrt{(-\frac{f}{g})} \tag{3.36}$$

It is not difficult to show that (3.7),(3.33) $\Rightarrow$

$$\nabla_\lambda \sigma_{\mu\nu} = 0 \tag{3.37}$$

Hence in view of the equivalence relation (3.22),

$$\{(3.36), (3.38)\} \Leftrightarrow \{(2.35)\} \tag{3.38}$$

In other words, application of the dynamic symmetries (3.12) and (3.13) to (3.29) recovers the kinematic (spinorial) symmetries (2.35) completely.

3.6. Field Equations

Rotating the indices λ, μ, ν in (3.33) and adding, taking into account (3.23) we have

$$\sum_{(\lambda,\mu,\nu)} \nabla_\lambda F_{\mu\nu} = \sum_{(\lambda,\mu,\nu)} \overset{\circ}{\nabla}_\lambda F_{\mu\nu} = 0 \tag{3.39}$$

i.e.

$$dF = 0 \tag{3.40}$$

where F is defined by (3.6). Again, (3.33) gives

$$d * F = J \tag{3.41}$$

where $*F$ is the dual of F and

$$J = \frac{1}{2!} S^{\cdot\cdot\rho}_{\lambda\mu} F_{\rho\nu} dx^\lambda \wedge dx^\mu \wedge dx^\nu \tag{3.42}$$

Further, in view of (3.25) and (3.30) equation (3.34) simplifies to

$$\nabla_\rho S^{\cdot\cdot\rho}_{\mu\nu} = -2\beta T_{[\mu\nu]} \tag{3.43}$$

where

$$T_{[\mu\nu]} = \frac{1}{L}[(\mathbf{F}^3)_{\mu\nu} + (H - 3G^2) F_{\mu\nu}] \tag{3.44}$$

In (4.9),

$$H \equiv \frac{1}{2} F^{\alpha\beta} F_{\alpha\beta} \tag{3.45}$$

and

$$(\mathbf{F}^3)_{\mu\nu} \equiv F_{\mu\alpha} F^{\alpha\beta} F_{\beta\nu} \tag{3.46}$$

The closed system {(3.36), (3.37), (3.45), (3.46), (3.48)} μ constitutes a set of 'field equations' and is a direct consequence of the set (3.12), (3.13), (3.29). It is a reflection of the dynamic symmetries and the Principle of Equivalence and contains in it kinematic (spinorial) symmetry (2.35). Nothing more can be said about it until the set of unknowns ($g_{\mu\nu}$, $F_{\mu\nu}$, h^λ) can be identified with the set of quantities of the physical universe. In order that the system μ constitutes a physical theory we must appeal to the Principle of Identification first formulated by Eddington[29]: 'The relation structure presents itself in our experience as a physical world consisting of space, time and things. The transition from the geometrical description to the physical description can only be made by identifying the tensors which measure physical quantities with tensors occurring in pure geometry; and we must proceed by inquiring first what experimental properties the physical tensor possesses, and then seeking a geometrical tensor which possesses these properties by virtue of mathematical identities.

If we can do this completely, we shall have constructed out of the primitive relation-structure a world of entities which behave in the same way and obey the same laws as the quantities recognised in physical experiments.' Hence, we suggest that in the scheme μ, $g_{\mu\nu}$ is in some way associated with the gravitational field and $F_{\mu\nu}$ the electromagnetic field. The new 4-vector h^λ is to be called the 'generalized momentum vector' of the 'field' and 'particle' - a name to be justified post-humously as the consequence of the scheme unfolds. A 'particle', according to Eddington[8], can be conceived as a carrier of 'variates'; in Eddington's scheme, a V_{16} particle, for instance, carries 16 variables: a linear momentum vector (4), an angular momentum bivector (6) and the electromagnetic field tensor (6).

In that sense, therefore, h^λ is to play the role of carrier of variates and thus admits various representations according to the contexts to which it is applied.

3.7. Dirac's Equation in Spinorial Space-time

On putting

$$\sqrt{12m}h_\mu = \overset{\star}{P}_\mu \tag{3.47}$$

in (3.32) gives

$$\overset{\circ}{R}_{\mu\nu} = \frac{1}{2m}\overset{\star}{P}_\mu \overset{\star}{P}_\nu - \frac{\alpha}{2L}(1 - G^2)g_{\mu\nu} + E_{\mu\nu} \tag{3.48}$$

where

$$E_{\mu\nu} = \frac{\alpha}{L}(F_{\mu\alpha}F^\alpha_\nu - \frac{1}{4}g_{\mu\nu}F_{\alpha\beta}F^{\alpha\beta}) \tag{3.49}$$

In the above m is called the 'mass parameter' associated with a particle and P_μ, its total momentum 4-vector such that

$$\overset{\star}{P}_\alpha \overset{\star}{P}^\alpha = -m^2 \tag{3.50}$$

Thus, contraction of (3.48) gives

$$\overset{\circ}{R} = \frac{1}{2}m + \frac{2\alpha}{L}(1 - G^2) \tag{3.51}$$

i.e.,

$$m = \frac{4\alpha}{L}(1 - G^2) + 2\hat{R} \tag{3.52}$$

It has been argued (Brans and Dicke[30]; Hoyle and Narlikar[31] and Ross[32]) that Mach's Principle implies that the mass of a particle is a function of its position in space-time. In the present theory, the very existence of space-time is a consequence of the existence of matter and hence it is natural that 'mass' of a 'particle', if such be identified, must be a property of space-time as demonstrated by (3.52).

In the presence of an electromagnetic field which is already there in the μ-scheme we may further put

$$\overset{\star}{P}_\mu = P_\mu - eA_\mu \tag{3.53}$$

where e is the charge of the 'particle' under consideration and A_μ the electromagnetic potential whose existence is quaranteed by (3.40) via Poincare's Lemma. In (3.53) p_μ is be interpreted as the inertial momentum 4-vector.

Substitution of (3.53) into (3.48) and multiplying the result by $\gamma^\mu\gamma^\nu$ we have

$$\gamma^\mu\gamma^\nu(P_\mu - eA_\mu)(p_\nu - eA_\nu) - 2m\gamma^\mu\gamma^\nu[\overset{\circ}{R}_{\mu\nu} + \frac{\alpha}{2L}(1-G^2)g_{\mu\nu} - E_{\mu\nu}] = 0 \tag{3.54}$$

If we postulate the quantum mechanical correspondence

$$P_\mu \to \hat{P}_\mu \equiv \hat{\nabla}_\mu \tag{3.55}$$

equation (3.54) corresponds to the wave equation

$$\{\gamma^\mu\gamma^\nu(\hat{P}_\mu - eA_\mu)(\hat{P}_\nu - eA_\nu) - 2m\gamma^\mu\gamma^\nu[\overset{\circ}{R}_{\mu\nu} + \frac{\alpha}{2L}(1-G^2)g_{\mu\nu} - E_{\mu\nu}]\}\Psi = 0 \tag{3.56}$$

where Ψ is the Dirac Bi-spinor. Now,

$$\begin{aligned}\gamma^\mu\gamma^\nu &= \frac{1}{2}(\gamma^\mu\gamma^\nu + \gamma^\nu\gamma^\mu) + \frac{1}{2}(\gamma^\mu\gamma^\nu - \gamma^\nu\gamma^\mu)\\ &\equiv g^{\mu\nu} + \sigma^{\nu\mu}\end{aligned} \tag{3.57}$$

which, when substituted into (3.56) gives

$$[(\hat{\mathbf{P}} - e\mathbf{A})^2 - m^2 - \frac{1}{2}eF_{\mu\nu}\sigma^{\mu\nu}]\Psi = 0 \tag{3.58}$$

where, using the notation of Landau and Lifshitz[46],

$$\sigma^{\mu\nu} = (\alpha, i\Sigma),\, F^{\mu\nu} = (-\mathbf{E}, \mathbf{H}) \tag{3.59}$$

Thus (3.58) becomes

$$[(\hat{\mathbf{P}} - e\mathbf{A})^2 - m^2 + e\Sigma\cdot H - ie\alpha\cdot\mathbf{E}]\psi = 0 \tag{3.60}$$

where the terms due to intrinsic spin are now made evident. The spinorial symmetry (2.35) is thus seen to be directly responsible for the appearance of intrinsic spin. Equation (3.60) is the second order Dirac Equation for the motion of an electron of mass m and charge e in an external field $(-\mathbf{E}, \mathbf{H})$. It can be factorized to give

$$[\gamma^\mu(\hat{\mathbf{P}}_\mu - e\mathbf{A}_\mu) + m][\gamma^\nu(\hat{\mathbf{P}}_\nu - e\mathbf{A}_\nu) - m]\psi = 0 \tag{3.61}$$

which is the sufficient condition for the existence of the Dirac Equation;

$$\left[\gamma^{\nu}(\hat{\mathbf{P}}_{\nu} - e\mathbf{A}_{\nu}) - m\right]\psi = 0 \tag{3.62}$$

Once the Dirac Equation is established, it can be shown (cf. Gurtler and Hestenes[22]) that the whole of Quantum Mechanics can be generated. To the extent that the correspondence (3.55) is permissible in the present scheme the Dirac equation is said to have been 'derived' from the non-dualistic Einstein equation (3.48). Equation (3.62) differs from the standard Dirac equation in that the definition of mass is given by (3.52) which, in principle depends on the solution of the μ-scheme. The definition (3.52) is in accord with Mach's Principle and couples the Dirac equation with the rest of the universe. The fact that the square of $\overset{\star}{P}_{\alpha}$ in (3.50) is a negative quantity shows that it is imaginary viz.,

$$\overset{\star}{P}_{\alpha} = i\pi_{\alpha} \tag{3.63}$$

where π_{α} is a real vector. The imaginary number i occurs because (3.27) gives

$$\mathbf{S}^2 = S_{\lambda\alpha\beta}S^{\lambda\alpha\beta} = -6h_{\alpha}h^{\alpha} = -6\mathbf{h}^2 \tag{3.64}$$

On substituting (3.64) into (3.47) one obtains

$$S^2 = -\frac{\overset{\star}{P}{}^2}{2m} \tag{3.65}$$

The all important imaginary number that characterizes QT as distinct from the standard classical field theory is thus seen to arise naturally from the space-time geometry and the Principle of Equivalence. It has been argued in the context of Einstein-Cartan Theory (Hehl, F.W. et al[33], V. De Sabbata et al[34,35]) that space-time torsion is linked to the intrinsic spin of elementary particles and here it has been shown that torsion also generates inertia of space-time and matter. As to be shown below, torsion is also seen to be responsible for the existence of the gravitational field as well as electric current and charge. Hence gravitation, electric charge and inertia are all unified, so to speak, in space-time torsion.

4. Gravitational Theories of the μ-Scheme

4.1. Space and Matter

'Matter' may be conceived of as being described by the spinorial symmetry (2.35) and the action density (3.9). The operations (3.12) and (3.13) on matter then produce the field equations (3.37), (3.46), (3.48), (3.67) which describe phenomena that are the manifestations of matter. Describing matter in terms of a relation-structure instead of miniscule golf balls is the essence of a field theory pioneered by the founders of GR. The same entries (2.35), (3.9) which describe matter can also be called a representation of 'space-time' which is otherwise labelled by four meaningless parameters; one could therefore call the μ-scheme manifestations of space-time - 'matter' and 'space-time' are just two different labels of the same reality. Hence, the concept of 'empty space' often employed in Einstein's Theory has no place in the present theory for, there is no space-time in the absence of matter; we do not have a universe situated 'in' space-time because universe is the space-time. One could, of course, have a situation in which the inertial manifestation or else the electromagnetic manifestation of matter is absent, but that does not represent the absence of matter altogether. Let us illustrate the point by a sub-theory of the μ-scheme in which $F_{\mu\nu} \to 0$. We then have the field equations

$$\overset{\circ}{R}_{\mu\nu} - \frac{1}{2}g_{\mu\nu}\overset{\circ}{R} = 6(h_\mu h_\nu - \frac{1}{2}g_{\mu\nu}h_\alpha h^\alpha) + \frac{\alpha}{2}g_{\mu\nu} \tag{4.1}$$

and

$$\nabla_\rho S_{\mu\nu}^{\cdot\cdot\rho} = 0 \tag{4.2a}$$

which is equivalent to

$$\frac{\partial h_\mu}{\partial x^\nu} - \frac{\partial h_\nu}{\partial x^\mu} = 0 \tag{4.2}$$

in view of (3.31). There are at least two solutions to equation (4.2) viz.,
(i)

$$h_\mu = 0 \tag{4.3}$$

(ii)

$$h_\mu = \pm\overset{\circ}{\nabla}_\mu\Omega, \quad h_\mu = \pm i\overset{\circ}{\nabla}_\mu\Omega \tag{4.4}$$

where Ω is a scalar field; The '+' or '-' sign in (4.4) correspond, as we shall see later to 'matter' or 'anti-matter'. A third solution is given by (5.1).
(i) Corresponding to (4.3) we have the standard de Sitter universe.

Whilst it largely gives a satisfactory account of the Hubble expansion of the universe it has been regarded, in the context of Einstein's GR, as unsatisfactory because it 'corresponds to a completely empty universe containing neither matter nor radiation' (Tolman, 1934)[36]. In the present theory, however, the universe is not 'empty' despite the absence of the inertial manifestation of matter. Indeed, it probably describes an early epoch of the universe before ponderable matter was formed when the universe was expanding at a fast rate giving rise to the 'inflationary' phase. A stage must have been reached during the expansion in which the universe cooled sufficiently for ponderable matter to form and we no longer have $h_\mu = 0$.

(ii) We here tentatively postulate that the subsequent evolution of the universe entails the solution

$$h_\mu = \pm\mathring{\nabla}_\mu\Omega \tag{4.5}$$

which, evidently can only give part pf the story since it has already been assumed that $F_{\mu\nu} \to 0$. Equation (4.1) now becomes

$$\mathring{R}_{\mu\nu}\frac{1}{2}g_{\mu\nu}\mathring{R} = 6(\partial_\mu\Omega\partial_\nu\Omega - \frac{1}{2}g_{\mu\nu}\partial_\alpha\Omega\partial^\alpha\Omega) + \lambda g_{\mu\nu} \tag{4.6}$$

On taking divergence of (4.6) we get

$$\mathring{\Box}\Omega = 0 \tag{4.7}$$

Weyl's postulate and the cosmological principle require that the space-time metric take the Robertson-Walker (R-W) form

$$ds^2 = dt^2 - a(t)[\frac{dr^2}{1-kr^3} + r^2(d\theta^2 + \sin^2\theta d\theta^2)] \tag{4.8}$$

The solutions of (4.6) corresponding to k = 0, 1, -1 are as follows.

(a) k=0

$$a = a_0\exp[\frac{\mu}{\sqrt{2}}(t-t_0)] \tag{4.9}$$

where a_0 =constant.

(b) k=1

$$a = \frac{1}{\mu\sqrt{2}}(A\exp[\frac{\mu}{\sqrt{2}}t] - A^{-1}\exp[-\frac{\mu}{\sqrt{2}}t]) \tag{4.10}$$

where A is an arbitrary constant.

(c) k=-1

$$a = \frac{1}{\mu\sqrt{2}}(B\exp[\frac{\mu}{\sqrt{2}}t] + B^{-1}\exp[-\frac{\mu}{\sqrt{2}}t]) \tag{4.11}$$

where B is an arbitrary constant.

With a positive real constant μ, the above solutions correspond to the different de Sitter universes but, as μ may be complex, it is also possible to obtain a solution corresponding to damped harmonic motion.

The physical significance of Ω in the above cosmological solution is not clear; its nature can be inferred, however, from the case of the radiation due to a radially pulsating star (Yu, 1989)[28] where it is shown that $(\frac{\partial\Omega}{\partial t})^2$ may be interpreted as the rate of energy radiation due to the rate of loss of mass of the central star. Hence this implies for the de Sitter solutions above the following conclusions:

(a) that gravitational energy is being continually converted into 'ponderable matter' as the universe expands and
(b) that the creation of mass is partially responsible for the Hubble expansion. The Ω on the other hand also takes on the meaning of static gravitation potential as demonstrated below.
(iii) The Ω-field Theory of Gravitation (Yu[28])

On neglecting the λ-term, the static spherically symmetric solution in isotropic coordinates for (4.6) can be shown to be and the corresponding solution for (4.7) is Thus, (4.12), (4.13)

$$g_{00} = 1 - \frac{2M}{r} + \frac{2M^2}{r^2} - .... \tag{4.14}$$

$$g_{11} = g_{22} = g_{33} = 1 + \frac{2M}{r} + ... \tag{4.15}$$

Hence if the above solution is taken to be the gravitational field of a central spherical star with mass M, then we have recovered the results for all the three classical tests for GR to the same degree of accuracy since a test particle follows the geodesic path in accordance with the PE which we have assumed.

Had we chosen the solution $h_\mu = 0$ for (4.2) we would have the field equation which gives the standard Schwarzschild solution for the gravitational field outside a spherical star which in turn established GR as the standard gravitational theory of Einstein. With the choice of (4.4) as the solution of (4.2) and the interpretation of Ω in (4.13) as the gravitational field, however, we have a possibly viable alternative gravitation theory in the same μ-scheme. This alternative theory called the 'Ω-field Theory of Gravitation' (Yu[28]) has certain theoretical advantages over the standard GR as will be demonstrated in Section VII (D).
(iv) Gravitational Field Equation with disordered Electromagnetic Radiation.

On appropriately applying th mean value theorem and Tolman's argument (R.C. Tolman, 1934)[36] equation (3.37) can be recast into the form

$$\overset{\circ}{R}_{\mu\nu} - \frac{1}{2} g_{\mu\nu} \overset{\circ}{R} = 6(h_\mu h_\nu - \frac{1}{2} g_{\mu\nu} h_\alpha h^\alpha) + \lambda g_{\mu\nu} + \frac{\alpha}{L_0}[(\rho_{00} + P_0) u_\mu u_\nu - P_0 g_{\mu\nu}] \tag{4.16}$$

with

$$\rho_{00} = 3P_0 \tag{4.17}$$

The last term on the RHS of (4.16) has been obtained by averaging the electromagnetic energy momentum tensor in (3.37). The density ρ_{00} and pressure P_0 of an equivalent perfect fluid are as defined in Tolman[36]. Moreover, if the electromagnetic field takes the particular form

$$F_{\mu\nu} = \phi^2 (P_\mu k_\nu - p_\nu k_\mu); \quad k_\mu k^\mu = P_\nu k^\mu = 0; \quad P_\mu P^\mu = 1 \tag{4.18}$$

then (3.48) reduces to

$$\frac{\partial h_\mu}{\partial x^\nu} - \frac{\partial h_\nu}{\partial x^\mu} = 0 \tag{4.19}$$

with a solution

$$h_\mu = i\partial_\mu \Omega \tag{4.20}$$

in which case (4.16) reduces to

$$\overset{\circ}{R}_{\mu\nu} - \frac{1}{2}g_{\mu\nu}\overset{\circ}{R} = 6(\partial_\mu\Omega\partial_\nu\Omega - \frac{1}{2}g_{\mu\nu}\partial_\alpha\Omega\partial^\alpha\Omega) + \lambda g_{\mu\nu} + \frac{\alpha}{L_0}[(\rho_{00} + P_0)u_\mu u_\nu - P_0 g_{\mu\nu}] \tag{4.21}$$

In the interpretation of an 'Ω-field Theory' (Yu[28]) the RHS of (4.21) is the EMT of the gravitational field and matter which EMT is convariantly conserved. Such a convariant conservation law for field and matter taken together was at one time regarded by Einstein[37] as essential to the completion of GR and that view was later echoed by Mercier et al[38] and Logunov et al[39] among others. All that having been said, however, it is essential to remember that whether the field equation takes the form (4.16) or (3.37) it is fundamentally the same at the source : the whole space-time is 'matter'. The difference lies with their different manifestations for the purposes of observation and experiment.

5. Spin Dynamics of Ponderable Matter and Gravitational Radiation-A Physical Postulate

5.1. *Matter Representation of the Gravitoinertial Vector h^λ.*

The above results have shown that the vector h^λ plays the multiple roles of a spin vector, an inertial vector as well as, in the case of the 'Ω-field Theory', the gravitational force field. Indeed, if h^λ is to represent matter in motion in some sense it must embody the concepts of particle and wave simultaneously; in particular, it should include tensors that represent waves, linear momentum as well as spin angular momentum. We suspect that it is the non-inclusion of a wave concept in the Newtonian momentum vector that causes that apparent discontinuity between classical dynamics and quantum mechanics and we here attempt to bridge the divide. In order to explicate the different manifestations, therefore, it admits the following representation:

$$h^\lambda = \overset{\circ}{\nabla}{}^\lambda\Omega + \epsilon^{\lambda\mu\nu\rho}\nabla_\rho A_{\mu\nu} + \gamma\epsilon^{\lambda\mu\nu\rho}M_{\mu\nu\rho} + \gamma P^\lambda \tag{5.1}$$

where $\gamma = \frac{8\pi}{c}\sqrt{G}$ is a physical constant related to the Blackett formula (V. de Sabbata[34]). The above decomposition is analogous to the theorem in vector anlaysis in E^3: in case a vector function has no potential but both its curl and divergence possess potentials, the vector function may be divided into three parts of which the first has no divergence; the second, no curl; the third, neither divergence nor curl. (Gibbs and Wilson, p.235)[40]. The difference here lies with the addition of an axial vector $\epsilon^{\lambda\mu\nu\rho}M_{\mu\nu\rho}$ which exists in 4-dimensional spacetime. The belief that such a decomposition exists in reality is strengthened by the fact that in Quantum Theory, there exists an angular momentum Helmholtz Theorem (Lomont and Moses[41], Moses[42]) and as h^λ is closely associated with spin angular momentum, the analogy is a natural one. Moreover, as (5.1) gives a sum of polar and axial vectors, chirality, which is important in microscopic theories is here naturally taken into account. Thus (5.1) gives an expression of the concept of an 'inertial wave vector' that includes the dual concepts of wave and particle if $(\Omega, A_{\mu\nu})$ are the wave tensors and $(\epsilon^{\lambda\mu\nu\rho}M_{\mu\nu\rho}, p^\lambda)$ particle vectors. Substitution of (5.1) into (3.48) gives

$$\mathring{\Box}A_{\mu\nu} - \gamma\nabla_\rho M^{\cdot\cdot\rho}_{\mu\nu} - \gamma(\frac{\partial P_\nu}{\partial x^\mu} - \frac{\partial P_\mu}{\partial x^\nu}) = -2\beta T_{[\mu\nu]} \tag{5.2}$$

Consider, in particular, the case in which $F_{\mu\nu} \to 0$ so that $T_{[\mu\nu]} \to 0$. The above equation then gives

$$\mathring{\Box}A_{\mu\nu} = \gamma\nabla_\rho M^{\cdot\cdot\rho}_{\mu\nu} + \gamma(\frac{\partial P_\nu}{\partial x^\mu} - \frac{\partial P_\mu}{\partial x^\nu}) \tag{5.3}$$

We then have the following physical postulate: if $M^{\cdot\cdot\rho}_{\mu\nu}$ is identified with the intrinsic spin angular momentum temsor of ponderable matter, p_μ with its linear momentum vector and $A_{\mu\nu}$ with the 'gravitational radiation tensor', then equation (5.3) states that the sum of the intrinsic spin angular momentum and orbital angular momentum of ponderable matter is conserved if and only if gravitational radiation $A_{\mu\nu}$ is absent. Conversely, the spacetime flux of the sum of intrinsic angular momentum and orbital angular momentum of ponderable matter is the source of gravitational radiation. The convariant radiation equation (5.3) is then exact. Equation (5.3), on the other hand, would appear to contradict the Law of Conservation of Angular Momentum but this is in fact not so; what it says is that angular momentum is conserved if and only if gravitational radiation is taken into account. Hence, the orbital decay of PSR1913+16 would indicate a decrease of angular momentum due to gravitational radiation and thus the total angular momentum of the binary system is not conserved in the Newtonian sense; but it is conserved if gravitational radiation is taken into account in the sense of (5.3). Conversely, should the angular momentum of the mass centr of the binary systme be conserved, its orbit would remain unchanged and no gravitational radiation can be expected.

It is here that the enclosed theory of gravitationbal radiation differs fundamentally from that in GR, where, by perturbation and weak field approximation one obtains the linear equation of gravitational radiation and the quadruple formula for the rate of energy loss. When (5.3) is coupled to the generalized Einstein equation (3.37) and the contracted Bianchi's second identity, one could in principle trace the decreasing orbit of a radiating planet as well as its gravitational radiation. It is not known yet whether this could also be done in the case of standard GR.

5.2. Angular Momentum and Electromagnetic Field

The full radiation equation is contained in (5.2) where

$$T_{[\mu\nu]} = \frac{1}{L}[(\mathbf{F}^3)_{\mu\nu} + (H - 3G^2)F_{\mu\nu}] \tag{5.4}$$

and hence in the absence of radiation we have

$$\gamma\nabla_\rho M_{\mu\nu}^{\cdot\cdot\rho} + \gamma(\frac{\partial P_\nu}{\partial x^\mu} - \frac{\partial P_\mu}{\partial x^\nu}) = -\frac{2\beta}{L}[(\mathbf{F}^3)_{\mu\nu} + (H - 3G^2)F_{\mu\nu}] \tag{5.5}$$

For astronomical objects such as stars and planets Blackett[43] proposed an empirical law relating the magnetic moment $\mathbf{M}$ and the intrinsic angular momentum $\mathbf{L}$ of the body and that relation has been determined rigorously on the basis of Einstein-Cartan Theory by de Sabbata et al[34] as :

$$\mathbf{M} = \frac{2\sqrt{G}}{C}\mathbf{L} \tag{5.6}$$

The same formula was later on developed by de Sabbata el al[34,35] into a relation between the magnetic moment $\mathbf{M}$ and the intrinsic spin σ of elementary particles :

$$\mathbf{M} = \frac{8\pi}{3C}\sqrt{\alpha G}\sigma \tag{5.7}$$

where α is the fine structure constant and G the universal gravitational constant. Equation (5.5) on the other hand expresses a general relationship between the intrinsic spin angular momentum $M_{\mu\nu}^{\cdot\cdot\rho}$, the orbital angular momentum $(\frac{\partial P_\nu}{\partial x^\mu} - \frac{\partial P_\mu}{\partial x^\nu})$ and the electromagnetic field $F_{\mu\nu}$ of elementary particles. For most continuous distribution of ponderable matter, taking the average of (5.5) over a given volume would see the vanishing of the intrinsic spin term (except, perhaps, at the surface of that volume) and we would have the relation

$$H_{\mu\nu} = K L_{\mu\nu} \tag{5.8}$$

where

$$H_{\mu\nu} =< [(\mathbf{F}^3)_{\mu\nu} + (H - 3G^2)F_{\mu\nu}]\frac{1}{L} > \tag{5.9}$$

may be taken to be the electromagnetic field inside macroscopic matter and $L_{\mu\nu}$ is the averaged angular momentum of the macroscipic body. To that extent (5.8) agrees with the Blackett formula (5.6) at least qualitatively. It is the existence of (5.6) and (5.8) that justifies the value of the physical constant γ.

5.3. Law of Conservation of Energy Momentum Tensor of Matter and Equation of Motion

The energy momentum tensor of matter (EMT) is by definition the RHS of equation (3.37) and hence by virture of the conttacted Bianchi's second identify, it is covariantly conserved. In order to illustrate the spirit of the present theory we shall here consider only the case for which $F_{\mu\nu} \to 0$ leaving the complete theory at a late stage. We have on neglecting the small cosmological constant α.

$$\overset{\circ}{\nabla}_\nu M^\nu_\mu = 0 \tag{5.10}$$

where

$$M^\nu_\mu \equiv h^\nu h_\mu - \frac{1}{2}\delta^\nu_\mu h_\rho h^\rho \tag{5.11}$$

And (5.10) gives

$$(\overset{\circ}{\nabla}_\nu h^\nu)h_\mu + (\overset{\circ}{\nabla}_\nu h_\rho - \overset{\circ}{\nabla}_\rho h_\mu)h^\rho = 0 \tag{5.12}$$

Hence for $F_{\mu\nu} = 0$, (3.48), (5.12) $\Rightarrow$

$$\overset{\circ}{\nabla}_\nu h^\nu = 0 \tag{5.13}$$

Substitution of (5.1) into (5.12) then gives

$$\overset{\circ}{\Box}\Omega = -\gamma \nabla_\lambda P^\lambda \tag{5.14}$$

In particular, if a test particle is defined by the conditions

$$P^\lambda = m u^\lambda \tag{5.15}$$

where u^λ is the velocity vector of the particle and on its constant mass,

$$\overset{\circ}{\nabla}_\lambda (m u^\lambda) = 0 \tag{5.16}$$

and

$$\overset{\circ}{\nabla}_\lambda \epsilon^{\lambda\mu\nu\rho} M_{\mu\nu\rho} = 0 \tag{5.17}$$

then (5.13) reduces to

$$\overset{\circ}{\Box}\Omega = 0 \tag{5.18}$$

Thus what (5.14) says is that the momentum vector is strictly conserved iff (5.17) and (5.18) are valid and in general it need not be the case. Now, for a test particle, (5.3) may be written as

$$B_{\mu\nu} = \gamma(\overset{\circ}{\nabla}_\mu u_\nu - \overset{\circ}{\nabla}_\nu u_\mu) \tag{5.19}$$

where

$$B_{\mu\nu} \equiv \frac{1}{m}(\overset{\circ}{\Box} A_{\mu\nu} - \gamma \nabla_\rho M^{\cdot\cdot\rho}_{\mu\nu}) \tag{5.20}$$

Thus, since

$$u_\mu u^\mu = 1 \tag{5.21}$$

(5.19) gives, after straight forward manipulation,

$$u^\mu B_{\mu\nu} = \gamma \frac{Du_\nu}{ds} \tag{5.22}$$

Hence the test particle follows a geodesic path iff

$$u^\mu B_{\mu\nu} = 0 \tag{5.23}$$

A sufficient condition for (5.23) to be valid is that the gravitational field is static with $A_{\mu\nu} = 0$ and that the particle does not possess spin. Such is the assumption used in the Ω-field Theory (Yu[27,44]). (5.21) otherwise gives an equation of motion (valid for a deformable body in general) that includes the effect of spin and radiation damping without imposition. In view of the difficulty in solving the full set of eqauations, approximation schemes would have to be proposed even for the simplest cases. For the motion of a planet round a central star, for instance, one could assume the static solutions in the Ω-field theory (Yu[28,45]), and then for radiation the radiatn field $A_{\mu\nu}$ field is to be imposed in a region far from the planetary system in a way to be illustrated in Section VI.

6. Gravitodynamics: Gravitational Radiation of a Closed Systems of Particles in Motion

6.1. Equations of Gravitational Radiation for a System of Accelerating Paticles in a Weak Gravitational Field

A system of neutral particles in motion induces gravitational radiation governed by

$$\mathring{\Box} A_{\mu\nu} = \gamma(\frac{\partial P_\mu}{\partial x^\nu} - \frac{\partial P_\nu}{\partial x^\mu}) \tag{6.1}$$

on neglecting the intrinsic spin at the interior of each macroscopic particle and the electromagnetic field. A question arises as to why the radiant screw tensor $A_{\mu\nu} \equiv (\mathbf{e}, \mathbf{b})$ should be crowned with the title 'dynamic gravitatinal field'. It is in reality immaterial whether or not $A_{\mu\nu}$ is entitled to that name as long as it represents radiation induced by accelerating neutral particles in the absence of any electromagnetic interaction. That radiation in turn influences and modifies the motion of these particles. As the radiation has been derived from (3.48) within the μ-structure it states in effect the unification of gravitation and inertia as a covariant condition of the world. This unification has hidden in it a Gravitodynamics just as elegant as the classical Electrodynamics as will be manifest below. Now, equation (6.1) can be decomposed into

$$\mathring{\Box} A_{0i} = \gamma(\frac{\partial P_0}{\partial x^i} - \frac{\partial P_i}{\partial x^0}) \tag{6.2}$$

and

$$\mathring{\Box} A_{jk} = \gamma(\frac{\partial P_i}{\partial x^k} - \frac{\partial P_k}{\partial x^i}) \quad i,j,k = 1,2,3 \tag{6.3}$$

In the usual weak field approximation, terms involving the connection coefficients are ignored and $g_{\mu\nu} \approx \eta_{\mu\nu}$ and we have

$$\mathring{\Box} \approx \nabla^2 - \frac{1}{c^2}\frac{\partial^2}{\partial t^2} \tag{6.4}$$

On putting

$$(A_{0i}) \equiv -\mathbf{e}, \tag{6.5}$$

$$(\epsilon_{ijk}A^{ik}) \equiv \mathbf{b}, \tag{6.6}$$

equations (6.2) and (6.3) take on the weak field form

$$\nabla^2 \mathbf{e} - \frac{1}{c^2}\frac{\partial^2 \mathbf{e}}{\partial t^2} = -\gamma(\nabla P_0 + \frac{1}{c}\frac{\partial \mathbf{P}}{\partial t}) \tag{6.7}$$

and,

$$\nabla^2 \mathbf{b} - \frac{1}{c^2}\frac{\partial^2 \mathbf{b}}{\partial t^2} = \gamma curl \mathbf{P} \tag{6.8}$$

which are the radiation equations for the radiant screw tensor $(\mathbf{e}, \mathbf{b})$ due to the inertial source. On putting $P_0 = \nabla \cdot \mathbf{e}$ and $\nabla \cdot \mathbf{b} = 0$, the system (6.7), (6.8) admits a solution set (6.9), (6.10):

$$curl \mathbf{e} = -\frac{1}{c}\frac{\partial \mathbf{b}}{\partial t} \tag{6.9}$$

and

$$curl \mathbf{b} = \frac{1}{c}\frac{\partial \mathbf{e}}{\partial t} - \gamma \mathbf{P} \tag{6.10}$$

which may be called the Gravito-Maxwell Equations and form the foundations of Gravitodynamics. Following the standard procedure (6.7), (6.8) may be deduced from (6.9), (6.10) which also produces the Gravito-Poynting Theorem:

$$\oint (\mathbf{e} \wedge \mathbf{b}) \cdot ds - \gamma \int \mathbf{e} \cdot \mathbf{P} dV = \frac{1}{2c}\frac{\partial}{\partial t}\int (e^2 + b^2) dV \tag{6.11}$$

That is, with the correspondence

$$(\mathbf{e}, \mathbf{b}) \leftrightarrow (\mathbf{E}, \mathbf{H}) \tag{6.12}$$

$$(-\gamma \mathbf{P}) \leftrightarrow (\mathbf{J}) \tag{6.13}$$

There is an isomorphism between classical Electrodynamics and the Gravitodynamics here established and everything that has been done for Electrodynamics can be uniquely translated into Gravitodynamics via the above correspondence. The negative sign associated with $\vec{P}$ in (6.13) and the positive sign associated with $\nabla\Omega$ in (6.21) ensure that gravitation 'attracts'. In particular, for a finite system of neutral particles in motion the intensity of gravitational radiation can be shown to be (Landau and Lifshitz[46])

$$-\frac{dE}{dt} = I = \frac{G}{45c^5}\dddot{D}^2_{\alpha\beta} + \frac{8G}{3c^3}\ddot{L}^2 \tag{6.14}$$

on putting

$$\text{on putting} \quad \gamma = \frac{8\pi}{c}\sqrt{G}, \quad G = \text{gravitational constant} \tag{6.15}$$

where

$$D_{\alpha\beta} = \sum \mu(3x_\alpha x_\beta - \delta_{\alpha\beta}r^2), \quad \mu = mass \tag{6.16}$$

is the total quadraupole mometn tensor and

$$\mathbf{L} = \frac{1}{2c}\sum \mu\mathbf{r}\wedge\mathbf{v} \tag{6.17}$$

is the total angular momentum vector of the system. Equation (6.14), having been translanted from equation (71.5) of Landau and Lifshitz[46] does not contain the dipole radiation term as it vanishes for neutral particles.

Although the μ-scheme contains Einstein's GR as a special case the energy radiation formula (6.14) evidently cannot be obtained for that case where $h^\lambda = 0$. That formula is, however, applicable to the Ω-field theory (Yu[28]) which arises as a first approximation to the representation (5.1) and in this case the Gravitodynamic System (6.9), (6.10) may be regarded as applicable to the far field zone where $A_{\mu\nu}$ becomes important as compared with Ω. In that zone, within the context of the Ω-field Theory, formula (6.14) is applicable.

Now the equation of motion of a test particle for the metric (4.12) in the Ω-field is (Yu[45])

$$\frac{d\mathbf{v}}{ds} + \frac{\mu\mathbf{r}}{r^3} = \frac{\mu\mathbf{r}}{r^3}[\mathbf{L}\wedge\mathbf{v} + (\mathbf{r}\cdot\mathbf{v})\mathbf{v}] \tag{6.18}$$

giving

$$\frac{d\mathbf{L}}{ds} = -2\frac{d\Omega}{ds}\mathbf{L} \neq 0 \tag{6.19}$$

However, in this case

$$\frac{\ddot{L}^2/c^3}{D^2_{\alpha\beta}/c^5} \sim \left(\frac{\mu v}{c}\right)^2 \tag{6.20}$$

and hence the $\ddot{\mathbf{L}}^2$ term is negligible for orbital speeds much less than that of the speed of light. Thus the Ω-field theory (Yu[28,44]) in combination with (6.14) predicts the result for the orbital decay of PSR1913+16 to the same degree of accuracy as Einstein's.

In general, however, the $\ddot{\mathbf{L}}^2$ term in (6.14) cannot in principle be ignored. Indeed, for a spinning body (e.g., a star) suddently brought to rest by e.g., collision, resulting in a jerk, the $\ddot{\mathbf{L}}^2$- term might not be negligible and experiments based on this may be devised for testing the validity of (6.14) versus sthe Einstein formula which does not contain $\ddot{\mathbf{L}}^2$. It must be added that whilst the derivation of (6.14) is here straight-forward within the context of Gravitodynamics here proposed, the standard derivation of the Einstein formula in standard text books is not without conceptual difficultiesa s has been shown by Wald[47] and Yu[48]. The refined and improved derivation by Damour[49] within the context of GR is rather elaborate involving many approximations and assumptions.

6.2. Quantum Gravitodynamics (QGD)

The verification of the radiation formula (6.14) by PRS1913+16 and the promise of (6.9), (6.10) to give an account of the Alfven universe without the assumption of the existence of large electric currents in space give hope for the physical reality of Gravitodynamics. Further, in view of its formal isomorphism with classical Lorentz-Maxwell Electrodynamics it is tempting to mimic QED directly via (6.9), (6.10) to obtain the corresponding relativistic quantum theory - the Quantum Gravitodynamics. For instance, one could write

$$\mathbf{e} = -\frac{1}{c}\frac{\partial \mathbf{G}}{\partial t} + \nabla\Omega \tag{6.21}$$

$$\mathbf{b} = curl\, G \tag{6.22}$$

where G is called the gravitational vector potential and Ω a scalar potential. The corresponding Dirac Equation is just

$$[\gamma^\nu(\hat{\mathbf{P}}_\nu - m\mathbf{G}_\nu) - m]\psi_0 = 0 \tag{6.23}$$

which can again be derived from the generalized Einstein equation (3.53) using an identical procedure for the derivation of (3.67). Should (6.23) correspond to physical reality, the rich store of experimental predictions is only to be expected. One must guard against such optimism, however, as there is a fundamental difference between electricity and gravitaion : electric charges can have both positive and negative values whilst gravitational charges (m) can only take on positive values. When and only when (6.9), (6.10) has been firmly established by observations and experiments can the QGD here suggested take on real meaning.

6.4. Quantization of Gravity

The parallel between Gravitodynamics and the Lorentz-Maxwell Electrodynamics makes the QGD as contained in (6.23) appear physically plausible and the quantization of gravity and mass could proceed much the same way as in teh case of electromagnetism and charge. Nonetheless, the old Bohr-Commerfeld quantization rules can be derived (and not postulated) from the $(\Omega, A_{\mu\nu})$-field theory and the Gravitodynamics as contained in (6.1). For, the orbital equation of a gravitationally charged particle in the gravitational field of a gravitational neucleus as given by (6.22) yields (Yu[44])

$$\frac{dE}{dS} = (\frac{e^{2\Omega}}{L_0}\frac{dL}{dS})v^2; \quad l = |\mathbf{L}| \tag{6.24}$$

where

$$E = (\frac{1}{2}v^2 - \frac{\mu}{r}); \quad v = |\mathbf{v}| \tag{6.25}$$

Equation (6.24) is then the equation of energy transport due to gravitational radiation if E is interpreted as the totla energy of the particle. We can then derive the Bohr-picture from (6.24): when a gravitational particle is executing a circular orbit in a spherically symmetric gravitational field it does not emit gravitational radiation since

$$\frac{dL}{dS} = 0 \tag{6.26}$$

for circular orbits and in this case the RHS of (6.24) vanishes. On the other hand, when it jumps from one (circular) orbit to another,

$$\frac{dL}{dS} \neq 0 \tag{6.27}$$

and gravitational radiation is emitted. It follows naturally that the possible orbits of the particle are restricted and we have

$$\oint P_\lambda dx^\lambda = 2\pi nh, \quad n = 0, 1, 2, ... \tag{6.28}$$

Thus (6.1) and (6.28) immediately give

$$\frac{1}{\gamma}\int \overset{\circ}{\Box} A_{\mu\nu} dx^\mu \wedge dx^\nu = 2\pi nh \tag{6.29}$$

which gives the quantization condition of the dynamic gravitational field $(\mathbf{e}, \mathbf{b})$ which is responsible for gravitational radiation induced by moving gravitational charges. As the dynamic gravity $A_{\mu\nu}$ is related to Torsion via (5.1) it may also be said that the quantization of spacetime has been effected through torsion, a scenario that has been pursued by de Sabbata et al[50,51].

ACKNOWLEDGEMENT

At the various stages of development of the present work I have benefited from the discussions with and communications from Sir H. Bondi and Prof. V. de Sabbata and their inspirations are here deeply acknowledged.

(1) Weyl, H., *'Space-time-Matter', Dover, N.Y.* (1950)
(2) Eddington A.S., *Proc. Roy. Soc. London A.*, **99**, (1921) p.104.
(3) Einstein, A., *Preuss. Akad. Wiss. Phys-Math. Kl. Sitz.*,, (1923) p.32, 76.
(4) Einstein, A., *Preuss. Akad. Wiss. Phys-Math. Kl. Sitz.*,, (1925) p.414.
(5) Einstein, A., *Preuss. Akad. Wiss. Phys-Math. Kl. Sitz.*,, (1928) p.217.
(6) Einstein, A., *'The Meaning of Relativity' Methuen, London*, (1956)
(7) Pauli, W., *'Theory of Relativity' Pergamon, Oxford*, (1958)
(8) Eddington A.S., *'Fundamental Theory' Cambridge Univ. Press*, (1948)
(9) Dirac, P.A.M., *Int. J. Theor. Phys.*, (1984) p.677.
(10) Penrose, R., *'Gen. Rel. Grav.'*, **7**, (1976) p.31.
(11) Penrose, R., *In 'Quantum theory of Gravity' ed. S.M. Christensen. Adam Hilger, Bristol*, (1983)
(12) Penrose, R., *In 'Quantum Concept in Space and Time' ed. C.J. Isham and R. Penrose, Oxford Univ. Press*, (1986)
(13) Stratton, J.A., *'Electromagnetic Theory', McGraw-Hill, N.Y.* (1941)
(14) Segal, I.E. *'Mathematical Cosmology and Extragalactic Astronomy' Academic Press, N.Y.* (1976)
(15) Sachs, M., *'General Relativity and Matter', D. Reidel, Dordrecht, Holland*, (1982)
(16) Kobayashi, S., and K. Nomizu, *'Foundations of Differential Geometry' Interscience, N.Y.*, **II**, (1969)
(17) Goldberg, S.I. *'Curvature and Homology' Academic Press, N.Y.*, (1962)
(18) Hestenes, D., *'Space-Time Algebra' Gorden and Breach, N.Y.*, (1966)
(19) Hestenes, D., *'Clifford Algebra to Geometric Calculus' D. Reidel Pub. Co., Dordrecht, Holland*, (1982)
(20) Hestenes, D., *J. Math. Phys.*, **8**, (1967) p.798.
(21) Hestenes, D., *J. Math. Phys.*, **14**, (1973) p.893.
(22) Hestenes, D., *J. Math. Phys.*, **16**, (1975) p.556.
(23) de Sabbata, V., *'Twistors, Torsion and Tensor-Spinor space-time'* in these proceedings.
(24) Bunge, M., *'Foundations of Physics', Springer, Berlin*, (1967)
(25) Tonnelat, M.A. *'The Principles of Electromagnetic Theory and of Relativity'*
(26) Schrodinger, E., 'Space-time Structure' Cambridge Univ. Press
(27) Schouten J.A., 'Ricci-Calculus' Spring, Berlin, (1954)
(28) X. Yu, 'Astrophysics & Space Science' **154** *(1989).*
(29) Eddington A.S., 'The Mathematical Theory of Relativity' Cambridge University Press, (1924)
(30) Brans and Dicke Phys. Rev, **124**, *(1961) p.925.*
(31) Hoyle, F. and V.J. Narlikar, Proc. Roy. Soc., **A282**, *(1964) p.191.*
(32) Ross D.K., Phys. Rev., **D5**, *(1972) p.284.*

(33) *Hehl, F.W., et al, Rev. Mod. Phys.,* **48**, *(1976) p.393.*

(34) *de Sabbata, V., and M. Gasperini, Lett. Nuovo Cimento,* **27**, *(1980) p.133.*

(35) *de Sabbata, V., and C. Sivaram, Nuovo Cimento,* **102B**, *(1988) p.107. de Sabbata, V. 'Torsion, String Tension and Quantum Gravity' in 'String Quantum Gravity and Physics at the Planck Energy Scale' ed. by N. Sanchez, World Scientific, Singapore, (1993) p.528. de Sabbata, V., and M.Gasperini, 'Introduction to Gravitation' World Scientific, Singapore (1985).*

(36) *Tolman, R.C. 'Relativity, Thermodynamics and Cosmology' Oxford Univ. Press, Oxford, (1934)*

(37) *Einstein, A., Phys. Z.,,* **15**, *(1914) p.108.*

(38) *Mercier, A. et at, "On General Relativity" Academic Verlag, Berlin, (1979)*

(39) *Logunov, A.A. (ed.), 'Gravitational and Elementary Particles' MIR publisher, Moscow (1989)*

(40) *Gibbs, J.W. and E.B. Wilson Vector Analysis" Dover, N.Y. (1960), (1901)*

(41) *Lomont, J.S. and H.E. Moses, Commun, Pure Appl. Maths.,* **14**, *(1961) p.69.*

(42) *Moses, H.E., J. Math. Phys.,* **17**, *(1976) p.1821.*

(43) *Blackett, P., Nature (London),* **159**, *(1947) p.658.*

(44) *X. Yu,Astrophysics and Space Science* **202** *(1993a) p.237*

(45) *X. Yu, J. Hong, Y.Y.Lam and W.T.Lee, Applied Mathematics and Mechanics,* **14**, *p.305*

(46) *Landau and Lifshitz, 'Course of Theoretical Physics, The Classical Theory Field' Pergamon, Oxford,* **2** *(1975)*

(47) *Wald R.M. General Relativity, The University of Chicago*

(48) *X. Yu,Astrophysics and Space Science (letts)***194** *(1992) p.159.*

(49) *Damour, T., in '300 years of Gravitation' Cambridge Univ. Press,* **128**, *(1989)*

(50) *de Sabbata, V., and C. Sivaram, in 'Modern Problem of Theoretical Physics', Festschrift for D. Ivanenko, World Scientific, Singapore, (1991), p.163. Nuovo Cimento,* **105B**, *(1990) p.603.*

(51) *de Sabbata, V., in 'The Earth and the Universe' ed. by W. Schroder, published by International Association of Geomagnetism and Aeronomy (Germany), (1993) p.332*

A DISCRETE APPROACH TO
THE QUANTIZATION OF SPACETIME*

ZHENJIU ZHANG

CCAST(World Laboratory), P.O.Box 8730, Beijing 100080, China
and
Centrer for Relativity Studies,
Department of Physics
*Huazhong Normal University, Wuhan, 430070, China***

ABSTRACT

Several approaches among others to quantum gravity reviewed, the discrete approach of lattice to quantum gravity, firstly proposed by T. D. Lee, reintroduced and refirmed in some sence. The foundation and the general background of both general relativity and quantum mechanics therefore modified. A foundamental constant introduced as a postulate from beginning. Locally Minkowski random lattice is the basic simplex. The concept of 'neighbor' described by the number of light cones crossed. 'Parallel transportation' along a closed path defined and connected to the curvature of the spacetime. The discrete quantum gravity action corresponding to the Hamiltonian formulation was given.

1. Introductory

It is general accepted that perturbative approaches to quantum gravity fail because they assume that spacetime geometry can be approximated by a smooth continuum at all scales. Ashtekar's non-perturbative canonical quantum gravity approach has been recently discussed extensively (for example, some at CCAST, and Lee Smolin at Erice School). Indeed, there exist a number of quantum gravity programs that begin by postulating discrete structures at the Planck scale (for example, Finkelstein,1969; Penrose,1971; 't Hooft,1979; Bombelli, et al.,1988; Barbour,1992). In this lecture, we will mainly concentrate on the discrete approach firstly proposed by T.D.Lee(1983) beginning with very foundamental concepts of physics, Hamiltonian formulation of general relativity, time as a dynamical variable, and 3+1 splitting of spacetime.

---

* invited lectures given to the Workshop of Quantum Gravity of China Center of Advanced Science and Technology (CCAST), April, 1995, Beijing, China; and to the 14th Course of the International School of Cosmology and Gravitation-Quantum Gravity, May, 1995, Erice, Italy

** mailing address

1.1. Spacetime

1. Newton's space–time: a 4–manifold with a preferred foliation by hyperplanes and a contravariant metric q^{ab} of signature (0,+,+,+) with an absolute observer–independent notion of distant simutaneity.

2. Special relativity's space–time: a 4–manifold with metric η^{ab} of signature (–,+,+,+) with a preferred light–cone whose normal is null with respect to η^{ab}. The speed of light is a universal constant, indepenedent of observers.

3. General reletivity's space–time: a 4–manifold with metric g^{ab}, which is sensitive to matter.

1.2. Internal Blemishes of General Relativity:

Perturbation expansions fail in gravitational case.

1. Einstein's theory is perturbatively non–renormalizable at two loops for pure gravity and at one loop for gravity interacting with matter.

2. The theories by adding suitable higher derivative terms is not unitary. In the perturbative treatment, the Hamiltonian is unbounded from below, signaling a dramatic instability.

1.3. Internal Blemishes of Quantum Field Theory:

1. Physical quantities such as scatering cross–sections are infinite when all radiative corrections are taken into account.

2. The renormalization procedure makes individual terms in perturbative expansions manageable, although the entire series almost certainly diverges.

3. When the renormalization works, it is not clear why it works. Dirac: "...How then do they manage with these incorrect equations? These equations lead to infinities when one tries to solve them; these infinities ought not to be there. They remove them artificially. . . . Just because the results happen to be in agreement with observations does not prove that one's theory is correct. After all, the Bohr theory (of the hydrogen atom) was correct in simple cases. It gave very good answers, but still the Bohr theory had very wrong concepts." (P.A.M.Dirac, The inadequacies of quantum field theory.)

1.4. Canonical Quantum Gravity

The Lagrangian formulation leads to the new canonical variables.
A concrete approach needs to find the inner product on the space of physical states to select a vacuum state for constrained system.
The new Hamiltonian framework was extended to allow for a cosmological constant and also matter sources including Klein– Gordon, Dirac and Yang–Mills fields.
The issue of time was analyzed using the connection representation. Non-perturbative canonlcal quantum gravity is promising (Ashtekar, Smolin et al. since 1986).

Example to show how infinity can arise:(Ashtekar,1991)
Consider the problem of self-energy of a point charge in classical field theory. We regard the point charge as the limit of a sequence of spherical shells of radius r, with uniform charge and mass density, as r tends to zero. Ignoring the gravity, the self-energy of the shell is given by

$$m(r) = m_o + e^2/r. \qquad (1)$$

It is infinite when $r \rightarrow 0$. Bring in Newtonian gravity,

$$m(r) = m_o + e^2/r - Gm_o^2/r. \qquad (2)$$

It is infinite when $r \rightarrow 0$, unless $m_o = e/\sqrt{G}$.

Example to show how infinity can be removed:
Bring in general relativity,

$$m(r) = m_o + e^2/r - Gm^2(r)/r. \qquad (3)$$

The solution of this equation, when $r \rightarrow 0$, is

$$m = e/\sqrt{G}. \qquad (4)$$

It is finite when $r \rightarrow 0$.

The calculation is meant to point out that there are certain mathematical mechanisms in general relativity which automatically regularize physical quantity. The result is essentially non-perturbative. This calculation is too naive, for it ignores all quantum effects as well as weak and strong interactions. A precise calculation was carried out by Arnowitt, et al. (1960,1965). The result is the same as (4).

1.5. Other Approaches

1. Feynman's theorem (Deser,1970): Fierz Pauli theory of a spin–2 particle in flat spacetime is inconsistent when coupled to matter, and the only self–consistent extention of it is the general relativity.

2. Quantum field theory in curved spacetime (Hawking 1974, Birrell and Davies 1982, Huang et al. 1994) : The Minkowski vacuum can be expressed as a coherent state of Cooper pairs of left and right quanta. Prodicted black hole radiation needs experimental confirmation. Neither the gravitational field, nor the spacetime is quantized.

3. Einstein gravity as a gauge theory (\`t Hooft, Veltman): Neither is pure gravity finite to two loops.

4. Supergravity fails at the third loop of renormalization.

5. Superstring theory fails for divergent sum of perturbative expansion.

6. A clear quantization program based on algebraic ideas (no need for geometrical quantum mechanics) is on the way for investigation.

7. 2–D dilaton gravity: This is exactly solvable quantum–corrected 2D dilaton gravity theories. The starting point is the classical action. Cone strings are equliblium configuretion of a cosmic string interacting with a black hole. In this sence, 2–D black hole can be created by a cosmic string in the presence of 4–D black hole. (Frolov and Novikov,1993; Frolov, 1995).

1.6. Arrangement of This Lecture

This lecture mainly investigate the discrete quantum gravity, firstly proposed by T.D. Lee and his coorperators (1983) based extensively on Regge's idea (1961), developed by them(1983), and Zhang et al.(1983,1991,1994).

As a preparation, in section 2 we give one motivation to the quantization of spacetime from the entropy of a spherically self–gravitating collapsing system; in section 3 we investigate the time as a dynamical variable, and give the 3+1 splitting of spacetime and the Hamiltonian formulation of general relativity; we introduce the discrete quantum gravity in section 4.

2. A Motivation From Black Hole Entropy

In the 1970's, Bekenstein (1973) and Hawking (1975) proposed that the relation between the entropy S and the area A of the event horizon of a Schwarzschild black hole is $S = A/4$. Later, Kundt (1976) and Bekenstein (1980) pointed out that the thermodynamical entropy of a baryon system with spherical symmetry is 20 orders less than the entropy of the Schwarzschild black hole with the same mass. This great difference makes the physical meaning of black hole entropy a particularly fascinating topic (Wilkins, 1979; Wald. 1979). Since the 1980's, much research on black hole entropy has been done in the areas of statistical physics (Zurek and Thorne, 1985), quantum physics (Bomebelli etc., 1988), the intrinsic entropy of the gravitational field, the self-gravitating entropy of thermodynamics (Sorkin, Wald and Zhang, 1981; Zhang et al. 1983; 1989; 1990,1991,1992), and others. It is shown that the black hole entropy in the general case to be connected with the Noether charge associated with the Killing horizon (Wald, 1993). Recently, the dynamical origin of black hole entropy has been investigated (Frolov et al. 1993, 1995).

We estimate the thermodynamical entropy of the collapsed star, and find the total entropy production during the entire collapsing process. For the estimate, we assume that the final state is at the instant of precollapse into a black hole, that it is in local thermodynamical equilibrium, and that its spacetime can be described by the time-symmetric metric. Under these assumptions, we show that the maximum total entropy of the final collapsed star is roughly (in the condition of Planck length quantum cutoff) that corresponding to a Schwarzschild black hole. This suggests that the self-gravitating collapse provides huge entropy production, and explains the great difference. It seems that in the content of classical general relativity, the process of gravitational collapse can increase the entropy of the system greatly, and can raise the maxima of the thermodynamical entropy near to the corresponding black hole entropy with the same mass. This suggests that the thermodynamical entropy production in the entire dissipative process of self-gravitating collapse might be one of the sources of the entropy increase of black holes.

The most important point here we must point out is that, if we do not apply the condition of Planck lenth quantum cutoff, when reaches the horizen, the maximum total entropy of the final collapsed star goes to infinite. This is the result of quantum field theory in unquantized curved spacetime. One of the ways to solve this problem is to quantize spacetime itself.

3. 3+1 Splitting and Hamiltonian Formulation of General Relativity

3.1. 3+1 Splitting of Spacetime (Wald,1984; Zhang,1992)

We decompose 4–spacetime into 3–space and 1–time. This is called 3+1 splitting. We choose a time function t and a "time flow" vector field t^a on M satisfying $t^a \nabla_a t = 1$,(Now, we cannot interpret t and t^a in terms of physical measurements using clocks until one knows the spacetime metric). Given a g^{ab}, we decompose t^a into its normal and tangent parts with respect to the surface Σ_t of constant t. We define a lapse function N as

$$N = -g_{ab} t^a n^b = (n^a \nabla_a t)^{-1}, \tag{1}$$

and shift vector N^a as

$$N^a = h^a_b t^b, \tag{2}$$

where n^a is unit normal to Σ_t and

$$h^a_b = g^a_b + n^a n_b \tag{3}$$

is the induced spatial metric on Σ_t. N measures the rate of flow of proper time τ, with respect to coordinate time t, as one moves normally to Σ_t. N^a measures the amount of "shift" tangent to Σ_t contained in the time flow vector field t^a. Therefore, we have

$$n^a = (t^a - N^a)/N \tag{4}$$

and

$$g^{ab} = h^{ab} - n^a n^b = h^{ab} - N^{-2}(t^a - N^a)(t^b - N^b). \tag{5}$$

Then we choose h^{ab}, N^a, N as our field variables rather than g^{ab}.

3.2. Hamiltonian Formulation of General Relativity

Now we turn to write the Hamiltonian for gravitational field (or curved spacetime). We write the Lagrangian density by scalar curvature R as

$$L_G = \sqrt{(-g)}R. \qquad (6)$$

In order to write L_G as a function of h^{ab}, N^a and N, we write the Einstein's equations as

$$G_{ab} = R_{ab} - (1/2)g_{ab}R, \qquad (7)$$

then contracting with n^a and n^b, We have

$$L_G = \sqrt{h}N\ {}^3R + K^{ab}K_{ab} - K^2, \qquad (8)$$

where K^{ab} is the extrinsic curvature of Σ_t, and $K = K^a_a$.

The conjugate momentum tensor corresponding to h^{ab} is

$$\Pi^{ab} = \sqrt{h}(K^{ab} - Kh^{ab}). \qquad (9)$$

Then the Hamiltonian density is

$$H_G = \int h^{1/2}N\ [\ -\ {}^3R + h^{-1}\ \Pi^{ab}\Pi_{ab} - (1/2)h^{-1}\Pi^2], \qquad (10)$$

where $\Pi = \Pi^a_a$.

From the variation of

$$H_G = \int (NH_o + N^a\ H_a) \qquad (11)$$

with respect to N and N_a, we have the initial value constraint equation

$$H_o = h^{1/2}\ [\ -\ {}^3R + h^{-1}\ \Pi^{ab}\Pi_{ab} - (1/2)h^{-1}\Pi^2] = 0. \qquad (12)$$

This is the Wheeler-DeWitt equation before quantization. It appears a serious obstacle to the formulation of a quantum theory of gravitation by the canonical quantization approach. We need to find a choice of configuration space for general relativity such that only the "true dynamical degrees of freedom" are present in its phase space, and

$$H^b = -2h^{1/2}D_a(h^{-1/2}\Pi^{ab}), \qquad (13)$$

where D_a is the derivative operator on Σ_t associated with h_{ab}. The equation (13) can be automatically satisfied in the superspace which is the configuration space diffeomorphic (with same causal structure) to our choice of configuration field h^{ab}.

Then from H_G and H_o we can get the Hamiltonian equations(Wald, 1984),(Zhang,1992). Equations (12), (13) and the Hamiltonian equation are equivalent to vacuum Einstein's equations $R_{ab} = 0$.

4. DISCRETE APPROACH TO QUANTUM GRAVITY

4.1. Basic Ideas

Gravitational physics is close related to the geometry of space and time. A primary goal of any quantum gravity should be to probe the quantum structure of space and time. As might be expected, a classical description of spacetime structure should break down and quantum gravity should be needed at scales of order of the Planck length l_p and smaller. Quantum gravity should be a logical development of theoretical physics. There are difficulties for formulating Quantum General Relativity. The key difficulty is the dual role of g_{ab}. g_{ab} describes the dynamical aspects of gravity; also describes the background spacetime structure. In order to "quantize" the dynamical degrees of freedom of the gravitational field, one must also give a quantum mechanical description of spacetime structure which has no analog in other quantum field theories. We can give an example to illustrate the difficulty. A fundamental property of quantum field theory in Minkowski spacetime is that the field operator Ψ, corresponding to a integer- spin classical field, evaluated at spacelike related points must commute with itself,

$$[\Psi(x), \ \Psi(x')] = 0, \qquad (1)$$

where x and x′ are spacelike related. This equation means that a measurement of Ψ at x cannot influence the value of Ψ at x′. But equation

$$[g_{ab}(x),\ g_{ab}(x')] = 0, \qquad (2)$$

for x and x′ spacelike related, makes nosense. Since: (1) we must know g_{ab}, then we can say x and x′ is or is not spacelike related; (2) But this operator equation, if valid, must hold independently of the state of the gravitational field, i.e., the value of g_{ab}; (3)If g_{ab} is a bandoned, causality becomes ill defined.

According to T.D.Lee(1983), at present, with QCD for the strong interaction, a unifying gauge theory for the electromagenetic and weak interactions and the Einstein theory of gravitation (plus possible further unification of some or all of these interactios), we may feel that we have arrived at a closed system of physical laws. Nevertheless, the fact that any of the unifying gauge theoretical models requires about twenty parameters suggests that our understanding of these forces most likely remains phenomenological, far from fundamental. This uneasy feeling is reinforced by the enormous technical difficulty of quantizing gravity. Attempts to improve this situation have so far invariably led back to the original unsatisfactory state, with only a change in scale. The underlying cause may be traced to the foundation of our relativistic quantum local field theory. In such a theory only fields are observables, represented by operator functions embedded in a space -time continuum. Hence we can conceive of performing an infinite sequence of observations, at x, x′, ⋯ , all lying infinitesimally close but with the result of the field measurements, say Φ(x), Φ(x′),⋯ ,fluctuating violently. The probability amplitude of such an unlikely event is proportional to exp(iA) with A as the corresponding action. Yet only the part of A that involves the gradient of the fields is effective in making such rapid fluctuations improbable. This then severly limits the candidates for an acceptable physical theory, and results in the present small set of renormalizable field theories: QED, QCD, Higgs, ⋯. All of these have been intensively investigated.

Let us suppose that the concept of locality is only an approximat one. After all, it is not difficult to imagine that the present quantizational field theory should break down at a distance l_p of the order of planck's length.

4.2. Time as a Dynamical Variable

Time has always been regarded as continuous parameter in physics. In general relativity, the metric is a dynamical variable, but the continuous four-dimensional space that the metric is embedded in is not. We wish to explore an altermative that time is a dynamical variable. Just like any orther dynamical variable, time is affected by material things, and in fact it is part of matter.

Let us consider a large time interval T. A fundamental postulate is that within such a time interval one can only assume N space- time positions (x_n, t_n), with $n=1,2,\cdots,N$. The ratio $T/N \equiv t_p$ is a fundamental constant of the theory. For convenience and without loss of generality we have arranged $t_1, t_2, \cdots, t_N$ in ascending order.

4.3. Discrete Quantum Gravity (T.D.Lee,1983)

The technical difficulties of quantizing gravity are well known. Among them the two most prominent are (i) nonrenormalizability; (ii) energy not positive definite.

Our strategy is to fomulate the discrete theory of gravity directly in the Minkowski space. Firstly, to set up a random lattice in the Minkowski space keeping the density of lattice sites $\rho=(1/l_p)^4$ finite, we avoid the ultraviolet divergence. Because we work directly in the Minkowski space, the path integrals become finite.

4.3.1. Random lattice

In a Minkowski Space, let x^μ be the Cartesian coordinates in a 4-dimensional flat Minkowski space where the superscript $\mu = 0, 1, 2, 3$ denotes the spacetime component with x^0=time t. The metric $\eta_{\mu\nu}$ is given by

$$\text{diag } \eta_{\mu\nu} = (-1,1,1,1). \tag{1}$$

The invariant volume element is

$$d^4x = dx^0 dx^1 \cdots dx^3. \tag{2}$$

the same as in a flat Euclidean space. Consider a random distribution of N lattice sites in a volume Ω. Clearly, N, Ω and the site density

$$\rho \equiv N/\Omega \tag{3}$$

are all Lorentz invariant.

The immediate problem with setting up a relativistic field theory on a random lattice in a Minkowski space is the concept of "neighbors". In a given random distribution of N lattice sites, we can envisage each site carrying its light cone. Between any two points p and q (which may or may not be the lattice sites), let us define

$$n_{pq} \equiv \text{number of light cones crossed by the straight line pq.} \tag{4}$$

Under a Lorentz transformation, a light cone remains a light cone, and a straight line remains a straight line. If pq crosses the light cone of, say, the i-th site in one Lorentz frame, this crossing must also be observed in any other Lorentz frame. Consequently, n_{pq} is invariant. As we shall see, this number n_{pq} can be used to establish the concept of "neighbors" in a Minkowski space. That is to use a number (not a distance) to distinguish the nearness or apartness of two points.

4.3.2. Simplex

First take a random lattice in a flat 4-dimensional Minkowski space. Label each site by i=1,2,⋯, and each 4-simplex by $\tau(a)$ with a=1,2,⋯. Let $x_\mu(i)$ be the μ-th coordinate of the i-th site. For every linked pair of sites i and j, the (link-length)^2 is given by the usual expression

$$l_{ij}^2 = \eta_{\mu\nu}\,[x(i) - x(j)]^\mu\,[x(i) - x(j)]^\nu \tag{1}$$

where $\eta_{\mu\nu}$ is defined by (4.3.1.1).

Consider now an arbitrary variation

$$l^2_{ij} \rightarrow l'^2_{ij}. \qquad (2)$$

Correspondingly, each 4–simplex changes from

$$\tau(a) \rightarrow \tau'(a) \qquad (3)$$

with the same vertices, but different link–lengths. Following Regge (1961), we assume that the interior of each 4–simplex remains flat. Hence, inside each simplex $\tau(a)$, we may set up an arbitrary Lorentz frame $\Sigma(a)$. A vector Φ^μ viewed in $\Sigma(a)$ can be written as a column matrix

$$\Phi(a) \equiv (\Phi^0(a), \Phi^1(a), \cdots, \Phi^4(a)). \qquad (4)$$

Under a Lorentz transformation of $\Sigma(a)$,

$$\varphi(a) \rightarrow u(a)\ \Phi(a) \qquad (5)$$

where, by definition, the real 4 x 4 matrix u(a) satisfies

$$u*(a)\ \eta\ u(a) = \eta \qquad (6)$$

with $*$ denoting the transpose and

$$\mathrm{diag}\ \eta = (-1,1,1,1) \qquad (7)$$

as before.

Each 4–simplex, say $\tau(a)$, has 4 vertices and 10 links; viewed in $\Sigma(a)$, apart from the absolute position of its center– of– mass, all the properties of $\tau(a)$ such as shape, volume, orientation, etc. are completely determined by the 4 independent link–vector: l^μ_{12}, l^μ_{13}, l^μ_{14},l^μ_{15}, where 1,2,⋯,5 refer to the vertices of $\tau(a)$. Under a Lorentz transformation of $\Sigma(a)$, these vectors transform according to (5). From these vectors, we can form a total of 10 independent invariants which may be represented by all the (link -length)^2 of $\tau(a)$.

So far, each $\tau(a)$ is separate from all other 4–simplices. Suppose that in the original flat Minkowski space, two 4–simplices, say $\tau(a)$ and $\tau(b)$, share a common 3–simplex s whose vertices are, say, sites 1,2,3,4. Under the link–length variation (2), these two 4–simplices transform from

$$\tau(a) \to \tau'(a) \qquad \text{and} \qquad \tau(b) \to \tau'(b); \tag{8}$$

viewed in $\Sigma(a)$ we have

$$s \to s'(a), \tag{9}$$

with each of its 6 link–vectors changing from

$$l^{\mu}{}_{ij} \to l'^{\mu}{}_{ij}(a) \tag{10}$$

where i and j=1,2,3,4. Likewise, viewed in $\Sigma(b)$ we have

$$s \to s'(b) \tag{11}$$

and

$$l^{\mu}{}_{ij} \to l'^{\mu}{}_{ij}(b) \tag{12}$$

Since in accordance with (2)

$$l^{\mu}{}_{ij}(a)^2 = l^{\mu}{}_{ij}(b)^2 = l^{\mu 2}_{ij} \tag{13}$$

for all i and j from 1,2,3,4, there exists a Lorentz transformation U(b,a) which can transform the set $l^{\mu}{}_{ij}(a)$ to $l^{\mu}{}_{ij}(b)$, and therefore s(a) to s(b); i.e.,

$$l^{\mu}{}_{ij}(b) = U(b,a)^{\mu}_{\nu}\ l^{\nu}{}_{ij}(a) \tag{14}$$

for all i and j=1,2,3,4. Under Lorentz transformations of both reference frames $\Sigma(a)$ and $\Sigma(b)$, we have on account of (5)

$$U(b,a) \to u(b)\ U(b,a)\ u(a)^{-1} \tag{15}$$

where u(a), u(b) and U(b,a), being Lorentz transformations, all satisfy (6), inside a 4-simplex τ (a). We can regard U(b, a) as connection. Viewed in Σ (a), we write $\Phi = \Phi$ (a), in accordance with (4). Because the interior of Σ (a) is flat, the parallel transport of Φ within τ (a) is trivial. When we transport Φ from τ (a) to one of its adjacent 4-simplices, say τ (b), Φ becomes Φ (b) [when viewed in Σ (b)] , where Φ (b) is defined to be

$$\Phi^{\mu}(b) = U(b,a)^{\mu}_{\nu}\ \Phi^{\nu}(a). \qquad (16)$$

Clearly, U(b,a) does not depend on which point in the interior of τ (a) we start from, nor on which point in the interior of τ (b) we end.

4.3.3. Parallel Transportation

In order to describe the curvature of the spacetime, we should establish the concept of "parallel transportation" and then the concept of "parallel transportation along a closed path". Continuing this way, we can parallel transport any vector Φ along a general path P. It is convenient to think first of the path P in the original flat Minkowski lattice. We assume P to be a continuous path starting from the interior of any 4-simplex, say τ (a_1), passing successively through a sequence of 4-simplices τ (a_2), τ (a_3),…, and finally arriving at the interior of τ (a_3). In going from one 4-simplex, say τ (a_i), to the next one τ (a_{i+1}), the path P always passes through their common 3- simplex (i. e. , P never intersects any 2-simplex, or 1-simplex, etc.) . After the link -length variation (2), P becomes a path from τ (a_1) through τ (a_2) , τ (a_3), … to τ (a_n), the parallel transport of any vector Φ along P is given by

$$\Phi^{\mu}(a_n) = U(P)^{\mu}_{\nu}\ \Phi^{\nu}(a_1) \qquad (1)$$

where

$$U(P) = U(a_n,\ a_{n-1}),\ \cdots\ U(a_2,\ a_1). \qquad (2)$$

When $a_n = a_1$, then P is a closed path C, in which case (2) becomes

$$U(C) = U(a_1,\ a_{n-1},\ \cdots\ U(a_2,\ a_1). \qquad (3)$$

Two different paths P(and therefore also two different C) that pass through the same sequence of $\tau(a_1)$, $\tau(a_2)$, $\cdots$ have same U(P)(or U(C)). Under arbitrary Lorentz transformations of $\Sigma(a_1)$, $\Sigma(a_2)$, $\cdots$ we have from (4-3-2-15)

$$U(P) \to u(a_n)\, U(P)\, u(a_1)^{-1} \quad \text{and} \quad U(C) \to u(a_1)\, U(C)\, u(a_1)^{-1} \qquad (4)$$

4.3.4. Curvature

The departure of U(C) from the unit matrix implies that the space is curved. Note that C is a closed path passing from one 4-simplex to the next 4-simplex by going through their common 3-simplex. Since the interiors of all 4-simplics are flat and since U(C) is independent of the precise passage of C through these 3- simplices, the curvature must rest solely in the 2-simplices that C encircles. In a 4-dimensional space, a 2-simplex, say γ, is shared by several 4-simplices, say

$$\tau(a_1), \tau(a_2), \cdots, \tau(a_m) \qquad (1)$$

in a sequential order; γ is also shared by an equal number of 3-simplices which are the common surfaces of these 4-simplices. For each γ, we introduce a closed path $C(\gamma)$ which encircles γ by following the order (1) and ending in $\tau(a_1)$. [As before, it is easiest to visualize the closed path $C(\gamma)$ as first drawn in the original flat Minkowski space, before the variation (4- 3- 2- 2)]. Define

$$U_\gamma \equiv U(C(\gamma)) = U(a_1,a_m)\, U(a_m, a_{m-1}) \cdots U(a_2,a_1) \qquad (2)$$

which depends only on γ and the reference system $\Sigma(a_1)$.In general, γ can be represented by a tensor of rank 2:

$$\gamma^{\mu_1 \mu_2}, \qquad (3)$$

whose components are referred to the reference frame $\Sigma(a_1)$.

Give γ and $\tau(a_1)$ in (1), there remains the question of the sense of $C(\gamma)$, defined by the sequence (1) versus its inverse order.

Without any loss of generality, we can set $C(\gamma)$ to be a planar figure; the two dimensional area that it encloses, when viewed in $\Sigma(a_1)$ and set in the order (1), can be represented by a tensor of second rank $C(\gamma)^{\alpha\beta}$. We require that the sense of the curve $C(\gamma)$ and the tensor γ satisfy the usual right-hand relation; i.e.,

$$\varepsilon_{\mu_1\mu_2\alpha\beta}\, C(\gamma)^{\alpha\beta}\, \gamma^{\mu_1\mu_2} \tag{4}$$

is positive for all γ, where $\varepsilon_{\mu_1\mu_2\mu_3\mu_4}$ is the total antisymmetric unit tensor. Let $J^{\rho\sigma} = -J^{\sigma\rho}$ be the 4- dimensional real matrix representation of the generator of the Lorentz group

$$\eta^{\sigma\nu}\, x^{\rho}(\partial/\partial x^{\nu}) - \eta^{\rho\nu}\, x^{\sigma}(\partial/\partial x^{\nu}) \tag{5}$$

where x^{ρ} is the coordinate of points inside $\tau(a_1)$, with respect to the frame $\Sigma(a_1)$.

4.3.5. Action

The action of the discrete gravity is defined to be, apart from a proportional constant,

$$\Sigma\ \mathrm{trace}(\varepsilon_{\mu_1\mu_2\rho\sigma}\, J^{\rho\sigma}\, \gamma^{\mu_1\mu_2}\, U_{\gamma}), \tag{1}$$

where the summation is over the γ. Because of the trace, the above expression is independent of the frame $\Sigma(a_1)$. Also, if the link -variation (4-3-2-2) is merely due to a coordinate change from $x(i) \rightarrow x'(i)$ with $l^{\mu}_{ij} \rightarrow l'^{\mu}_{ij}$, then $U_{\gamma} = 1$ and (1) remains zero. The invariances of the action under arbitrary independent Lorentz transformations of all $\Sigma(a)$ and under arbitrary independent translations (when U_{γ} is near 1) of all sites are the discrete equivalent of Einstein's arbitrary coordinate transformations in the continuum theory. We note that the discrete QCD action in a flat Minkowski space (i.e., without gravity) is similar to (1).

When we include gravity, the action A and the generating function Z become

$$A = A_{\mathrm{gravity}} + A_{\mathrm{QCD}} \tag{2}$$

and

$$Z = \int e^{iA} [dl^2_{ij}] \ [du_{ij}] \tag{3}$$

where $A_{gravity}$ is equal to the production of (1/G) to (1).

4. Concluding Remarks

In this lecture, we have tried to show how by viewing time as a dynamical variable, a fundamental length l_p can be introduced which removes all ultra-violet divergences and thereby makes quantization of gravity possible. Because all symmetries, unitary and spacetime transformations, are respected, even for renormalizable theories such as QCD the discrete mechanics gives new insight into their solutions. At the minimum this represents an advance in theoretical techniques; at the maximum it may change our basic concept of spacetime.

Of course, we do not know how nature really works, and there is no evidence that the discrete mechanics is more appropriate than the usual continuum mechanics. However, once a theoretical possibility of such scope is open, it seems difficult not to face the intellectual challenge. Even at the present stage of enormous imperfection, we find aspects of this new mechanics appealing.

Acknowledgements

We wish to give thanks to the Wrokshop of Quantum Gravity of The China Center of Advanced Science and Technology(CCAST), to Professor Antonino Zichichi, the President of the World Lab and the director of the Ettore Majorana Centre for Scientific Culture, to Professor Venzo de Sabbata, the director of the Internationl School of Cosmology and Gravitation, and to the Bologna University for hospitalities.

References

Ashteckar, A., 1986, Phys. Rev. Lett. 57, 2244
1987, Phys. Rev. D36, 1587
1991, Non-Perturbative Canonical Quantum Gravity, (World Scientific, Singapore)

Arnowitt, et al., Phys. Rev. 120, 313(1960); Ann. Phys.(N.Y.), 33, 88(1965)

Barbour,J. and L. Smolin, Syracuse University Report No. SU-GP-92/2-4, 1992

Barvinsky, A.O., V. P. Frolov and A. I. Zelnikov, 1995, Phys.Rev. D51,1741

Bekenstein, J.D., 1973, Phys. Rev. D7, 2333;
1980, Phys. Today, 33, 24,
1982, Gen. Relativ.& Gravitation, 14, 355

Birrell, N. D. and P. C. W. Davies, Quantum Fields in Curved Spacetime, 1982, Cambridge University Press

Bomebelli, L. et al., 1988, Phys.Rev.Lett. 60, 655

Christ,N.H., R. Friedberg and T.D.Lee, Nucl.Phys. B202, 89, (1982).

Christ,N.H., and A.E. Terrano, CU- TP- 259 (Nuclear Methods and Instrumentation).

Finkelstein,D., Phy.Rev. 184,1261(1969)

Friedberg,R., and T.D.Lee, Nucl. Phys. B225 [F59] ,(1983).

Frolov, V. P., 1995, Phys. Rev. Lett. 74, 3319
1995, this volume

Frolov, V. P. and Thorne, K. S., 1989, Phys. Rev., D39, 2125

Frolov, V. P. and Novikov, I., 1993, Phys.Rev. D48, 1607

Gardner,E.J., C. Itzykson and B.Derrida, J.Phys.A; C.Itzykson, "Fields on a Random Lattice," Saclay Dph. G/SPT/83/148.

Hawking, S., 1975, Commun. Math. Phys. 43, 199;
1976, Phys. Rev. D14, 2460
1988, A Brief History of Time, Bantam, New York.

Huang, H., B. Zhou and Z. Zhang, 1994, Quantum Field Theory in Curved Spacetime, Huazhong Normal University Press, (in Chinese)

Itzykson,C., in Proceedings of the Trieste Workshop, eds. R.Lengo et al. (Singapore, World Scientific, (1983);

Kundt, W., 1976, Nature, 259, 30

Lee,T.D., Phys.Lett. 122B,217(1983),

Maxwell,J.C., in "A Treatise on Electricity and Magnetism" (New York, Dover Publications, 1954).

Penrose, R., in Quantum Theory and Beyond, edited by T. Bastin(Cambridge University Press, New York, 1971

Regge,C., Nuovo Cimento 19,558 (1961).

Ren,H.C., CU-TP-272 (Nucl.Phys.); R.Friedberg and H.C.Ren, CU- TP- 268 (Nucl.Phys).

Rocek,M., and R.M.Williams, Phys.Lett. 104 B,31(1981) , in "Quantum Structure of Space and Time", eds. M. J. Duff and C. J. Isham (Cambridge, Cambridge Unviersity Press, 1981).

Sorkin, R., Wald, R. and Zhang, Z., 1981, Gen. Relativ. & Gravitation, 13, 1127

't Hooft, G., in Recent Developments in Gravitation, edited by M.Levy and S.Deser (Plenum, New York, 1979)

Wald, M. R., 1977, Commun. Math. Phys. 54, 1;
1978, Phys. Rev. D17, 1477;
1979, Phys. Rev. D20, 1271;
1984, General Relativity, University of Chicago Press, Chicago.
1993, Phys. Rev. D48, R3423

Wilkins, D., 1979, Gen. Relativ. Gravitation., 11, 45

Zhang, Z., 1983, in Proceedings of The Third Marcel Grossmann Meeting on General Relativity}, ed. Hu, N., 1289, Science Press and North-Holand.
1986, Modern Relativity and Black Hole Physics, Huazhong Normal University Press, Wuhan. (in Chinese)
1992, in Black Hole Physics, eds. V de Sabbata and Z. Zhang, Kluwer Academic, Dordrecht/Boston/London, 1-53

Zhang, Z., Huang, H., Bao, G. and He, C., 1990, in Quantum Mechanics in Curved Spacetime}, eds.J. Audrestch and V. de Sabbata, Plenum, New York.

Zhang, Z. and Shi, J., 1986, Science Bulletin of Academia Sinica, (English Edition), Beijing. 32, 1007

Zhang, Z. and Wang, X., 1989, in Proceedings of The Fifth Marcel Grosmann Meeting on General Relativity, eds.B.G. Blair et al., World Scien. Singapore, A. 633

Zhang, Z., X.Wang et al., 1991, Nuovo Cimento, 106B, 1189

Zurek, W.H. and Page, D.N., 1984, Phys. Rev. D29, 628

Zurek, W.H. and Thorne, K.S., 1985, Phys. Rev. Lett., 54, 2171

SUBJECT INDEX

LESSONS BEGIN
9.00 a.m.
LESSONS BEGIN
9.30 a.m.
San Francesco
Lecture Hall
TONIGHT